HANDBOOK OF FINANCIAL ENGINEERING

Springer Optimization and Its Applications

VOLUME 18

Managing Editor
Panos M. Pardalos (University of Florida)

Editor—Combinatorial Optimization
Ding-Zhu Du (University of Texas at Dallas)

Advisory Board
J. Birge (University of Chicago)
C.A. Floudas (Princeton University)
F. Giannessi (University of Pisa)
H.D. Sherali (Virginia Polytechnic and State University)
T. Terlaky (McMaster University)
Y. Ye (Stanford University)

Aims and Scope
Optimization has been expanding in all directions at an astonishing rate during the last few decades. New algorithmic and theoretical techniques have been developed, the diffusion into other disciplines has proceeded at a rapid pace, and our knowledge of all aspects of the field has grown even more profound. At the same time, one of the most striking trends in optimization is the constantly increasing emphasis on the interdisciplinary nature of the field. Optimization has been a basic tool in all areas of applied mathematics, engineering, medicine, economics and other sciences.

The series *Optimization and Its Applications* publishes undergraduate and graduate textbooks, monographs and state-of-the-art expository works that focus on algorithms for solving optimization problems and also study applications involving such problems. Some of the topics covered include nonlinear optimization (convex and nonconvex), network flow problems, stochastic optimization, optimal control, discrete optimization, multi-objective programming, description of software packages, approximation techniques and heuristic approaches.

HANDBOOK OF FINANCIAL ENGINEERING

Edited By

CONSTANTIN ZOPOUNIDIS
Technical University of Crete, Chania, Greece

MICHAEL DOUMPOS
Technical University of Crete, Chania, Greece

PANOS M. PARDALOS
University of Florida, Gainesville, Florida

Constantin Zopounidis
Department of Production
  Engineering and Management
Financial Engineering Laboratory
Technical University of Crete
University Campus
731 00 Chania
Greece

Michael Doumpos
Department of Production
  Engineering and Management
Technical University of Crete
University Campus
731 00 Chania
Greece

Panos M. Pardalos
Department of Industrial
  and Systems Engineering
University of Florida
303 Weil Hall
P.O. Box 116595
Gainesville FL 32611-6595
USA

Managing Editor:
Panos M. Pardalos
Department of Industrial
  and Systems Engineering
University of Florida
Gainesville, FL 32611
pardalos@cao.ise.ufl.edu

Editor/ Combinatorial Optimization:
Ding-Zhu Du
Department of Computer Science
  and Engineering
University of Texas at Dallas
Richardson, TX 75083
dzdu@utdallas.edu

ISSN: 1931-6828
ISBN: 978-0-387-76681-2   e-ISBN: 978-0-387-76682-9
DOI: 10.1007/978-0-387-76682-9

Library of Congress Control Number: 2008923930

© 2008 by Springer Science+Business Media, LLC
All rights reserved. This work may not be translated or copied in whole or in part without the written permission of the publisher (Springer Science+Business Media, LLC, 233 Spring Street, New York, NY 10013, USA), except for brief excerpts in connection with reviews or scholarly analysis. Use in connection with any form of information storage and retrieval, electronic adaptation, computer software, or by similar or dissimilar methodology now know or hereafter developed is forbidden.
The use in this publication of trade names, trademarks, service marks and similar terms, even if they are not identified as such, is not to be taken as an expression of opinion as to whether or not they are subject to proprietary rights.

Printed on acid-free paper

springer.com

To our families

Preface

From the beginning of the 20th century, finance has consolidated its position as a science of major practical importance for corporate entities, firms, organizations, and investors. Over this period finance has undergone significant change keeping pace with the technological innovations that occurred after the World War II and the socio-economic changes that affected the global business environment. The changes in the field of finance resulted in a transformation of its nature from a descriptive science to an analytic science, involved with the identification of the relationship among financial decisions and the decision environment and ultimately to an engineering science, involved with the design of new financial products and the development of innovations with regard to financial instruments, processes, and solutions.

This transformation began in the late 1950s with the work of Markowitz on portfolio selection and later, during the 1970s, with the work of Black and Scholes on option pricing. These pioneering works have demonstrated that the descriptive character of financial theory was gradually progressing toward a more analytic one that ultimately led to the engineering phase of finance by the late 1980s.

Several financial researchers and practitioners consider the engineering phase as a new era in finance. This led to the introduction of the term "financial engineering" to describe the new approach to the study of financial decision-making problems. Since the late 1980s, financial engineering has consolidated its position among financial researchers and practitioners, referring to the design, development, and implementation of innovative financial instruments and processes and the formulation of innovative solutions to financial decision-making problems.

For an analyst (practitioner or academic researcher) to be able to address the three major aspects of financial engineering (design, development, implementation) in an innovative way, the knowledge of financial theory is not enough. While the financial theory constitutes the underlying knowledge required to address financial engineering problems, some synthesis and analysis are also necessary for innovation. Thus, the major characteristic of this new context is the extensive use of advanced decision analysis and modeling tools to manage the increasing complexity of the financial and business environments. These tools originate from a variety of different disciplines including statistical analysis, econometrics, artificial intelligence, and operations research. Thus, the role of financial decision makers (financial engineers) within the financial engineering context becomes even more complex. They are involved not only with the application of the financial theory, but also with the implementation of advanced methodological tools and quantitative analysis techniques in order to address effectively financial decision problems.

Over the past decade the financial and business environments have undergone significant changes. During the same period several advances have been made within the field of financial engineering, involving both the methodological tools used as well as the application areas. These findings motivate the preparation of a book with the aim to present, in a comprehensive way, the most recent advances within the field of financial engineering, focusing not only on the description of the existing areas in financial engineering research, but also on the new methodologies that have been developed for modeling and addressing financial engineering problems more efficiently.

The objective for the preparation of this book has been to address this requirement through the collection of up-to-date research and real-world applications of financial engineering, in a comprehensive edited volume.

The book is organized into four major parts, each covering different aspects of financial engineering and modeling.

Part I is devoted to portfolio management and trading. It covers several important issues, including portfolio optimization, efficiency analysis, financial trading, and technical analysis.

The first paper in this part, by Steuer, Qi, and Hirschberger, discusses the role of multicriteria optimization in portfolio selection. In the traditional portfolio theory the expected return and the associated variance are the only two criteria used to select efficient portfolios. The authors extend this setting to a multicriteria context. This seems to be a more realistic setting, because beyond the random variable of portfolio return, an investor's utility function can take additional stochastic and deterministic arguments, such as dividends, liquidity, the number of securities in a portfolio, turnover, and the amount of short selling. The authors discuss the stochastic and deterministic nature of portfolio selection, illustrate how multiple-argument utility functions lead to multiple criteria portfolio selection formulations, analyze the mean-variance nondominated frontier and the nondominated sets of multiple criteria portfolio selection problems, and discuss issues involved in their computation.

The second paper, by Konno and Yamamoto, involves the recent developments in integer programming formulations for financial optimization problems. In the context of financial engineering optimization plays a major role, especially in large-scale and difficult problems. Konno and Yamamoto show that nonconvex financial optimization problems can be solved within a practical amount of time using state-of-the-art integer programming methodologies. The authors focus on three classes of problems. The first involves mean-variance portfolio optimization subject to nonconvex constraints related to the minimal transaction unit and the number of assets in the portfolio. Nonconvex transaction costs are also discussed. It all cases it is shown that these problems can be formulated as integer linear programming problems and solved to optimality using a convex piecewise linear risk measure such as absolute deviation instead of variance. A similar approach is also used in a second class of problems related to "maximal predictability portfolio optimization" and a solution algorithm is given. The third class of problems is related to attribute selection in discriminant analysis for failure prediction.

The third paper, by Hall and Satchell, considers the trade-off between portfolio return and portfolio downside risk that has arisen because of inadequacies of the mean-variance framework and regulatory requirements to calculate value at risk and related measures by banks and other financial institutions. Analytical results are given that allow one to better understand the form of mean-risk frontiers. The authors show that the set of minimum risk portfolios are essentially the same under ellipticity for a wide class of risk measures. The authors also derive explicit expressions for mean-value at risk, mean-expected loss and mean-semivariance frontiers under normality and propose extensions for portfolio simulation and for the analysis of fairly arbitrary risk measures with arbitrary return distributions.

In the fourth chapter of Part I, Deville provides a complete overview of exchange traded funds (ETFs), which have emerged as major financial innovation since the early 1990s. The chapter begins with the history of ETFs, from their creation in North American markets to their more recent developments in the U.S. and European markets. The mechanics of ETFs are then presented along with the current status of the ETF industry. The chapter also covers several important aspects of ETFs, including their pricing efficiency compared to closed-end funds, the relative performance of ETFs over conventional index mutual funds, the impact of ETFs on the market quality of the stock components of the underlying indices, as well as the efficiency of index derivatives markets and the pricing discovery process for index prices.

In the final chapter of Part I, Chen, Kuo, and Hoi investigate the development of technical trading rules using a computational intelligence methodology. Technical analysis is widely used by practitioners in securities trading, but the development of proper trading rules in a dynamic and evolving environment is cumbersome. Genetic programming (GP) techniques is a promising methodology that can be used for this purpose. The chapter presents a thorough examination of the applicability and performance of GP in this context using data from different markets. This extensive study enriches our understanding of the behavior of GP in financial markets, and the robustness of their results.

The second part of the book is devoted to risk management, which is an integral part of financial engineering. In the first chapter of this part, Ioannidis,

Miao, and Williams provide a review of interest rate models. Interest rate modeling has been a major research topic in financial engineering with direct applications in the pricing of rate-sensitive instruments. This review addresses three general approaches to interest rate modeling, namely single and multi-factor models of the short rate, models of forward rates, and finally LIBOR models. The chapter focuses on key results and pertinent pricing formulas and discusses several practical approaches to implementing short rate models.

In the next chapter, Dash and Kajiji illustrate the contribution of an artificial intelligence methodology in modeling volatility spillovers. The analysis is implemented in two stages. The first stage focuses on the development of a radial basis function neural network mapping of government bond excess returns. The second stage establishes the overall effectiveness of the model to control for the known conditional volatility properties that define transmission linkages among government bond excess returns. The developed network model provides helpful policy inferences and research findings and it proves to be extremely efficient in the separation of global, regional and local volatility effects.

In the third chapter of Part II, MacLean, Zhao, Consigli, and Ziemba develop a model of market returns that, under certain conditions, determines the emergence of a speculative bubble in the economy and drives bond and equity returns more generally. The model has diffusion parameters, which are random variables plus shock/jump terms with random coefficients. The model incorporates both over-and under-valuation of stocks, and an algorithm is proposed for the estimation of its parameters. Empirical results indicate that the procedures are able to accurately estimate the parameters and that there is a dependence of shock intensity on the state of returns.

The last chapter of Part II, by Topaloglou, Vladimirou, and Zenios, analyzes alternative means for controlling currency risk exposure in actively managed international portfolios. Multistage stochastic programming models are extended to incorporate decisions for optimal selection of forward contracts or currency options for hedging purposes, and a valuation procedure to price currency options is presented with discrete distributions of exchange rates. The authors also provide an empirical analysis of the effectiveness of alternative decision strategies through extensive numerical tests. Individual put options strategies as well as combinations of options are considered and compared to optimal choices of forward contracts, using static tests and dynamic backtesting. The results show that optimally selected currency forward contracts yield superior results in comparison to single protective puts. Moreover, it is shown that a multistage stochastic programming model consistently outperforms its single-stage counterpart and yields incremental benefits.

The third part of the book is devoted to the applications of operations research methods in financial engineering. This part consists of three chapters. In the first one, Kosmidou and Zopounidis give an overview of the main methodologies that have been used for asset liability management, focusing on banking institutions. Different types of models are covered, which are categorized into two major groups: deterministic mathematical programming models and stochastic models. An example of a deterministic model is also given, in the form of a goal programming formulation, which uses Monte Carlo simulation for handling the interest rate risk.

The next chapter, by Kunsch, is devoted to capital budgeting. Kunsch gives a detailed tutorial overview of modern operations research methods applied in this field. The chapter begins with an introduction to the traditional capital budgeting process and then presents several extensions, including the application of multicriteria decision aid methods, the treatment of uncertainty in the context of the fuzzy sets theory (e.g., fuzzy arithmetic and fuzzy rule systems), and the use of real options in the investment process. All these issues are discussed in a unified context and their connections are highlighted.

In the last chapter of Part III, Nagurney overviews some of the major developments in financial engineering in the context of financial networks. The chapter begins with a discussion of financial optimization problems within a network context. Then, financial network equilibrium problems that involve more than a single decision maker are analyzed. The presentation is based on the extension of the classical mean-variance portfolio optimization to multiple sectors. The dynamics of the financial economy are also explored with the discussion of dynamic financial networks with intermediation and the integration of social networks with financial networks is explored. Optimality conditions, solution algorithms, and examples are also presented.

The last part of the handbook includes three chapters on mergers/acquisitions and credit rating models. The first paper, by Chevalier and Redor, surveys the theories on the choice of the payment method in mergers and acquisitions and the empirical studies on this topic. Initially, asymmetric information models are discussed, which assume that both sides (i.e., the target and the bidder) have private information on their own value. Then, the impact of taxation is reviewed, followed by the presentation of theories related to managerial ownership and outside control. Additional issues discussed include past performances, investment opportunities, business cycles, capital structure, the delay of completion theory, as well as acquisitions for non-public firms.

The next chapter, by Pasiouras, Gaganis, Tanna, and Zopounidis, is also related to mergers and acquisitions. The authors present empirical results on the potential of developing reliable models for identifying acquisition targets, focusing on the European banking industry. The relevant literature on this topic is reviewed and a methodology is developed based on nonparametric pattern recognition approach, namely support vector machines. The methodology is applied to a sample of European commercial banks and issues such as the selection of proper explanatory variables and the performance of the resulting models are discussed.

The last chapter of the handbook, by Papageorgiou, Doumpos, Zopounidis, and Pardalos, is devoted to the development of credit rating systems, which have become an integral part of the risk management process under the new Basel capital adequacy accord. The authors discuss the requirements of the new regulatory environment, the role of credit rating systems as well as their specifications and development procedure, along with their use for estimating the minimum capital requirements. A review of the current rating systems and methodologies is also presented together with a survey of comparative studies. On the empirical side, the chapter presents detailed results on the relative performance of several methodologies for the development of credit rating systems, covering issues such as variable and sample selection as well as model stability.

Sincere thanks must be expressed to all the authors, whose contributions have been essential in creating this high-quality volume. We hope that this volume will be of great help to financial engineers/analysts, bank managers, risk analysts, investment managers, pension fund managers, and of course to financial engineering researchers and graduate students, as reference material to the recent advances in the different aspects of financial engineering and the existing methodologies in this field.

Constantin Zopounidis
Michael Doumpos
Panos M. Pardalos
May 2008

Contents

Preface VII

Part I Portfolio Management and Trading

Portfolio Selection in the Presence of Multiple Criteria
Ralph E. Steuer, Yue Qi, Markus Hirschberger 3

Applications of Integer Programming to Financial Optimization
Hiroshi Konno, Rei Yamamoto 25

Computing Mean/Downside Risk Frontiers: The Role of Ellipticity
Antony D. Hall, Steve E. Satchell 49

Exchange Traded Funds: History, Trading, and Research
Laurent Deville 67

Genetic Programming and Financial Trading: How Much About "What We Know"
Shu-Heng Chen, Tzu-Wen Kuo, Kong-Mui Hoi 99

Part II Risk Management

Interest Rate Models: A Review
Christos Ioannidis, Rong Hui Miao, Julian M. Williams 157

Engineering a Generalized Neural Network Mapping of Volatility Spillovers in European Government Bond Markets
Gordon H. Dash, Jr., Nina Kajiji 201

**Estimating Parameters in a Pricing Model
with State-Dependent Shocks**
Leonard MacLean, Yonggan Zhao, Giorgio Consigli, William Ziemba ... 231

Controlling Currency Risk with Options or Forwards
Nikolas Topaloglou, Hercules Vladimirou, Stavros A. Zenios 245

Part III Operations Research Methods in Financial Engineering

Asset Liability Management Techniques
Kyriaki Kosmidou, Constantin Zopounidis 281

**Advanced Operations Research Techniques in Capital
Budgeting**
Pierre L. Kunsch 301

Financial Networks
Anna Nagurney 343

Part IV Mergers, Acquisitions, and Credit Risk Ratings

**The Choice of the Payment Method in Mergers
and Acquisitions**
Alain Chevalier, Etienne Redor 385

**An Application of Support Vector Machines in the Prediction
of Acquisition Targets: Evidence from the EU Banking Sector**
*Fotios Pasiouras, Chrysovalantis Gaganis, Sailesh Tanna,
Constantin Zopounidis* 431

**Credit Rating Systems: Regulatory Framework
and Comparative Evaluation of Existing Methods**
*Dimitris Papageorgiou, Michael Doumpos, Constantin Zopounidis,
Panos M. Pardalos* 457

Index 489

List of Contributors

Shu-Heng Chen
AI-ECON Research Center
Department of Economics
National Chengchi University
Taipei, Taiwan 11623
chchen@nccu.edu.tw

Alain Chevalier
ESCP EAP
79 Avenue de la République
75543 Paris Cedex 11, France
chevalier@escp-eap.net

Giorgio Consigli
University of Bergamo
Via Salvecchio 19
24129, Bergamo, Italy
giorgio.consigli@unibg.it

Gordon H. Dash, Jr.
777 Smith Street
Providence, RI 02908, USA
GHDash@uri.edu

Laurent Deville
Paris-Dauphine University
CNRS, DRM-CEREG
Paris, France
Laurent.Deville@dauphine.fr

Michael Doumpos
Technical University of Crete
Department of Production
Engineering and Management
Financial Engineering Laboratory
University Campus
73100 Chania, Greece
mdoumpos@dpem.tuc.gr

Chrysovalantis Gaganis
Technical University of Crete
Department of Production
Engineering and Management
Financial Engineering Laboratory
University Campus
73100 Chania, Greece
bgaganis@yahoo.com

Antony D. Hall
University of Technology Sydney
School of Finance and Economics
Sydney, Australia
tony.hall@uts.edu.au

Markus Hirschberger
Department of Mathematics
University of Eichstätt-Ingolstadt
Eichstätt, Germany

Kong-Mui Hoi
AI-ECON Research Center
Department of Economics

National Chengchi University
Taipei, Taiwan 11623
92258038@nccu.edu.tw

Christos Ioannidis
School of Management
University of Bath
Bath, BA2 7AY, UK
C.Ioannidis@bath.ac.uk

Nina Kajiji
National Center on Public Education
and Social Policy
University of Rhode Island
80 Washington Street
Providence, RI 02903, USA
nina@uri.edu

Hiroshi Konno
Department of Industrial and
Systems Engineering
Chuo University
Tokyo, Japan
konno@indsys.chuo-u.ac.jp

Kyriaki Kosmidou
Technical University of Crete
Department of Production
Engineering and Management
Financial Engineering Laboratory
University Campus
73100 Chania, Greece
kikikosmidou@yahoo.com

Pierre L. Kunsch
Vrije Universiteit Brussel
MOSI Department
Pleinlaan 2
BE-1050 Brussels, Belgium
pikunsch@ulb.ac.be

Tzu-Wen Kuo
Department of Finance
and Banking
Aletheia University
Tamsui, Taipei, Taiwan 25103
kuo@aiecon.org

Leonard MacLean
School of Business Administration
Dalhousie University
Halifax, Nova Scotia, B3H 3J5
Canada
lmaclean@mgmt.dal.ca

Rong Hui Miao
School of Management
University of Bath
Bath, BA2 7AY, UK

Anna Nagurney
Department of Finance and
Operations Management
Isenberg School of Management
University of Massachusetts at
Amherst
Amherst, MA 01003, USA
nagurney@gbfin.umass.edu

Dimitris Papageorgiou
Technical University of Crete
Department of Production
Engineering and Management
Financial Engineering Laboratory
University Campus
73100 Chania, Greece
papageorgiou.d@gmail.com

Panos M. Pardalos
Department of Industrial and
Systems Engineering
Center for Applied Optimization
University of Florida
303 Weil Hall
PO Box 116595
Gainesville, FL 32611-6595, USA
pardalos@cao.ise.ufl.edu

Fotios Pasiouras
School of Management
University of Bath
Claverton Down
Bath, BA2 7AY, UK
f.pasiouras@bath.ac.uk

List of Contributors   XVII

Yue Qi
Hedge Fund Research Institute
International University of Monaco
Principality of Monaco

Etienne Redor
UESCP EAP
79 Avenue de la République
75543 Paris Cedex 11, France
etienne.redor@escp-eap.net

Steve E. Satchell
University of Cambridge
Faculty of Economics
Sidgwick Avenue
Cambridge, CB3 9DD, UK
steve.satchell@econ.cam.ac.uk

Ralph E. Steuer
Terry College of Business
University of Georgia
Athens, GA 30602-6253, USA
rsteuer@uga.edu

Sailesh Tanna
Department of Economics, Finance
and Accounting
Faculty of Business, Environment
and Society
Coventry University
Priory Street
Coventry, CV1 5FB, UK
s.tanna@coventry.ac.uk

Nikolas Topaloglou
HERMES European Center of
Excellence on Computational
Finance and Economics
School of Economics and
Management
University of Cyprus
PO Box 20537
CY-1678 Nicosia, Cyprus

Hercules Vladimirou
HERMES European Center of
Excellence on Computational
Finance and Economics
School of Economics and
Management
University of Cyprus
PO Box 20537
CY-1678 Nicosia, Cyprus
hercules@ucy.ac.cy

Julian M. Williams
School of Management
University of Bath
Bath, BA2 7AY, UK

Rei Yamamoto
Department of Industrial and
Systems Engineering
Chuo University
Tokyo, Japan
yamamoto@mtec-institute.co.jp

Stavros A. Zenios
HERMES European Center of
Excellence on Computational
Finance and Economics
School of Economics and
Management
University of Cyprus
PO Box 20537
CY-1678 Nicosia, Cyprus

Yonggan Zhao
RBC Centre for Risk Management
Faculty of Management
Dalhousie University
Halifax, Nova Scotia, B3H 3J5,
Canada
Yonggan.Zhao@dal.ca

William Ziemba
Sauder School of Business
University of British Columbia
Vancouver, BC, V6T 1Z2, Canada
ziemba@interchange.ubc.ca

Constantin Zopounidis
Technical University of Crete
Department of Production
Engineering and Management
Financial Engineering Laboratory
University Campus
73100 Chania, Greece
kostas@dpem.tuc.gr

Part I

Portfolio Management and Trading

Portfolio Selection in the Presence of Multiple Criteria

Ralph E. Steuer[1], Yue Qi[2], and Markus Hirschberger[3]

[1] Terry College of Business, University of Georgia, Athens, GA 30602-6253, USA
rsteuer@uga.edu
[2] Hedge Fund Research Institute, International University of Monaco, Principality of Monaco
[3] Department of Mathematics, University of Eichstätt-Ingolstadt, Eichstätt, Germany

1 Introduction

There has been growing interest in how to incorporate additional criteria beyond "risk and return" into the portfolio selection process. In response, our purpose is to describe the latest in results that have been coming together under the topic of multiple criteria portfolio selection. Starting with a review of conventional portfolio selection from a somewhat different perspective so as better to lead into the topic of multiple criteria portfolio selection, we start from the basics as follows.

In portfolio selection, two vectors are associated with each portfolio. One is used to "define" a portfolio. The other is used to "describe" the portfolio. The vector used to define a portfolio is an *investment proportion vector*. It specifies the proportions of an amount of money to be invested in different securities, thereby defining the composition of a portfolio. The length of an investment proportion vector is the number of securities under consideration.

The other of a portfolio's two vectors is a *criterion vector*. A portfolio's criterion vector contains the values of measures used to evaluate the portfolio. For instance, in mean-variance portfolio selection, criterion vectors have two components. One is for specifying the expected value of the portfolio's return random variable. The other is for specifying the variance of the random variable. The idea is that the variance of the random variable is a measure of risk. In reality, investors may have additional concerns.

To accommodate multiple criteria in portfolio selection, we no longer call an "efficient frontier" by that name. Instead we call it a "nondominated frontier" or "nondominated set." Terminologically, criterion vectors are now either *nondominated* or *dominated*. This does not mean that the term "efficiency" has been discarded. Efficiency is simply redirected to apply only to investment proportion vectors in the following sense. An investment proportion vector is *efficient* if and only if its criterion vector is nondominated, and an investment proportion vector is *inefficient* if and only if its criterion vector is dominated.

A portfolio selection problem is a *multiple criteria portfolio selection* problem when its criterion vectors have three or more components. In conventional portfolio selection, with criterion vectors of length two, the nondominated set is typically a curved line in two-dimensional space, which, when graphed, usually has expected value of the portfolio return random variable on the vertical axis and variance (or more commonly, standard deviation) of the same random variable on the horizontal axis. But, when criterion vectors are of length three or more, the nondominated set is best thought of as a surface in higher-dimensional space. Because of the increased difficulties involved in computing nondominated surfaces and communicating them to investors, multiple criteria portfolio selection problems can be expected to be much more difficult to solve than the types of problems we are used to seeing in conventional portfolio selection.

Multiple criteria portfolio selection problems normally stem from multiple-argument investor utility functions but can stem from a single-argument utility function.[1] While portfolio selection problems with criterion vectors of length two are the usual case with single-argument utility functions (when the argument is stochastic), it is possible for a multiple criteria portfolio selection problem to result from a single-argument utility function when the investor's nondominated set is a consequence of three or more measures derived from the same single stochastic argument. An example of this is when a mean-variance portfolio selection problem (which revolves around the single random variable of portfolio return) is extended to take into account additional measures, such as skewness, based upon the same random variable.

Despite the above, we will primarily focus on the more general and interesting cases of multiple criteria portfolio selection problems resulting from multiple-argument utility functions. Beyond the random variable of portfolio return, utility functions can take additional stochastic and deterministic arguments. Additional stochastic arguments might include dividend, liquidity, and excess return over of a benchmark random variables. Deterministic arguments might include the number of securities in a portfolio, turnover, and the amount of short selling.

Conventional mean-variance portfolio analysis (described as "modern portfolio analysis" in Elton et al., 2002) dates back to the papers of Roy (1952) and Markowitz (1952). In addition to introducing new ways to think about finance, the papers are important because they symbolize different strategies for solving portfolio selection problems. In computing his "safety first" point, Roy's paper symbolizes approaches that attempt to directly compute portfolios whose criterion vectors possess prechosen characteristics.

---

[1] Although we use the term *utility function* throughout, *preference function* or *value function* could just as well have been used.

On the other hand, Markowitz's approach is more reflective. It recognizes that there are likely to be differences among investors. It essentially eschews preconceived notions, preferring to compute the entire nondominated set first. Then, only after studying the nondominated set should an investor attempt to identify a most preferred criterion vector. Overall, Markowitz's solution approach consists of the following four stages:

1. Compute the nondominated set.
2. Communicate the nondominated set to the investor.
3. Select the most preferred of the points in the nondominated set.
4. Working backwards, identify an investment proportion vector whose image is the nondominated point (i.e., criterion vector) selected in stage 3.

Under assumptions generally accepted in portfolio selection, these four stages, when properly carried out, will lead to an investor's optimal portfolio. Because of the widespread acceptance of Markowitz's approach, his name is virtually synonymous with portfolio selection, although Markowitz (1999) has tried to see that Roy also receives credit.

Despite the degree to which mean-variance portfolio selection dominates the landscape, there has almost always been a slight undercurrent of multiple objectives in portfolio selection. However, this undercurrent has become more pronounced of late. For instance, Steuer and Na (2003), the number of papers reported as dealing with multiple criteria in portfolio selection has increased from about 1.5 to 4.5 per year over the period from 1973 to 2000. Such papers can be grouped into three categories.

In the first category we have overview articles such as those by Colson and DeBruyn (1989), Spronk and Hallerbach (1997), Bana e Costa and Soares (2001), Hallerbach and Spronk (2002a, 2002b), Spronk et al. (2005), and Steuer et al. (2005, 2006a, 2006b).

In the second category, in the spirit of Roy, are articles that attempt to directly compute points on the nondominated surface that possess certain characteristics. Papers in this category include Lee and Lerro (1973), Hurson and Zopounidis (1995), Ballestero and Romero (1996), Dominiak (1997a, 1997b), Doumpos et al. (1999), Arenas Parra et al. (2001), Ballestero (2002), Bouri et al. (2002), Ballestero and Plà-Santamaría (2004), Bana e Costa and Soares (2004), and Aouni et al. (2006).

In the third category, in the spirit of Markowitz, are articles that attempt to compute, or at least interactively search or sample, the nondominated set before selecting a "final" portfolio. Here, a *final* solution is a portfolio that is either optimal or sufficiently close to being optimal to terminate the decision process. Contributions in this category include those by Spronk (1981), Konno et al. (1993), L'Hoir and Teghem (1995), Chow (1995), Tamiz et al. (1996), Korhonen and Yu (1997), Yu (1997), Ogryczak (2000), Xu and Li (2002), Lo et al. (2003), Ehrgott et al. (2004), Fliege (2004), and Kliber (2005).

The organization of the rest of this chapter is as follows. Sections 2 and 3 discuss the initially stochastic, and then deterministic, nature of portfolio selection. Section 4 discusses single- and multiple-argument utility functions and shows the natural way multiple-argument utility functions lead to multiple criteria portfolio selection formulations. After a careful study of the mean-variance nondominated frontier in Section 5, the nondominated sets of multiple criteria portfolio selection problems, and issues involved in their computation, are discussed in Section 6. Section 7 concludes the chapter.

2 Initial Stochastic Programming Problem

In its most basic form, the problem of portfolio selection is as follows. Consider a fixed sum of money to be invested in securities selected from a universe of n securities. Let there be a beginning of a holding period and an end of the holding period. Also, let x_i be the proportion of the fixed sum to be invested in the ith security. Being proportions, the sum of the x_i equals 1.

Let r_i denote the random variable for the ith security's return over the holding period. While the realized values of the r_i are not known until the end of the holding period, it is nevertheless assumed that all means μ_i, variances σ_{ii}, and covariances σ_{ij} of the distributions from which the r_i come are known at the beginning of the holding period.

Letting r_p denote the random variable for the return on a portfolio defined by the r_i and some set of x_i over the holding period, we have

$$r_p = \sum_{i=1}^{n} r_i x_i,$$

Under the assumption that investors are only interested in maximizing the uncertain objective of return on a portfolio, the problem of portfolio selection is then to maximize r_p as in

$$\max\{ r_p = \sum_{i=1}^{n} r_i x_i \}, \tag{1}$$

$$\text{s.t.} \quad \mathbf{x} \in S = \{\mathbf{x} \in \mathbb{R}^n \mid \sum_{i=1}^{n} x_i = 1, \ \alpha_i \leq x_i \leq \omega_i\},$$

where S as above is a typical feasible region. While (1) may look like a linear programming problem, it is not a linear programming problem. Since the r_i are not known until the end of the holding period, but the x_i must be determined at the beginning of the holding period, (1) is a *stochastic* programming problem. For use later, let (1) be called the investor's *initial stochastic programming problem*. As stated in Caballero et al. (2001), if in a problem some parameters take unknown values at the time of making a decision, and these parameters are random variables, then the resulting problem is called a *stochastic programming problem*. Since S is deterministic, problem (1)'s stochastic nature only derives from random variable elements being present in the objective function portion of the program. Interested readers might also wish to consult Ziemba (2003) for additional stochastic discussions about portfolio selection.

3 Equivalent Deterministic Formulations

The difficulty with a stochastic programming problem is that its solution is not well defined. Hence, to solve (1) requires an interpretation and a decision. The approach taken in the literature (for instance, in Stancu-Minasian, 1984; Slowinski and Teghem, 1990; Prékopa, 1995) is to ultimately transform the stochastic problem into an *equivalent deterministic problem* for solution. Equivalent deterministic problems typically involve the utilization of some statistical characteristic or characteristics of the random variables in question. For problems with a single stochastic objective as in (1), Caballero et al. (2001) discuss the following five equivalent deterministic possibilities:

(a) $\max\{E[r_p]\}$
    s.t. $\mathbf{x} \in S$

(b) $\min\{\text{Var}[r_p]\}$
    s.t. $\mathbf{x} \in S$

(c) $\max\{E[r_p]\}$
    $\min\{\text{Var}[r_p]\}$
    s.t. $\mathbf{x} \in S$

(d) $\max\{P(r_p) \geq u\}$ for some chosen level of u
    s.t. $\mathbf{x} \in S$

(e) $\max\{u\}$
    s.t. $P(r_p \geq u) \geq \beta$ for some chosen level of β
    $\mathbf{x} \in S$

If there is a question about how any of the above can be deterministic, recall that from the previous section all means μ_i, variances σ_{ii}, and covariances σ_{ij} of the r_i are assumed to be known at the beginning of the holding period. But with a list of choices, how is one to know which should replace (1) for a given investor? At this point it is illuminating to take a step back and delve into the rationale that leads from the investor's initial stochastic programming problem to equivalent deterministic possibilities **(a)** to **(e)**.

Early 17th-century mathematicians assumed that a gambler would be indifferent between receiving the uncertain outcome of a gamble and receiving its expected value in cash. In the context of portfolio selection, the gambler would be an investor, the gamble would be the return on a portfolio, and the *certainty equivalent* would be

$$CE = E[r_p].$$

Given that an investor would want to maximize the amount of cash received for certain, this rationale leads directly to equivalent deterministic possibility **(a)**. However, Bernoulli (1738) discovered what has become known as the St. Petersburg paradox.[2] A coin is tossed until it lands "heads." The gambler receives one ducat if it lands "heads" on the first throw, two ducats if it first lands "heads" on the second throw, four ducats if it first lands "heads" on the third throw, and so on (2^{h-1} ducats on the hth throw). The expected value of the gamble is infinite, but in reality many gamblers would be willing to accept only a small number of ducats in exchange for the gamble. Hence, Bernoulli suggested not to compare cash outcomes, but to compare the "utilities" of cash outcomes. With the utility of a cash outcome given by a $U : \mathbb{R} \to \mathbb{R}$, we thus have:

$$U(CE) = E[U(r_p)].$$

That is, the utility of CE equals the expected utility of the uncertain portfolio return.

With an investor wishing to maximize $U(CE)$, this leads to the problem of Bernoulli's principle of maximum expected utility:

$$\max\{E[U(r_p)]\} \qquad (2)$$
$$\text{s.t.} \quad \mathbf{x} \in S.$$

With U obviously increasing with r_p, this means that any $\mathbf{x}$ that solves (2) solves (1), and vice versa. Although Bernoulli's maximum expected utility problem (2) is a deterministic equivalent to (1), we call it an equivalent "undetermined" deterministic problem. This is because it is not fully *determined* in that it contains unknown utility function parameters and cannot be solved in its present form. However, with investors assumed to be *risk-averse* (i.e., the expected value $E[r_p]$ is always preferred over the uncertain outcome r_p), we at least know that in (2) U is concave.

Two schools of thought have evolved for dealing with the undetermined nature of U. One, in the spirit of Roy, involves attempting to ascertain aspects of an investor's preference structure for the purpose of using them to solve (2) directly for an optimal portfolio. The other, in the spirit of Markowitz, involves parameterizing U and then attempting to solve (2) for all possible values of its unknown parameters. With this at the core of contemporary portfolio theory, Markowitz considered a parameterized *quadratic* utility function[3]

$$U(x) = x - (\lambda/2)x^2. \qquad (3)$$

---

[2] Because it was published in the *Commentaries from the Academy of Sciences of St. Petersburg*.

[3] There is an anomaly with quadratic utility functions since they decrease from a certain point on. Instead of quadratic utility, an alternative argument (not shown) leading to the same result can be made by assuming that U is concave and increasing and that $\mathbf{r} = (r_1, \ldots, r_n)$ follows a multinormal distribution.

Since $U(x)$ above is normalized such that $U(0) = 0$ and $U'(0) = 1$, this leaves exactly one parameter λ, the *coefficient of risk aversion*. By this parameterization, Markowitz showed that precisely all potentially maximizing solutions of the equivalent "undetermined" deterministic problem (2) for a risk-averse investor can be obtained by solving equivalent deterministic possibility (c):

$$\max\{E[r_p]\}$$
$$\min\{\text{Var}[r_p]\}$$
$$\text{s.t.} \quad \mathbf{x} \in S$$

for all $\mathbf{x} \in S$ from which it is not possible to increase the expected portfolio return without increasing the portfolio variance, or to decrease the portfolio variance without decreasing the expected portfolio return. In accordance with terminology introduced earlier, the set of all such $\mathbf{x}$-vectors constitutes the *efficient set* (in investment proportion space) and the set of all images of the efficient points constitutes the *nondominated set* (in criterion space). Thus, with U as in (3), (c) is the most appropriate equivalent deterministic problem among the five. Note that with respect to the extreme values of $\lambda = 0$ (risk neutrality) or $\lambda \to \infty$ (extreme risk aversion), we obtain possibility (a) or (b), respectively, as special cases of (c). It should be noted that since the limit function of U does not exist for $\lambda \to \infty$, (b) is not directly obtained as an expected utility maximizing solution. It is only obtained as the limit of expected utility solutions for increasing risk aversion.

Should we consider another extreme situation in which

$$U(x) = \begin{cases} 1, & c + \varepsilon \leq x \\ (x-c)/\varepsilon, & c \leq x < c + \varepsilon, \\ 0, & x < c \end{cases}$$

with an unknown parameter c and an $\varepsilon > 0$, we could observe that

$$P(r \geq c) \geq E[U(r)] \geq P(r \geq c + \varepsilon).$$

For a continuous random variable r, we obtain $E[U(r)] = P(r \geq c)$ for $\varepsilon \to 0$, which would lead to candidates (d) and (e). For instance, let c be the risk-free rate of return. Then candidate (d) would mean that the probability to receive at least the risk-free rate of return on a portfolio is maximized. If $r = (r_1, \ldots, r_n)$ follows a multinormal distribution, in the case of c equaling the risk-free rate, solving (d) then yields Roy's "safety first" portfolio. Again, it should be noted that (d) and (e) are not obtained as expected utility maximizing solutions,[4] but only as the limit of expected utility solutions for an increasing focus on c.

---

[4] While the limit function of U does exist for $\varepsilon \to 0$, it contradicts the Archimedean axiom of von Neumann and Morgenstern (1947), i.e., the function is discontinuous.

Although not mentioned in Caballero et al. (2001), a sixth equivalent deterministic possibility stemming from (1) is

(f) $\max\{E[r_p]\}$
$\min\{\text{Var}[r_p]\}$
$\max\{\text{Skew}[r_p]\}$
s.t. $\mathbf{x} \in S,$

where Skew stands for skewness. With criterion vectors of length three, **(f)** is a multiple criteria portfolio selection problem. This formulation is probably the only multiple criteria formulation that is not totally unfamiliar to conventional portfolio selection as a result of the interest taken in skewness by authors such as Stone (1973), Konno and Suzuki (1995), Chunhachinda et al. (1997), and Prakash et al. (2003). However, we will not dwell on **(f)**, as this formulation, as a result of the severe nonlinearities of its third criterion, has not gained much traction in practice. Instead, we will concentrate on the newer types of multiple criteria portfolio selection problems that have begun to appear as a result of the more sophisticated purposes of many investors.

4 Portfolio Selection with Multiple-Argument Utility Functions

Whereas multiple criteria formulations are little more than a curiosity in conventional portfolio selection, multiple criteria formulations are mostly appropriate when attempting to meet the modeling needs of investors with multiple-argument utility functions. Two situations in which multiple-argument utility functions are likely to occur are as follows.

One is that in addition to portfolio return, an investor has other considerations, such as to maximize social responsibility or to minimize the number of securities in a portfolio, that are also important to the investor. That is, instead of being interested in solely maximizing the stochastic objective of portfolio return, the investor can be viewed as being interested in optimizing some combination of several stochastic and several deterministic objectives.

A second situation in which a multiple-argument utility function might pertain is when an investor is unwilling to accept the assumption that all means μ_i, variances σ_{ii}, and covariances σ_{ij} can be treated as known at the beginning of the holding period. In response, an investor might wish to monitor the construction of his or her portfolio with the help of additional measures such as dividends, growth in sales, amount invested in R&D, and so forth, to guard against relying on any single measure that might have imperfections associated with it.

Let z_1 be alternative notation for r_p. Then, a list of z_i criterion values, from which arguments might be selected to staff an investor's multiple-argument utility function, is as follows:

max{ z_1 = portfolio return}
max{ z_2 = dividends}
max{ z_3 = growth in sales}
max{ z_4 = social responsibility}
max{ z_5 = liquidity}
max{ z_6 = portfolio return over that of a benchmark}

max{ z_7 = amount invested in R&D}

min{ z_8 = deviations from asset allocation percentages}
min{ z_9 = number of securities in portfolio}
min{ z_{10} = turnover (i.e., costs of adjustment)}
min{ z_{11} = maximum investment proportion weight}
min{ z_{12} = amount of short selling}
min{ z_{13} = number of securities sold short}

Of course, other z_i can be imagined. Note the differences between the first six and last six of the z_i. For the first six, it is not possible to know the realized values of the z_i until the end of the holding period. Depending in turn upon random variables associated with each of the n securities, these z_i, like z_1, are themselves random variables. Thus, the first six are stochastic objectives.

For the last six z_i, the actual values of these z_i, for any investment proportion vector **x**, are available at the beginning of the holding period. For example, for any investment proportion vector **x**, z_9 is given by the number of nonzero components in **x**. With the last six z_i known in this way at the beginning of the holding period, they are deterministic objectives.

As for z_7 in the middle, it is an example of a measure that could be argued either way. It could be argued that only the most recent amounts invested in R&D are relevant to the situation at the end of the holding period, thus enabling the objective to be treated deterministically.

One might ask why extra objectives can't be handled by means of constraints. The difficulty is in the setting of the right-hand sides of the constraints. In general, for a model to produce a mean-variance nondominated frontier that contains the criterion vector of an optimal portfolio, one would need to know the optimal value of each objective modeled as a constraint prior to computing the frontier. It is not likely that this would be possible in many situations.

With z_1 almost certainly an argument of every investor's utility function, additional arguments depend upon the investor. For instance, one investor's set of arguments might consist of $\{z_1, z_2, z_{10}\}$, and another's might consist of $\{z_1, z_5, z_7, z_8, z_{11}\}$. The point is that all investors need not be the same. If we let k be the number of selected objectives, in the case of the first investor, $k = 3$, and in the case of the second investor, $k = 5$. Of course, a conventional mean-variance investor's set of arguments would only be $\{z_1\}$, in which case $k = 1$.

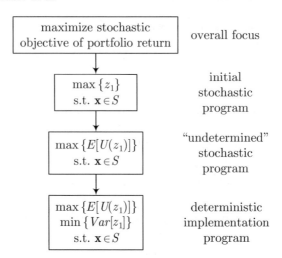

Fig. 1. Hierarchical structure of the overall focus, initial stochastic program, equivalent "undetermined" deterministic program, and equivalent deterministic implementation program of conventional portfolio selection.

The differences between conventional portfolio selection and multiple criteria portfolio selection are highlighted in Figures 1 and 2. At the top of each, as in Saaty's Analytic Hierarchy Process (1999), is the investor's *overall focus*. In Figure 1, the overall focus is to maximize the portfolio return random variable. In Figure 2, the overall focus is to optimize some combination of stochastic and deterministic objectives. In the second box of each is the investor's initial stochastic programming problem. Note that the initial stochastic programming problem in Figure 2 reflects the multiple stochastic and deterministic objectives involved in the investor's overall focus and hence is a *multiobjective* stochastic program. As for notation in the second, third, and fourth boxes of Figure 2, η specifies the number of stochastic objectives of concern and $D_{i_{\eta+1}}(\mathbf{x})$ represents the first of the $k - \eta$ deterministic objectives of concern. For instance, if $D_{13}(\mathbf{x})$ were included, then $D_{13}(\mathbf{x})$ would represent a function that returns the number of negative x_i.

In the third box of Figure 2 is the equivalent "undetermined" deterministic problem

$$\max\{E[U(z_{i_1}, \ldots, z_{i_\eta}, z_{i_{\eta+1}}, \ldots, z_{i_k})]\} \qquad (4)$$
$$\text{s.t.} \quad \mathbf{x} \in S,$$

which shows the multiple-argument utility function that follows from the investor's overall focus. Employing a mean-variance pair for each stochastic argument of the utility function, we have the equivalent deterministic implementation program of the bottom box. We use the term "implementation" because this is the actual deterministic problem that is implemented. Note

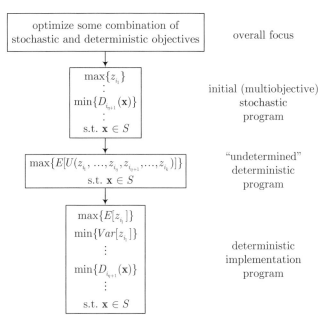

Fig. 2. Hierarchical structure of the overall focus, initial (multiobjective) stochastic program, equivalent "undetermined" deterministic program, and equivalent deterministic implementation program of multiple criteria portfolio selection.

that all deterministic objectives of the initial (multiobjective) stochastic program are repeated unchanged in the equivalent deterministic implementation program.

As a practical matter, for stochastic objectives in which variation is small (or not) of noteworthy importance, it may be possible to represent them in the equivalent deterministic implementation program of the bottom box of Figure 2 by **(a)** instead of **(c)**. This would be very advantageous when possible. For example, suppose an investor's set is $\{z_1, z_2, z_5\}$. Since these objectives are linear in the portfolio weights, the investor's initial (multiobjective) stochastic program would be

$$\max\{ z_1 = \sum_{j=1}^{n} r_j x_j \}$$

$$\max\{ z_2 = \sum_{j=1}^{n} d_j x_j \}$$

$$\max\{ z_5 = \sum_{j=1}^{n} \ell_j x_j \}$$

$$\text{s.t.} \quad \mathbf{x} \in S$$

in which d_j is the random variable for the dividends, and ℓ_j is the random variable for the liquidity, of the jth security. Should variations in portfolio dividends and portfolio liquidity be much less important than variations in portfolio return, then it may well be acceptable to use **(a)** instead of **(c)** for each of the dividends and liquidity. Then the resulting equivalent deterministic implementation program would be

$$\max\{E[z_1]\} \tag{5}$$
$$\min\{\text{Var}[z_1]\}$$
$$\max\{E[z_2]\}$$
$$\max\{E[z_5]\}$$
$$\text{s.t.} \quad \mathbf{x} \in S.$$

The advantage of being able to use **(a)** instead of **(c)** with stochastic objectives beyond portfolio return is, of course, that a Var objective for each such objective can be eliminated from the equivalent deterministic implementation program. This not only simplifies data collection requirements (as it is necessary to know only the means of the relevant random variables) but also lessens the burden on computing the nondominated set.

5 Mean-Variance Nondominated Sets

We now utilize matrix notation when convenient. To prepare for the application of the four stages of the Markowitz solution procedure to multiple criteria portfolio selection problems, it is useful to study in a little greater detail the mean-variance formulation in the bottom box of Figure 1:

$$\max\{E[z_1] = \boldsymbol{\mu}^T \mathbf{x}\} \tag{6}$$
$$\min\{\text{Var}[z_1] = \mathbf{x}^T \boldsymbol{\Sigma} \mathbf{x}\}$$
$$\text{s.t.} \quad \mathbf{x} \in S$$

in which $\boldsymbol{\mu} \in \mathbb{R}^n$ is the expected value vector of the r_i and $\boldsymbol{\Sigma} \in \mathbb{R}^{n \times n}$ is the covariance matrix of the σ_{ij}. In this problem, the efficient set is a piecewise linear path in S. The nondominated set, being the set of images of all efficient points, is piecewise parabolic in $(\text{Var}[z_1], E[z_1])$ space. This means that when portrayed in $(\text{Stdev}[z_1], E[z_1])$ space, the nondominated set is piecewise hyperbolic. Although theory and computation are customarily carried out in $(\text{Var}[z_1], E[z_1])$ space, we mention $(\text{Stdev}[z_1], E[z_1])$ space, as most nondominated sets are communicated to investors in this space.

When the feasible region is the $\mathbf{1}^T \mathbf{x} = 1$ hyperplane as in

$$S = \{\mathbf{x} \in \mathbb{R}^n \mid \mathbf{1}^T \mathbf{x} = 1\,\}, \tag{7}$$

the efficient and nondominated sets are straightforward. The efficient set is a (single) straight line in the hyperplane, bounded at one end and unbounded at the other. The nondominated set is the top half of a (single) hyperbola. And, as a consequence of the $\mathbf{1}^T\mathbf{x} = 1$ nature of S, the efficient and nondominated sets can, after taking the Lagrangian, be obtained by formula (see, for instance, Campbell et al., 1997).

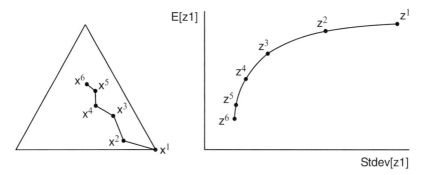

Fig. 3. Piecewise linear efficient set in S (left) and piecewise hyperbolic nondominated set in $(\text{Stdev}[z_1], E[z_1])$ space (right).

However, as soon as additional constraints become involved, as in

$$S = \{\mathbf{x} \in \mathbb{R}^n \mid \mathbf{1}^T\mathbf{x} = 1,\ \alpha_i \leq x_i \leq \omega_i\}, \tag{8}$$

thereby making S a subset of the $\mathbf{1}^T\mathbf{x} = 1$ hyperplane, the situation becomes more complicated. To illustrate, consider Figure 3 (which might correspond to a problem with about eight securities). With a little poetic license mandated by the fact that it is not possible to draw a graph in 8-space, the graph on the left is intended to portray (a) the subset of the $\mathbf{1}^T\mathbf{x} = 1$ hyperplane that is S and (b) the efficient set, which is normally a piecewise linear path. On the right in $(\text{Stdev}[z_1], E[z_1])$ space is the nondominated set (or frontier). Corresponding to the five segments of the piecewise linear path, the nondominated frontier consists of five hyperbolic segments. Note that the inverse images of the endpoints of a given nondominated hyperbolic segment are the endpoints of the efficient line segment that generates the hyperbolic segment. For instance, the inverse images of $\mathbf{z}^1$ and $\mathbf{z}^2$ are $\mathbf{x}^1$ and $\mathbf{x}^2$, respectively.

A property of a nondominated hyperbolic segment is that along the segment excluding its endpoints, the securities in a portfolio remain the same. Only their proportions change as we move along the segment. Securities can only leave a nondominated portfolio at an endpoint, and securities can only enter a nondominated portfolio if we cross over an endpoint (such as $\mathbf{z}^2$) to an adjacent nondominated hyperbolic segment. As for the number of nondominated hyperbolic segments, the larger the problem, the greater the number of nondominated hyperbolic segments. For instance, a problem with 100 securities might have 30 to 60 nondominated hyperbolic segments. Apart from when S is the entire $\mathbf{1}^T\mathbf{x} = 1$ hyperplane, mathematical programming is now the tool for obtaining information about efficient and nondominated sets.

What about software? In the past there was the IBM (1965) code. Early computer codes suffered from two problems. One was speed and the other was core (i.e., memory). Because of the amount of core required for storing a dense covariance matrix, methods for "sparsifying" or "simplifying" the covariance matrix structure all but dominated portfolio optimization research for the next 20 years. Also, there was debate about whether a portfolio code should be "parametric" or "one-at-a-time." A *parametric* code is one that is able to define the nondominated frontier as a function of some single parameter. A *one-at-a-time* code simply computes points, one at a time, on the nondominated frontier, for instance, by repetitively solving the "e-constraint" formulation[5]

$$\min\{\mathbf{x}^T \mathbf{\Sigma} \mathbf{x}\} \tag{9}$$
$$\text{s.t.} \quad \boldsymbol{\mu}^T\mathbf{x} \geq \rho$$
$$\mathbf{x} \in S$$

for different values of ρ. Then, with the points obtained, representations of the nondominated frontier as in Figure 4 can be prepared.

In the 1980s there was the Perold code. For achieving a breakthrough with large-scale problems (500 securities was considered large-scale at the time), the code was predicated upon a covariance matrix structure sparsified according to the techniques in Markowitz and Perold (1981a, 1981b) and Perold (1984). Algorithmically drawing upon Markowitz (1956), the code was not one at a time, but parametric as it was able to compute parametrically the nondominated frontier. Having been programmed on older platforms, neither the IBM code nor Perold's code is in distribution today. A paper describing the latest developments in portfolio optimization up until the early to mid 1990s is by Pardalos et al. (1994).

The current situation is eclectic. On the one hand, there are "proprietary systems," which are not really intended for university use. Rather, they are intended for integration into the computing systems of (mainly large) firms in the financial services industry. Depending upon the modifications necessary to fit into a client's computer system, the number of users, and the amount of training involved, such systems can easily run into the tens of thousands of dollars. Two such proprietary systems are FortMP, as alluded to in Mitra et al. (2003), and QOS which developed out of the works of Best (1996), Best and Kale (2000), and Best and Hlouskova (2005). Each has its own features, is designed for large-scale problems, and performs at a high speed. However, they do not parametrically specify the nondominated frontier. Rather, they

---

[5] In multiple criteria optimization, programs with all objectives but one converted to constraints are often called *e-constraint* formulations.

compute points on the nondominated frontier, thus resulting in both being classified as one-at-a-time.

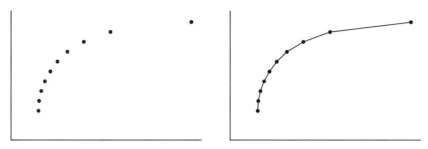

Fig. 4. Dotted (left) and piecewise linear (right) representations of a nondominated frontier.

As for codes more suitable for university use, there is the public domain code Optimizer given in the appendix of Markowitz and Todd (2000). Optimizer is parametric, as it implements the critical line algorithm of Markowitz (1956). It is written in VBA (Visual Basic for Applications). However, as of this writing, it is limited to 248 securities.

One might think that commands for computing the linear segments of the efficient set and the hyperbolic segments of the nondominated frontier of (6) would be included in packages such as Cplex, Mathematica, MATLAB, LINGO, SAS, and premium versions of Solver, but this is not the case. Other than for the simplistic case when S is the $\mathbf{1}^T\mathbf{x} = 1$ hyperplane, the best that can be done with the packages is to write routines within them to compute points on the nondominated frontier utilizing formulations such as (9), thus consigning us, with the packages, to an essentially one-at-a-time world.

6 Solving a Multiple Criteria Portfolio Selection Problem

Building upon knowledge gained in the previous section, we are now able to discuss the task of solving a multiple criteria portfolio selection problem. While the protocol of computing the nondominated set, communicating it to the investor, searching the nondominated set for a most preferred point, and then taking an inverse image of the selected point still remains intact, the first three stages present much greater difficulties. As for the equivalent deterministic implementation program of the bottom box of Figure 2, different types of formulations may result. For the purposes of our discussion, we divide them into three categories: (1) those with one quadratic and two or more linear objectives, (2) those with two or more quadratic and one or more linear objectives, and (3) those with one or more non-smooth objective functions [for instance, $D_9(\mathbf{x})$, which is to minimize the number of securities, is non-smooth].

6.1 One Quadratic and Two or More Linear Objectives

Although many of the problems that can emerge in the bottom box of Figure 2 cannot, given where we are in the development of multiple criteria portfolio selection, yet be effectively addressed, progress is being made on 1-quadratic 2-linear and 1-quadratic 3-linear problems in Hirschberger, et al. (2007), and on this we comment. For instance, whereas the nondominated frontier of a mean-variance problem is piecewise hyperbolic in $(\text{Stdev}[z_1], E[z_1])$ space, the nondominated set (surface) of a 1-quadratic multilinear multiple criteria portfolio selection is platelet-wise *hyperboloidic* in $(\text{Stdev}[z_1], E[z_1], E[z_2], \ldots)$ space. That is, the nondominated set is composed of patches, with each patch coming from the surface of a different hyperboloid. Also, whereas the efficient set in the mean-variance case is a path of linear line segments, the efficient set of a 1-quadratic multilinear problem is a connected union of low-dimensional polyhedra in S.

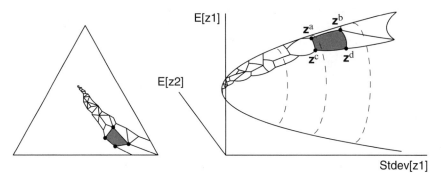

Fig. 5. Efficient set in S (left) and platelet-wise hyperboloidic nondominated set of a 1-quadratic 2-linear multiple criteria portfolio selection problem in $(\text{Stdev}[z_1], E[z_1], E[z_2])$ space (right).

Consider Figure 5. On the left, the efficient set is portrayed showing the polyhedral subsets of which it is composed. On the right, the nondominated set is portrayed showing the hyperboloidic platelets of which it is composed. Note that not all platelets are of the same size, and they generally decrease in size the closer we are to the minimum standard deviation point. This is normal. Also, it is normal for not all platelets to have the same number of (platelet) *corner points*. Note the correspondence between the nondominated hyperboloidic platelets and the efficient polyhedral subsets. For instance, the platelet defined by corner points $\mathbf{z}^a$, $\mathbf{z}^b$, $\mathbf{z}^c$, $\mathbf{z}^d$ would be associated with a polyhedral subset such as the shaded one with four extreme points on the left. And as in the mean-variance case, all portfolios in the relative interior of a platelet contain the same securities, just in different proportions, and for a security to leave or for a new one to enter, one would need to cross the boundary to another platelet.

With regard to computing the nondominated set, we are able to report limited computational results using a code under development in Hirschberger et al. (2007). The results obtained are for 1-quadratic 2-linear problems whose covariance matrices are 100% dense. With $n = 200$ securities, 1-quadratic 2-linear problems were found to have in the neighborhood of about 1,000 nondominated platelets, taking on average about 10 seconds to compute. The computer used was a Dell 2.13GHz Pentium M Centrino laptop. With $n = 400$ securities, 1-quadratic 2-linear problems were found to have in the neighborhood of about 2,000 nondominated platelets, taking on average about one minute to compute. These encouraging results lead us to believe that the nondominated sets of larger problems (in terms of either the number of securities or the number of linear objectives) are computable in a reasonable time.

Unfortunately, it is not as easy to display the nondominated set in multiple criteria portfolio selection as in mean-variance portfolio selection. In problems with criterion vectors of length three, 3D graphics can be used, but in problems with more objectives, probably about the best that can be done is to enable the investor to learn about the nondominated set while in the process of searching for a most preferred point.

As for searching the nondominated set of a problem such as in Figure 5, one approach is to discretize the nondominated set to some desired level of resolution. This can be accomplished as follows. For each polyhedral subset of the efficient set, take convex combinations of its extreme points. Because platelet size tends to increase the more distant the platelet is from the minimum standard deviation point, one would probably want to increase the number of convex combinations the farther the platelet is away. Then with perhaps tens of thousands, if not hundreds of thousands, of nondominated points generated in this way, the question is how to locate a most preferred. Four strategies come to mind. One is to employ interactive multiple probing as in the Tchebycheff method described in Steuer et al. (1993). Another is to pursue a projected line search strategy as in Korhonen and Wallenius (1988) and Korhonen and Karaivanova (1999). A third is to utilize an interactive criterion vector component classification scheme as, for instance, in Miettinen (1999). And a fourth might involve the utilization of some of the visualization techniques described in Lotov et al. (2004).

6.2 Two or More Quadratic and One or More Linear Objectives

To illustrate what can be done in this category, consider the 2-quadratic 2-linear problem

$$\min\{\mathbf{x}^T\mathbf{\Sigma}_1\mathbf{x}\} \qquad (10)$$
$$\min\{\mathbf{x}^T\mathbf{\Sigma}_2\mathbf{x}\}$$
$$\max\{\boldsymbol{\mu}^T\mathbf{x}\}$$

$$\max\{\boldsymbol{\nu}^T\mathbf{x}\}$$
$$\text{s.t.} \quad \mathbf{x} \in S,$$

where $\boldsymbol{\Sigma}_1$ and $\boldsymbol{\Sigma}_2$ are positive definite and S is defined, for instance, as in (8). Given that $\boldsymbol{\Sigma}_1$ and $\boldsymbol{\Sigma}_2$ are positive definite, any convex combination

$$\bar{\boldsymbol{\Sigma}} = \lambda\boldsymbol{\Sigma}_1 + (1-\lambda)\boldsymbol{\Sigma}_2$$

(where $\lambda \in [0,1]$) renders $\bar{\boldsymbol{\Sigma}}$ positive definite. This means that the nondominated set of the 1-quadratic 2-linear

$$\min\{\mathbf{x}^T\bar{\boldsymbol{\Sigma}}\mathbf{x}\} \tag{11}$$
$$\max\{\boldsymbol{\mu}^T\mathbf{x}\}$$
$$\max\{\boldsymbol{\nu}^T\mathbf{x}\}$$
$$\text{s.t.} \quad \mathbf{x} \in S$$

is a subset of the nondominated set of (10). Thus, by solving (11) for a series of different λ-values, we should be able to obtain a covering of the nondominated set of the original 2-quadratic 2-linear. The covering can then be handled in the same way as in the previous subsection.

6.3 One or More Non-Smooth Objectives

When a multiobjective equivalent deterministic problem possesses a non-smooth objective (such as to minimize the number of securities), we face major difficulties in that the problem is no longer continuous. Moreover, the nonpositive hull of the nondominated set might not even be convex. In addition, a problem might possess non-smooth constraints, for instance, in the form of semicontinuous variables (variables that are either zero or in some interval $[a,b]$ where a is materially greater than zero). Possibly, the only way to attack such problems is to utilize evolutionary algorithms such as set forth in Deb (2001).

7 Conclusions

For investors with additional concerns, portfolio selection need not only be looked at within a mean-variance framework. Steps can now be taken to integrate additional concerns into the portfolio optimization process more in accordance with their criterion status. Instead of attempting to interject additional concerns into portfolio selection by means of constraints — an *ad hoc* process that often ends prematurely because of losses in user patience — the methods that have been outlined form the basis for a new era of solution methodologies whose purposes are to converge to a final portfolio that more formally achieves optimal trade-offs among all of the criteria that the investor wishes to deem important. Of course, as with any area that is gaining momentum, more work needs to be done.

References

1. Aouni, B., Ben Abdelaziz, F., and El-Fayedh, R. *Chance constrained compromise programming for portfolio selection*. Laboratoire LARODEC, Institut Superieur de Gestion, La Bardo 2000, Tunis, Tunisia, 2006.
2. Arenas Parra, M., Bilbao Terol, A., and Rodríguez Uría, M. V. A fuzzy goal programming approach to portfolio selection. *European Journal of Operational Research*, 133(2):287–297, 2001.
3. Ballestero, E. Using compromise programming in a stock market pricing model. In Y. Y. Haimes and R. E. Steuer, Editors, *Lecture Notes in Economics and Mathematical Systems*, vol. 487. Springer-Verlag, Berlin, 2002, pages 388–399.
4. Ballestero, E., and Plà-Santamaría, D. Selecting portfolios for mutual funds. *Omega*, 32:385–394, 2004.
5. Ballestero, E., and Romero, C. Portfolio selection: A compromise programming solution. *Journal of the Operational Research Society*, 47(11):1377–1386, 1996.
6. Bana e Costa, C. A., and Soares, J. O. Multicriteria approaches for portfolio selection: An overview. *Review of Financial Markets*, 4(1):19–26, 2001.
7. Bana e Costa, C. A., and Soares, J. O. A multicriteria model for portfolio management. *European Journal of Finance*, 10(3):198–211, 2004.
8. Bernoulli, D. Specimen theoria novae de mensura sortis. *Commentarii Academiae Scientarum Imperialis Petropolitnae*, 5(2):175–192, 1738. Translated into English by L. Sommer, Exposition of a new theory on the measurement of risk, *Econometrica*, 22(1):23–26, 1954.
9. Best, M. J. An algorithm for the solution of the parametric quadratic programming problem. In B. Riedmüller H. Fischer and S. Schäffler, Editors, *Applied Mathematics and Parallel Computing: Festschrift for Klaus Ritter*. Physica-Verlag, Berlin, 1996, pages 57–76.
10. Best, M. J., and Hlouskova, J. An algorithm for portfolio optimization with transaction costs. *Management Science*, 51(11):1676–1688, 2005.
11. Best, M.J., and Kale, J. Quadratic programming for large-scale portfolio optimization. In J. Keyes, Editor, *Financial Services Information Systems*. CRC Press, Boca Raton, FL, 2000, pages 513–529.
12. Bouri, G., Martel, J. M., and Chabchoub, H. A multi-criterion approach for selecting attractive portfolio. *Journal of Multi-Criteria Decision Analysis*, 11(3):269–277, 2002
13. Caballero, R., Cerdá, E., Muñoz, M. M., Rey, L., and Stancu-Minasian, I. M. Efficient solution concepts and their relations in stochastic multiobjective programming. *Journal of Optimization Theory and Applications*, 110(1):53–74, 2001
14. Campbell, J. Y., Lo, A. W., and Mackinlay, A. C. *The Econometrics of Financial Markets*. Princeton University Press, Princeton, NJ, 1997.
15. Chow, G. Portfolio selection based on return, risk, and relative performance. *Financial Analysts Journal*, 54–60, March-April 1995.
16. Chunhachinda, P., Dandapani, K., Hamid, S., and Prakash, A. J. Portfolio selection and skewness evidence from international stock markets. *Journal of Banking & Finance*, 21:143–167, 1997.
17. Colson, G., and DeBruyn, C. An integrated multiobjective portfolio management system. *Mathematical and Computer Modelling*, 12(10-11):1359–1381, 1989

18. Deb, K. *Multi-Objective Optimization Using Evolutionary Algorithms*. John Wiley, New York, 2001.
19. Dominiak, C. An application of interactive multiple objective goal programming on the Warsaw stock exchange. In R. Caballero, F. Ruiz, and R. E. Steuer, Editors, *Lecture Notes in Economics and Mathematical Systems*, vol. 455. Springer-Verlag, Berlin, 1997a, pages 66–74.
20. Dominiak, C. Portfolio selection using the idea of reference solution. In G. Fandel and T. Gal, Editors, *Lecture Notes in Economics and Mathematical Systems*, vol. 448. Springer-Verlag, Berlin, 1997b, pages 593–602.
21. Doumpos, M., Spanos, M., and Zopounidis, C. On the use of goal programming techniques in the assessment of financial risks. *Journal of Euro-Asian Management*, 5(1):83–100, 1999.
22. Ehrgott, M., Klamroth, K., and Schwehm, C. An MCDM approach to portfolio optimization. *European Journal of Operational Research*, 155(3):752–770, 2004.
23. Elton, E. J., Gruber, M. J., Brown, S. J., and Goetzmann, W. *Modern Portfolio Theory and Investment Analysis*, 6th edition. John Wiley, New York, 2002.
24. Fliege, J. Gap-free computation of Pareto-points by quadratic scalarizations. *Mathematical Methods of Operations Research*, 59(1):69–89, 2004
25. Hallerbach, W. G., and Spronk, J. A multidimensional framework for financial-economic decisions. *Journal of Multi-Criteria Decision Analysis*, 11(3):111–124, 2002a.
26. Hallerbach, W.G., and Spronk, J. The relevance of MCDM for financial decisions. *Journal of Multi-Criteria Decision Analysis*, 11(4-5):187–195, 2002b.
27. Hirschberger, M., Qi, Y., and Steuer, R. E. Tri-criterion quadratic-linear programming. Working paper, Department of Banking and Finance, University of Georgia, Athens, 2007.
28. Hurson, C., and Zopounidis, C. On the use of multi-criteria decision aid methods to portfolio selection. *Journal of Euro-Asian Management*, 1(2):69–94, 1995.
29. IBM 1401 Portfolio Selection Program (1401-F1-04X) Program Reference Manual. IBM, New York, 1965.
30. Kliber, P. A three-criteria portfolio selection: Mean return, risk and costs of adjustment. Akademia Ekonomiczna w Poznaniu, Poznan, Poland, 2005.
31. Konno, H., and Suzuki, K.-I. A mean-variance-skewness portfolio optimization model. *Journal of the Operations Research Society of Japan*, 38(2):173–187, 1995.
32. Konno, H., Shirakawa, H., and Yamazaki, H. A mean-absolute deviation-skewness portfolio optimization model. *Annals of Operations Research*, 45:205–220, 1993.
33. Korhonen, P., and Karaivanova, J. An algoithm for projecting a reference direction onto the nondominated set of given points. *IEEE Transactions on Systems, Man, and Cybernetics*, 29(5):429–435, 1999.
34. Korhonen, P., and Wallenius, J. A Pareto race. *Naval Research Logistics*, 35(6):615–623, 1988.
35. Korhonen, P., and Yu, G.-Y. A reference direction approach to multiple objective quadratic-linear programming. *European Journal of Operational Research*, 102(3):601–610, 1997
36. Lee, S. M., and Lerro, A. J. Optimizing the portfolio selection for mutual funds. *Journal of Finance*, 28(5):1087–1101, 1973.

37. L'Hoir, H., and Teghem, J. Portfolio selection by MOLP using interactive branch and bound. *Foundations of Computing and Decision Sciences*, 20(3):175–185, 1995
38. Lo, A. W., Petrov, C., and Wierzbicki, M. It's 11pm - Do you know where your liquidity is? The mean-variance-liquidity frontier. *Journal of Investment Management*, 1(1):55–93, 2003.
39. Lotov, A.B., Bushenkov, V.A., and Kamenev, G.K. *Interactuve Decision Maps: Approximation and Visualization of Pareto Frontier.* Kluwer Academic Publishers, Boston, 2004.
40. Markowitz, H. M. Portfolio selection. *Journal of Finance*, 7(1):77–91, 1952.
41. Markowitz, H. M. The optimization of a quadratic function subject to linear constraints. *Naval Research Logistics Quarterly*, 3:111–133, 1956.
42. Markowitz, H. M. The early history of portfolio selection: 1600-1960. *Financial Analysts Journal*, 5–16, July 1999.
43. Markowitz, H.M., and Perold, A. Portfolio analysis with factors and scenarios. *Journal of Finance*, 36(14):871–877, 1981a.
44. Markowitz, H. M., and Perold, A. Sparsity and piecewise linearity in large scale portfolio optimization problems. In I. Duff, Editor, *Sparse Matrices and Their Use,* (The Institute of Mathematics and Its Applications Conference Series). Academic Press, New York, 1981b, pages 89–108.
45. Markowitz, H. M., and Todd, G. P. *Mean-Variance Analysis in Portfolio Choice and Capital Markets.* Frank J. Fabozzi Associates, New Hope, PA, 2000.
46. Miettinen, K. M. *Nonlinear Multiobjective Optimization.* Kluwer, Boston, 1999.
47. Mitra, G., Kyriakis, T., Lucas, C., and Pirbhai, M. A review of portfolio planning: models and systems. In S. E. Satchell and A. E. Scowcroft, Editors, *Advances in Portfolio Construction and Implementation.* Butterworth-Heinemann, Oxford, 2003, pages 1–39.
48. Ogryczak, W. Multiple criteria linear programming model for portfolio selection. *Annals of Operations Research*, 97:143–162, 2000.
49. Pardalos, P. M., Sandström, M., and Zopounidis, C. On the use of optimization models for portfolio selection: A review and some computational results. *Computational Economics*, 7(4):227–244, 1994.
50. Perold, A. Large-scale portfolio optimization. *Management Science*, 30(10):1143–1160, 1984.
51. Prakash, A. J., Chang, C. H., and Pactwa, T. E. Selecting a portfolio with skewness: Recent evidence from US, European, and Latin American equity markets. *Journal of Banking & Finance*, 27:1375–1390, 2003.
52. Prékopa, A. *Stochastic Programming.* Kluwer Academic Publishers, Dordrecht, 1995.
53. Roy, A. D. Safety first and the holding of assets. *Econometrica*, 20(3):431–449, 1952.
54. Saaty, T. L. *Decision Making for Leaders.* RWS Publications, Pittsburgh, 1999.
55. Slowinski, R., and Teghem, J., Editors *Stochastic Versus Fuzzy Approaches to Multiobjective Mathematical Programming Under Uncertaintyd.* Kluwer Academic Publishers, Dordrecht, 1990.
56. Spronk, J. *Interactive Multiple Goal Programming: Applications to Financial Management.* Martinus Nijhoff Publishing, Boston, 1981.
57. Spronk, J., and Hallerbach, W.G. Financial modelling: Where to go? With an illustration for portfolio management. *European Journal of Operational Research*, 99(1):113–127, 1997.

58. Spronk, J., Steuer, R. E., and Zopounidis, C. Multicriteria decision analysis/aid in finance. In J. Figuiera, S. Greco, and M. Ehrgott, Editors, *Multiple Criteria Decision Analysis: State of the Art Surveys*. Springer Science, New York, 2005, pages 799–857.
59. Stancu-Minasian, I. *Stochastic Programming with Multiple-Objective Functions*. D. Reidel Publishing Company, Dordrecht, 1984.
60. Steuer, R. E., and Na, P. Multiple criteria decision making combined with finance: A categorized bibliography. *European Journal of Operational Research*, 150(3):496–515, 2003
61. Steuer, R. E., Qi, Y., and Hirschberger, M. Multiple objectives in portfolio selection. *Journal of Financial Decision Making*, 1(1):5–20, 2005.
62. Steuer, R. E., Qi, Y., and Hirschberger, M. Developments in multi-attribute portfolio selection. In T. Trzaskalik, Editor, *Multiple Criteria Decision Making '05*. Karol Adamiecki University of Economics in Katowice, 2006a, pages 251–262.
63. Steuer, R. E., Qi, Y., and Hirschberger, M. Suitable-portfolio investors, nondominated frontier sensitivity, and the effect of multiple objectives on standard portfolio selection. *Annals of Operations Research*, 2006b. forthcoming.
64. Steuer, R. E., Silverman, J., and Whisman, A. W. A combined Tchebycheff/aspiration criterion vector interactive multiobjective programming procedure. *Management Science*, 39(10):1255–1260, 1993.
65. Stone, B. K. A linear programming formulation of the general portfolio selection problem. *Journal of Financial and Quantitative Analysis*, 8(4):621–636, 1973
66. Tamiz, M., Hasham, R., Jones, D. F., Hesni, B., and Fargher, E. K. A two staged goal programming model for portfolio selection. In M. Tamiz, Editor, *Lecture Notes in Economics and Mathematical Systems,* vol. 432. Springer-Verlag, Berlin, 1996, pages 386–399.
67. Von Neumann, J., and Morgenstern, O. *Theory of Games and Economic Behavior*. 2nd Edition, Princeton, NJ, 1947.
68. Xu, J., and Li, J. A class of stochastic optimization problems with one quadratic & several linear objective functions and extended portfolio selection model. *Journal of Computational and Applied Mathematics*, 146:99–113, 2002
69. Yu, G. Y. A multiple criteria approach to choosing an efficient stock portfolio at the Helsinki stock exchange. *Journal of Euro-Asian Management*, 3(2):53–85, 1997.
70. Ziemba, W. T. *The Stochastic Programming Approach to Asset, Liability, and Wealth Management*. Research Foundation of the AIMR, Charlottesville, VA, 2003.

Applications of Integer Programming to Financial Optimization

Hiroshi Konno[1] and Rei Yamamoto[1,2]

[1] Department of Industrial and Systems Engineering, Chuo University, Tokyo, Japan konno@indsys.chuo-u.ac.jp
[2] Mitsubishi UFJ Trust Investment Technology Institute Co., Ltd., Tokyo, Japan yamamoto@mtec-institute.co.jp

1 Introduction

The problems to be discussed in this chapter make up a class of nonconvex financial optimization problems that can be solved within a practical amount of time using the state-of-the-art integer programming methodologies.

We will first discuss mean-risk portfolio optimization problems (Elton and Gruber, 1998; Konno and Yamazaki, 1991; Markowitz, 1959) subject to nonconvex constraints such as minimal transaction unit constraints and cardinality constraints on the number of assets to be included in the portfolio (Konno and Yamamoto, 2005b). Also, we will discuss problems with piecewise linear nonconvex transaction costs (Konno and Wijayanayake, 2001, 2002; Konno and Yamamoto, 2005a, 2005b). It will be shown that fairly large-scale problems can now be solved to optimality by formulating the problem as a mixed 0−1 integer linear programming problem if we use convex piecewise linear risk measure such as absolute deviation instead of variance.

The second class of problems are so-called maximal predictability portfolio optimization problems (Lo and MacKinlay, 1997), where we maximize the coefficient of determination of the portfolio using factor models. This model, though very promising, was set aside long ago, since we need to maximize the ratio of convex quadratic functions, which is not a concave function. This problem can be solved to optimality by a hyper-rectangular subdivision algorithm (Gotoh and Konno, 2001; Phong et al., 1995) or by 0−1 integer programming approach (Yamamoto and Konno, to appear; Yamamoto et al., to appear) if the number of assets is relatively small.

To solve larger problems, we employ absolute deviation as a measure of variation and define the coefficient of determination as the ratio of functions defined by the sum of absolute values of linear functions. The resulting nonconvex minimization problem can be reformulated as a linear complementarity problem that can be solved by using 0−1 integer programming algorithms (Konno et al., 2007, to appear).

The third class of problems is the choice of financial attributes to be included in failure discriminant analysis (Galindo and Tamayo, 2000; Konno and Yamamoto, 2007). Which and how many financial attributes out of 100 candidates should be included in the model to achieve the best performance in failure prediction of a large number of small-to medium-scale enterprises?

We pose this combinatorial optimization problem as a 0−1 integer linear programming problem using the absolute deviation as the measure of variation. It has been demonstrated by Konno and Yamamoto (2005b, 2007) that medium-scale problems can be solved to optimality within a practical amount of time.

Our success is due to the remarkable developments of integer programming in the past decade. In 2002, Bixby reported that some class of 0−1 integer programming problems can be solved 2 million times faster than 15 years ago. This trend continues, and we can now solve almost 200 million times faster, so that very large scale-problems in scheduling, distribution, and supply chain management are solved to optimality.

Unfortunately, however the relatively few people in finance are aware of these remarkable developments. The important thing is that this trend is expected to continue for at least another decade. Therefore, more difficult and important financial optimization problems would be solved through a combination of integer programming, global optimizations, and heuristic algorithms.

2 Mean-Risk Portfolio Optimization Problems

2.1 Mean-Absolute Deviation Model

Let there be n assets $S_j, j = 1, 2, \ldots, n$, and let R_j be the random variable representing the rate of return of S_j. Then the rate of return $R(\boldsymbol{x})$ of portfolio $\boldsymbol{x} = (x_1, x_2, \ldots, x_n)^T$ is given by

$$R(\boldsymbol{x}) = \sum_{j=1}^{n} R_j x_j. \tag{1}$$

Let us define the absolute deviation of $R(\boldsymbol{x})$ as follows:

$$W(\boldsymbol{x}) = E[|R(\boldsymbol{x}) - E[R(\boldsymbol{x})]|]. \tag{2}$$

Theorem 2.1. If $R(\boldsymbol{x})$ follows a normal distribution with mean $r(\boldsymbol{x})$ and variance $\sigma^2(\boldsymbol{x})$, then

$$W(\boldsymbol{x}) = \sqrt{2/\pi}\sigma(\boldsymbol{x}). \tag{3}$$

Proof. See Konno and Koshizuka (2005). □

We will assume in the sequel that $\boldsymbol{R} \equiv (R_1, R_2, \ldots, R_n)$ is distributed over a finite set of points $(r_{1t}, r_{2t}, \ldots, r_{nt}), t = 1, 2, \ldots, T$, and that

$$p_t = Pr\{(R_1, R_2, \ldots, R_n) = (r_{1t}, r_{2t}, \ldots, r_{nt})\}, \quad t = 1, 2, \ldots, T, \quad (4)$$

is known in advance.

Let x_j be the proportion of the fund to be invested into S_j. Then the absolute deviation $W(\boldsymbol{x})$ of the portfolio $\boldsymbol{x} = (x_1, x_2, \ldots, x_n)^T$ is defined as follows:

$$W(\boldsymbol{x}) = \sum_{t=1}^{T} p_t \left| \sum_{j=1}^{n} (r_{jt} - r_j) x_j \right|, \quad (5)$$

where $r_j = \sum_{t=1}^{T} p_t r_{jt}$ is the expected value of R_j.

The mean-absolute deviation (MAD) model is defined as follows:

$$\begin{aligned} \text{maximize} \quad & \sum_{j=1}^{n} r_j x_j \\ \text{s.t.} \quad & W(\boldsymbol{x}) \leq w, \\ & \boldsymbol{x} \in X, \end{aligned} \quad (6)$$

where w is a constant representing the acceptable level of risk and $X \subset R^n$ is an investable set defined by

$$X = \left\{ \boldsymbol{x} \in R^n \;\middle|\; \sum_{j=1}^{n} x_j = 1, \; 0 \leq x_j \leq \alpha_j, \; j = 1, 2, \ldots, n, \right.$$

$$\left. \sum_{j=1}^{n} a_{ij} x_j \geq b_i, \; i = 1, 2, \ldots, m \right\}, \quad (7)$$

where $\sum_{j=1}^{n} a_{ij} x_j \geq b_i, \; i = 1, 2, \ldots, m$, are usually called institutional constraints. Readers are referred to Konno and Koshizuka (2005) and Ogryczak and Ruszczynski (1999) for basic properties of the MAD model.

Let us introduce a set of nonnegative variables ϕ_t, ψ_t satisfying the condition

$$\phi_t - \psi_t = p_t \sum_{j=1}^{n} (r_{jt} - r_j) x_j, \quad t = 1, 2, \ldots, T,$$

$$\phi_t \psi_t = 0, \; \phi_t \geq 0, \; \psi_t \geq 0, \quad t = 1, 2, \ldots, T.$$

The absolute deviation $W(\boldsymbol{x})$ is then represented as follows:

$$W(\boldsymbol{x}) = \sum_{t=1}^{T} (\phi_t + \psi_t).$$

Hence, the MAD model (6) reduces to

$$\begin{aligned}
\text{maximize} \quad & \sum_{j=1}^{n} r_j x_j \\
\text{s.t.} \quad & \sum_{t=1}^{T} (\phi_t + \psi_t) \leq w, \\
& \phi_t - \psi_t = p_t \sum_{j=1}^{n} (r_{jt} - r_j) x_j, \quad t = 1, 2, \ldots, T, \\
& \phi_t \psi_t = 0, \ \phi_t \geq 0, \ \psi_t \geq 0, \quad t = 1, 2, \ldots, T, \\
& x \in X.
\end{aligned} \qquad (8)$$

By using the standard result in linear programming (Chvatal, 1983; Konno and Yamazaki, 1991), complementarity conditions $\phi_t \psi_t = 0$, $t = 1, 2, \ldots, T$, can be removed, so that the problem becomes a linear programming problem:

$$\begin{aligned}
\text{maximize} \quad & \sum_{j=1}^{n} r_j x_j \\
\text{s.t.} \quad & \sum_{t=1}^{T} (\phi_t + \psi_t) \leq w, \\
& \phi_t - \psi_t = p_t \sum_{j=1}^{n} (r_{jt} - r_j) x_j, \quad t = 1, 2, \ldots, T, \\
& \phi_t \geq 0, \ \psi_t \geq 0, \quad t = 1, 2, \ldots, T, \\
& \sum_{j=1}^{n} x_j = 1, \\
& 0 \leq x_j \leq \alpha_j, \quad j = 1, 2, \ldots, n.
\end{aligned} \qquad (9)$$

2.2 Minimal Transaction Unit

Associated with a real transaction is a minimal transaction unit (MTU) one can include in the portfolio. This minimal unit is usually 1,000 stocks in the Tokyo Stock Exchange. Let γ_j be the minimal proportion of the fund. Then x_j has to satisfy the following constraint:

$$x_j = \gamma_j z_j, \ z_j \in Z_+, \qquad (10)$$

where Z_+ is the set of nonnegative integers.

Once this constraint is added, the constraint $\sum_{j=1}^{n} x_j = 1$ may not be satisfied exactly. Hence, we relax this constraint as follows:

$$1 - \varepsilon \leq \sum_{j=1}^{n} x_j \leq 1 + \varepsilon, \tag{11}$$

where ε is a small positive constant. Also, we adjust the upper bound constraint $x_j \leq \alpha_j$ in such a way that α_j is an integer multiple of γ_j.

2.3 Transaction Cost

The typical cost function $c(\cdot)$ applied to a real transaction in the Tokyo Stock Exchange is either piecewise linear concave [Figure 1(a)] or piecewise constant [Figure 1(b)] function with up to 8 linear pieces.

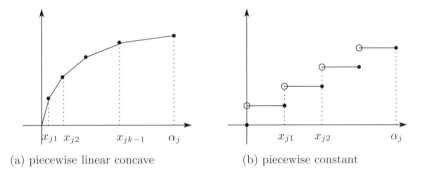

(a) piecewise linear concave (b) piecewise constant

Fig. 1. Transaction cost function.

The net return is given by

$$\sum_{j=1}^{n} \{r_j x_j - c(x_j)\}. \tag{12}$$

It is well known that these cost functions can be converted to a linear function by introducing a number of 0–1 integer variables.

(a) Piecewise Linear Concave Case
Let $0 = x_{j0} < x_{j1} < \cdots < x_{jk} = \alpha_j$ be nondifferentiable points. Then the piecewise linear concave cost function can be represented as follows (Konno and Yamamoto, 2005b; Wolsey, 1998):

$$c(x_j) = \sum_{l=0}^{k} c_{jl} \lambda_{jl}, \tag{13}$$

where

$$x_j = \sum_{l=1}^{k} \lambda_{jl} x_{jl}, \quad j = 1, 2, \ldots, n,$$

$$\sum_{l=0}^{k} \lambda_{jk} = 1, \quad j = 1, 2, \ldots, n,$$

$$\sum_{l=1}^{k} y_{jl} = 1, \quad j = 1, 2, \ldots, n, \quad (14)$$

$$\lambda_{j0} \leq y_{j1}, \quad j = 1, 2, \ldots, n,$$
$$\lambda_{jl} \leq y_{jl} + y_{jl-1}, \quad j = 1, 2, \ldots, n; l = 1, 2, \ldots, k-1,$$
$$\lambda_{jk} \leq y_{jk}, \quad j = 1, 2, \ldots, n,$$
$$y_{jl} = 0 \text{ or } 1, \quad l = 1, 2, \ldots, k; j = 1, 2, \ldots, n.$$

(b) Piecewise Constant Case

Let $0 = x_{j0} < x_{j1} < \cdots < x_{jk} = \alpha_j$ be the jump points and let $c(x_j) = c_{jl}$ for $x_j \in [x_{jl-1}, x_{jl}), l = 1, 2, \ldots, k$.

Then the piecewise constant cost function can be represented as follows:

$$c(x_j) = \sum_{l=1}^{k}(c_{jl} - c_{jl-1})y_{jl}, \quad (15)$$

where

$$\frac{x_j - x_{jl}}{\alpha_j} \leq y_{jl} \leq 1 + \frac{x_j - x_{jl}}{\alpha_j},$$

$$y_{jl} = 0 \text{ or } 1.$$

Note that this constraint ensures that

$$x_j > x_{jl} \Rightarrow y_{jl} = 1,$$
$$x_j \leq x_{jl} \Rightarrow y_{jl} = 0.$$

2.4 Cardinality Constraints

It is widely believed among fund managers that one has to include a significant number of assets in a portfolio to achieve a specified risk-return structure. In fact, a typical mutual fund consists of over 100 assets.

Also, when a fund manager constructs a portfolio simulating an index, he or she often purchases virtually all assets included in the index with the same weight as the index. This will certainly guarantee the same performance as the associated index.

However, this strategy will lead to a significant amount of transaction cost and management cost. When the amount of the fund is relatively small, the net performance would not be satisfactory due to these costs. Therefore, it would be nice to find a portfolio satisfying the specified risk-return condition with a smaller number of assets.

Let p be the maximal number of assets to be included in the portfolio. Then the cardinality constraint can be represented by adding the following constraints:

$$\gamma_j y_j \leq x_j \leq \alpha_j y_j, \qquad j = 1, 2, \ldots, n, \tag{16}$$

$$\sum_{j=1}^{n} y_j = p,\ y_j \in \{0,\ 1\}, \qquad j = 1, 2, \ldots, n, \tag{17}$$

where γ_j is the minimal transaction unit.

A number of other nonconvex portfolio optimization problems such as index tracking problems under nonconvex constraints (Konno and Wijayanayake, 2001) can be solved using a similar approach.

2.5 Computational Experiments

In this section, we will report computational results on the mean-absolute deviation model under piecewise constant transaction cost and cardinality constraints (Konno and Yamamoto, 2005b):

$$\begin{aligned}
\text{maximize} \quad & \sum_{j=1}^{n} r_j x_j - \sum_{j=1}^{n}\sum_{l=1}^{k}(c_{jl} - c_{jl-1})y_{jl} \\
\text{s.t.} \quad & \sum_{t=1}^{T}(\phi_t + \psi_t) \leq w, \\
& \phi_t - \psi_t = p_t \sum_{j=1}^{n}(r_{jt} - r_j)x_j, \qquad t = 1, 2, \ldots, T, \\
& 1 - \varepsilon \leq \sum_{j=1}^{n} x_j \leq 1 + \varepsilon, \\
& \frac{x_j - x_{jl}}{\alpha_j} \leq y_{jl} \leq 1 + \frac{x_j - x_{jl}}{\alpha_j}, \\
& \gamma_j y_{j1} \leq x_j \leq \alpha_j y_{j1}, \qquad j = 1, 2, \ldots, n, \\
& \sum_{j=1}^{n} y_{j1} = p, \\
& y_{jl} = 0\ \text{or}\ 1, \qquad l = 1, 2, \ldots, k;\ j = 1, 2, \ldots, n, \\
& \phi_t \geq 0,\ \psi_t \geq 0, \qquad t = 1, 2, \ldots, T, \\
& x_j = \gamma_j z_j,\ z_j \in Z_+, \qquad j = 1, 2, \ldots, n.
\end{aligned} \tag{18}$$

This problem contains kn 0–1 variables and n integer variables.

We used historical data collected in the Tokyo Stock Exchange and solved the problem using CPLEX7.1 on an Athlon XP 2000 (1.67 GHz, 512 MB) PC.

Table 1 shows the CPU time for various values of p and n where $\varepsilon = 0.01, w = 3.5\%$. The computation time increases sharply as we increase n, as expected, while it is not sensitive to the magnitude of p.

Table 1. CPU Time (sec)

$p \backslash n$	200	400	600	800
55	33	52	705	1,754
70	13	79	124	996
100	102	158	99	639
130	61	265	50	7,908
160	43	45	89	2,639
190	21	90	216	5,926

Table 2 shows the computation time as a function of M (the amount of the fund). Note that k, the number of steps of a piecewise constant function, increases as we increase the amount of the fund.

Table 2. CPU Time (sec)

$p \backslash M$ (million)	300($k=2$)	450($k=3$)	600($k=4$)	750($k=5$)
55	33	81	51	151
70	13	43	56	70
100	102	63	572	289
130	61	279	99	341
160	43	106	370	197
190	21	59	12,534	336

The number of 0–1 variables increases linearly as a function of k. When $n = 200$ and $k = 5$, the problem (18) contains 1,200 integer variables. Surprisingly, the number of computations increases more or less linearly as a function of k.

Table 3 shows the computation time as a function of ε. We see that the computation time increase sharply as we decrease ε, while the value of the objective function remains stable. We conclude from this that it is appropriate to choose $\varepsilon = 0.01 \sim 0.001$.

Figure 2 shows the maximal return under the constraint on the maximal number of assets in the portfolio, i.e., under the constraint $\sum_{j=1}^{n} y_{j1} \leq p$, with MTU constraints but without transaction cost.

We see from this figure that the maximal return under a specified level of risk is attained by a relatively small number of assets. In fact, when the level of allowable risk is larger, the best portfolio is attained when $p = 10$, i.e., a portfolio consisting of 10 assets. This number gradually increases as we decrease w. However, even when w is close to the globally minimal risk point,

Table 3. CPU Time (sec)

$p \backslash \varepsilon$	0.01	0.001	0
55	33	57	805
70	13	30	356
100	102	56	858
130	61	23	327
160	43	95	735
190	21	74	301

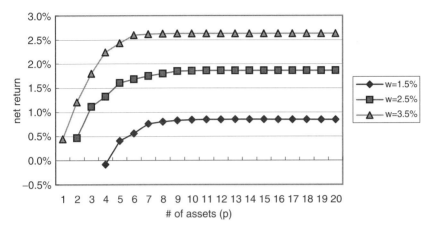

Fig. 2. Maximal return as a function of p.

the best value for p is only 20. This means that we do not have to purchase more than 20 assets to construct a portfolio to achieve a specified risk-return structure.

3 Maximal Predictability Portfolio Optimization Problems

3.1 Definitions

Let r_{jt} be the rate of return of the jth asset during time period t. Let us consider the following factor representation of r_{jt}:

$$r_{jt} = \beta_{j0} + \beta_{j1}F_{1t-1} + \beta_{j2}F_{2t-1} + \cdots + \beta_{jK}F_{Kt-1} + \varepsilon_{jt},$$
$$j = 1, 2, \ldots, n;\ t = 1, 2, \ldots, T, \qquad (19)$$

where F_{kt} is the value of factor k during period t, the ε_{jt} are independent and $E[\varepsilon_{jt}] = 0$, and β_{jk} is a constant called the *factor loading*.

Let us denote

$$r_t = (r_{1t}, r_{2t}, \ldots, r_{nt})^T \in R^n, \tag{20}$$

$$F_t = (F_{1t}, F_{2t}, \ldots, F_{Kt})^T \in R^K, \tag{21}$$

$$\varepsilon_t = (\varepsilon_{1t}, \varepsilon_{2t}, \ldots, \varepsilon_{nt})^T \in R^n. \tag{22}$$

Let x be a portfolio and let $R^2(x)$ be the coefficient of determination of portfolio x defined by

$$R^2(x) = \frac{\mathrm{Var}[\tilde{r}_t^T x]}{\mathrm{Var}[r_t^T x]}, \tag{23}$$

where $\tilde{r}_t = E[r_t | F_{t-1}]$.

Let us denote the investable set by

$$X' = \left\{ x \in R^n \mid \tilde{r}_t^T x \geq \rho,\ e^T x = 1,\ A^T x \geq b,\ 0 \leq x \leq \alpha \right\}, \tag{24}$$

where ρ is a constant, $e = (1, 1, \ldots, 1)$, α is the vector of upper bounds, and $A^T x \geq b$ is a set of institutional constraints. Lo and MacKinlay (1997) defined a maximal predictability portfolio (MPP) as a portfolio x that maximizes $R^2(x)$ over X, which is claimed to achieve the best predictive power in terms of the expected rate of return.

$R^2(x)$ is the ratio of two convex quadratic functions, so that the problem becomes a nonconcave maximization problem. In Konno and Yamamoto (2005a, 2005b), we proposed an algorithm for solving the problem using the Dinkelbach transformation (Dinkelbach, 1967) and 0−1 integer programming method. This algorithm can successfully solve problems with n up to 200, and the resulting portfolio performs better than index. Results should be better if we solve problems with a larger universe.

To solve large-scale problems, we replace the variance by the absolute deviation in the definition of $R^2(x)$ as follows:

$$Q(x) = \frac{E[|\tilde{r}_t^T x - E[\tilde{r}_t^T x]|]}{E[|r_t^T x - E[r_t^T x]|]}. \tag{25}$$

It is usually assumed in a multifactor approach that F_{t-1} and ε_{t-1} are both normally distributed. Then r_t and $\tilde{r}_t$ are also normally distributed, so that maximizing $Q(x)$ is equivalent to maximizing $R^2(x)$ by Theorem 2.1.

Let us assume that $(F_1, F_2, \ldots, F_K)$ are distributed over a finite set of points $(f_{1t}, f_{2t}, \ldots, f_{Kt}), t = 1, 2, \ldots, T$. Then

$$W[r_t^T x] = \sum_{t=1}^{T} p_t \left| \sum_{k=1}^{K} \beta_k(x)(f_{kt-1} - \hat{f}_k) + \varepsilon_t(x) \right|, \tag{26}$$

$$W[\tilde{r}_t^T x] = \sum_{t=1}^{T} p_t \left| \sum_{k=1}^{K} \beta_k(x)(f_{kt-1} - \hat{f}_k) \right|, \tag{27}$$

where $\hat{f}_k = \sum_{t=1}^{T} p_t f_{kt-1}$, $\beta_k(\boldsymbol{x}) = \sum_{j=1}^{n} \beta_{jk} x_j$, and $\varepsilon_t(\boldsymbol{x}) = \sum_{j=1}^{n} \varepsilon_{jt} x_j$. Then the maximal predictability portfolio construction problem in terms of the absolute deviation is defined as follows:

$$
\begin{aligned}
\text{minimize} \quad & \frac{\sum_{t=1}^{T} p_t \left| \sum_{k=1}^{K} \beta_k(\boldsymbol{x})(f_{kt-1} - \hat{f}_k) + \varepsilon_t(\boldsymbol{x}) \right|}{\sum_{t=1}^{T} p_t \left| \sum_{k=1}^{K} \beta_k(\boldsymbol{x})(f_{kt-1} - \hat{f}_k) \right|} \\
\text{s.t.} \quad & \beta_k(\boldsymbol{x}) = \sum_{j=1}^{n} \beta_{jk} x_j, \quad k = 1, 2, \ldots, K, \\
& \varepsilon_t(\boldsymbol{x}) = \sum_{j=1}^{n} \varepsilon_{jt} x_j, \quad t = 1, 2, \ldots, T, \\
& \tilde{r}_1 x_1 + \tilde{r}_2 x_2 + \cdots + \tilde{r}_n x_n \geq \rho, \\
& x_1 + x_2 + \cdots + x_n = 1, \\
& 0 \leq x_j \leq \alpha, \quad j = 1, 2, \ldots, n.
\end{aligned} \quad (28)
$$

3.2 0−1 Mixed Integer Programming Formulation

The first step to solve (28) is to apply the Charnes–Cooper transformation (Charnes and Cooper, 1962). Let

$$
y_0 = \frac{1}{\sum_{t=1}^{T} p_t \left| \sum_{k=1}^{K} \beta_k(\boldsymbol{x})(f_{kt-1} - \hat{f}_k) \right|}. \quad (29)
$$

Then problem (28) can be reformulated as follows:

$$
\begin{aligned}
\text{minimize} \quad & \sum_{t=1}^{T} p_t \left| \sum_{k=1}^{K} \beta_k(\boldsymbol{x})(f_{kt-1} - \hat{f}_k) + \varepsilon_t(\boldsymbol{x}) \right| \cdot y_0 \\
\text{s.t.} \quad & \sum_{t=1}^{T} p_t \left| \sum_{k=1}^{K} \beta_k(\boldsymbol{x})(f_{kt-1} - \hat{f}_k) \right| \cdot y_0 = 1, \\
& \beta_k(\boldsymbol{x}) = \sum_{j=1}^{n} \beta_{jk} x_j, \quad k = 1, 2, \ldots, K, \\
& \varepsilon_t(\boldsymbol{x}) = \sum_{j=1}^{n} \varepsilon_{jt} x_j, \quad t = 1, 2, \ldots, T, \\
& \tilde{r}_1 x_1 + \tilde{r}_2 x_2 + \cdots + \tilde{r}_n x_n \geq \rho, \\
& x_1 + x_2 + \cdots + x_n = 1, \\
& 0 \leq x_j \leq \alpha, \quad j = 1, 2, \ldots, n.
\end{aligned} \quad (30)
$$

Let $\boldsymbol{y} = y_0 \cdot \boldsymbol{x}$. Then problem (30) is equivalent to

$$\text{minimize} \quad \sum_{t=1}^{T} p_t \left| \sum_{k=1}^{K} \beta_k(\boldsymbol{y})(f_{kt-1} - \hat{f}_k) + \varepsilon_t(\boldsymbol{y}) \right|$$

$$\text{s.t.} \quad \sum_{t=1}^{T} p_t \left| \sum_{k=1}^{K} \beta_k(\boldsymbol{y})(f_{kt-1} - \hat{f}_k) \right| = 1,$$

$$\beta_k(\boldsymbol{y}) = \sum_{j=1}^{n} \beta_{jk} y_j, \quad k = 1, 2, \ldots, K,$$

$$\varepsilon_t(\boldsymbol{y}) = \sum_{j=1}^{n} \varepsilon_{jt} y_j, \quad t = 1, 2, \ldots, T, \qquad (31)$$

$$\sum_{j=1}^{n} \tilde{r}_j y_j \geq \rho y_0,$$

$$y_1 + y_2 + \cdots + y_n = y_0,$$

$$0 \leq y_j \leq \alpha y_0, \quad j = 1, 2, \ldots, n; \ y_0 \geq 0.$$

Let $(\boldsymbol{y}^*, y_0^*)$ be an optimal solution of this problem. It is straightforward to see that $\boldsymbol{x}^* = \boldsymbol{y}^*/y_0^*$ is an optimal solution of the original problem (28).

Let

$$\boldsymbol{u} = (u_1, u_2, \ldots, u_T), \qquad \boldsymbol{v} = (v_1, v_2, \ldots, v_T), \qquad (32)$$

$$\boldsymbol{\xi} = (\xi_1, \xi_2, \ldots, \xi_T), \qquad \boldsymbol{\eta} = (\eta_1, \eta_2, \ldots, \eta_T). \qquad (33)$$

Then problem (31) is reformulated as follows:

$$\text{minimize} \quad \sum_{t=1}^{T} p_t (u_t + v_t)$$

$$\text{s.t.} \quad \sum_{t=1}^{T} p_t (\xi_t + \eta_t) = 1,$$

$$u_t - v_t = \sum_{k=1}^{K} \beta_k(\boldsymbol{y})(f_{kt-1} - \hat{f}_k) + \varepsilon_t(\boldsymbol{y}), \quad t = 1, 2, \ldots, T,$$

$$\xi_t - \eta_t = \sum_{k=1}^{K} \beta_k(\boldsymbol{y})(f_{kt-1} - \hat{f}_k), \quad t = 1, 2, \ldots, T, \qquad (34)$$

$$u_t v_t = 0, \quad t = 1, 2, \ldots, T,$$

$$\xi_t \eta_t = 0, \quad t = 1, 2, \ldots, T,$$

$$u_t \geq 0, \ v_t \geq 0, \quad t = 1, 2, \ldots, T,$$

$$\xi_t \geq 0, \ \eta_t \geq 0, \quad t = 1, 2, \ldots, T,$$

$$(\boldsymbol{y}, y_0) \in Y.$$

Let

$$Y = \left\{(\boldsymbol{y}, y_0) | \beta_k(\boldsymbol{y}) = \sum_{j=1}^n \beta_{jk} y_j, \ k = 1, 2, \ldots, K; \right.$$

$$\varepsilon_t(\boldsymbol{y}) = \sum_{j=1}^n \varepsilon_{jt} y_j, \ t = 1, 2, \ldots, T,$$

$$\sum_{j=1}^n \tilde{r}_j y_j \geq \rho y_0; \ \sum_{j=1}^n y_j = y_0;$$

$$\left. 0 \leq y_j \leq \alpha y_0, \ j = 1, 2, \ldots, n; \ y_0 \geq 0 \right\}. \quad (35)$$

The following theorem shows that we can eliminate the complementarity conditions $u_t v_t = 0, t = 1, 2, \ldots, T$.

Theorem 3.1. Let $(\boldsymbol{y}^*, y_0^*, \boldsymbol{u}^*, \boldsymbol{v}^*, \boldsymbol{\xi}^*, \boldsymbol{\eta}^*)$ be an optimal basic feasible solution to problem (34) without the complementarity conditions. Then $u_t^* v_t^* = 0$, for all t. Also, there exists at most one t such that $\xi_t^* > 0, \eta_t^* > 0$.

Proof. See Konno et al. (2005). □

What makes our problem difficult is a set of complementarity conditions $\xi_t \eta_t = 0, t = 1, 2, \ldots, T$. As is well known, these conditions can be represented as a system of linear inequalities by introducing 0–1 integer variables $z_t, t = 1, 2, \ldots, T$.

Let us consider a pair of linear inequalities:

$$\xi_t \leq a_t z_t, \quad t = 1, 2, \ldots, T, \quad (36)$$
$$\eta_t \leq b_t(1 - z_t), \quad t = 1, 2, \ldots, T, \quad (37)$$

where

$$a_t = \max\{\max\{\sum_{k=1}^K \beta_k(\boldsymbol{y})(f_{kt-1} - \hat{f}_k) \,|\, (\boldsymbol{y}, y_0) \in Y\}, 0\}, \quad (38)$$

$$b_t = -\min\{\min\{\sum_{k=1}^K \beta_k(\boldsymbol{y})(f_{kt-1} - \hat{f}_k) \,|\, (\boldsymbol{y}, y_0) \in Y\}, 0\}, \quad (39)$$

and $z_t = \{0, 1\}$. When $z_t = 0$, then $\xi_t = 0$ and η_t is unconstrained. Also, when $z_t = 1$, then $\eta_t = 0$ and ξ_t is unconstrained.

Problem (34) can now be formulated as a 0–1 integer programming problem:

$$\text{minimize} \quad \sum_{t=1}^{T} p_t(u_t + v_t)$$

$$\text{s.t.} \quad \sum_{t=1}^{T} p_t(\xi_t + \eta_t) = 1,$$

$$u_t - v_t = \sum_{k=1}^{K} \beta_k(\boldsymbol{y})(f_{kt-1} - \hat{f}_k) + \varepsilon_t(\boldsymbol{y}), \quad t = 1, 2, \ldots, T, \tag{40}$$

$$\xi_t - \eta_t = \sum_{k=1}^{K} \beta_k(\boldsymbol{y})(f_{kt-1} - \hat{f}_k), \quad t = 1, 2, \ldots, T,$$

$$0 \leq \xi_t \leq a_t z_t, \quad 0 \leq \eta_t \leq b_t(1 - z_t), \quad t = 1, 2, \ldots, T$$
$$z_t \in \{0, 1\}, \quad t = 1, 2, \ldots, T,$$
$$u_t \geq 0, \; v_t \geq 0, \quad t = 1, 2, \ldots, T,$$
$$(\boldsymbol{y}, y_0) \in Y.$$

3.3 Computational Experiments

Our target is to solve problems with up to $(T, n, K) = (36, 1500, 5)$ on an Xeon (3.73 GHz, 2.00 GB) PC using CPLEX10.1. Factors to be included in the model are chosen in accordance with Fama and French (1993). The computation time is expected to increase sharply as we increase T. A preliminary test shows that the maximal size of the problem solvable within 1,500 CPU seconds is $(T, n, K) = (36, 600, 3)$.

To solve larger problems, we introduce several constraints to reduce the feasible region of the associated linear relaxation problems.

Among several schemes, the following two are very useful:

(a) Bound on y_0
Let y_0^* be the minimal value of y_0 over the feasible region of problem (28), and let us add the following constraint:

$$y_0 \geq y_0^*. \tag{41}$$

To calculate y_0^*, we need to maximize a piecewise linear convex function over a polytope. This procedure requires additional computation time but results in a significant reduction of the total computation time.

(b) Tighter bounds on ξ_t and η_t
Upon a close look at the optimal solution of (34), we find that the ξ_t and η_t in the optimal solution attain a much smaller value than its a priori bounds a_t and b_t. In fact, they are usually less than 0.3 times of its bounds or even much smaller.

Applications of Integer Programming to Financial Optimization

Therefore, we will introduce tighter bounds on these variables, i.e.,

$$\xi_t \leq \delta a_t z_t, \quad \eta_t \leq \delta b_t (1 - z_t), \quad t = 1, 2, \ldots, T, \tag{42}$$

where δ is some constant less than 1.0.

Table 4 shows the effect of δ on the CPU time. Blanks in Table 4 represent that an optimal solution was not obtained in 1,500 CPU seconds. We compared 27 cases for different T, n, K by varying δ and found that the solution remains the same for all problems down to $\delta = 0.3$. When $\delta = 0.2$, however, we observe around a 3.0% relative error or the average.

Table 4. CPU Time (sec)

T	K	n	Original	With constraint (39) and (40)	
				$\delta = 1$	$\delta = 0.2$
12	3	225	1.2	1.8	1.7
		1,500	9.6	6.5	6.8
	5	225	19.7	4.2	2.8
		1,500	21.4	23.3	15.0
24	3	225	2.1	7.1	3.4
		1,500		82.2	20.4
	5	225		137.7	49.7
		1,500		674.4	851.2
36	3	225	3.1	5.7	5.1
		1,500		46.7	31.2
	5	225			989.2
		1,500			

Finally, Figures 3 and 4 show the ex-post performance of the Nikkei Index, MAD portfolio, and MPP for $(T, K) = (36, 3)$ and $n = 225$ and 1,500, respectively, using an annual rebalancing strategy. We see that MPP significantly outperforms the Index and MAD portfolio. The difference tends to become larger as we increase n.

4 Choosing the Best Set of Financial Attributes in Failure Discriminant Analysis

4.1 The Problem

Let there be n companies $E_t, t = 1, 2, \ldots, T$, and let $X_k, k = 1, 2, \ldots, K$, be the kth financial attribute.

Let us define the failure intensity f as a linear combination of x as follows:

$$f = \alpha_0 + \alpha_1 x_1 + \alpha_2 x_2 + \cdots + \alpha_K x_K. \tag{43}$$

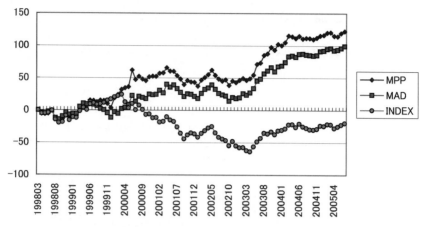

Fig. 3. Ex-post performance ($n = 225$).

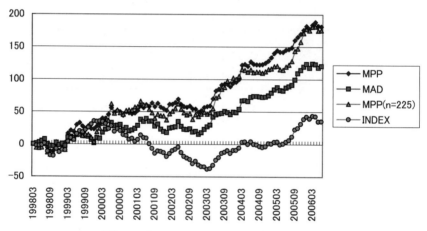

Fig. 4. Ex-post performance ($n = 1,500$).

In failure discriminant analysis, we determine the threshold value f^* and separate n companies into failing group F_1 and nonfailing group F_0 depending upon whether the failure intensity is above and below f^*.

Let $y_t = 1$ if E_t failed during the last period and $y_t = 0$ if E_t did not fail. To estimate $(\alpha_0, \alpha_1, \alpha_2, \ldots, \alpha_K)$, we introduce the error term

$$\varepsilon_t = y_t - (\alpha_0 + \alpha_1 x_{1t} + \alpha_2 x_{2t} + \cdots + \alpha_K x_{Kt}), \qquad t = 1, 2, \ldots, T, \quad (44)$$

where x_{jk} is the magnitude of X_k of company E_t, and we apply the least-squares method. However, if there are too many attributes, we have to choose a certain portion of variables that achieve the required quality of fitting. Akaike's Information Criterion (AIC) is a very popular method for this purpose (Akaike, 1974). To apply this criterion, however, we need to have prior information about the statistical properties of residual variables. Many other criteria have been proposed in the past, but they all inherit the advantages and disadvantages of AIC (Akaike, 1974).

Problems to be considered here are:

Problem 1. Given a positive integer s, find the set of attributes $X_{i_1}, X_{i_2}, \ldots, X_{i_s}$ such that the total amount of residual error is minimal.

Problem 2. Find the set of variables $X_{i_1}, X_{i_2}, \ldots, X_{i_s}$ whose (freedom adjusted) coefficient of determination is maximal.

These are difficult combinatorial optimization problems for which no exact and efficient algorithm has been proposed until recently. To find an optimal combination, we had to use an enumeration approach of some sort.

The number of possible explanatory variables k in the traditional field is not very large, say less than 20 or 30. However, much larger problems are under consideration. For example, k is around 100 in the failure discriminant analysis (Judge et al., 1988; Konno et al., 2004) and is sometimes larger than 1,000 in bio-informatics (Pardalos, et al., 2007).

When $k = 100$ and $s = 20$ as in the case of failure discriminant analysis (Konno et al., 2004), a possible combination is $_{100}C_{20} \sim 10^{21}$, so that total enumeration is completely out of reach.

Therefore, people use some sort of heuristic approach (Burnkam and Andersen, 2002; Galindo and Tamayo, 2000; Miller, 1990; Osborne, 1976). One commonly used method is to sequentially introduce s "important" variables one at a time. When the residual error is small enough, we are done. Otherwise, we eliminate a certain variable from the model, add a new one in its place, and continue until the fitting is satisfactory enough. This procedure usually leads to a good solution, but it may not generate the best combination. The purpose of this paper is to propose an efficient method to solve Problems 1 and 2 above.

Though this procedure need not always generate an optimal solution to Problem 1, it usually generates such a solution as demonstrated by a series of numerical experiments to be presented later. Once Problem 1 is solved, then Problem 2 can be solved by solving Problem 1 by increasing s.

4.2 Least Absolute Deviation Fitting

Given T sets of data $(y_t, x_{1t}, x_{2t}, \ldots, x_{kt})$, $t = 1, 2, \ldots, T$, let us define

$$f(\alpha_0, \alpha_1, \ldots, \alpha_k) = \sum_{t=1}^{T} \left\{ y_t - \left(\alpha_0 + \sum_{j=1}^{k} \alpha_j x_{jt}\right) \right\}^2. \tag{45}$$

Then Problem 1 can be formulated as the following constrained minimization problem:

minimize $f(\alpha_0, \alpha_1, \ldots, \alpha_k)$ (46)

s.t. at most s components of $(\alpha_1, \alpha_2, \ldots, \alpha_k)$ are nonzero.

By introducing 0–1 integer variables $z_j, j = 1, 2, \ldots, k$, this problem can be formulated as a quadratic 0–1 integer programming problem:

$$P_k(s) \begin{cases} \text{minimize} & f(\alpha_0, \alpha_1, \ldots, \alpha_k) \\ \text{s.t.} & \sum_{j=1}^{k} z_j = s, \\ & 0 \leq \alpha_j \leq \bar{\alpha}_j z_j, \quad j = 1, 2, \ldots, k, \\ & z_j \in \{0, 1\}, \quad j = 1, 2, \ldots, k, \end{cases} \quad (47)$$

where $\bar{\alpha}_j > 0$ is the largest attainable value of α_j. If $z_j = 0$, then $\alpha_j = 0$, so that at most s components of α_j can be positive, as required. Algorithmic research for solving a quadratic 0–1 integer programming problem is now under way (Wolsey, 1998). However, to date there exists no efficient algorithm.

The key idea is to replace variance $f(\alpha_0, \alpha_1, \ldots, \alpha_k)$ by absolute deviation:

$$g(\alpha_0, \alpha_1, \ldots, \alpha_k) = \sum_{t=1}^{T} \left| y_t - \left(\alpha_0 + \sum_{j=1}^{k} \alpha_j x_{jt} \right) \right|, \quad (48)$$

and consider the following problem:

$$Q_k(s) \begin{cases} \text{minimize} & g(\alpha_0, \alpha_1, \ldots, \alpha_k) \\ \text{s.t.} & \sum_{j=1}^{k} z_j = s, \\ & 0 \leq \alpha_j \leq \bar{\alpha}_j z_j, \quad j = 1, 2, \ldots, k, \\ & z_j \in \{0, 1\}, \quad j = 1, 2, \ldots, k. \end{cases} \quad (49)$$

Let us note that least absolute deviation estimator is more robust than least-squares estimator (Judge et al., 1988). Also, problems of $P_k(s)$ and $Q_k(s)$ are equivalent under normality assumption (Theorem 2.1).

The least absolute deviation estimator is not a linear estimator, but it shares some nice properties as stated in the following theorem:

Theorem 4.1. If the ε_i are i.i.d. and symmetrically distributed with $E[\varepsilon_i] = 0$, $V[\varepsilon_i] = \sigma^2$, then an optimal solution of $Q_k(s)$ is an unbiased estimator of $(\alpha_0, \alpha_1, \ldots, \alpha_k)$.

Proof. See Judge et al. (1988). □

Therefore, $Q_k(s)$ can serve as a proper alternative to $P_k(s)$. More importantly, problem $Q_k(s)$ can be reduced to a 0–1 mixed integer programming problem that can be solved by the state-of-the-art algorithm (Cplex, 2006; Wolsey, 1998).

Let $\bar{y} = \sum_{t=1}^{T} y_t / T$ and let $(\hat{\alpha_0}, \hat{\alpha_1}, \ldots, \hat{\alpha_k})$ be an optimal solution of $P_k(s)$. Then the quality of the cardinality constrained least-squares fitting may be measured by the adjusted R^2 defined below:

$$\bar{R}^2 = 1 - \frac{T-1}{T-s-1}\left\{1 - \frac{\sum_{t=1}^{T}(\hat{y}_t - \bar{\hat{y}})^2}{\sum_{t=1}^{T}(y_t - \bar{y})^2}\right\}, \tag{50}$$

where

$$\hat{y}_t = \hat{\alpha_0} + \sum_{j=1}^{k} \hat{\alpha_j} x_{jt}, \tag{51}$$

$$\bar{\hat{y}} = \sum_{t=1}^{T} \hat{y}_t / T. \tag{52}$$

4.3 A Two-Step Algorithm

Those who are familiar with linear programming (Chvatal, 1983) should know that $Q_k(s)$ can be rewritten as a 0–1 linear integer programming problem:

$$\begin{aligned}
\text{minimize} \quad & \sum_{t=1}^{T}(u_t + v_t) \\
\text{s.t.} \quad & u_t - v_t = y_t - \left(\alpha_0 + \sum_{j=1}^{k} \alpha_j x_{jt}\right), \quad t = 1, 2, \ldots, T, \\
& u_t \geq 0,\ v_t \geq 0, \quad t = 1, 2, \ldots, T, \\
& \sum_{j=1}^{k} z_j = s, \\
& 0 \leq \alpha_j \leq \bar{\alpha}_j z_j, \quad j = 1, 2, \ldots, k, \\
& z_j \in \{0, 1\}, \quad j = 1, 2, \ldots, k.
\end{aligned} \tag{53}$$

Problem (53) has an optimal solution since it is feasible and the objective function is bounded below.

Theorem 4.2. Let $(\alpha_0^*, \alpha_1^*, \ldots, \alpha_k^*, u_1^*, u_2^*, \ldots, u_T^*, v_1^*, v_2^*, \ldots, v_T^*, z_1^*, z_2^*, \ldots, z_k^*)$ be an optimal solution (53). Then $(\alpha_0^*, \alpha_1^*, \ldots, \alpha_k^*, z_1^*, z_2^*, \ldots, z_k^*)$ is an optimal solution of (49).

Proof. See Konno and Yamamoto (2007). □

When k, s, T are not very large, problem (53) can be solved by state-of-the-art software such as CPLEX10.1.

Let us now propose a two-step algorithm for solving $P_k(s)$. The first step is to solve the least absolute deviation fitting problem $Q_k(s+r)$:

$$Q_k(s+r) \begin{cases} \text{minimize} & g(\alpha_0, \alpha_1, \ldots, \alpha_k) \\ \text{s.t.} & \sum_{j=1}^{k} z_j = s + r, \\ & 0 \leq \alpha_j \leq \bar{\alpha}_j z_j, \quad j = 1, 2, \ldots, k, \\ & z_j \in \{0, 1\}, \quad j = 1, 2, \ldots, k, \end{cases} \quad (54)$$

where r is some positive integer. Let $(\alpha_0^*, \alpha_1^*, \ldots, \alpha_k^*)$ be an optimal solution to (54). Let $J_1 = \{j \mid z_j^* = 1\}$ and let $(\hat{\alpha}_0, \hat{\alpha}_1, \ldots, \hat{\alpha}_k)$ be an optimal solution to $P_k(s)$. Then it is quite likely that $\hat{\alpha}_j = 0$ for almost all j such that $j \notin J_1$ for large enough r since absolute deviation and standard deviation are similar measures of variation.

To recover an optimal solution of $P_k(s)$, we solve

$$\begin{aligned} \text{minimize} \quad & f(\alpha_0, \alpha_1, \ldots, \alpha_k) \\ \text{s.t.} \quad & \sum_{j \in J_1} z_j = s, \\ & 0 \leq \alpha_j \leq \bar{\alpha}_j z_j, \quad j \in J_1, \\ & z_j \in \{0, 1\}, \quad j \in J_1. \end{aligned} \quad (55)$$

If r is not large, this problem can be solved by solving $_{s+r}C_s$ least-squares subproblems associated with all possible s out of $s+r$ combinations.

4.4 Results of Computational Experiments

We compare a two-step algorithm above and S-plus (Furnival and Wilson, 1974; SPLUS, 2001) using data associated with failure discriminant analysis, where $y_t = 1$ or 0 depending upon whether or not the tth company failed and x_{jt} is the jth financial attribute of the tth company.

We prepared four data sets $(T, k) = (200, 50)$, $(200, 70)$, $(1000, 50)$, $(1000, 70)$ randomly chosen from 6,556 corporate data among which 40% failed.

Figure 5 shows the distribution of correlation coefficients of 50 financial attributes. We see that the majority of financial attributes are not strongly correlated, with a correlation coefficient between -0.3 and 0.3, but there are a nonnegligible number of highly correlated pairs.

Figures 6 and 7 show the quality of fitting. We see that a two-step algorithm generates a significantly better solution (see Figure 7). In fact, when $s + r$ is over 15, a two-step algorithm outperforms S-plus.

Table 5 shows the CPU time for a two-step algorithm. The number in the bracket shows the CPU time required for the first step of a two-step algorithm.

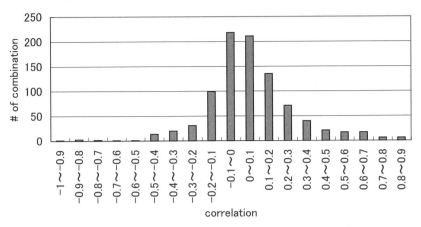

Fig. 5. Distribution of correlation coefficients ($k = 50$).

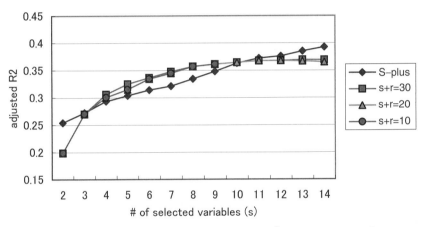

Fig. 6. Comparison of freedom adjusted R^2 ($T = 200, k = 50$).

Table 5. CPU Time (sec)

k	$s+r$	CPU	k	$s+r$	CPU
50	10	1.20 (0.65)	50	10	11.53 (10.17)
	20	1.51 (0.65)		20	13.32 (10.31)
	30	1.98 (0.65)		30	4.14 (0.53)
	S-plus	6.20		S-plus	11.54
70	10	136.13(135.52)	70	10	1,501.34(1,500.00)
	20	26.47 (24.78)		20	1,502.98(1,500.00)
	30	19.02 (1.08)		30	32.23 (23.23)
	S-plus	9.65		S-plus	18.63

Finally, Figure 8 shows the magnitude of the absolute deviation calculated by solving $Q_k(s)$ and the sequential method similar to the one used in S-plus

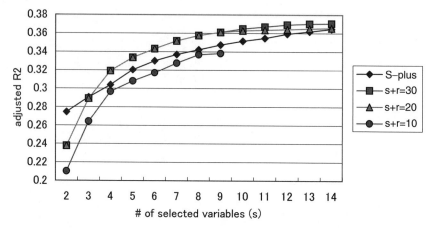

Fig. 7. Comparison of freedom adjusted R^2 ($T = 1000, k = 70$).

for least-squares fitting. We see that our method generates significantly better results than the sequential method adopted in S-plus at the expense of more CPU time.

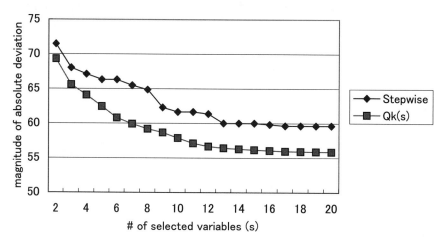

Fig. 8. Comparison of the magnitude of absolute deviation ($T = 200, k = 50$).

Acknowledgments

The first author's research was supported in part by the Grant-in-Aid for Scientific Research B18310109 of the Ministry of Education, Science, Sports and Culture of Government of Japan.

References

1. Akaike, H. A new look at the statistical model identification. *IEEE Transactions on Automatic Control*, 19:716–723, 1974.
2. Bixby, R. Solving real-world linear programs: A decade and more of progress. *Operations Research*, 50:3–15, 2002.
3. Burnkam, K., and Andersen, D., *Model Selection and Multimodel Inference: A Practical Theoretic Approach*, 2nd edition. Springer, New York, 2002.
4. Charnes, A., and Cooper, W. W. Programming with linear fractional functionys. *Naval Research Logistics Quarterly*, 9:181–186, 1962.
5. Chvatál, V., *Linear Programming*. Freeman and Co., New York, 1983.
6. CPLEX10.1 User's Manual, ILOG, 2006.
7. Dinkelbach, W. On nonlinear fractional programming. *Management Science*, 13:492–498, 1967.
8. Elton, J. E., and Gruber, M. J., *Modern Portfolio Theory and Investment Analysis*, 6th edition. John Wiley & Sons, New York, 1998.
9. Fama, F., and French, K. Common risk factors in the returns on stock and bonds. *Journal of Financial Economics*, 33:3–56, 1993.
10. Furnival, G. M., and Wilson, R. W. Jr. Regressions by leaps and bounds. *Technometrics*, 16:499–511, 1974.
11. Galindo, J., and Tamayo, P. Credit risk assessment using statistical and machine learning: Basic methodology and risk modeling applications. *Computational Economics*, 15:107–143, 2000.
12. Gotoh, J., and Konno, H. Maximization of the ratio of two convex quadratic functions over a polytope. *Computational Optimization and Applications*, 20:43–60, 2001.
13. Judge, G. R., Hill, R. C., Griffiths, W. E., Lutkepohl, H., and Lee, T.-C., *Introduction to the Theory and Practice of Econometrics*. John Wiley & Sons, New York, 1988.
14. Konno, H., Kawadai, N., and Wu, D. Estimation of failure probability using semi-definite logit model. *Computational Management Science*, 1:59–73, 2004.
15. Konno, H., Koshizuka, T., and Yamamoto, R. Mean-variance portfolio optimization problems under short sale opportunity. *Dynamics of Continuous Descrete and Impulsive Systems*, 12:483–498, 2005.
16. Konno, H., and Koshizuka, T. Mean-absolute deviation model. *IIE Transactions*, 37:893–900, 2005.
17. Konno, H., Morita, Y., and Yamamoto, R. Large scale maximal predictability portfolio optimization using absolute deviation as the measure of deviation. *ISE07-03*, Department of Industrial and Systems Engineering, Chuo University, 2007.
18. Konno, H., Tsuchiya, K., and Yamamoto, R. Minimization of the ratio of functions defined as a sum of absolute values of affine functions. To appear in *Journal of Optimization Theory and Applications*.
19. Konno, H., and Wijayanayake, A. Portfolio optimization under D.C. transaction costs and minimal transaction unit constraints. *Journal of Global Optimization*, 22:137–154, 2002.
20. Konno, H., and Wijayanayake, A. Minimal cost index tracking under concave transaction costs. *International Journal of Theoretical and Applied Finance*, 4:939–957, 2001.

21. Konno, H., and Yamamoto, R. Global optimization versus integer programming in portfolio optimization under nonconvex transaction costs. *Journal of Global Optimization*, 32:207–219, 2005a.
22. Konno, H., and Yamamoto, R. Integer programming approaches in mean-risk models. *Computational Management Science*, 4:339–351, 2005b.
23. Konno, H., and Yamamoto, R. Choosing the best set of variables in regression analysis using integer programming. *ISE07-01*, Department of Industrial and Systems Engineering, Chuo University, 2007.
24. Konno, H., and Yamazaki, H. Mean-absolute deviation portfolio optimization model and its applications to Tokyo stock market. *Management Science*, 37:519–531, 1991.
25. Lo, A., and MacKinlay, C. Maximizing predictability in the stock and bond markets. *Macroeconomic Dynamics*, 1:102–134, 1997.
26. Markowitz, H., *Portfolio Selection: Efficient Diversification of Investment*. John Wiley & Sons, New York, 1959.
27. Miller, A. J., *Subset Selection in Regression*. Chapman and Hall, New York, 1990.
28. Ogryczak, O., and Ruszczynski, A. From stochastic dominance to mean-risk model. *European Journal of Operational Research*, 116:33–50, 1999.
29. Osborne, M. R. On the computation of stepwise regression. *Australia Computer Journal*, 8:61–68, 1976.
30. Pardalos, P., Boginski, V., and Vazacopoulos, A., *Data Mining in Biomedicine*. Springer, New York, 2007.
31. Phong, T. Q., An, L. T. H., and Tao, P. D. Decomposition branch and bound method for globally solving linearly constrained indefinite quadratic minimization problems. *Operations Research Letters*, 17:215–220, 1995.
32. S-PLUS 6 for Windows Guide to Statistics, Volume 1, Insightful Corporation, 2001.
33. Wolsey, L. A., *Integer Programming*. John Wiley & Sons, New York, 1998.
34. Yamamoto, R., and Konno, H. An efficient algorithm for solving convex-convex quadratic fractional programs. To appear in *Journal of Optimization Theory and Applications*.
35. Yamamoto, R., Ishii, D., and Konno, H. A maximal predictability portfolio model: Algorithm and performance evaluation. To appear in *International Journal of Theoretical and Applied Finance*.

Computing Mean/Downside Risk Frontiers: The Role of Ellipticity

Antony D. Hall[1] and Steve E. Satchell[2]

[1] University of Technology, Sydney, School of Finance and Economics, Sydney, Australia `tony.hall@uts.edu.au`
[2] University of Cambridge, Faculty of Economics, Sidgwick Avenue, Cambridge, CB3 9DD, UK `steve.satchell@econ.cam.ac.uk`

1 Introduction

There is currently considerable interest in the trade-off between portfolio return and portfolio downside risk. This has arisen due to inadequacies of the mean/variance framework and regulatory requirements to calculate the value at risk and related measures by banks and other financial institutions.

Much of the literature focuses on the case where asset and portfolio returns are assumed to be normally or elliptically distributed, see Alexander and Bapista (2001) and Campbell et al. (2002). In both cases the problem of a mean/value at risk frontier differs only slightly from a classic mean/variance analysis. However, the nature of the mean-risk frontiers under normality is not discussed in any detail, and we present some new results in Section 2. These are a consequence of two results we present, Proposition 1 and a generalization, Proposition 2, which prove that the set of minimum risk portfolios are essentially the same under ellipticity for a wide class of risk measures. In addition to these results, we present three extensions. The extensions we propose in this paper are threefold. First, we consider extensions for portfolio simulation of those advocated for value at risk simulation by Bensalah (2002). Second, under normality we compute explicit expressions for mean/value at risk, mean/expected loss and mean/semivariance frontiers in the two asset case and in the general $N-$asset case, complementing the results for mean/value at risk under normality provided by Alexander and Bapista (2001). Finally, our framework allows us to consider fairly arbitrary risk measures in the two asset case with arbitrary return distributions; in particular, some explorations under bivariate log-normality are considered. In Section 6, we present issues with the simulation of portfolios, pointing out some of the limitations of our proposed methodology. These methodologies are applied to general downside risk frontiers for general distributions. Conclusions follow in Section 7.

2 Main Proposition

It is worth noting that although it is well known that normality implies mean/variance analysis for an arbitrary utility function (see, for example, Sargent, 1979, page 149) it is not clear what happens to the mean/downside risk frontier under normality. What is known is that the (μ_p, θ_p) frontier should be concave under appropriate assumptions for θ_p and that the set of minimum downside risk portfolios should be the same as those that make up the set of minimum variance portfolios. We note that Wang (2000) provided examples that show when returns are not elliptical, the two sets do not coincide. A proof of this second assertion is provided in Proposition 1. We shall initially assume that the $(N \times 1)$ vector of returns is distributed as $N_N(\mu, \Sigma)$ where μ is an $(N \times 1)$ vector of expected returns and Σ is an $(N \times N)$ positive definite covariance matrix. We define the scalars $\alpha = \mu'\Sigma^{-1}\mu$, $\beta = \mu'\Sigma^{-1}e$, and $\gamma = e'\Sigma^{-1}e$, where e is an $(N \times 1)$ vector of ones.

Proposition 1. *For any risk measure $\phi_p = \phi(\mu_p, \sigma_p^2)$, where $\mu_p = \mu'x$, $\sigma_p^2 = x'\Sigma x$, $\phi_1 = \partial\phi/\partial\mu_p$, $\phi_2 = \partial\phi/\partial\sigma_p^2$, and ϕ_2 is assumed nonzero, the mean minimum risk frontier (μ_p, ϕ_p) is spanned by the same set of vectors for all ϕ, namely, $x = \Sigma^{-1}E(E'\Sigma^{-1}E)^{-1}\psi_p$, where $\psi_p' = (\mu_p, 1)$, $E = (\mu, e)$.*

Proof. Our optimization problem is to minimize ϕ_p subject to $E'x = \psi_p$. That is,

$$\min_x \quad \phi(\mu_p, \sigma_p^2) - \lambda'(E'x - \psi_p), \qquad (1)$$

where λ is a (2×1) vector of Lagrangians. The first-order conditions are

$$\frac{\partial\phi}{\partial\mu_p}\frac{\partial\mu_p}{\partial x} + \frac{\partial\phi}{\partial\sigma_p^2}\frac{\partial\sigma_p^2}{\partial x} - E\lambda = 0 \qquad (2)$$

and

$$E'x - \psi_p = 0. \qquad (3)$$

Here

$$\frac{\partial\mu_p}{\partial x} = \mu \qquad (4)$$

and

$$\frac{\partial\sigma_p^2}{\partial x} = 2\Sigma x. \qquad (5)$$

This implies that

$$\phi_1\mu + 2\phi_2\Sigma x = E\lambda \qquad (6)$$

or

$$\phi_1\Sigma^{-1}\mu + 2\phi_2 x = \Sigma^{-1}E\lambda \qquad (7)$$

or
$$\hat{\lambda} = (E'\Sigma E)^{-1}(\phi_1 E'\Sigma^{-1}\mu + 2\phi_2\psi_p). \tag{8}$$

Thus, the general "solution" x satisfies
$$\phi_1\Sigma^{-1}\mu + 2\phi_2 x = \Sigma^{-1}E(E'\Sigma^{-1}E)^{-1}(\phi_1 E'\Sigma^{-1}\mu + 2\phi_2\psi_p). \tag{9}$$

We now show that
$$x = E(E'\Sigma^{-1}E)^{-1}\psi_p \tag{10}$$

satisfies Equation (9) for any ϕ_1 and ϕ_2. This is because the right-hand side of (9) can be written as
$$\phi_1\Sigma^{-1}E(E'\Sigma^{-1}E)^{-1}E'\Sigma^{-1}\mu + 2\phi_2\Sigma^{-1}E(E'\Sigma^{-1}E)^{-1}\psi_p. \tag{11}$$

But
$$(E'\Sigma^{-1}E)^{-1}E'\Sigma^{-1}\mu = \begin{pmatrix}1\\0\end{pmatrix}, \tag{12}$$

so that
$$\phi_1\Sigma^{-1}E(E'\Sigma^{-1}E)^{-1}E'\Sigma^{-1}\mu = \phi_1\Sigma^{-1}\mu, \tag{13}$$

and Equation (9) then simplifies to
$$2\phi_2 x = 2\phi_2\Sigma^{-1}E(E'\Sigma^{-1}E)^{-1}\psi_p \tag{14}$$

or
$$x = \Sigma^{-1}E(E'\Sigma^{-1}E)^{-1}\psi_p. \tag{15}$$

Corollary 1. *Our result includes as a special case the value at risk calculations of Alexander and Bapista (2001) since $\phi(\mu_p, \sigma_p^2) = t\sigma_p - \mu_p$ for $t > 0$.*

We note that Alexander and Bapista (2001) are more concerned with the efficient set than the minimum risk set. The distinction between the efficient set and the minimum risk set is addressed by finding the minimum point on the mean minimum risk frontier. Nevertheless, their Proposition 2 (page 1168), which implicitly relates normality, goes to some length to show that the mean/value at risk efficient set of portfolios is the same as the mean minimum variance portfolios. This follows as a consequence of our Proposition 1 and its corollaries.

We now turn to the question as to whether the mean minimum risk frontier is concave (i.e. $\partial^2 v/\partial \mu^2 > 0$). Suppose that returns are elliptical, so that our risk measure v can be expressed as

$$v = \varphi(\mu, \sigma_p), \quad \mu = \mu_p, \tag{16}$$

where

$$\sigma_p^2 = \frac{\mu^2 \gamma - 2\beta\mu + \alpha}{\Delta}, \tag{17}$$

and

$$\Delta = \alpha\gamma - \beta^2, \tag{18}$$

so that

$$\frac{\partial v}{\partial \mu} = \varphi_1 + \frac{\varphi_2 (\mu\gamma - \beta)}{\sigma_p \Delta}, \tag{19}$$

so that $\partial v / \partial \mu \geq = \leq 0$ when

$$\varphi_1 + \frac{\varphi_2 (\mu\gamma - \beta)}{\sigma_p \Delta} \geq = \leq 0. \tag{20}$$

We shall assume the existence of a unique minimum risk portfolio; if it exists, the minimum risk portfolio occurs when $\partial v / \partial \mu = 0$ or when

$$\mu^* = \varphi_1 + \frac{\beta}{\gamma} - \frac{\sigma_p \varphi_1 \Delta}{\gamma \varphi_2}. \tag{21}$$

We note that β/γ is the expected return of the global minimum variance portfolio, so the global minimum "v" portfolio is to the right or to the left depending on the signs of φ_1 and φ_2.

For value at risk, $\varphi_2 = t > 0$ and $\varphi_1 = -1$, whereas for variance, $\varphi_2 = 2\sigma_p$ and $\varphi_1 = 0$. In general, one may wish to impose the restriction that $\varphi_2 > 0$ and $\varphi_1 \leq 0$; then $\mu^* \geq \frac{\beta}{\gamma}$ if $\sigma_p \Delta \geq \varphi_2 \gamma$. Other cases can be elaborated. Multiple solutions may be possible, but we shall ignore these. Consider now the second derivatives:

$$\frac{\partial^2 v}{\partial \mu^2} = \varphi_{11} + \frac{2\varphi_{12}(\mu\gamma - \beta)}{\sigma_p \Delta} + \frac{\varphi_{22}(\mu\gamma - \beta)^2}{\sigma_p^2 \Delta^2} + \frac{\varphi_2 \gamma}{\sigma_p \Delta}. \tag{22}$$

For $\partial^2 v / \partial \mu^2 > 0$ as required, we need the matrix of second derivatives $\begin{pmatrix} \varphi_{11} & \varphi_{12} \\ \varphi_{21} & \varphi_{22} \end{pmatrix}$ to be positive definite and $\varphi_2 > 0$. This condition is satisfied for variance and for value at risk as in this case $\varphi_{ij} = 0$ while $\varphi_2 = t$.

Corollary 2. *The above generalizes to a large family of risk measures for a range of distributions that reduce to the "mean-variance" analysis, namely the class of distributions as outlined in Chamberlain (1983).*

To clarify Chamberlain's class we need to define elliptical distributions.

Ingersoll (1987, page 104) defined a vector of n random variables to be elliptically distributed if its density (*pdf*) can be written as

$$pdf(y) = |\Omega|^{-1/2} g((y - \mu)' \Omega^{-1} (y - \mu)). \tag{23}$$

If means exist, then $E[y] = \mu$; if variances exist then $\text{cov}(y)$ is proportional to Ω. The characteristic function of y is

$$\varphi_n(t) = E[\exp(it'y)] = e^{it'\mu}\psi(t'\Omega t) \qquad (24)$$

for some function ψ that does not depend on n. It is apparent from (24) that if $\omega'y = z$ is a portfolio of elliptical variables from (24), then

$$E[\exp(isz)] = E[\exp(is\omega'y)] = e^{is\omega'\mu}\psi(s^2\omega'\Omega\omega), \qquad (25)$$

and all portfolios from the joint *pdf* given by (23) will have the same marginal distribution, which can be obtained by inverting (24). Furthermore, the distribution is location scale, in the sense that all portfolios differ only in terms of $\omega'\mu$ and $\omega'\Omega\omega$, and it is for this reason that mean-variance analysis holds for these families. A particular subfamily base of elliptical distributions is called the *family of spherical distributions*, namely those joint distributions invariant under orthogonal transformation.

Chamberlain proved (see Theorem 2) that for mean/variance analysis to be valid when there is no riskless asset, the case we are considering, a necessary and sufficient condition for $pdf(y)$, $E(y) \neq 0$, is that $\omega'y$ is determined by its mean and variance for every ω iff there is a nonsingular matrix T such that $Ty = \binom{m}{v}$, where, conditional on m, v is spherically distributed around the origin.

This condition is rather cumbersome. However, we note that for mean/variance analysis to be valid, the family of distributions needs to be identical if and only if they have the same mean and variances.

Consider, therefore, an important example. Let

$$\underset{\sim}{y} = \mu s_1 + s_2 \sum\nolimits^{\frac{1}{2}} \underset{\sim}{z}, \quad \text{where } z \sim N(0, I_N)$$

and $\sum^{\frac{1}{2}}$ is the square set matrix of the positive definite matrix $\sum$. The variables s_1 and s_2 are arbitrary and independent of z; furthermore, s_2 is assumed positive. Thus,

$$pdf(y/s_1, s_2) \sim N(\mu s_1, \sum s_2).$$

Furthermore, we see that $r_p = \omega'y$, and then $pdf(r_p|s_1, s_2) \sim N(\mu_p s_1, \sigma_p^2 s_2)$, where $\mu_p = \omega'\mu$ and $\sigma_p^2 = \omega'\sum\omega$. Clearly, if s_1 and s_2 are common to all portfolios then two portfolios p and p' will have the same pdfs if $\mu_p = \mu_{p'}$ and $\sigma_p^2 = \sigma_{p'}^2$. This can be reparameterized in terms of Chamberlain's condition as his theorem asserts.

Ingersoll demonstrated (1987, page 108) that this class, which he calls *jointly subordinated normal*, is compatible with mean/variance analysis in the case of no riskless asset; the proof amounts to noting that if the joint characteristic function of s_1 and s_2 is $\phi(t_1, t_2)$, then any portfolio r_p will have characteristic function $\phi(t) = \mu_p t \phi(\mu_p t, t^2 \sigma_p^2/2)$ and so distributions will be equalized iff $\mu_p = \mu_{p'}$ and $\sigma_p = \sigma_{p'}$.

The above model is especially important in that it provides a tractable case where we can have mean-variance analysis and distributions that have both skewness and kurtosis. To see this consider the case where s_1 and s_2 are independent. Then $k(t) = \ln \phi(t) = \ln \phi_1(\mu_p t) + \ln \phi_2 \left(t^2 \sigma_p^2/2\right)$, so that the $\phi_2(.)$ term, being even, contributes to the even moments while the $\phi_1(.)$ term contributes to both odd and even moments. In particular, the third central moment is the third central moment of the first term, so the model will capture skewness.

3 The Case of Two Assets

We now consider the nature of the mean/semivariance frontier. When $N = 2$, some general expressions can be calculated for the frontier.

In particular, for any distribution, for $N = 2$, $r_p = w r_1 + (1-w) r_2$ and

$$\mu_p = w\mu_1 + (1-w)\mu_2. \tag{26}$$

Two special cases arise: If $\mu_1 = \mu_2 = \mu$, then μ_p always equals μ, and the (μ_p, θ_p^2) frontier is degenerate, consisting of a single point. Otherwise, assume that $\mu_1 \neq \mu_2$, and then Equation (26) can be solved for w^*, so that

$$w^* = \frac{\mu_p - \mu_2}{\mu_1 - \mu_2}. \tag{27}$$

If we assume, without loss of generality, that $\mu_1 > \mu_2$, so that when there is no short selling, $\mu_1 \geq \mu_p \geq \mu_2$, for $0 \leq w^* \leq 1$. We shall concentrate on this part of the frontier, but we could consider extensions for $w^* > 1$ or $w^* < 0$.

Now define $\theta_p^2(\tau)$ as the lower partial moment of degree 2 with truncation point τ. Then

$$\theta_p^2(\tau) = \int_{-\infty}^{\tau} (\tau - r_p)^2 pdf(r_p) dr_p. \tag{28}$$

However, an alternative representation in terms of the joint *pdf* of r_1 and r_2 is available. Namely,

$$\theta_p^2(\tau) = \int_{\Re} (\tau - r_p)^2 pdf(r_1, r_2) dr_1 dr_2, \tag{29}$$

where $\Re = \{(r_1, r_2); r_1 w^* + r_2(1 - w^*) \leq \tau\}$.

We can now change variables from (r_1, r_2) to (r_1, r_p) by the (linear) transformations

$$r_p = w^* r_1 + (1 - w^*) r_2 \quad \text{and} \quad r_1 = r_1. \tag{30}$$

Therefore,
$$dr_1 dr_2 = \frac{1}{(1-w^*)} dr_1 dr_p \quad \text{if} \quad 0 \le w^* < 1. \tag{31}$$

If $w^* = 1$, then the transformation is $(r_1, r_2) \mapsto (r_2, r_p)$.

Now,
$$\theta_p^2(\tau) = \int_{-\infty}^{\infty} \int_{-\infty}^{\tau} \frac{(\tau - r_p)^2}{(1-w^*)} pdf\left(r_1, \frac{r_p - w^* r_1}{1 - w^*}\right) dr_1 dr_p. \tag{32}$$

This equation gives us the mean/semivariance locus for **any** joint *pdf*, i.e., $pdf(r_1, r_2)$. As μ_p changes, w^* changes and so does $\theta_p^2(\tau)$.

In certain cases, i.e., ellipticity, we can explicitly compute $pdf(r_p)$ and we can directly use Equation (28). In general, however, we have to resort to Equation (32) for our calculations. For the jointly subordinated normal class described earlier, we can compute $pdf(r_p)$, but it would involve integrating over s_1 and s_2 and, except in some special cases, will not lead to tractable solutions.

In what follows, we present two results, first, for $N = 2$ under normality, where Equation (28) can be applied directly and a closed-form solution derived. Second, we assume joint log-normality, where we use either Equation (28) or Equation (32) and numerical methods.

We see, by manipulating the above, that if we can compute the marginal *pdf* of r_p, the results will simplify considerably. For the case of normality, and for general elliptical distributions, the *pdf* of r_p is known. We proceed to compute the (μ_p, θ_p^2) frontier under normality.

Proposition 2. *Assuming that*

$$\begin{pmatrix} r_1 \\ r_2 \end{pmatrix} \sim N\left[\begin{pmatrix} \mu_1 \\ \mu_2 \end{pmatrix}, \begin{pmatrix} \sigma_1^2 & \sigma_{12} \\ \sigma_{12} & \sigma_2^2 \end{pmatrix}\right], \tag{33}$$

the mean/semivariance frontier can be written as

$$\theta_p^2 = \left(\sigma_p^2 + (t - \mu_p)^2\right) \Phi\left(\frac{t - \mu_p}{\sigma_p}\right) + (t - \mu_p)\sigma_p \phi\left(\frac{t - \mu_p}{\sigma_p}\right), \tag{34}$$

where ϕ and Φ are the standard normal density and distribution functions, respectively, μ_p is given by Equation (26), and $\sigma_p^2 = w^2 \sigma_1^2 + 2w(1-w)\sigma_{12} + (1-w)^2 \sigma_2^2$. Moreover, if r_1 and r_2 are any two mean/variance efficient portfolios, the above result will hold for any $N > 2$.

Proof. Consider the integral, [letting $\theta_p^2(t) = I(t)$],

$$I(t) = \int_{-\infty}^{t} (t - r_p)^2 pdf(r_p) dr_p, \tag{35}$$

where $r_p \sim N(\mu_p, \sigma_p^2)$, so that

$$pdf(r_p) = \frac{1}{\sigma_p\sqrt{2\pi}} \exp\left(-\frac{(r_p - \mu_p^2)}{2\sigma_p^2}\right). \tag{36}$$

Transform $r_p \to y = t - r_p \Rightarrow r_p = t - y, |dr_p| = |dy|$, so

$$I(t) = \int_0^\infty y^2 \frac{1}{\sigma_p\sqrt{2\pi}} \exp\left(-\frac{(y + t - \mu_p)^2}{2\sigma_p^2}\right) dy, \tag{37}$$

$$I(t) = \frac{\partial^2}{\partial q^2}\left[\int_0^\infty e^{qy} \frac{1}{\sigma_p\sqrt{2\pi}} \exp\left(-\frac{(y + t - \mu_p)^2}{2\sigma_p^2}\right) dy\right]_{q=0}. \tag{38}$$

So examine the integral in brackets:

$$J = \int_0^\infty e^{qy} \frac{1}{\sigma_p\sqrt{2\pi}} \exp\left(-\frac{(y + t - \mu_p)^2}{2\sigma_p^2}\right) dy. \tag{39}$$

Let $t - \mu_p = -\mu_t$, so

$$J = \int_0^\infty \frac{1}{\sigma_p\sqrt{2\pi}} \exp\left(-\frac{(y^2 - 2y\mu_t - 2\sigma_p^2 qy + \mu_t^2)}{2\sigma_p^2}\right) dy, \tag{40}$$

$$J = \exp\left(-\frac{\mu_t^2}{2\sigma_p^2}\right)\exp\left(\frac{(\mu_t^2 + q\sigma_p^2)^2}{2\sigma_p^2}\right)\int_0^\infty \frac{1}{\sigma_p\sqrt{2\pi}}\exp\left(-\frac{(y-(q\sigma_p^2-\mu_t))^2}{2\sigma_p^2}\right)dy. \tag{41}$$

Transform $y \to z = (y - (q\sigma_p^2 - \mu_t))/\sigma_p \Rightarrow y = z\sigma_p + q\sigma_p^2 - \mu_t$, so

$$J = \exp\left(\frac{-2q\mu_t + q^2\sigma_p^2}{2}\right)\int_{-(q\sigma_p^2-\mu_t)/\sigma_p}^\infty \frac{1}{\sqrt{2\pi}}\exp\left(-\frac{z^2}{2}\right)dz, \tag{42}$$

$$J = \exp\left(\frac{-2q\mu_t + q^2\sigma_p^2}{2}\right)\left(1 - \Phi\left(-\frac{q\sigma_p^2 - \mu_t}{\sigma_p}\right)\right). \tag{43}$$

That is,

$$I(t) = \frac{\partial^2}{\partial q^2}\left[\exp\left(\frac{-2q\mu_t + q^2\sigma_p^2}{2}\right)\Phi\left(\frac{q\sigma_p^2 - \mu_t}{\sigma_p}\right)\right]_{q=0}, \tag{44}$$

$$I(t) = \frac{\partial}{\partial q}\left[(q\sigma_p^2 - \mu_t)\exp\left(\frac{-2q\mu_t + q^2\sigma_p^2}{2}\right)\Phi\left(\frac{q\sigma_p^2 - \mu_t}{\sigma_p}\right)\right.$$
$$\left. + \sigma_p \exp\left(\frac{-2q\mu_t + q^2\sigma_p^2}{2}\right)\Phi'\left(\frac{q\sigma_p^2 - \mu_t}{\sigma_p}\right)\right]_{q=0}, \tag{45}$$

$$I(t) = \left[\sigma_p^2 \exp\left(\frac{-2q\mu_t + q^2\sigma_p^2}{2}\right) \Phi\left(\frac{q\sigma_p^2 - \mu_t}{\sigma_p}\right)\right.$$

$$+ (q\sigma_p^2 - \mu_t)^2 \exp\left(\frac{-2q\mu_t + q^2\sigma_p^2}{2}\right) \Phi\left(\frac{q\sigma_p^2 - \mu_t}{\sigma_p}\right)$$

$$+ 2\sigma_p(q\sigma_p^2 - \mu_t) \exp\left(\frac{-2q\mu_t + q^2\sigma_p^2}{2}\right) \Phi'\left(\frac{q\sigma_p^2 - \mu_t}{\sigma_p}\right)$$

$$\left. + \sigma_p^2 \exp\left(\frac{-2q\mu_t + q^2\sigma_p^2}{2}\right) \Phi''\left(\frac{q\sigma_p^2 - \mu_t}{\sigma_p}\right)\right]_{q=0}, \qquad (46)$$

$$I(t) = \sigma_p^2 \Phi\left(\frac{-\mu_t}{\sigma_p}\right) + \mu_t^2 \Phi\left(\frac{-\mu_t}{\sigma_p}\right) - 2\mu_t \sigma_p \Phi'\left(\frac{-\mu_t}{\sigma_p}\right) + \sigma_p^2 \Phi''\left(\frac{-\mu_t}{\sigma_p}\right). \qquad (47)$$

Now

$$\Phi(x) = \int_{-\infty}^{x} \frac{1}{\sqrt{2\pi}} \exp\left(-\frac{z^2}{2}\right) dz, \qquad (48)$$

$$\Phi'(x) = \frac{1}{\sqrt{2\pi}} \exp\left(-\frac{x^2}{2}\right) = \phi(x), \qquad (49)$$

$$\Phi''(x) = -x \frac{1}{\sqrt{2\pi}} \exp\left(-\frac{x^2}{2}\right) = -x\phi(x). \qquad (50)$$

Thus,

$$I(t) = (\sigma_p^2 + \mu_t^2)\Phi\left(\frac{-\mu_t}{\sigma_p}\right) - 2\mu_t \sigma_p \phi\left(\frac{\mu_t}{\sigma_p}\right) + \sigma_p^2 \frac{\mu_t}{\sigma_p} \phi\left(\frac{\mu_t}{\sigma_p}\right) \qquad (51)$$

or

$$I(t) = (\sigma_p^2 + \mu_t^2)\Phi\left(\frac{-\mu_t}{\sigma_p}\right) - \mu_t \sigma_p \phi\left(\frac{\mu_t}{\sigma_p}\right). \qquad (52)$$

Finally, if r_1 and r_2 are any two mean/variance efficient portfolios, then they span the set of minimum risk portfolios as described in Proposition 1, and the result thus follows.

Corollary 3. *If we wish to consider expected loss under normality, which we denote by L, where $L = E[r_p$ and $r_p < t]$, then*

$$L = (t - \mu_p)\Phi\left(\frac{t - \mu_p}{\sigma_p}\right) + \sigma_p \phi\left(\frac{t - \mu_p}{\sigma_p}\right). \qquad (53)$$

Proof. The same argument as before.

The above results can be generalized to any elliptical distribution with a finite second moment since in all cases we know, at least in principle, the marginal distribution of any portfolio. Let $f()$ and $F()$ be the *pdf* and *cdf* of any portfolio return r_p, and let μ_p and σ_p be the relevant mean and scale parameters.

Then it is a consequence of ellipticity that a density (if it exists) $f(z)$ has the following property. For any (μ_p, σ_p) and $r_p = \mu_p + \sigma_p z$,

$$f(z) = \sigma_p pdf\left(\frac{r_p - \mu_p}{\sigma_p}\right). \tag{54}$$

Furthermore, there are incomplete moment distributions

$$F^k(x) = \int_{-\infty}^{x} z^k f(z) dz, \qquad F^0(x) = F(x) \tag{55}$$

for k a positive integer [the existence of the kth moment is required for $F^k(x)$ to exist]. So

$$L = (t - \mu_p) F\left(\frac{t - \mu_p}{\sigma_p}\right) - \sigma_p F^1\left(\frac{t - \mu_p}{\sigma_p}\right) \tag{56}$$

and letting $w = (t - \mu_p)/\sigma_p$,

$$\text{Semivariance} = (t - \mu_p)^2 F(w) - 2(t - \mu_p)\sigma_p F^1(w) + \sigma_p^2 F^2(w). \tag{57}$$

To illustrate the problems that arise when the *pdf* of r_p is not available in closed form, we consider the case of bivariate log-normality; as we see below, the previous simplifications no longer occur. Suppose that

$$r_1 = \left(\frac{P_1}{P_0} - 1\right); \tag{58}$$

thus,

$$(1 + r_1) = \left(\frac{P_1}{P_0}\right) = \exp(y_1), \tag{59}$$

so that

$$(1 + r_p) = 1 + \sum w_i r_i = w \exp(y_1) + (1 - w) \exp(y_2) \tag{60}$$

and

$$r_p = w \exp(y_1) + (1 - w) \exp(y_2) - 1. \tag{61}$$

Therefore,

$$sv(r_p) = \int_{-\infty}^{t} (t - r_p)^2 pdf(r_p) dr_p. \tag{62}$$

If

$$r_p < t, \quad \text{then} \quad w \exp(y_1) + (1 - w) \exp(y_2) < 1 + t. \tag{63}$$

The above transforms to a region $\mathbf{R}$ in (y_1, y_2) space. Hence,

$$sv(r_p) = \int_{\mathbf{R}} (1 + t - w \exp(y_1) - (1 - w) \exp(y_2))^2 pdf(y_1, y_2) dy_1 dy_2 \tag{64}$$

or

$$sv(r_p) = \int_{\mathbf{R}} (1+t)^2 pdf(y_1, y_2) dy_1 dy_2 + c_1 \int_{\mathbf{R}} \exp(y_1) pdf(y_1, y_2) dy_1 dy_2$$
$$+ c_2 \int_{\mathbf{R}} \exp(y_1 + y_2))^2 pdf(y_1, y_2) dy_1 dy_2, \tag{65}$$

where c_1 and c_2 are some constants. None of the above integrals can be computed in closed form, although they can be calculated by numerical methods.

4 Conic Results

Using our definition of value at risk as $VaR_p = t\sigma_p - \mu_p$ with $t > 0$, and noting from Proposition 1 that σ_p^2 must lie on the minimum variance frontier so that

$$\sigma_p^2 = \frac{(\mu_p^2 \gamma - 2\beta\mu_p + \alpha)}{(\alpha\gamma - \beta^2)}, \tag{66}$$

we see that

$$(\text{Var}_p + \mu_p)^2 = \frac{(t^2)(\mu_p^2 \gamma - 2\beta\mu_p + \alpha)}{(\alpha\gamma - \beta^2)}. \tag{67}$$

Equation (67) is a general quadratic (conic) in μ_p and Var_p and we can apply the methods of analytical geometry to understand what locus it describes.

The conic for v and u is

$$v^2 + 2vu + u^2(1 - \gamma\theta) + 2\beta\theta u - \alpha\theta = 0, \tag{68}$$

where $\theta = t^2/(\alpha\gamma - \beta^2)$ and $u = \mu_p$, $v = \text{Var}_p$, α, γ, and θ are always positive.

Following standard arguments we can show that this conic must always be an hyperbola since $\gamma\theta > 0$ (see Brown and Manson, 1959, page 284). Furthermore, the center of the conic in (u, v) space is $(\beta/\gamma, -\beta/\gamma)$, so that the center corresponds to the same expected return as the global minimum variance portfolio. The center of the hyperbola divides the (u, v) space into two regions.

We now consider implicit differentiation of (68) for the region where $\mu \geq \beta/\gamma$, which corresponds to the relevant region for computing our frontier:

$$2v\frac{\partial v}{\partial u} + 2\frac{\partial v}{\partial u}u + 2v + 2u(1 - \gamma\theta) + 2\beta\theta = 0. \tag{69}$$

So

$$\frac{\partial v}{\partial u} = \frac{u(\gamma\theta - 1) - v - \beta\theta}{u + v}. \tag{70}$$

Thus,

$$v = u(\gamma\theta - 1) - \beta\theta \quad \text{when} \quad \frac{\partial v}{\partial u} = 0, \tag{71}$$

or
$$u = \frac{(v+\beta\theta)}{(\gamma\theta-1)}. \tag{72}$$

Substituting into Equation (68),

$$v^2 + 2v\frac{(v+\beta\theta)}{(\gamma\theta-1)} + \frac{(v+\beta\theta)^2}{(\gamma\theta-1)^2}(1-\gamma\theta) + 2\beta\theta\frac{(v+\beta\theta)}{(\gamma\theta-1)} - \alpha\theta = 0. \tag{73}$$

Simplifying, letting $\Delta = \alpha\gamma - \beta^2 > 0$,

$$v^2(\gamma\theta-1) + 2v^2 + 2\beta\theta v - v^2 - 2\beta\theta v - \beta^2\theta^2 + 2\beta\theta v + 2\beta^2\theta^2 - \alpha\theta(\gamma\theta-1) = 0 \tag{74}$$

or
$$v^2\gamma + 2v\beta + (\alpha - \Delta\theta) = 0, \tag{75}$$

so that
$$v = -\frac{\beta}{\gamma} \pm \frac{\sqrt{\beta^2 - \alpha\gamma + \Delta\theta\gamma}}{\gamma}. \tag{76}$$

Since
$$\beta^2 - \alpha\gamma + \Delta\theta\gamma \geq 0, \tag{77}$$

or
$$-\Delta + \Delta\theta\gamma \geq 0, \tag{78}$$

or
$$\Delta(\theta\gamma - 1) \geq 0, \tag{79}$$

we have
$$\theta\gamma \geq 1. \tag{80}$$

The solution for the upper part of the hyperbola is where $v > -\beta/\gamma$, which corresponds to

$$v = -\frac{\beta}{\gamma} + \frac{\sqrt{\beta^2 - \alpha\gamma + \Delta\theta\gamma}}{\gamma}. \tag{81}$$

To check that this is a minimum, we compute

$$u = \frac{(-\frac{\beta}{\gamma} + \frac{1}{\gamma}\sqrt{\Delta(\theta\gamma-1)}) + \beta\theta}{(\gamma\theta-1)}, \tag{82}$$

$$u = \frac{(-\beta + \sqrt{\Delta(\theta\gamma-1)} + \beta\theta\gamma)}{\gamma(\gamma\theta-1)}, \tag{83}$$

$$u = \frac{(\beta(\theta\gamma-1) + \sqrt{\Delta(\theta\gamma-1)})}{(\gamma\theta-1)}, \tag{84}$$

$$u = \frac{\beta}{\gamma} + \frac{1}{\gamma}\frac{\sqrt{\Delta(\theta\gamma-1)}}{(\theta\gamma-1)}, \tag{85}$$

$$u = \frac{\beta}{\gamma} + \frac{1}{\gamma}\sqrt{\frac{\Delta}{(\theta\gamma - 1)}}, \qquad (86)$$

$$u > \frac{\beta}{\gamma}. \qquad (87)$$

Now considering the second-order conditions, first we have

$$u + v = \frac{\beta}{\gamma} + \frac{1}{\gamma}\sqrt{\frac{\Delta}{(\theta\gamma - 1)}} - \frac{\beta}{\gamma} + \frac{1}{\gamma}\sqrt{\Delta(\theta\gamma - 1)}, \qquad (88)$$

$$u + v = \frac{1}{\gamma}[\sqrt{\frac{\Delta}{(\theta\gamma - 1)}} + \sqrt{\Delta(\theta\gamma - 1)}] > 0. \qquad (89)$$

Differentiating Equation (69) gives

$$\left(\frac{\partial v}{\partial u}\right)^2 + v\frac{\partial^2 v}{\partial u^2} + u + v\frac{\partial^2 v}{\partial u^2} + \frac{\partial v}{\partial u} + \frac{\partial v}{\partial u} + (1 - \gamma\theta) = 0. \qquad (90)$$

Since $\partial v/\partial u = 0$ at the minimum, then

$$\frac{\partial^2 v}{\partial u^2} = \frac{(\gamma\theta - 1)}{(u+v)}, \qquad (91)$$

and since $\gamma\theta > 1$ and $(u+v) > 0$ at (u^*, v^*) from (89), the minimum point is established.

We note that the condition $\gamma\theta > 1$ corresponds to the condition given in Proposition 1 of Alexander and Bapista (2001). By substituting (69) back into (73), we can recover the mean/variance portfolio, which is the minimum value at risk portfolio as described by Equation (10).

5 Simulation Methodology

We consider simulation of portfolios $\underline{w}$ of length N subject to symmetric linear constraints, i.e., $\sum w_i = 1$ and $a \leq w_i \leq b$.

We assume w_1 is distributed as $uniform[a, b]$ and let w_1^* be the sampled value. Then w_2 is also sampled from a $uniform[a, b]$ distribution with sampled value w_2^*. The procedure is repeated sequentially as long as $\sum_{j=1}^{m} w_j^* \leq (N-m)a$. If a value w_{m+1}^* is chosen such that $\sum_{j=1}^{m+1} w_j^* \geq (N-m-1)a$, we set w_{m+1}^* so that $\sum_{j=1}^{m+1} w_j^* + (N-m-1)a = 1$. This is always feasible.

Because the sequential sampling tends to make the weights selected early in the process larger, for any feasible $\underline{w}^*$ we consider all $n!$ permutations of $\underline{w}^*$. From the symmetric constraints, these will also be feasible portfolios. Unfortunately, with $N = 8$, then $N! = 40,320$, so it may be feasible, but with

$N = 12$ then $N! = 47,9001,600$ and it may no longer be a feasible approach! We have to rely on random sampling in the first instance.

If we have a history of N stock returns for T periods, then for any vector of portfolio weights, we can calculate the portfolio returns for the T periods. These values can then be used to compute an estimate of the expected return μ_i and the risk measure ϕ_i, which can be mapped into points on a diagram as in Figure 1. If we believe the data to be elliptically generated, then, following Proposition 2, we can save the weights of the set of minimum risk portfolios.

We amend an algorithm suggested by Bensalah (2000). If the risk-return value of individual portfolios can be computed, then an intuitive procedure is to draw the surface of many possible portfolios in a risk-return framework and then identify the optimal portfolio in a mean-minimum risk sense. In the case of no short selling, an algorithm for approximating any frontier portfolios of N assets each with a history of T returns can be described as follows:

Step 1: Define the number of portfolios to be simulated as M.
Step 2: Randomize the order of the assets in the portfolio.
Step 3: Randomly generate the weight of the first asset $w_1 \sim U[0, 1]$ from a Uniform distribution, $w_2 \sim U[0, 1 - w_1]$, $w_3 \sim U[0, 1 - w_1 - w_2]$, $\ldots$, $w_N = 1 - w_1 - w_2 - \ldots - w_{N-1}$.
Step 4: Generate a history of T returns for this portfolio, and compute the average return and risk measure.
Step 5: Repeat steps 2 to 4, M times.
Step 6: From the M sets of risk-return measures, rank the returns in ascending order and allocate the returns and allocate each pair to B buckets equally spaced from the smallest to the largest return. Within each bucket determine the portfolio with the minimum (or maximum as required) risk measure. Across the B buckets these portfolios define the approximate risk-return frontier.

We illustrate the feasibility and accuracy of this algorithm using data from the Australian Stock Exchange. Daily returns were obtained for the trading days from June 1, 1995, to August 30, 2002, on the eight largest capitalization stocks, giving a history of 1,836 returns. Summary statistics for the daily percentage return on these stocks are reported in Table 1. They display typical properties of stock returns, in particular, with all stocks displaying significant excess kurtosis. A total of 128,000 random portfolios were generated. For each portfolio the average return, the sample standard deviation, the sample semi-standard deviation, the 5% value at risk and the sample expected loss below zero were computed. Figure 1 illustrates the surface of the mean/standard deviation and mean/value at risk pairs obtained from 8,000 random portfolios. The frontiers for these risk measures are readily identified. From the 128,000 random portfolios, the approximated risk frontiers are presented in Figure 2 for the mean/standard deviation, the mean/semistandard deviation, the mean/value at risk (5%) and the mean/expected loss. We represent the frontiers when it is assumed that the returns are elliptically distributed. In this case, a quadratic programming algorithm has been solved for the optimal mean/standard deviation portfolio with no short selling for a number of target returns. The portfolio weights were then used to compute a history of returns to obtain the three other risk measures. The other lines represent the frontiers identified by the algorithm. In this case we use the portfolio weights for the optimal portfolios in each bucket found by minimizing the sample standard deviation. The identified frontiers for the mean/standard deviation and mean/semistandard deviation are very close to that identified by the quadratic programming algorithm. The two expected loss frontiers are superimposed and cannot be distinguished on this diagram. By contrast, the deviations between the two value at risk frontiers are relatively larger. The quality of the approximations for the 128,000 random portfolios are summarized in Table 2, where 100 buckets of portfolios have been used. In Panel A, we chose the optimal portfolio in each bucket, is that portfolio with the smallest standard deviation. The metric here is returns measured in daily percentages. For the mean/standard deviation frontier we measure the error as the distance of the approximation from the quadratic programming solution, and across 100 points on the frontier, the average error is 1.05 basis points with a standard deviation of 0.49 basis points and the largest error is 2.42 basis points. For the mean/semistandard deviation frontier, the average error is 0.72 basis points with a standard deviation of 0.36 basis points. The expected loss frontiers almost coincide, with the average error of 0.01 basis points with a standard deviation of 0.08 basis points. The value at risk frontiers show the largest discrepancies, with an average error of 1.75 basis points, a standard deviation of 2.84 basis points, and a range in values from -5.08 up to 9.55 basis points. Panels B, C, and D report similar measures when the optimal portfolios in each bucket are chosen by minimizing semistandard deviation,

Table 1. Summary Statistics for Eight Stocks

Rank	Stock	Average	StDev	Min	Max	Skewness	Kurtosis
1	NAB	0.058	1.407	-13.871	4.999	-0.823	9.688
2	CBA	0.066	1.254	-7.131	7.435	-0.188	5.022
3	BHP	0.008	1.675	-7.617	7.843	0.107	4.205
4	ANZ	0.073	1.503	-7.064	9.195	-0.122	4.617
5	WBC	0.059	1.357	-6.397	5.123	-0.181	3.985
6	NCP	0.013	2.432	-14.891	24.573	0.564	11.321
7	RIO	0.035	1.659	-12.002	7.663	0.069	5.360
8	WOW	0.078	1.482	-8.392	11.483	0.025	6.846

Notes: Stocks are ranked by market capitalization on August 30, 2002. The statistics are based on 1,836 daily returns and are expressed as percentage per day. Average is the sample mean; StDev is the sample standard deviation; Min is the minimum observed return; Max is the maximum observed return; Skewness is the sample skewness; and Kurtosis is the sample kurtosis.

maximizing value at risk (5%), and maximizing the associated expected loss. The results are qualitatively the same as in Panel A, but it is notable that the portfolios chosen using value at risk result in the largest deviations from the quadratic programming solutions.

This example does illustrate that it is feasible to approximate the frontiers for a variety of risk measures using this intuitive simulation methodology, but in this case there is little to be gained over the frontiers identified from assuming that the returns are elliptically distributed. Finally, by using jointly subordinated normal returns, we can have mean risk frontiers while using data that are skewed and kurtotis.

Table 2. Summary Statistics: Distance Between Frontiers, 8 Assets, 128,000 Simulated Portfolios

		StDev	Semi-StDev	VaR	Exp. Loss
		Panel A			
StDev	Avg	0.0105	0.0072	0.0175	0.0001
	StDev	0.0049	0.0036	0.0284	0.0008
	Min	0.0004	0.0003	−0.0508	−0.0015
	Max	0.0242	0.0158	0.0955	0.0020
		Panel B			
Semi-StDev	Avg	0.0111	0.0069	0.0185	0.0000
	StDev	0.0055	0.0035	0.0295	0.0008
	Min	0.0004	0.0003	−0.0508	−0.0015
	Max	0.0270	0.0166	0.0981	0.0023
		Panel C			
VaR	Avg	0.0218	0.0148	−0.0106	0.0008
	StDev	0.0123	0.0093	0.0187	0.0012
	Min	0.0001	0.0001	−0.0508	−0.0013
	Max	0.0614	0.0481	0.0817	0.0048
		Panel D			
Exp. Loss	Avg	0.0125	0.0082	0.0122	−0.0003
	StDev	0.0064	0.0050	0.0258	0.0007
	Min	0.0004	0.0003	−0.048	−0.0015
	Max	0.0391	0.0276	0.0955	0.0014

Note: The units of measurement are percent per day. For Standard Deviation and semistandard deviation, we measure the deviation as the simulated portfolio value minus the quadratic programming value. For value at risk and expected loss, we measure the quadratic programming value minus the simulated portfolio value. The statistics reported here are based on 100 points equally spaced along the respective frontiers from the minimum to the maximum observed sample portfolio returns.

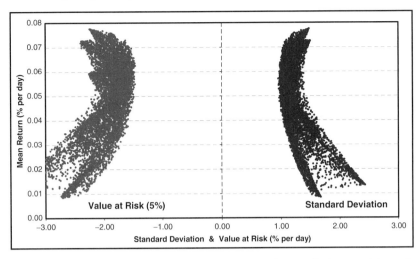

Fig. 1. Illustration of mean/standard deviation and mean/value at risk surfaces using 8,000 portfolios.

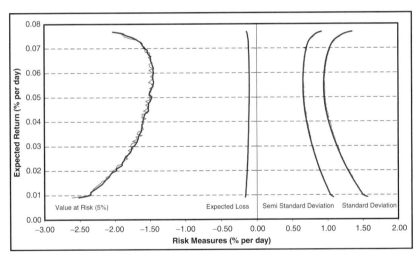

Fig. 2. Risk frontiers obtained using quadratic programming and 128,000 simulated portfolios (*Smooth*: frontiers using weights from solving minimum standard deviation frontiers with no short selling, *Wavy*: frontiers obtained using the simulated portfolios where portfolio weights are derived by minimising standard deviation).

6 Conclusion

We have presented analytical results that allow us to understand better what mean/risk frontiers look like. For elliptical, and related, returns these simplify to explicit formulas and we present closed-form expressions for mean/value at risk frontiers under ellipticity and mean/expected loss and mean/semivariance frontiers under normality. For nonelliptical distributions, a simulation methodology is presented that can be applied easily to historical data. We do not consider the case of a riskless asset since this only has relevance when index-linked bonds are available. However, our results could be extended to this case.

Acknowledgments

We thank John Knight and Peter Buchen for many helpful comments.

References

1. Alexander, G. J., and Bapista, A. M. Economic implications of using a mean-VaR model for portfolio selection: A comparison with mean-variance analysis. *Journal of Economic Dynamics and Control*, 26:1159–1193, 2001.
2. Bensalah, Y. Asset allocation using extreme value theory. Bank of Canada Working paper 2002-2.
3. Bradley, B. O., and Jaggia, M. S. Financial risk and heavy tails. In F. Comroy and S. T. Rader, Editors, *Heavy-tailed Distributions in Finance*. North Holland, Amsterdam, 2002.
4. Brown, J. T., and Manson, C. W. M. *The Elements of Analytical Geometry*. Macmillan, New York, 1959.
5. Campbell, R., Huisman, R., and Koedijk, K. Optimal portfolio selection in a value at risk framework. *Journal of Banking and Finance*, 25:1789–1804, 2002.
6. Chamberlain, G. A characterisation of the distributions that imply mean-variance utility functions. *Journal of Economic Theory*, 29: 185–201, 1983.
7. Embrecht, P., McNeil, A. J., and Straumann, D. Correlation and dependence in risk management: Properties and pitfalls. In M. Dempster and H. K. Moffatt, Editors, *Risk Management: Value at Risk and Beyond*. Cambridge University Press, Cambridge, 2002.
8. Ingersoll, J. E. *Theory of Financial Decision Making*. Rowman & Littlefield, Totowa, NJ, 1987.
9. Sargent, T. J. *Macroeconomic Theory*. Academic Press, New York, 1979.
10. Wang, J. Mean-variance-VaR based portfolio optimization. Mimeo, Valdosta State University, 2000.

Exchange Traded Funds: History, Trading, and Research

Laurent Deville

Paris-Dauphine University, CNRS, DRM-CEREG, Paris, France
Laurent.Deville@dauphine.fr

1 Introduction

One of the most spectacular successes in financial innovation since the advent of financial futures is probably the creation of exchange traded funds (ETFs). As index funds, they aim at replicating the performance of their benchmark indices as closely as possible. Contrary to conventional mutual funds, however, ETFs are listed on an exchange and can be traded intradaily. Issuers and exchanges set forth the diversification opportunities they provide to all types of investors at a lower cost, but also highlight their tax efficiency, transparency, and low management fees. All of these features rely on a specific "in-kind" creation and redemption principle: New shares can continuously be created by depositing a portfolio of stocks that closely approximates the holdings of the fund; similarly, investors can redeem outstanding ETF shares and receive the basket portfolio in return. Holdings are transparent since fund portfolios are disclosed at the end of the trading day.

ETFs were introduced to U.S. and Canadian exchanges in the early 1990s. In the first several years, they represented a small fraction of the assets under management in index funds. However, the 132% average annual growth rate of ETF assets from 1995 through 2001 (Gastineau, 2002) illustrates the increasing importance of these instruments. The launching of Cubes in 1999 was accompanied by a spectacular growth in trading volume, making the major ETFs the most actively traded equity securities on the U.S. stock exchanges. Since then, ETF markets have continued to grow, not only in the number and variety of products, but also in terms of assets and market value. Initially, they aimed at replicating broad-based stock indices; new ETFs extended their fields to sectors, international markets, fixed-income instruments, and, lately, commodities. By the end of 2005, 453 ETFs were listed around the world, for assets worth $343 billion. In the United States, overall ETF assets totaled $296.02 billion, compared to $8.9 trillion in mutual funds.[1]

---
[1] Data from the Investment Company Institute; http://www.ici.org/.

C. Zopounidis, M. Doumpos, and P.M. Pardalos (eds.) *Handbook of Financial Engineering,* doi: 10.1007/978-0-387-76682-9_4,
© Springer Science+Business Media, LLC 2008

ETFs were initially developed in the United States by the American Stock Exchange (AMEX) but soon faced competition for trading. Before the NYSE ventured into ETFs, these securities were already traded on the Nasdaq InterMarket, regional exchanges, and the Island Electronic Crossing Network. Though long opposed to this practice, for the first time in its history the NYSE began trading the three most active ETFs under Unlisted Trading Privileges on July 31, 2001. Moreover, the different trading venues also competed for listings. On December 1, 2004, Nasdaq-100 Index Tracking Stocks, more commonly known as "Cubes," changed listing from AMEX to Nasdaq. More recently, on July 20, 2005, Barclays Global Investors announced the transfer of 61 iShares ETFs to the NYSE from the AMEX.

Several questions arise:

- Does the ETF-specific structure allow for more efficient index fund pricing?
- Do ETFs represent a performing alternative to conventional index mutual funds?
- What impact does the advent of ETFs have on trading and market quality with regard to index component stocks and index derivatives?

Other empirical studies also focus on ETFs and investigate diverse topics, such as competition between trading venues, the shape of the demand curve, or the use of ETFs. Even though they are only loosely related, we will discuss these studies under the heading "more studies devoted to ETFs".

In the following section, we start by providing an overview of the history of ETFs, from their creation in North American markets to their more recent developments in the U.S. and European markets. In Section 3, we detail the mechanics of ETFs with a special focus on creation and redemption and present the ETF industry. The next four sections are devoted to the survey itself. In Section 4, we look at the pricing efficiency of ETFs and compare it to that of closed-end funds, while in Section 5 we examine the relative performance of ETFs over conventional index mutual funds. In Section 6, we explore the impact the arrival of ETFs has on the market quality of the stock components of the underlying indices, the efficiency of index derivatives markets and the pricing discovery process for index prices. In Section 7, we discuss other, less studied ETF-related issues. Section 8 concludes and presents directions for further research.

2 The History of ETFs

2.1 The Birth and Development of ETFs in North America

Depending on how restrictive the authors are in their definition, ETFs as we now know them were first introduced in the early 1990s, either in Canada (with the TIPs that were first traded in 1990) or three years later in the United States (with the SPDRs). However, the ability to trade a whole stock basket in a single transaction dates farther back. Major U.S. brokerage firms provided such program trading facilities as early as the late 1970s, particularly for the S&P 500 index. With the introduction of index futures contracts, program trading became more popular. As such, the opportunity to develop a suitable instrument allowing index components to be negotiated in a single trade became increasingly interesting.

In 1989, the American Stock Exchange and the Philadelphia Stock Exchange started trading Index Participation Shares (IPS). These synthetic instruments were aimed at replicating the performance of the S&P 500 index, among others, but they had characteristics similar to those of futures contracts. Despite significant interest from investors, IPS had to stop trading after the lawsuit by the Chicago Mercantile Exchange and the Commodity Futures Trading Commission (CFTC) was won. As futures contracts, IPS had to be traded on a futures exchange regulated by the CFTC.

The first equity-like index fund, the Toronto Index Participation units (TIPs), was introduced to the Toronto Stock Exchange on March 9, 1990. Tracking the Toronto 35, they were traded on the stock exchange and were characterized by extremely low management fees, given that the fund manager was authorized to loan the stocks held by the fund, for which demand was usually high. This product was followed in 1994 by HIPs, based on the broader TSE-100 index. Despite the huge success of these securities, their very low expense ratios finally made them too costly for the exchange and its members. TIPs and HIPs were terminated in 2000.[2] In 1993, after three years of dispute with the SEC, the American Stock Exchange (AMEX) began trading Standard & Poor's 500 Depositary Receipt (SPDR, popularly known as "Spider"; ticker SPY), which is often referred to as the world's first ETF. The fund was sponsored by PDR Services Corporation, an AMEX subsidiary, with State Street Bank and Trust as trustee. Its specific trust structure and trading process then constituted a model for the next ETFs introduced, such as MidCap SPDRs, Diamonds (ticker DIA), based on the Dow Jones Industrial Average, or Select Sector SPDRs. In 1996, Barclays Global Investors preferred a mutual fund structure for their WEBS (World Equity Benchmark Shares), ETFs that track the performance of foreign markets indices. Despite a growing interest, it took a few years for these funds to really take off.

The ETF marketplace experienced its effective boom in March 1999 with the launch of the Nasdaq-100 Index Tracking Stock, popularly known as Cubes or Qubes in reference to its initial ticker, QQQ, recently changed to QQQQ. In

---

[2] In 1999, S&P acquired the rights to the TSE index family and the TSE-30 and TSE-100 indices were combined into the S&P/TSE-60. Owners of TIPs and HIPs then had to choose between redemption and conversion into the i60 Fund based on this new index, the first ETF managed by Barclays Global Investors. Biktimirov (2004) analyzed the conversion of the remaining assets to examine the effect of demand on stock prices.

its second year of trading, a daily average of 70 million shares was being traded in Cubes, which is roughly 4% of the Nasdaq trading volume. The popularity of this specific fund increased market awareness for the other ETFs and the total assets under management more than doubled in 2000, up to $70 billion at the end of December (Frino and Gallagher, 2001). Since then, growth in ETF assets has shown no signs of slowing in the United States: 27% in 2001, 23% in 2002, 48% in 2003, 50% in 2004, even remaining high at 31% in 2005.[3] Over the years, ETFs progressively became an alternative to traditional non-traded index mutual funds which led their major competitors such as Vanguard or Fidelity to lower their fees by up to 10 basis points or less.

By the end of 2002, there were 113 ETFs in the United States with about $102.14 billion in assets under management. At the end of April 2006, with new cash invested in the existing ETFs and new ETFs based on still more diverse types of indices launched, the ETF marketplace consisted of four stock exchanges listing 216 ETFs with $335 billion in assets. The iShares (sponsored by Barclays Global Investors) and StreetTracks (sponsored by State Street Global Advisors) series present an extremely diversified offer among sectors and/or countries, but ETF assets are dominated by Spider, Cube, and Diamond, which are based on relatively broad market indexes. Trading volume concentrates on the two most popular ETFs, Cubes and Spiders, with annual turnovers as high as 3,700% for the former and 2,400% for the latter, according to Bogle (2004). This made Cubes, a passive investment instrument, the most actively traded listed equity security in the United States in 2005, with a daily average of 97 million shares traded.

2.2 The Market for ETFs in Europe

European stock exchanges started listing their first ETFs in 2000, while they had already gained popularity in the United States. The first exchanges to quote ETFs in Europe were the Deutsche Börse and the London Stock Exchange in April 2000 with the opening of the XTF and extraMARK specific market segments. Competition rapidly intensified with the entry of the Stockholm Stock Exchange at the end of October 2000, Euronext in January 2001 when NextTrack began trading ETFs first in Paris and Amsterdam marketplaces (trading in Brussels began in October 2002), and of the Swiss Stock Exchange in March 2001. In February of 2002, the Helsinki Stock Exchange listed its first ETF, the IHEX 35, whereas the Borsa Italiana opened the MTF segment dedicated to ETFs in September. More recently, ETFs were launched in the Icelandic market (December 2004), the Norwegian market (March 2005), the Irish market (April 2005), and the Austrian market (November 2005).

As of the end of 2005, 11 exchanges listed more than 160 ETFs, with assets growing at an annual rate of 60% up to €45 billion. Following the same trend

---

[3] Data source: Investment Company Institute.

as the one observed in the United States, exchanges began by quoting broad-based national and regional equity index ETFs. They then quickly diversified the benchmarks to a variety of underlying indices. For example, after only six and five years, respectively, Euronext and the Deutsche Börse listed 95 and 77 ETFs. This included ETFs based on eurozone or European indices, emerging country indices, style (socially responsible, growth, value, small caps, mid caps, etc.), or sectors indices. Besides these equity-based ETFs, sponsors launched fixed-income ETFs, ETFs based on precious metals and, lastly, those based on commodities.

Table 1 reports ETF trading on European marketplaces for year 2005. The Deutsche Börse and Euronext account for more than 70% of the total amount traded in ETFs in Europe. A monthly average of €3,842 million was traded on the Deutsche Börse in 2005 versus €1,481 million on Euronext, although fewer ETFs were listed on the dominant exchange at the time. Despite continuous growth, these figures are still far from those observed in the United States. Surprisingly, the leader in the number of trades is the Borsa Italiana, with almost twice as many transactions a month as the Deutsche Börse and Euronext, but worth only €0,524 million. This highlights the difference in types of investors in the European ETF markets. In the first two markets, the trading volume essentially stems from institutional investors posting large orders, whereas the Italian market is characterized by a higher proportion of individual investors posting significantly smaller orders.

Table 1. Overview of the European ETF Markets, 2005

	ETFs			Monthly Average Trading Volume	
Exchange	No. of ETFs	No. of Under. Indices	No. of Issuers	No. of Trades	Amount Traded (K€)
Deutsche Börse	77	68	9	18,787	3,842.1
Euronext	95	68	10	14,434	1,481.9
London Stock Exchange	28	28	1	-	770.2
Borsa Italiana	30	29	5	29,964	727.1
SWX Swiss Exchange	34	26	8	6,383	524.3
Virt-X	17	17	4	552	59.8
OMX	11	11	2	744	28.7
Wiener Börse	11	10	2	119	19.6
Oslo Børs	2	2	1	45	1.9

Data source: FESE and Deutsche Börse

Table 1 also illustrates the competition that exists between exchanges concerning the order flow in of ETFs and between issuers for the attraction of new cash invested. The London Stock Exchange is the only European marketplace with a single ETF series, the iShares sponsored by Barclays GI. In every other exchange, there are multiple issuers managing ETFs based either on specific "home" indices or under licence from index providers. The latter represent most of the ETFs listed in Europe, with indices from STOXX, FTSE, MSCI, or iBoxx, who sometimes grant multiple licences to competing issuers. For example, the 95 ETFs issued by 10 sponsors that are traded on Euronext track the performance of only 68 different underlying indices. As in the United States, the major national (the French CAC 40, the English FTSE 100) and regional (Dow Jones EURO STOXX 50, Dow Jones STOXX 50) indices concentrate most of the assets under management as well as the trading volume. Typically, several ETFs use these indices as benchmark, and are either listed on different European exchanges or on the same exchange. Table 2 reports basic information on the ETFs tracking the CAC 40 and the DJ Euro STOXX 50 indices competing on NextTrack as of December 31, 2005. It appears that even if those ETFs are the most traded on Euronext, the average daily number of transactions is low and highly concentrated on a single ETF for each index, namely those issued by Lyxor AM. If the same observation applies to the assets under management for the CAC 40 index with more than €3 billion, it does not apply to the DJ Euro STOXX 50. For this eurozone index, three ETFs, issued by Lyxor, Barclays, and IndExchange, have assets greater than €3 billion under management. Nonetheless, the trading volume still mostly concentrates on a single ETF. This situation is typical of the cross-listing of ETFs in Europe (DJ Euro STOXX 50-based ETFs are listed on seven different exchanges) where issuers benefit both from their nationality on their home market and, more importantly, from anteriority. Investors appear to keep trading on the same ETF even when competitors are launched on the same indices.

3 ETF Trading

ETFs are hybrid instruments combining the advantages of both open-end unit trusts and closed-end funds. They combine the creation and redemption process of the former with the continuous stock market tradability of the latter. Conventional mutual funds must typically buy back their units for cash, with the disadvantage that investors can only trade once a day at the net asset value (NAV) computed after the close.[4] Moreover, the trustee needs to keep a fraction of the portfolio invested in cash to meet the possible redemption outflows. Closed-end funds avoid this so-called cash drag as investors who wish to exit the fund can trade it throughout the day on exchanges. However, as no further creations and redemptions are allowed, excess offer or demand for closed-end funds may result in significant premiums or discounts with respect to their NAV. An innovative structure has been set up for ETFs. They trade on the stock market on a continuous basis, but shares can also be

---

[4] NAV is defined as the market value of the securities held less liabilities, all divided by the number of shares outstanding.

Table 2. CAC 40 and DJ Euro STOXX 50 Competing ETFs Listed on NextTrack

ETF	Issuer	Manag. Fees	Div. Freq.	Trading Volume			Assets Under Management	
				# Trades	# Shares	Amount (K€)	# shares (K)	Amount (K€)
Underlying index: CAC 40								
Lyxor ETF CAC 40	Lyxor AM	0.25%	Annual	242	466,805	19,857	67,832	3,233,539
CAC40 indexes (02/03/05)	Credit Agricole AM	0.25%	Annual	9	45,375	1,883	14,227	680,170
EasyETF CAC40 (17/03/05)	AXA IM, BNP Paribas	0.25%	Annual	12	107,997	4,719	17,800	840,516
Underlying index: DJ Euro STOXX 50								
Lyxor ETF DJ Euro STOXX 50	Lyxor AM	0.25%	Annual	85	544,386	17,426	108,128	3,906,665
iShares DJ Euro STOXX 50	Barclays GI	0.15%	Quarterly	20	149,637	4,830	92,900	3,368,483
Dow Jones Euro STOXX 50 EX	IndExchange Investment AG	0.15%	Annual	3	18,135	598	94,398	3,460,613
EasyETF Euro STOXX 50 A	AXA IM, BNP Paribas	0.45%	Annual (cap.)	2	43,197	146	20,340	76,477
EasyETF Euro STOXX 50 B	AXA IM, BNP Paribas	0.25%	Annual	2	19,623	684	1,118	40,801
UBS ETF DJ EURO STOXX 50 I	UBS ETF	0.10%	Half-yearly					

Data source: NextTrack. Reported trading volumes are computed as daily averages for 2005. Complete figures for the ETF DJ EURO STOXX 50 I were not available.

created or redeemed directly from the fund. The efficiency of the ETF specific dual trading system essentially relies on the in-kind creation and redemption process that is only available to institutional investors. We will first describe the ETF trading structure and then present the different players in the ETF marketplace.

3.1 The ETF Trading Process

ETF trading in the major marketplaces around the world closely resembles the system that was set up in the AMEX for SPDRs. The basic idea the original designer of ETFs, Nathan Most, had was to organize ETFs as commodity warehouse receipts with the physicals delivered and stored, whereas only the receipts are traded, although holders of the receipt can take delivery. This "in-kind" creation and redemption principle has been extended from commodities to stock baskets. Market makers and institutional investors can deposit the stock basket underlying an index with the fund trustee and receive fund shares in return. The shares thus created can then be traded on an exchange as simple stocks or later redeemed for the stock basket then making up the underlying index. The interesting feature in this process is that the performance earned by an investor who creates new shares and redeems them later is equal to the index return less fees even if the composition of the index has changed in the meantime.

Figure 1 illustrates the dual structure of the ETF trading process with a primary market open to institutional investors for the creation and redemption of ETF shares in lots directly from the fund, and a secondary market where ETF shares can be traded with no limitation on order size. The conditions for the creation and redemption of shares, such as the size of creation units, can vary from one fund to another, but the equity ETF process is typically as follows.

Creation of New Shares

Only authorized participants (APs), typically large institutional investors who have an agreement with the fund sponsor, are allowed to create new shares, in blocks of specified minimal amounts called *creation units*. Creation units vary in size from one fund to another, ranging from 25,000 up to 300,000 shares. Most ETFs have creation units of 50,000 shares, which represents an amount 500 times the dollar value of the index underlying the ETF. APs deposit the corresponding prespecified stock basket plus an amount of cash into the fund and receive the corresponding number of shares in return.[5] For some ETFs,

---

[5] The cash component is equal to the difference between the NAV and the value of the stock basket. It accounts for the dividends cumulated by the funds, the management fees, and adjustments due to rounding. This cash component may be negative.

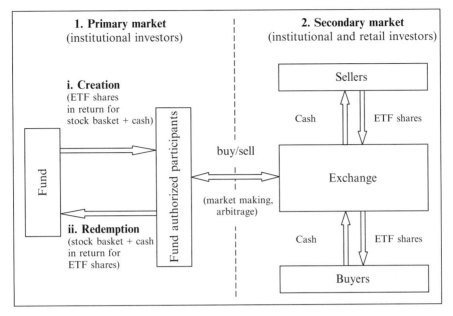

Fig. 1. Primary and secondary ETF market structure.

creation is allowed in cash but the APs then incur higher creation fees to account for the additional cost of the transactions that the replication of the index requires. Consequently, ongoing shareholders do not bear the cost of the entry (or exit) of new shareholders.

Redemption of Outstanding Shares

Shares are not individually redeemable. Investors can ask for redemption only by tendering to the trust shares in creation units. Typically, the operation is done "in-kind." Redeemers are offered the portfolio of stocks that make up the underlying index plus a cash amount in return for creation units. As is the case with creation, some funds may redeem ETF units in cash under specific terms, such as delays or costs.

The number of outstanding shares tradable on the secondary market varies over time according to creation and redemption operations carried out on the primary market. Both institutional and individual investors can buy and sell shares in the secondary market like ordinary stocks at any time during the trading day. As such, there is no fee payable for secondary market purchases or sales, but secondary market transactions are subject to regular brokerage commissions. Negotiating on the secondary market is subject to local exchange regulations. However, as index funds, ETFs typically need to receive a number of exemptions to trade like common stocks, and the launch of ETFs has generally been accompanied by the creation of dedicated market segments with their own specific rules.

In the United States, the ETF structure could not exist under the Investment Company Act of 1940. Gastineau (2002) reviewed the exemptions necessary for ETFs to exist and operate. The major exemptions are related to the permission for the "in-kind" creation and redemption process to occur only in creation units and the permission for shares to trade throughout the day at a price different from its NAV. Generally, ETFs also receive exemptions from the Securities and Exchange Act of 1934 so as to permit short selling on a down tick, for example. In European markets, exemptions are generally embedded in the dedicated market segment regulations. On NextTrack, Euronext's dedicated market segment, besides the conventional information referring to the fund, admission to trading new ETFs is essentially subject to the nomination of at least two liquidity providers, although Euronext is organized as a pure order book market.[6] Moreover, specific trading halts have been required for ETFs listed on Euronext Paris since French laws stipulate that Index Funds must trade at a price that does not deviate from their NAV by more than 1.5%.[7]

3.2 The Importance of "In-Kind" Creations and Redemptions

Since an ETF may be negotiated on two markets, it has two prices: the NAV of the shares and their market price. The first price is the value per share of the fund holdings computed at the end of each trading day. The second depends on the supply and demand for shares on the exchange. If selling or buying pressure is high, these two prices may deviate from each other. The possibility of "in-kind" creation and redemption helps market makers absorb the liquidity shocks that might occur on the secondary market, either by redeeming outstanding shares or by creating new shares directly from the fund. Moreover, the process ensures that departures are not too large. Indeed, APs could arbitrage any sizable differences between the ETF and the underlying index component stocks. If the ETF market price fell below the indicative NAV, it could be profitable for APs to buy ETFs in the secondary market, take on a short position in the underlying index stocks and, then ask the fund manager to redeem the ETFs for the stock basket before closing the short position at a profit.

Another major advantage of the "in-kind" process relies on the receipt and delivery of the stock basket with its weightings specified so as to replicate the underlying index. As they need not sell any stocks on the exchange to meet redemptions, ETF fund managers can fully invest their portfolio. Moreover, creations do not yield any additional costly trading within the fund. In the United States, "in-kind" operations are a nontaxable event, making the ETF structure seem particularly tax-efficient. When confronted with massive redemptions, which often occur in bull markets, classical funds must sell their

---

[6] See Euronext Instruction N3-09, Avis 2000-5271.
[7] Decree no. 89-624, September 6, 1989.

stock, resulting in taxable capital gains. When requesting redemption, APs are indifferent to the cost basis of the stocks they receive in return for the shares since their basis is the price at which they first delivered the stocks for the creation of the ETF shares. The ETF sponsor thus has the ability to deliver the stocks with the largest embedded capital gain. Historically, Dellva (2001) reported almost insignificant capital gains delivered by ETFs with respect to conventional mutual funds.

As efficient as it may be, this process is not sufficient to ensure a perfect replication of the underlying index. Changes in the composition of the index and constraints on the use of dividends and management fees induce some tracking error. Constraints depend on the legal structure chosen for the fund, but they generally remain low. Some structures allow the use of derivatives to ensure replication, whereas others restrict the holdings to the stocks that make up the index.[8] Loaning securities held by the fund might be permitted and is all the more profitable as the fund turnover is low and demand for its constituting stocks is high. This may help reduce management fees and expense ratios. Dividends are generally paid quarterly or half-yearly. Their value includes the cumulated dividends delivered by the underlying stocks, less management fees, bringing the NAV back to the initially specified multiple of the index (usually 1/10th or 1/100th).

3.3 ETF Market Participants

Besides the index providers that develop and provide licences for existing or new indices, ETF players are the stock exchanges, sponsors, and trustees, ETF-authorized participants, market makers, and investors on the secondary market.

Stock Exchanges

A stock exchange's first task upon entering the ETF business is to define admission to trading conditions and trading rules in conjunction with market authorities and regulators. The rules depend on local regulations and often require exemptions to the existing security laws and regulations. Its second task is to provide information. Stock exchanges disseminate classical intraday and daily data on market activity such as trades and quotes, trading volume, and so on. More specific information including assets under management or the number of outstanding shares is also made available. More importantly, exchanges compute and disclose indicative NAVs on a frequent basis. They are updated throughout the day so as to reflect changes in the underlying index. Investors can thus assess how far from their value ETFs trade on the marketplace. Historical data on the mean premium and discount values may

---

[8] The major differences between the two structures used in the United States, namely the mutual fund or the unit investment trust, will be reviewed later.

also be available for each ETF. That is the case with the AMEX, for example. Usually, effective NAVs are computed and disclosed at the end of the trading day, along with the composition of the creation and redemption stock baskets.

Moreover, exchanges usually undertake marketing and educational activities to benefit investors. However, the role of stock exchanges is not limited to these regulatory and operating aspects. More specifically, the exchanges generally select which ETFs will be listed in the last resort. For example, Euronext explicitly states that it "may reject an application for admission to trading on NextTrack if the applicable conditions are not met or if Euronext believes that admission would not be in the interests of NextTrack or investors." It is unclear whether listings of competitor ETFs based on the same or close indices improve or deteriorate the market quality of their underlying stocks and other ETFs. In this respect, European exchanges have very different strategies. Contrary to Euronext and the Deutsche Börse, the Italian and English exchanges follow very restrictive listing strategies, limiting listings to very specific indices for the first and to one single sponsor for the second. Finally, exchanges can also influence the offer by providing and licensing their own indices as benchmarks for ETFs.

Sponsors and Trustees

Sponsors and trustees issue ETFs and manage the fund's holdings so as to replicate their underlying index or benchmark as closely as possible. However, in the United States, before an ETF is admitted to trading on a stock exchange, it must pass through the SEC's exemptive process since no set of rules exists to allow firms to launch such an instrument. When a sponsor wants to cross-list his ETFs in multiple markets, the full prospectus may eventually be rewritten since, even though regulations are similar, as in Europe, different information may be needed or different presentation formats may prevail according to the country. The prospectuses provide information on the risks associated to the index replicating scheme. They also contain various information, including the list of shareholders, legal representatives, and directors of the ETF's management company, the terms and conditions of the product, and the way it operates. More specifically, the creation and redemption conditions are fully detailed. Replicating the performance of the underlying index is an objective but not a mandatory one. Prospectuses include a tracking error objective but specify that it may not be achieved. Holdings management is broadly limited to adjustments caused by changes to the index, managing dividends, and creating new shares or redeeming outstanding shares. ETFs are extremely transparent since the information on the holdings and their value as well as the number of outstanding shares must be reported to the exchange and then made public.

Fund-Authorized Participants and Market Makers

Although theoretically opened to all investors, the ETF primary market practically aims at the fund managers and authorized participants. Fund managers, whose role has already been briefly described, are responsible for issuing and redeeming trackers. Authorized participants have the fund manager's permission to request share creation and redemption, generally in multiples of the creation units. All investors requesting that creation units be created or redeemed must place an order with an AP. APs may be simple investors in the ETF fund or act as market makers on the secondary market. As with the AMEX, most ETF marketplaces have specialists or market makers. One major difference with stock market specialists is their ability to create or redeem shares to manage their inventory risk. They play an essential role in the efficient pricing of ETFs through possible arbitrage between the ETF primary and secondary markets as well as with the underlying index futures and options markets.

Retail and Institutional Investors on the Secondary Market

Most ETF trading occurs in the secondary market. That is one major advantage ETFs have over classical mutual funds. Investors need not redeem their shares to exit the fund; they can simply sell them on the market. Depending on the market and the ETF, the secondary market may be dominated either by institutional investors and APs or by retail investors. Trading ETF shares on the secondary market is organized in the same way as regular stocks, with the possible difference that there are specialists and market markers posting bid and offering prices even on order-driven markets. Short selling, even on a down tick, and margin buying are usually allowed and ETFs may be eligible to block trades and other trading facilities.

3.4 ETFs of Different Kinds

Differences in Legal Structure

The ETF legal structure primarily depends on which exchange it is listed on. Security laws and stock exchange regulations differ from country to country. Sponsors who want to cross-list their ETFs have to accommodate multiple legal regimes. Even in the United States, three main legal structures co-exist: open-ended index mutual funds, unit investment trusts, and exchange-traded grantor trusts. ETFs are regulated by the SEC as mutual funds, but, as we discussed earlier, their structure is subject to a number of exemptions and there is still no set of rules that would allow new ETFs to be listed directly. Historically, the first ETFs were initially designed as Unit Investment Trusts (UIT) for simplicity and cost-saving reasons (Gastineau, 2002). Followers and most of the new ETFs preferred the more flexible structure provided by mutual funds. The main difference between the two structures is the use of dividends and the securities the fund holds. Unlike open-end index mutual funds, UITs cannot reinvest the dividends delivered by the underlying stocks and must cumulate them in cash.[9] Mutual funds are also allowed to use derivatives such as futures, which allows them to equitize their dividend stream, and, finally, unlike UITs, they can generate income from loaning the securities they hold. QQQs (Qubes), DIAMONDS, and S&P 500 SPDRs are structured as UITs while Select Sector SPDRs and iShares are open-end index mutual funds.

Although HOLDRs are sometimes referred to as ETFs, such exchange-traded grantor trusts cannot be considered as such according to strict definitions of the term. They are more similar to owning the underlying shares, since investors keep the right to vote shares and to receive dividends. However, such funds do not track independent indices, given that the stocks to be included in the fund are selected based on objective criteria once the industry sector, or more generally the group of securities, has been chosen. New shares can then be created and outstanding shares can be cancelled against the delivery of the stock portfolio. The included stocks are fixed and cannot be changed even though some of the basket components are acquired by other companies.

Differences in the Underlying Indices

ETFs were initially meant to replicate broad-based stock indices. However, as the instrument became more familiar to investors, the universe of ETFs expanded progressively to replicate indices built around sectors, countries, or styles. The process continued with the launch of fixed-income, commodity, and finally currency ETFs.

Broad-based stock indices measure the performance of companies that represent a market. The number of stocks included in the index, and therefore the diversification of the associated fund, varies from one index to another, from 30 stocks in the case of the Dow Jones Industrial Average to as much as 3,000 for the Russell 3000 Index (which measures the performance of the largest U.S. companies based on total market capitalization) or almost 5,000 for the Dow Jones Wilshire 5000 Composite Index. Major ETFs based on broad-based stock indices include SPDRs, QQQQs, DIAMONDS, or iShares Russell 2000 that replicate the S&P 500, the Nasdaq 100, the DJIA, and the Russell 2000 indices, respectively. Specific ETF series usually break down broad-based indices into "growth" and "value" management styles and small, medium, or large capitalization stock sizes. With broad-based ETFs, it is possible to establish positions in global markets very quickly and equitize temporary cash positions. They may be used for long-term investing, as a

---

[9] This UIT feature results in a so-called dividend-drag during rising markets. As we will discuss more extensively later, it partially explains the poor performance of UITs, such as the S&P 500 SPDR, in comparison with some of their major mutual fund competitors. See Elton et al. (2002).

tool to hedge well-diversified portfolios or to implement multiple strategies. Core-satellite strategies typically use such broad-based ETFs to build the core allocation. Capitalization size trading strategies can be implemented with large, medium, and small capitalization ETFs.

Country and regionally based ETFs also generally replicate broad-based foreign equity market indices. Country ETFs replicate indices that focus on a single country, whereas regional ETFs track an index that focuses on a geographical or monetary zone such as Asia, Europe, or the Eurozone. They provide easy and rapid international diversification but, since ETFs and their underlying stocks need not be traded synchronously, deviations to the NAV may be larger for these instruments. WEBs were the first country ETFs to be launched in 1996, and the iShares MSCI series offer worldwide country ETFs.

Broad-based indices are also generally broken down into sectors. In some instances, sector indices are designed specifically for the funds. For example, select sector SPDRs break down the overall S&P 500 index into industry components that differ from classical S&P sector indices (an individual security cannot account for more than 25% of the index in order to comply with the Internal Revenue Code). Other examples of sector ETF series are iShares Dow Jones, SPDR Sector Series, or Merrill Lynch HOLDRs insofar as they can be considered ETFs. Sector ETFs appear to be particularly useful in implementing sector rotation strategies since they make it is easy to overweigh or underweigh sectors in a single transaction.

The first fixed-income ETFs appeared in Canada in 2000; to date, there are only six of these products on the U.S. market, all iShares funds issued by Barclays in 2002. These were meant to replicate Goldman Sachs and Lehman Brothers bond indices, which measure the performance of obligations with different maturities and issuers, both public and corporate. Fixed-income ETFs do not mature but maintain a portfolio that reflects the underlying bond index's target maturity. As of April 2006, fixed-income ETFs represented $16.14 billion in assets out of a total $334.87 billion in assets under management for the whole ETF industry. Fixed-income ETFs are used for portfolio diversification, core holding for bond portfolios, or transition management, among others. European marketplaces list country-specific fixed-income ETFs.

Commodity ETFs were first launched in the U.S. with StreetTracks Gold Shares, whose objective is to reflect the performance of the gold bullion. To create new shares, APs must deposit a specified gold amount plus cash. They are redeemed in the same in-kind basket in return for shares. Commodity ETFs give investors exposure to a variety of commodities such as gold, silver, oil, or broad-based index commodities that include commodities from sectors as diverse as energy, metals, agriculture, and livestock. Examples of such broad-based commodity ETFs include EasyETF GSCIs based on the Goldman Sachs Commodity Index (listed on the Deutsche Börse and the Swiss Stock Exchange), Lyxor ETF Commodities CRBs based on the Reuters/Jefferies CRB index (listed on Euronext), and the Deutsche Bank Commodity Index Tracking Fund and based on the Deutsche Bank Liquid Commodity Index (listed on the AMEX). Unlike classical ETFs, these funds invest in futures contracts on the underlying commodities.

To date, the last type of ETFs to be created is currency ETFs, with the launch of Euro CurrencyShares sponsored by Rydex Investments in December 2005. It is listed on the NYSE, and its objective is to reflect the price of the euro. In this last case, no derivatives are used and the Trust's assets consist only of euros on demand deposit. APs can issue and redeem shares in euro-based creation units.

4 ETFs' Pricing Efficiency

The specificity of ETF trading is based on the creation and redemption process we presented in the previous section. Exchanges and sponsors claim that this structure necessarily brings a high pricing efficiency to the ETF market. Pricing efficiency is a major concern since trading in index funds has long been at the root of the most intriguing puzzles in finance: the closed-end fund discount. Although fund holdings are made public and the NAV is disclosed at least daily, closed-end funds generally trade at a discount to NAV. Conventional explanations for the closed-end fund puzzle include biases in NAV calculation, agency costs, tax inefficiency, and market segmentation.[10] However, none of these theories can explain the full set of anomalies associated with the pricing of closed-end funds. One must forego the rational expectation framework to encompass these anomalies in a single theory. The limited rationality model developed by Lee et al. (1991) shows how the behavior of individual investors can explain the puzzle. Misperception leads to overreaction, and the unpredictability of variations in investor sentiment makes arbitrage risky. Most of the empirical tests support this investor sentiment theory.

In contrast to closed-end funds, whose capitalization is fixed, ETFs are characterized by a variable number of shares in issue. APs can ask the fund to create new shares or redeem outstanding shares with no impact on market prices and thus should be able to quickly arbitrage any deviation of the price to the NAV. No specific model that integrates the ETF arbitrage process has yet been developed. However, some empirical studies test the ability of the ETF structure to ensure efficient pricing in the U.S. ETF market. Using closing data, Ackert and Tian (2000) show that discounts on the price of SPDRs had no economic significance between 1993 and 1997, even though individual investors were the primary investors in the fund. They measure larger discounts for the MidCap SDPR based on the S&P 400 index. This confirms the hypothesis that limits to arbitrage cause deviations given that this ETF is likely to have higher arbitrage costs due to higher fundamental risk, transactions costs, and lower dividend yield associated with its benchmark index.

---

[10] For a review on the closed-end fund puzzle, see Dimson and Minio-Kozerski (1999).

Discounts remain very low compared to those observed on closed-end funds, and excessive volatility is only observed for MidCap SDPRs. Hence, the ETF specific structure lessens the impact of noise traders since rational traders can more easily arbitrage deviations to the NAV.

These results are confirmed by the empirical studies carried out by Elton et al. (2002) on SPDRs, Engle and Sarkar (2006) on a sample of 21 ETFs listed on the AMEX, and Curcio et al. (2004) on Cubes or Cherry (2004) on 73 iShares ETFs. Elton et al. (2002) showed that deviations to the NAV do not persist from day to day. The fact that trading volume is linked to premium and discount values supports the claim that the arbitrage mechanism is responsible for this efficiency. Engle and Sarkar (2006) examined the magnitude and persistence of discounts both daily and intradaily. On average, they found that ETFs are efficiently priced since only small deviations were seen, lasting for only a few minutes. The daily results of Curcio et al. (2004) confirmed those of Ackert and Tian (2000). From an intradaily perspective, their study of transaction size proves that, even if individual investors seem to be the primary holders of SPDRs and Cubes, they account for less than one-half of the trading volume. Some economically significant discounts are found, but these are very short-lived and can be attributed to institutional arbitrage activity.

The structure is the same for all ETFs, but, in the case of country ETFs, the arbitrage mechanism is somewhat inhibited by nonoverlapping trading hours between ETFs and their underlying index component stocks. Engle and Sarkar (2006) found that deviations to the NAV are greater and more persistent for the 16 country ETFs sample compared to the 21 domestic ETFs sample. Though imperfect, the existence of the creation/redemption process along with the high transparency of the funds holdings appears to enhance price efficiency. In effect, deviations remain smaller in magnitude (around 100 basis points on average with a maximum of 211 bps) than those generally observed for comparable closed-end country funds (often greater than 10%). Jares and Lavin (2004) studied this issue for Japan and Hong Kong WEBs that trade on the AMEX. Nontradability of the underlying stocks is an especially meaningful concern in this case since Asian markets are closed for the day before U.S. markets open. For these ETFs, an Indicated Optimized Portfolio Value serves as the indicative NAV and is disclosed throughout the day. It is based on stale stock prices and accounts solely for changes in exchange rates. Jares and Lavin (2004) found frequent discounts and premiums for the period ranging from 1996 to 2001. Moreover, there is predictability in returns giving rise to highly profitable trading rules. These results are confirmed in Madura and Richie (2004), who found reversals in prices that support the hypothesis that informed traders arbitrage overreacting investors. The measured reversals are insignificant for broad-based ETFs, but are more pronounced for international ETFs. Simon and Sternberg (2004) also found significant premiums and discounts at the end of the day and overreaction for European ETFs traded on the AMEX. Hence, if the trading system appears to enhance pricing efficiency for traded funds, some inefficiency seems to remain for ETFs replicating illiquid or foreign benchmarks.

5 ETF Performance

Marketing for ETFs, presented as a low-cost alternative to traditional mutual funds, has always focused on their low management fees and expense ratios. As ETFs attracted more and more cash, fierce competition between ETFs and mutual funds led to the fee war described in Dellva (2001) and Bogle (2004). Fidelity and Vanguard progressively lowered their fees; after an almost 10-year fall in expense ratios, they are now at a historical low with 10 basis points and still less for Vanguard major funds. Broad-based ETFs generally display annual expense ratios of 20 basis points or less. Recently, the expense ratio for SPDRs was lowered from 0.12% to 0.10% while Barclays' iShares S&P 500 fees are set at 0.09%. The expense ratio comparisons used as a competitive tool by issuers are obviously in favor of ETFs. However, such direct comparisons are too simplistic since they omit ETF trading costs and relative tracking performance over mutual funds.

Dellva (2001) and Kostovetsky (2003) compared both types of funds based on total costs supported by investors. ETFs generally have lower expense ratios. Investors incur transaction costs when they buy and sell ETFs, while there is no supplementary cost for trading no-load mutual funds. Taxes are also of importance to taxable investors. As registered investments companies, mutual fund and ETFs must both distribute capital gains to their shareholders. If mutual funds are considered tax-friendly investments, this is even truer of ETFs. Actually, ETF managers do not need to sell shares to meet redemptions as creations/redemptions are done in-kind. Moreover, they can also redeem shares with the higher tax basis. ETFs distribute almost no capital gains, but, overall, Dellva (2001) found that trading costs are typically higher than expense ratios and tax savings for small investors. However, as the invested amount increases, ETFs become more profitable than mutual funds, even for short-term investment of two or three years. Kostovetsky (2003) goes one step further by quantitatively modeling the difference in cost in both a single and multiple-period setting. He also found that there is a threshold in the amount invested over which ETFs dominate mutual funds. However, both studies assume that there is no tracking error for both types of funds, although it is well known that the replication of the benchmark index is rarely, if ever, perfect.

Replication strategies cannot always be perfect. Even if most times fund holdings mimic the index composition, when it changes, fund managers must trade to adjust their holdings. The related transaction costs and possible flaws in the replication strategies induce tracking error. Elton et al. (2002) evidenced an average 0.28% annual underperformance for SPDRs relative to the S&P 500 index over the 1993–1998 period. Moreover, SPDRs do not compare favorably with major index mutual funds: annually, the Vanguard mutual fund that replicates the S&P 500 index yields on average 0.18% more than the SPDRs. These results are confirmed in the study by Poterba and Shoven (2002) over the 1994–2000 period, even when taxes on capital gains delivered by both funds are taken into account. Although differences in performance are reduced, they remain economically significant. For Elton et al. (2002), 9.95 basis points are lost due to the SPDR structure. As a Unit Investment Trust, the dividends received on the fund holdings have been kept in a noninterest-bearing account until distributed to shareholders. The authors claim that investors are still investing in SPDRs rather than in relatively outperforming mutual funds because they assign value to the ability to trade their shares intradaily. The value of immediacy is 9.95 basis points. In addition to comparing SPDRs and Vanguard performances, Gastineau (2004) investigated the difference in returns between iShares Russell 2000 ETFs and Vanguard Small Cap Investor Shares over the 1994–2002 period. Irrespective of the underlying index, ETFs underperform the corresponding mutual fund. However, Gastineau attributed these differences to passiveness from ETF managers when faced with changes in index composition. Mutual fund managers typically anticipate upcoming events to reduce transaction costs embedded in the index modification process while ETF managers wait until the announcement.

To date, only a few studies deal with this issue concerning ETFs that are not based on major broad-based indices or listed in the United States. Harper et al. (2006) extended the performance issue to country funds. Due to significantly lower expense ratios, iShares country ETFs offer higher returns than corresponding closed-end country funds. ETFs also have higher Sharpe ratios. In this case, the ETF cost-efficient structure proves decisive. On the younger Australian ETF market, Gallagher and Segara (2004) did not find evidence that the street TRACKS S&P/ASX 200 ETF over or underperformed the mutual funds tracking the same index in 2002 and 2003. To our best knowledge, despite the growing success of ETFs in Europe, only one study is dedicated to the performance of European ETFs to date. For the seven most important ETFs traded on the Italian market, Zanotti and Russo (2005) showed that risk-adjusted returns are higher on average than those observed for traditional mutual funds. It therefore seems that, contrary to what is observed in U.S. major broad-based indices, ETFs based on less liquid indices or listed on less mature exchanges might outperform their mutual fund counterparts.

6 The Impact of the Introduction of ETFs on Trading and Efficiency of Related Securities

Before the introduction of ETFs, investors could already trade stock indices intradaily through their component stocks and index derivatives. The advent of ETFs offers a new means to take quick and inexpensive positions in indices.

Given their specific characteristics and organization, ETFs have attracted a significant portion of index-based trading. Either ETF investors are new to indexing or they come from the other pre-existing index markets. However, it is not very clear whether the arrival of new investors and the possible migration of existing investors from one market to another alter the mix between liquidity and informed traders for the basket and underlying stock components.

The spectacular growth of index futures markets in the 1980s had already raised the question of what impact the introduction of a basket market would have on market quality. Depending on the assumptions made about the integration of the different markets, theoretical models predict opposite effects. In the framework of perfectly integrated markets, Subrahmanyam (1991) and Gorton and Pennacchi (1993) modeled the strategic behavior of traders who can choose to trade either in the basket stock market or in the underlying stocks market. Subrahmanyam (1991) demonstrated that the basket security market most probably serves as the lowest-cost market for the index. Adverse selection costs are lower on the market for the basket in which the firm-specific private information is diversified. In Gorton and Pennacchi's (1993) model, liquidity traders will prefer the basket market, as it enables them to build their portfolios at a lower cost. Hence, the proportion of informed traders negotiating the individual securities increases, which results in higher adverse selection costs.

In Fremault (1991) and Kumar and Seppi (1994), markets are assumed to be imperfectly integrated. The introduction of a basket instrument removes some of the obstacles that limited arbitrageurs from establishing profitable portfolios. Information asymmetry across markets and arbitrage costs will tend to decrease, attracting new arbitrageurs. Arbitrage activity and competition between informed traders will increase and result in higher liquidity in the individual securities market.

Hedge and McDermott (2004) transposed these predictions to the introduction of ETFs. On the one hand, the migration of liquidity trading from the stock market to the ETF market could deter the liquidity of individual securities. On the other hand, if ETFs facilitate arbitrage trading, their introduction would increase arbitrage activity and enhance both the liquidity of the underlying stocks and the efficiency of the derivatives markets. Two contrasting theories on how the introduction of ETFs modifies the established equilibriums can be tested: the "adverse selection hypothesis" and the "arbitrage hypothesis." The number of studies on this issue is still limited, but we will nonetheless divide the discussion into three parts. We will first review the studies that analyze to what extent ETF trading affects the quality of the underlying index component stocks. Then we will look at the research that tests what impact the introduction of ETFs has on derivatives markets efficiency. Finally, we will consider works that measure how the index pricing discovery process is influenced by the high ETF trading levels.

6.1 ETFs and the Market Quality of Their Underlying Stocks

The advent of ETF trading is likely to have modified the mix of informed and liquidity traders on the market for individual securities. To test to what extent this is the case, empirical studies measure the importance of information asymmetries both in ETFs and individual stock markets. This research typically relies on the analysis of bid-ask spreads and trading volumes and measures the evolution of market quality after trading in ETFs becomes possible. Overall, the ETF market is found to attract very few informed trading and to be more liquid than the individual stocks market. However, there is no clear consensus on whether ETFs enhance liquidity in the underlying stocks.

Hedge and McDermott (2004) provided an in-depth analysis of the market liquidity of Diamonds and the stocks that constitute the DJIA around the launch of this ETF. Empirical pre- and post-ETF comparisons of various liquidity measures computed over two 50-day periods were used to test the impact of the ETF's introduction. Their results support the arbitrage hypothesis. The different measures of the individual stocks' liquidity improve, appearing to be mostly due to a decrease in adverse selection costs as measured with the price formation model developed by Madhavan et al. (1997). Moreover, the trading volume and open interest of DJIA futures contracts increase over the sample period. Similar, but less significant, results are obtained for the introduction of Cubes.

The study of Madura and Richie (2005) on the introduction of Cubes supported the arbitrage hypothesis. First, there is evidence of a decrease in the spreads of Nasdaq 100 index components over the three months following the introduction of Cubes compared to the preceding three-month period. Second, the decrease in the spreads is all the more significant as the weight of the stock is low. This result supports the role of ETFs in the measured decrease. Passive fund managers need not invest in all securities to replicate the index. Rather, they use sampling techniques and limit their activity to the top holdings. In contrast, arbitrageurs investing in ETFs through the in-kind creation/redemption process must transact the full 100-stock portfolio. Among these stocks, the less weighted stocks experience the largest increase in liquidity. Third, the introduction of Cubes was followed by an increase in the pricing efficiency of the individual stock and a significant decline in systematic risk. Yu (2005) also found more efficient pricing and a decline in the trading costs of component stocks following the introduction of the basket security for a sample of 63 ETFs and 15 HOLDRs listed on the AMEX.

On the contrary, the results found by Van Ness et al. (2005) and Ascioglu et al. (2006) supported the adverse selection hypothesis drawn from Subrahmanyam's (1991) model. Over the two-month period that brackets the introduction of Diamonds, Van Ness et al. (2005) claim that liquidity traders move to the ETF market since the spreads measured for the DJIA component stock experience a smaller decline than those of the control sample. However, no significant modification is found in the adverse selection components of the individual stocks, but the authors argue that this may simply be due to the poor performance of adverse selection models in general. Ascioglu et al. (2006) broke down the spread of 64 broad-based ETFs listed on Nasdaq using Madhavan et al.'s (1997) methodology. In this preliminary study, tests are carried out over two months, March and April 2005. They showed that information asymmetry is less severe for ETFs than it is for comparable stocks, insofar as it is possible to match the most traded ETFs with stocks.

Nonetheless, whatever the measured effect the introduction of ETFs has on individual stocks, all studies find that the ETF liquidity is higher than that of the underlying stock portfolio. For liquidity traders, ETFs appear to be a cheaper vehicle for building a diversified index portfolio compared to investing in the individual stocks directly. The study of Bennett and Kerins, Jr. (2003) confirmed this last point for 92 ETFs listed on the AMEX over the last quarter of 2000. This remains true even though some ETFs exhibit a lower trading volume than the underlying stocks.

6.2 ETFs and the Efficiency of the Underlying Index Derivatives

In complete and perfect markets, arbitrage relationships tightly constrain the price of derivatives with respect to their underlying asset. On real markets, with the existence of friction and trading constraints, futures (Chung, 1991; Miller et al., 1994) and options (Kamara and Miller, 1995) prices can fluctuate around their theoretical value without giving rise to arbitrage opportunities. Arbitrage relationships only impose bounds that widen with the prevalence of friction. As Ackert and Tian (2000) noted, the advent of ETFs removes some of the obstacles that prohibited arbitrageurs to enter in efficiency-creating trades in index derivatives markets. Besides the possibility of shorting the index, even on a down tick, ETFs should lower both trading costs and the liquidity risk of building an index position. Moreover, in the imperfectly integrated market framework of Fremault (1991) and Kumar and Seppi (1994), the advent of ETFs should increase intermarket arbitrage activity. As a first hint in favor of these predictions, Hedge and McDermott (2004) found a significant increase in the daily average DJIA and Nasdaq-100 futures trading volume and open interest over the 101 trading days surrounding the introduction of Diamonds and Cubes, respectively.

Empirical tests that study what impact the introduction of ETFs have on the efficiency of derivatives markets first rely on the computation of arbitrage profits. The frequency and values of arbitrage opportunities measured prior to the advent of ETFs are then compared to those measured after. Though early studies use daily data, most recent works use tick-by-tick data, which eventually allows differences in the persistence of efficient value distortions to be tested. Futures markets distortions are defined with respect to cost-of-carry prices, whereas the put-call parity relationship is the main benchmark for theoretical option values, even though other arbitrage relationships such as the lower boundary or constraints on spreads may set efficiency boundaries.

Overall, though futures market studies highlight an improvement in intermarket efficiency, evidence for a similar pattern in the options market is mixed.

Park and Switzer (1995) tested how TIPs, the very first ETF listed on the Toronto Stock Exchange, impacted the efficiency of the Toronto 35 index futures market. Using closing data, they found a reduction in arbitrage opportunities in terms of both frequency and value. The authors interpreted this result as evidence that the TIPs lowered arbitrage costs and thus attracted more arbitrage activity. Switzer et al. (2000) drew the same conclusion from the reduction in mispricings measured after SPDRs were introduced. Nonsynchronous prices do not explain the improved efficiency observed since the pattern is obtained with both daily and hourly data. As for the advent of Cubes, Kurov and Lasser (2002) worked with one year of transaction data concerning the near maturity of Nasdaq-100 futures. Whatever the assumed transaction cost levels, both the size and frequency of deviations decrease once Cubes are traded. Kurov and Lasser also conducted ex ante tests that consisted in building the arbitrage portfolios only after an ad-hoc period had elapsed. They documented faster market reactions to observed deviations since a larger percentage of opportunities disappear within two minutes in the post-ETF period. On the French index futures market, Deville et al. (2006) investigated the impact the introduction of the Lyxor CAC 40 ETF had on the pricing efficiency of the French broad-based CAC 40 index futures over a two-year period. Even after controlling for liquidity of futures contracts and individual stocks and market volatility, ETFs appear to enhance intermarket efficiency. However, further analysis shows that this improvement cannot be directly attributed to ETF trading. Rather, the introduction of ETFs increased the liquidity of the underlying stocks, which may have attracted new arbitrage activity, thus tightening the spot-futures pricing relationship.

In contrast, there is no clear evidence of improved efficiency in the options markets. Their efficiency seems to improve over time, as evidenced by Ackert and Tian (1998) on the Toronto Stock Exchange when TIPs were launched or by Ackert and Tian (2001) on the CBOE when SPDRs were launched. However, they found no clear effect on the link between stock and index options markets as measured by sole deviations to relationships that require trading in the index. No significant improvement is found in the compatibility of their closing price samples with these relationships. Deville (2003, 2005) obtained opposite results with tick-by-tick data on the French market for the launch of CAC 40 index ETFs. The improvement of all market efficiency measures that rely on put-call parity supports the notion that ETFs improve the efficiency of the options market. Moreover, the duration of deviations drops twofold with the introduction of ETFs. Deville and Riva (2006) confirmed the importance of ETFs in enhancing intermarket efficiency through a survival analysis approach. The existence of ETFs is found to be a major determinant of the process that drives prices back to values compatible with efficiency.

6.3 ETFs and Price Discovery

With the creation and development of index futures, the cash market for component stocks has gradually lost its prominence in index trading. Empirically, for U.S. broad-based indices, studies that explore the dynamics of index prices show that the futures markets incorporate information more rapidly than the stock markets. However, significant but weaker effects are measured from the latter to the first market. ETFs allow indices to be traded throughout the day at low cost and may appear to be more convenient trading vehicles than futures for smaller orders and liquidity traders. A question that naturally arises from this is whether futures contracts remain the lead instrument in the price discovery process. A byproduct of the studies on price discovery is the insightful information they provide on where uninformed and informed traders trade.

Despite the introduction of SPDRs, Chu et al. (1999) showed in a vector error correction framework that price discovery still takes place on S&P 500 futures. SPDRs only make a small contribution to the common factor, but more so than the spot market. Since the study is based on the ETFs' first year of trading, it is necessary to view these results with some caution. SPDRs only began to exhibit a high trading volume years later. Over the March–May 2000 period, Hasbrouck (2003) analyzed the price discovery process using the information share approach of Hasbrouck (1995) for three major U.S. indices. Investors can take positions on the S&P 500 and Nasdaq-100 indices through individual stocks, floor-traded futures contracts, electronically traded E-mini futures contracts, options, or ETFs. The largest informational contributions come from the futures market, with the ETF market playing a minor, though significant, role. Interestingly enough, there was no E-mini contract for the S&P MidCap 400 over the sample period, and the ETF information share is the most important for this last index.

Recent work by Tse et al. (2006) showed that although the E-mini DJIA futures contracts dominate price discovery, Diamonds also play a very significant part in the process. Their results for the S&P 500 highlighted a contribution of about 49% for the ETF. However, this does not cast doubt on Hasbrouck's (2003) results, since they were based on floor-based quotes and trades from the AMEX, whereas Tse et al. used quotes from the ArcaEx Electronic Crossing Network.[11] The anonymous and immediate trading execution obtained on electronic trading platforms may indeed attract informed trading.

The results obtained by Henker and Martens (2004) contrast with the view that derivatives and ETFs are the leading instruments. They follow Hasbrouck's (1995) methodology to assess the discovery process for two liquid HOLDRs from January to July 2003. Although there are no futures contracts that could attract most of the informed trading activity, HOLDRs are dominated in the pricing discovery by component stocks. This evidence is in line

---

[11] This confirms what Tse and Erenburg (2003) found concerning the Nasdaq-100 index in that trading in ECNs contributes the most to price discovery.

with the predictions of the Subrahmanyam (1991) model, in that the underlying stocks will lead the basket instrument.

Each stock in the S&P 500 is also assigned to one of the nine Select Sector indices. Sector ETFs may be of interest to liquidity traders looking for specific diversification as well as to investors trading on private information at the sector level. Even though some Sector SPDRs such as the XLK (technology) exhibit significant trading, Hasbrouck (2003) showed that their information share is limited. In the period running from July 1, 2002, to September 20, 2002, the results obtained by Yu (2005) in a VAR framework are consistent with the view that low information production occurs at the sector level. One explanation is that the high trading costs and low liquidity that characterize these ETFs might deter liquidity trading. Consequently, Sector SPDRs are unattractive to informed traders.

7 More Studies Devoted to ETFs

ETFs are often presented as an alternative, either interesting or not, to other index instruments, mutual funds, and derivatives. Literature on ETFs mostly takes the same perspective. ETF performance is compared to that of index mutual funds and their efficiency to that of closed-end funds. Their trading is essentially analyzed for the impact the advent of ETFs has on the efficiency of the related index markets. However, ETFs trade like stocks and a few studies started to transpose security market issues to ETFs. In particular, Boehmer and Boehmer (2003) and Tse and Erenburg (2003) studied the influence the NYSE ETF listing has on the competition for order flow and market quality with regard to ETFs primarily traded on the AMEX. Furthermore, the specific ETF structure may shed new light on other classical questions. Arshanapalli et al. (2002) measured the impact SPDR creations and redemptions have on the SPDR market price and index component stocks. Biktimirov (2004) studied the conversion of TIPs to the i60 Fund to assess the shape of the demand curve for equities. Finally, ETFs were initially designed to offer low-cost diversification in a single trade, but little is known about their real use. The capacity of country ETFs to enhance international diversification is questioned by Pennathur et al. (2002) and Miffre (2004). Amenc et al. (2004) illustrated the potential use of fixed-income ETFs in core-satellite portfolio management.

On July 31, 2001, for the first time in its history, the NYSE exercised Unlisted Trading Privileges and began trading the three majors ETFs, namely QQQQs, SPDRs, and DIAs, that were then primarily listed on the AMEX.[12] On April 15, 2002, the process was continued with the addition of 27 new AMEX-listed ETFs and HOLDRs, mostly based on sector indices. As of April

---

[12] An Unlisted Trading Privilege (UTP) is a right, provided by the Securities Exchange Act of 1934, that permits securities listed on any U.S. securities exchange to be traded by other such exchanges.

2006, 270 ETFs now trade on the NYSE on a UTP basis along with 94 primary listed ETFs. The considerable ETF trading volume made the NYSE decide to join in the competition.[13] Before the NYSE entered into ETFs, QQQQs, SPDRs, and DIAs were already traded on the Nasdaq InterMarket, regional exchanges, and the Island ECN. Tse and Erenburg (2003) focused on QQQQs to investigate to what extent NYSE trading influenced the ETF market quality as measured by liquidity, efficiency, and price discovery. They found evidence that trading on the NYSE has increased competition for order since spread declined in all trading venues and the information shares of QQQQs relative to Nasdaq-100 futures increased following competition from NYSE. However, this accrued competition between different trading centers did not result in market fragmentation or increased trading costs. Boehmer and Boehmer (2003) confirmed these results for the entire 30-ETF set that began trading on the NYSE. Post-NYSE liquidity is higher compared to pre-NYSE figures both in the entire market and in different market centers. Further analysis supports the hypothesis that ETF market makers earned significant rents prior to the NYSE entry. However, in his discussion of Boehmer and Boehmer (2003), Peterson (2003) suggested that these results are also consistent with a segmentation hypothesis in which traders migrate to the market offering the best liquidity for their trades. Nonetheless, competition appears to enhance overall market liquidity without impeding the price discovery process. Competition between exchanges for ETF listings caused Cubes to switch their listing from AMEX to Nasdaq on December 1, 2004, with a change in ticker from QQQ to QQQQ. Broom et al. (2006) showed that even when trading already takes place in different market venues, the location of the primary listing is an important determinant of trading activity since the move resulted in a decline in trading costs, a consolidation of order flow, and a less fragmented market.

ETFs are of particular interest in the study of the shape of the stock demand curve since noninformational events regarding individual stocks are likely to occur for these securities. Such events may be regular, as is the case for in-kind creations and redemptions studied by Arshanapalli et al. (2002) or exceptional like the conversion of TIPs into a new fund that is the central point of Biktimirov (2004). The findings of Arshanapalli et al. (2002) concerning the impact of index composition changes, and SPDR creations and redemptions from January 29, 1993, to September 29, 2001, support the downward sloping demand curve concept. Biktimirov (2004) made use of a more specific event to examine the effect of demand on stock prices: the conversion of unredeemed TIPs 35 and TIPs 100 shares into new S&P/TSE 60 Index Participation Fund (i60 Fund) shares that occurred on the Toronto Stock Exchange on March 6, 2000. The 40 stocks in the Toronto 100 index that were not included in the S&P/TSE 60 index (which served as a benchmark for the i60 Fund) had to be sold. Biktimirov (2004) claimed that this event is completely noninforma-

---

[13] Boehmer and Boehmer (2003) noted that, in early 2001, Cubes, Spiders, and Diamonds generated an average daily trading volume of about $5 billion all together.

tional since it has been long anticipated and is not associated with a change in the composition of the index. Selling pressure results in a decline in value both the day before conversion and the day of conversion, with abnormal trading volumes. There is no change in liquidity and the price decline is permanent. All this evidence is consistent with the downward-sloping demand curve hypothesis.

The natural properties of country ETFs for international diversification were studied by Pennathur et al. (2002). They found that the international iShares series efficiently replicates its foreign index benchmarks. However, its potential for diversification is limited due to of a high degree of exposure to U.S. equity markets. Miffre (2004) nonetheless insists on the specific advantages country ETFs have over conventional mutual and closed-end country funds: short selling on a down tick, low costs, and tax efficiency, to name but a few. Investors are thus able to achieve superior diversification with ETFs as long as they invest significant amounts. Amenc et al. (2004) measured the performance of a dynamic core-satellite approach based on fixed-income ETFs. However, ETFs only serve illustrative purposes and no empirical comparison with other investment vehicles is provided.

8 Conclusion and Perspectives

ETFs are open-end index funds that trade like regular stocks on exchanges. They combine the features of conventional mutual funds and closed-end funds since new shares can be continuously created or redeemed and outstanding shares trade throughout the day on exchanges. They were initially launched in North American markets in the early 1990s, and new listings on exchanges led to more than 450 different ETFs being traded around the world with steadily increasing assets under management. What is even more spectacular is the growth in trading volume these instruments have generated. In the United States, major ETFs are more traded than any other security. European ETF markets are younger, but they exhibit similar tendencies, with fierce competition both between issuers for new cash and between exchanges for order flow. Their success raises the issue of the organization of mutual fund trading.

Research on ETFs mostly focuses on their efficiency and performance as well as on their impact on the other index markets. Compared to closed-end funds, the specific in-kind creation and redemption process ensures a higher degree of pricing efficiency. Nonetheless, the advantages inherent to the in-kind process do not help ETF managers provide higher performance over the least-cost no-load index mutual funds. Overall, the advent of ETFs enhances the liquidity of the individual stock making up the benchmark indices and the efficiency of index derivatives markets. Finally, ETFs play a significant, though not prominent, role in the price discovery process.

Despite the increasing importance of ETFs markets, literature on these topics is still scarce, although research perspectives are promising. For example, European and Asian ETFs markets are very active but remain an almost untouched research field. The empirical but also theoretical questions of competition between marketplaces and between ETFs tracking the same index still need to be investigated. Regulatory issues should also be included in future research as the evolution of ETF markets may lead markets and regulators to adopt new rules. This has already been the case for the so-called trade-through rule exemption implemented by SEC for ETFs studied by Hendershott and Jones (2005). Finally, new types of ETFs, such as the recent commodity ETFs, are launched on a regular basis and a study has yet to examine their specificities, trading, or uses for fixed-income ETFs and ETF derivatives.

References

1. Ackert, L. F., and Tian, Y. S. The introduction of Toronto index participation units and arbitrage opportunities. *Journal of Derivatives*, 5(4):44–53, 1998.
2. Ackert, L. F., and Tian, Y. S. Arbitrage and valuation in the market for Standard and Poor's depositary receipts. *Financial Management*, 29:71–87, 2000.
3. Ackert, L. F., and Y. S. Tian 2001. Efficiency in index options markets and trading in stock baskets. *Journal of Banking and Finance*, 25:1607–1634.
4. Amenc, M., Malaise, P., and Martellini, L. Revisiting core-satellite investing. *Journal of Portfolio Management*, 31(1):64–75, 2004.
5. Ascioglu, A., Aydogdu, M., Chou, R. K., and Kugele, L. P. An analysis of intraday patterns in ETF and common stock spreads. Working paper, 2006.
6. Arshanapalli, B., Switzer, L. N., and Abersfeld, J. Index participation units and the structure of equity Market demand: Evidence from new issues and redemption of SPDRs. Working paper, 2002.
7. Bennet, J., and Kerins, F. J. Jr. Exchange traded funds: Liquidity and informed trading levels. Working paper, 2003.
8. Biktimirov, E. N. The effects of demand on stock prices: Evidence from index fund rebalancing. *Financial Review*, 39:455–472, 2004.
9. Blank, H., and Lam, P. The development and evolution of ETFs in North America. Investment & Pensions Europe, 2002.
10. Boehmer, B., and Boehmer, E. Trading your neighbor's ETFs: Competition or fragmentation? *Journal of Banking and Finance*, 27:1667–1703, 2003.
11. Bogle, J. C. Convergence: The Great Paradox? Remarks before The Art of Indexing Conference, 2004.
12. Broom, K. D., Van Ness, R. A., and Warr, R. S. Cubes to Quads: The move of the QQQ ETF from AMEX to NASDAQ. Working paper, 2006.
13. Cherry, J. The limits of arbitrage: Evidence from exchange traded funds. University of California, Berkeley, Working paper, 2004.
14. Chu, Q. C., Hsieh, W.-L. G., and Tse, Y. Price discovery on the S&P 500 index markets: An analysis of spot index, index futures, and SPDRs. *International Review of Financial Analysis*, 8:21–34, 1999.

15. Chung, P. A transactions data test of stock index futures market efficiency and index arbitrage profitability. *Journal of Finance*, 46(5):1791–1809, 1991.
16. Curcio, R. J., Lipka, J. M., and Thornton, J. H. Jr. Cubes and the individual investor. *Financial Services Review*, 13:123–138, 2004.
17. Dellva, W. L. Exchange-traded funds not for everyone. *Journal of Financial Planning*, 14(4):110–124, 2001.
18. Deville, L. Impact de l'introduction du tracker master share CAC40 sur la relation de parité call-put. *Banque et Marchés*, 62:50–57, 2003.
19. Deville, L. Time to efficiency in options markets and the introduction of ETFs: Evidence from the French CAC40 index. Paris Dauphine University, Working paper, 2005.
20. Deville, L., and Riva, F. The efficiency process in options markets: A survival analysis approach. Paris Dauphine University, Working paper, 2006.
21. Deville, L., Gresse, C., and de Séverac, B. The introduction of the CAC 40 master unit and the CAC 40 index spot-futures pricing relationship. Paris Dauphine University, Working paper, 2006.
22. Dimson, E., and Minio-Kozerski, C. Closed-end funds: A survey. *Financial Markets, Institutions & Instruments*, 8(2):1-41, 1999.
23. Elton, E. J., Gruber, M. J., Comer, G., and Li, K. Spiders: Where are the bugs? *Journal of Business*, 75(3):453–472, 2002.
24. Engle, R., and Sarkar, D. Premiums-discounts and exchange traded funds. *Journal of Derivatives*, 13(4):27–45, 2006.
25. Fremault, A. Stock index futures and index arbitrage in a rational expectations model. *Journal of Business*, 64:523–547, 1991.
26. Gallagher, D. R., and Segara,R. The performance and trading characteristics of exchange-traded funds. University of New South Wales, Working paper, 2004.
27. Gastineau, G. L. Exchange traded funds: An introduction. *Journal of Portfolio Management*, 27(3):88–96, 2001.
28. Gastineau, G. L. *The Exchange-traded Funds Manual*. John Wiley & Sons, New York, 2002.
29. Gastineau, G. L. The benchmark index ETF performance problem. *Journal of Portfolio Management*, 30(2):96–103, 2004.
30. Gleason, K. C., Mathur, I., and Peterson, M. A. Analysis of intraday herding behaviour among the sector ETFs. *Journal of Empirical Finance*, 11(5):681–694, 2004.
31. Gorton, G., and Pennacchi, G. Security baskets and index-linked securities. *Journal of Business*, 66:1–27, 1993.
32. Harper, J. T., Madura, J., and Schnusenberg, O. Performance comparison between exchange traded funds and closed-end country funds. *Journal of International Markets, Institutions, and Money*, 16(2):104–122, 2006.
33. Hasbrouck, J. One security, many markets: Determining the contributions to price discovery. *Journal of Finance*, 50(4):1175–1199, 1995.
34. Hasbrouck, J. Intraday price formation in U.S. equity index markets. *Journal of Finance*, 58(6):2375–2399, 2003.
35. Hegde, S. P., and McDermott, J. B. The market liquidity of DIAMONDS, Q's and their underlying stocks. *Journal of Banking and Finance*, 28:1043–1067, 2004.
36. Hendershott, T., and Jones, C. M. Trade-through prohibitions and market quality. *Journal of Financial Markets*, 8:1–23, 2005.

37. Henker, T., and Martens, M. Price discovery in HOLDR security baskets and the underlying stocks. University of New South Wales, Working paper, 2004.
38. Jares, T. E., and Lavin, A. M. Japan and Hong Kong exchange traded funds (ETFs): Discounts, returns and trading strategies. *Journal of Financial Services Research*, 25(1):57–69, 2004.
39. Kostovetsky, L. Index mutual funds and exchange-traded funds. *Journal of Portfolio Management*, 29(4):80–92, 2003.
40. Kumar, P., and Seppi, D. Information and index arbitrage. *Journal of Business*, 67:481–509, 1994.
41. Kurov, A. A., and Lasser, D. J. The effect of the introduction of cubes on the Nasdaq-100 index spot-futures pricing relationship. *Journal of Futures Markets*, 22(3):197–218, 2002.
42. Lee, C. M. C., Shleifer, A., and Thaler, R. Investor sentiment and the closed-end fund puzzle. *Journal of Finance*, 46(1):75–109, 1991.
43. Madhavan, A., Richardson, M., and Roomans, M. Why do security prices change? A transaction-level analysis of NYSE stocks. *Review of Financial Studies*, 10:1035–1064, 1997.
44. Madura, J., and Richie, N. Overreaction of exchange-traded funds during the bubble of 1998–2002. *Journal of Behavioral Finance*, 5(2):91–104, 2004.
45. Madura, J., and Richie, N. Impact of the QQQ on liquidity, pricing efficiency, and risk of the underlying stocks. Susquehanna University, Working paper, 2005.
46. Miffre, J. Country specific ETFs: An efficient approach to global asset allocation. Cass Business School, Working paper, 2004.
47. Miller, M. H., Muthuswamy, J., and Whaley, R. E. Mean reversion of S&P 500 index basis changes: Arbitrage induced or statistical illusion? *Journal of Finance*, 49(2):479–513, 1994.
48. Mussavian, M., and Hirsch, L. European exchange-traded funds: An overview. Journal of Alternative Investments, 5:63–77, 2002.
49. Park, T. H., and Switzer, L. N. Index participation units and the performance of index futures markets: Evidence from the Toronto 35 index participation units market. *Journal of Futures Markets*, 15:187–2000, 1995.
50. Pennathur, A. K., Delcoure, N., and Anderson, D. Diversification benefits of iShares and closed-end country funds. *Journal of Financial Research*, 25:541–557, 2002.
51. Peterson, M. Discussion of trading your neighbor's ETFs: Competition or fragmentation? *Journal of Banking and Finance*, 27:1705–1709, 2003.
52. Poterba, J. M., and Shoven, J. B. Exchange-traded funds: A new investment option for taxable investors. *American Economic Review*, 92(2):422–427, 2002.
53. Simon, D. P., and Sternberg, J. S. Overreaction and trading strategies in European iShares. Working paper, 2004.
54. Subrahmanyam, A. A theory of trading in stock index futures. *Review of Financial Studies*, 4:17–51, 1991.
55. Switzer, L. N., Varson, P. L., and Zghidi, S. Standard and Poor's depository receipts and the performance of the S&P 500 index futures market. *Journal of Futures Markets*, 20(8):705–716, 2000.
56. Thirulamai, R. S. Active vs. passive ETFs. Working paper, 2003.
57. Tse, Y., Bandyopadhyay, P., and Shen, Y.-P. Intraday price discovery in the DJIA index markets. Working paper, 2006.

58. Tse, Y., and Erenburg, G. Competition for order flow, market quality, and price discovery in the Nasdaq 100 index tracking stock. *Journal of Financial Research*, 26(3):301–318, 2003.
59. Van Ness, B. F., Van Ness, R. A., and Warr, R. S. The impact of the introduction of index securities on the underlying stocks: The case of the diamonds and the Dow 30. *Advances in Quantitative Finance and Accounting*, 2:105–128, 2005.
60. Yu, L. Basket securities, price formation and informational efficiency. Notre Dame University, Working paper, 2005.
61. Zanotti, G., and Russo, C. Exchange traded funds versus traditional mutual funds: A comparative analysis on the Italian market. Bocconi University, Working paper, 2005.

Genetic Programming and Financial Trading: How Much About "What We Know"

Shu-Heng Chen[1], Tzu-Wen Kuo[2], and Kong-Mui Hoi[1]

[1] AI-ECON Research Center, Department of Economics, National Chengchi University, Taipei, Taiwan 11623 {chchen; 92258038}@nccu.edu.tw
[2] Department of Finance and Banking, Aletheia University, Tamsui, Taipei, Taiwan 25103 kuo@aiecon.org

1 Motivation and Literature Review

The relevance of genetic programming (GP) to technical analysis is quite obvious. Technical analysis is mainly built upon some mathematical manipulations of the historical data of prices and volumes. The mathematical manipulations can be regarded as the function set of GP, whereas the signals of prices and volumes can be taken as the terminal set of GP. The interest in using GP to study technical analysis is, therefore, motivated by the two following concerns.[1] First, technical analysis generally does not refer to a fixed set of trading rules. They are evolving and changing over time. Many of them are still not even known to the public. However, for some time academic studies seem to have overlooked this property and have tended to study them as if they are fixed over time.[2] It is, therefore, not surprising to see the diversity of the results: They are profitable in some markets some of the time, while they fail in other markets at other times, and so they are very inconclusive.[3] A more systematic way to study this evolving subject is to place it in a dynamic and evolving environment. As a tool for simulating the evolution of trading rules in response to the changing environment, genetic programming can then serve this purpose well. Second, technical analysis usually involves quite complicated transformations and combinations of price and volume signals, which is too demanding to be harnessed by the human mind. GP, as a rules-generating machine, can better facilitate our travel through this jungle.

---

[1] See also Granville (1976).
[2] Brock et al. (1992), as a celebrated work in this area, is a case in point. This paper motivated a series of follow-ups to test the financial market's efficiency by testing the profitability of the simple trading strategies. Trading strategies, later on, have been parameterized as a parametric model. Either the parameters have to be inferred from the sample, or they are simply open to many possible values for us to play with.
[3] This phenomenon is sometimes connected to the *selection bias*.

Given these two concerns, the financial application of GP to trading rules is distinguished from the use of other computational intelligence tools. The point of interest here is to see how well we are able to simulate the rule-discovery process without assuming the size and the shape of trading rules. Needless to say, other computational intelligence tools can help us with the market-timing decision, but either they do not provide us with trading rules, such as the linear perceptron neural networks and support vector machine, or they assume a fixed size or shape of trading rules, such as the decision trees, self-organizing maps, and genetic algorithms. In fact, GP allows us to work with an issue of academic interest similar to automatic-theorem proofing, which is not shared by other competing tools.[4]

Given this academic uniqueness, it is then interesting to inquire whether this idea actually works. The answer that we have from the literature seems to be mixed. For example, it seems to work well in the foreign exchange markets (Bhattacharyya et al., 2002; Neely and Weller, 1999; Neely, et al. 1997) and has succeeded in some stock markets (O'Neill et al., 2002), but it has failed in some other markets (Allen and Karjalainen, 1999; Potvin et al., 2004) and has failed in the future markets (Wang, 2000).[5] Nonetheless, the real issue is that research in this area is so limited that we are far from concluding anything firm. In particular, GP is notorious for its large number of user-supplied parameters, and the current research is not rich enough to allow us to inquire whether these parameters may impact the performance of GP. Obviously, in order to better understand the present and the future of GP in evolving trading rules, more research needs to be done.

The purpose of this paper is to give a more thorough examination of what was found earlier in the stock markets (Allen and Karjalainen, 1999) and foreign exchange markets (Neely and Weller, 1999; Neely et al., 1997). To do so, we test the performance of GP more extensively with many more markets. In this chapter a total of eight stock markets and eight foreign exchange markets are tested. This scale has not been attempted in the past.[6] Even for the same market, our test period differs from earlier ones by including the most recent data after the year 2000, which features significant changes in trends within this extended period. Putting these efforts together, this extensive study may enrich our understanding of the behavior of GP in financial markets, and in particular whether those successes or failures are robust or are general enough.

Furthermore, motivated by Potvin et al. (2004), this chapter also considers the use of short selling. This consideration has its value. For instance, when the market experiences a long downward tendency, it will make GP essentially

---

[4] Of course, it is always interesting to compare different tools on the same application domain based on their performance. However, that should not be the only thing to look at, and it is beyond the scope of this paper.

[5] See Chen and Kuo (2003a) for a survey.

[6] Allen and Karjalainen (1999) only tested the series for the S&P 500, Neely et al. (1997) tested six foreign exchanges, and Neely and Weller (1999) tested four foreign exchanges.

very inactive; in this case, staying out of the market is the best strategy. However, when short selling is admissible, GP can become more active by learning to sell short. We therefore test the performance of GP by allowing traders to sell short in the stock market.[7]

To choose a benchmark with which GP is compared, most earlier studies selected the buy-and-hold strategy (B&H). Few have taken into account other practical trading rules used by practitioners. In this paper, we have the chance to work with an investment firm and to use the 21 trading strategies supplied by them as alternative benchmarks. As a result, GP is not just compared with B&H, but also competes with many other, more active trading strategies.

The rest of the chapter is organized as follows. Section 2 gives a brief introduction to genetic programming (GP) and the experimental designs in terms of the setting of the control parameters. One essential part of GP, namely, the fitness function, is fully discussed in Section 3. The data used in this paper associated with its preprocessing are detailed in Section 4. The experimental results together with an analysis are given in Section 5 and are followed by the concluding remarks in Section 6.

2 Genetic Programming

Genetic programming applies the ideas of biological evolution to a society of computer programs.[8] From one generation to another generation, it follows the survival of the fittest principle to select from the existing programs and then to operate these selected programs via some familiar genetic operators, such as reproduction, crossover, and mutation. To facilitate the genetic operation, each computer program is visualized as a manifestation of the computer language LISP (List Programming), known as the S-expression (symbolic expression). Each S-expression can be expanded into a parse tree. GP implements crossover and mutation operations by taking advantage of this parse-tree structure.

The parse-tree structure and its alteration or recombination make genetic programming able to demonstrate and operate the idea of modularity and complexity, which are the two essential ingredients of modern economics. In light of this virtue, Chen (2001) pointed out the significant relevance of genetic programming to economics, and Chen and Chie (2005) further used it to demonstrate how economic activities of discovery or innovation can be modeled and simulated by GP. Chen and Kuo (2003a) reviewed literature on the financial applications of GP by emphasizing its use as a tool for discovering hidden financial patterns.

---

[7] There is no appropriate counterpart for short selling in the foreign exchange market, since our position has to be either one of the possible currencies: the host currency and the foreign currency.

[8] For the readers who are not familiar with GP, John Koza's series of studies provides a systematic treatment (Koza, 1992, 1994; Koza et al., 1999, 2005).

Specifically, in financial trading, each computer program represents a trading program, and the society of computer programs represents a collection of trading programs. The population size, denoted by *Pop*, is a key control parameter of GP. The evolution of the population of the trading program proceeds in a cycle. Each cycle is counted as one generation. The maximum number of generations (*Gen*) combined with *Pop*, usually as a product of the two, gives the maximum search resources to be used in the discovery process. There are some studies on the choices of these two parameters (Chen et al., 2002). However, studies using empirical data shed little light on the choice of these two parameters (Chen and Kuo, 2003b). In this chapter, *Pop* is set to 500, and *Gen* is set to 100 (Table 1).[9]

The content of each trading program is determined by the associated LISP S-expression, i.e., a list composed of atoms. The atoms can be functions (operators), variables, or constants (operands) and can be a list as well. The functions and variables that can be used to span a possible LISP S-expression (trading program) must be declared at the beginning and are known as the function set and the terminal set. Our choice of the function set and the terminal set is basically in line with Allen and Karjalainen (1999), Neely et al. (1997), and Neely and Weller (1999). The idea is to satisfy the closure property (Koza, 1992). We first have a collection of simple technical trading rules, such as the moving-average rule, the filter rule, the trading range break-out rule, etc., and we thus see what operators and operands are required to make them reachable. A collection of these functions becomes our function set, and a collection of these variables or constants becomes our terminal set (see Table 1). By the closure property, these functions and terminals are sufficient enough to span the above-mentioned simple technical trading rules; but, more than that, they also have the potential to generate complex rules.

The functions and terminals used in Table 1 are all very primitive and simple. However, they can be enriched in several directions; for example, in addition to the price, we can include other variables in the terminal set such as volatility, volume, and technical indexes; we can also add other functions to the terminal set, such as some well-known trading rules. The significance of enriching the terminal set and the function set has been discussed in Chen et al. (2002), and we regard it as a direction for further research.

With the function set and terminal set specified in Table 1, we start the evolution by initializing a population of *Pop* random trading programs, called the initial population. A standard method for this initialization is the *ramp-half-and-half* method, which is a combination of the growth method and the full method. Given a size limit or a depth ceiling, the growth method grows

---

[9] In pilot experiments, we have also tried a larger *Pop* and *Gen*, such as 1,000 and 200 for each. However, there is no significant difference. In fact, as we shall discuss below, an issue arising here is the possibly low efficiency in searching: GP repeatedly discovers something we already knew, such as the buy-and-hold strategy.

Table 1. Control Parameters of GP

Population Size (Pop)	500
Initialization	ramp-half-and-half
Offspring trees created	
by crossover	50%
by point mutation	20%
by tree mutation (grow method)	20%
by elite	0.2%
by reproduction	9.8%
Function set	$+, -, \times, \div,$
	norm, average, max, min, lag
	and, or, not, $>, <$
	if-then-else, true, false
Terminal set	price, constants
Selection scheme	Tournament selection (size $= 2$)
Termination criterion	
number of generations (Gen)	100
stagnation tolerance (g)	50
Validation	
number of best models saved (k)	1
Fitness	
Fitness function	Equation (2) (stock market)
	Equation (10) (foreign exchange market)
Transaction cost (c)	0.5% (stock market)
	0.05% (foreign exchange market)

the tree randomly up to or below the size limit, whereas the full method grows the tree exactly to the size limit.[10]

Each trading program is then applied to the market and gives the trading recommendation. Based on the recommendation and the data, we measure its fitness.[11] We then rank all trading programs of the current generation by their fitness. The top-ranked k programs are selected and are further tested over another unseen data set, called the *validation set* (see Section 4). Their fitness is calculated, and they are saved in a list of winners.

The standard genetic operators, including reproduction, crossover, and mutation, are then applied to the current generation of the programs, in order to generate a new generation. In addition to these three operators, as a device for disturbance avoidance (Chen et al., 2002), the elitism operator, which is designed to keep the few best programs of the current generation, is also applied. The share of the total programs assigned to each of these operators is another control parameter in GP and is given in Table 1. Chen et al. (2002)

---

[10] For details, see Koza (1992).
[11] The use of the data is detailed in Section 4, and the exact fitness measure is detailed in Section 3.

thoroughly discussed the setting of these control parameters, and their device, which may not be optimal, is followed here.

After the emergence of the new generation, we then follow the same procedure to measure the fitness of each trading program of the new generation. The best k models are selected, and their fitness over the validation set is calculated and further compared with that of the k incumbents in the list of winners. The best k of these $2k$ programs are then selected as the new list of winners. This finishes the evolution for one generation, and the cycle starts over again by generating the next new generation of trading programs, and this cycle goes on and on until it meets a termination criterion.

There are two termination criteria. One is the maximum number of generations (Gen), and the other is the maximum number of consecutive stagnations (g). A stagnation is defined as a state in which none of the k incumbents in the winners list has been replaced by any from the new generation of programs. A stagnation implies that we have not learned anything new from the last generation to this generation. Therefore, to save computational resources, we may stop the cycle if the stagnating state consecutively lasts for a number of times. In this paper, k is set to 1, the same as Allen and Karjalainen (1999) and Neely et al. (1997), whereas g is set to 50, which is twice the size of Allen and Karjalainen (1999) and Neely et al. (1997). The idea of using a validation set of data to watch the progress in learning and to determine the point to terminate the evolution cycle is a device to avoid overfitting or overlearning. However, the effectiveness of this design has been questioned (Chen and Kuo, 2003c). In Appendix A, we shall revisit this issue.

3 Fitness Function

While the choice of the fitness function is usually not a trivial task for evolutionary computation, there is a generally acceptable one for the purpose of evaluating investment strategy, i.e., the *investment return*, e.g., Allen and Karjalainen (1999) and Neely et al. (1997). For the purpose of comparing our results with those of earlier studies, we shall also employ this standard fitness function.[12]

3.1 Return of Investment in the Stock Markets

There are three steps in computing the return. First, the GP program is applied to the time series to produce a sequence of trading decisions: *True* means to enter/stay in the market, and *False* means to exit/stay out of the market. Second, this decision sequence is executed based on the original stock price series and the daily interest rate in order to calculate the compounded

---

[12] To be more practical, it is definitely more interesting to take also risk and, more importantly, the downside risk, into account.

return. Finally, each transaction (buy or sell) is charged with a transaction cost, which is deducted from the compounded return to give the final fitness.

Let P_t be the S&P 500 index on day t, i_t be the interest rate on day t, and the return for day t be π_t:

$$\pi_t = \begin{cases} \ln(P_t) - \ln(P_{t-1}), & \text{in the market,} \\ \ln(1 + i_t), & \text{out of the market.} \end{cases} \quad (1)$$

Let n denote the total number of transactions, i.e., the number of times a *True* (in the market) is followed by a *False* (out of the market) plus the number of times a *False* (out of the market) is followed by a *True* (in the market). Also, let c be the one-way transaction cost. The *rate of return* over the entire period of T days, as an *arithmetic sum*, is

$$\Pi = \sum_{t=1}^{T} \pi_t + n * \log \frac{1-c}{1+c}. \quad (2)$$

However, it can be shown that the *total return*, based on a continuously compounded return (*geometric sum*), is[13]

$$R = \exp^{\Pi} - 1. \quad (3)$$

When a short sale is allowed, the investor may not have to leave the market.[14] In this case, a *False*, a signal to sell, implies not only just to sell, but also to *sell short*. Similarly, a *True*, a signal to buy, not only indicates to buy, but further implies to *recover short*. Figure 1(a), a three-state automaton, illustrates the operation of the transaction direction when a short sale is allowed. The three states are denoted by "0," "1," and "-1." "1" refers to being *long* on one unit of stock, and "-1" refers to being *short* on one unit of stock. "0" serves as both the initial state and the terminal state. In addition to these three states, there are three actions that will determine the transition to the next state, namely, "B" (**BUY**), "S" (**SELL**), and "H" (**Hold**). As Figure 1(b) shows, at any point in time, a trader can be either *long* or *short* on one unit of stock. Therefore, if a trader is already in the long (short) position, then any "BUY" ("SELL") action will be ignored. Finally, the terminal state "0" will not be reached until the clearance date, i.e., the period T.

---

[13] The *total return* means the return earned over the entire investment horizon $[0, T]$ per unit of investment. So, say $R = 0.05$ means that a one-dollar investment will earn five cents as a total over the entire investment horizon $[0, T]$. For the derivation of Equation (2), see Allen and Karjalainen (1999).

[14] While a short sale is implementable in individual stocks, the financial product corresponding to the short sale of a stock index does not exist. The closest product that we have is the index futures. Given this circumstance, what we are testing is not a real implementation. The purpose of doing this is to gain some insights into GP behavior, when trading can take place in both directions. After all, this mechanism is crucial when later on we wish to extend our applications to individual stocks.

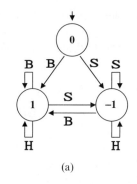

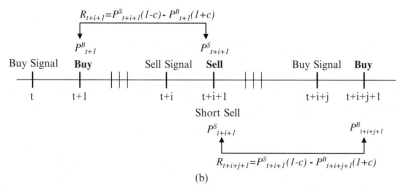

Fig. 1. The use of data: training, validation, and testing.

Based on this three-state automaton, one calculates the sequence of accumulated profits π_t $(0 \leq \tau < T)$ as follows [see Figure 1(b)],

$$\pi_0 = 0, \qquad (4)$$

and, for $0 < t \leq T$,

$$\pi_t = \begin{cases} 0 & \text{if } I_{t-1} = 0, I_t = 1, -1, \\ \pi_{t-1} & \text{if } I_t = I_{t-1}, \\ \pi_{t-1} + P_t(1-c) - P_{\lambda_t}(1+c) & \text{if } I_{t-1} = 1, I_t = -1, 0, \\ \pi_{t-1} + P_{\lambda_t}(1-c) - P_t(1+c) & \text{if } I_{t-1} = -1, I_t = 1, 0, \end{cases} \qquad (5)$$

where I_t is the state at time period t. As depicted in Figure 1(a), $I_t \in \{1, 0, -1\}$. c is the transaction cost. λ_t is an index function:

$$\lambda_t = \max\{\lambda \mid 0 \leq \lambda < t, I_{\lambda-1}I_\lambda = -1 \text{ or } 0\}. \qquad (6)$$

3.2 Return of Investment in Foreign Exchange Markets

In the foreign exchange markets, the investment return is again employed as the fitness function. As to a specific functional form, we follow Neely et al. (1997) to distinguish the return associated with long positions from that associated with short positions. Let us denote the exchange rate at date t (dollars per unit of foreign currency) by S_t and the domestic (foreign) interest rate by i_t (i_t^*).[15] A long position can be thought of as borrowing S_t dollars at the interest rate i_t at date t, and converting the dollars into one unit of foreign currency, reinvesting it at the interest rate i_t^* and, in the end, converting the principal plus the interest back to S_{t+1} dollars. Let

$$r_{t+1} = \frac{S_{t+1}(1+i_t^*)}{S_t(1+i_t)}; \tag{7}$$

the return of a long position, π_{t+1}, is, therefore,

$$\pi_{t+1} = \ln r_{t+1}. \tag{8}$$

Similarly, a short position can be thought of as borrowing one unit of the foreign currency at the interest rate i_t^*, converting it to S_t dollars, reinvesting it at the interest rate i_t, and, in the end, paying back the borrowing foreign currency of $1+i_t^*$, which is equal in value to $S_{t+1}(1+i_t^*)$ dollars. So, the return of a short position is

$$\ln \frac{S_t(1+i_t)}{S_{t+1}(1+i_t^*)} = \ln \frac{1}{r_{t+1}} = -\pi_{t+1}. \tag{9}$$

The *rate of return* of a trading rule over the period from time zero to time T (the arithmetic sum), with a one-way proportional transaction cost c, is given by

$$\Pi = \sum_{t=1}^{T} z_t \pi_t + n * \log \frac{1-c}{1+c}, \tag{10}$$

where z_t is an indicator variable taking the value "+1" for a long position and "-1" for a short position, and n is the number of round-trip trades. The continuously compounded return (geometric sum) is simply

$$R = \exp^{\Pi} - 1. \tag{11}$$

3.3 Transaction Cost

The transaction cost is another variable that may affect the results we have. Needless to say, transaction costs can be high to a degree in that there does not exist any profitable trading strategy, and B&H would be the only survivable strategy. To react to the sensitivity to this arbitrariness, the break-even transaction cost has been proposed in the literature and used in some related studies (Pereira, 2002). However, this idea is very difficult to implement in the

---

[15] In this study, we shall assume that USD is the domestic currency, and the other currencies are the foreign currencies.

context of the evolutionary cycle; therefore, we just follow what the standard literature has done to enhance the comparability of our results.

For the case of the stock market (Allen and Karjalainen, 1999), from low to high, considered three levels of c, namely $c = 0.1\%, 0.25\%$, and 0.5%. We choose the highest one, $c = 0.5\%$.[16] For the case of the foreign exchange market, we follow Neely et al. (1997) to choose $c = 0.05\%$.[17]

4 Data and Data Preprocessing

The stock indexes from eight stock markets are used in this paper. The eight stock indexes are, respectively, the S&P 500 (US), FTSE 100 (UK), TSE 300 (Canada), Frankfurt Commerzbank (Germany), Madrid-SE (Spain), Nikkei Dow Jones (Japan), Straits Times (Singapore), and the Capitalization Weighted Stock Index (Taiwan). For each stock index, we use the daily data from January 1988 to December 2004, except that, for Madrid-SE, due to the lack of appropriate interest rate data, the data start only from January 1992. In addition to the stock markets, data from eight foreign exchange markets are also used in this study. These are the U.S. dollar (USD) to the British pound (GBP), Canadian dollar (CAD), Deutsche mark (DEM), Italian lira (ITL), Spanish peseta (ESP), Japanese yen (JPY), Singapore dollar (SGD), and Taiwan dollar (TWD). For each foreign exchange rate, we use the average of the daily U.S. dollar bid and ask quotations from 1992 to 2004.

As to the riskless interest rate, we mainly consider the short-term treasury bills. Given the data availability, for the United States and Singapore, we use three-month Treasury bills, and for the UK and Canada, we employ one-month Treasury bills. However, for Germany, Spain, Japan, and Taiwan, whose Treasury bill rates are not available, we take the rate of the interbank overnight loan.

Data for stock indexes are made available from the AREMOS databank,[18] whereas the data for the foreign exchange rates are obtained from the Data-stream databank.[19] Most interest rate data can be downloaded from the Data-stream databank, except that the interest rate for Germany is directly downloaded from the Bundesbank.

The right halves of Figures 10 and 11 in Appendix D depict the time series of the eight indexes. It is quite evident that the United States, UK, Canada, Germany, and Spain share a similar pattern. The stock prices in these five

---

[16] We also try the zero transaction cost ($c = 0$). As expected, the performance gets better, but it does not enable us to reach qualitatively different conclusions. We will leave some details in Appendix B, for interested readers.

[17] Actually, Neely et al. (1997) also considered a different c, a higher c for the training period, as a device to avoid overfitting. We also try this device, but doubt its effectiveness. More discussion appears around Table 8.

[18] http://140.111.1.22/moecc/rs/pkg/tedc/bank/tse/c4t4.htm.

[19] http://www.datastream.com.mt/.

stock markets experienced a long upward tendency from 1989 to 2000 and then dropped quite significantly over the next two years (2001–2002) before they turned up again (after 2003). This pattern is, however, not shared by the other three Asian countries. Instead of the long upward tendency, Japan experienced a long declining tendency for almost the entire sample horizon: The Nikkei dropped from 32,000 to just around 10,000. On the other hand, Taiwan and Singapore, after a series of fluctuations, did not grow or decline much. The TAIEX (Taiwan) started at 6,000 and ended about the same. The Straits index (Singapore) began at 1,200 and remained almost the same in the year 2003, before it bloated again to 1,800.

The discussion above highlights the nonstationary nature of the data.[20] It is therefore desirable to transform the original series into a stationary series.[21] In this chapter, we follow what was conventionally done in the literature by dividing each day's price by a 250-day moving average (Allen and Karjalainen, 1999; Neely et al., 1997).[22] The data after normalization are shown in parallel to the original series and are depicted on the right halves of Figures 10 and 11 (see Appendix D). The normalized prices have an order of magnitude of around 1. Since we use the first year of data for normalization (taking the 250-day moving average), the usable data set starts from 1989 or 1993 (Spain).

Like most similar studies conducted before, we adopt a very standard way of decomposing the whole data set into three sections, namely, the training set, the validation set, and the testing set. To guard against potential data snooping in the choice of time periods, we use three successive training periods, validation periods, and test periods, as shown in Figure 2.[23] The five-year training and five-year validation periods start in 1989, 1991, and 1993, with the out-of-sample test periods starting in 1999, 2001, and 2003.[24] For instance, the first trial uses the years 1989 to 1993 as the training period, 1994 to 1998 as the selection period, and 1999 to 2000 as the test period. For each of the three training periods, we carry out 50 trials.

---

[20] More technically, this can be rigorously examined by the standard *unit-root test*. The test applied to these series of indexes does show that all series have a unit root, which means that they are nonstationary.

[21] The hypothesis that using original data without transformation may adversely affect the performance of GP has been widely acknowledged by researchers, while there is no formal test to show how big the difference can be. In Appendix C, we actually conduct such a test and find that the hypothesis is largely true.

[22] Alternatively, one can also take the log and the difference to obtain the the daily return series.

[23] The idea of the rolling-window design was originally proposed by Allen and Karjalainen (1999).

[24] The window size of the three sections of data is another issue for which we do not yet have a clear answer. We are aware that our result is not independent of the choices of the sizes. As we will discuss in Section 6, a more flexible way is to allow the size to be endogenously determined (Lanquillon, 1999).

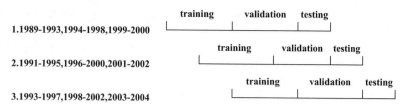

Fig. 2. The use of stock data: Training, validation, and testing.

The left halves of Figures 12 and 13 in appendix D depict the time series of the foreign exchange rates. Due to the nonstationary nature of these series, we also smooth the data by dividing them by a 250-day moving average, which we show in the right halves of Figures 12 and 13.

A similar way of dividing the data into three sections is also applied to the foreign exchange market, which is shown in Figure 3. As in Neely et al. (1997), both the training and validation periods are shortened to only three years. Neely et al. (1997) also considered a testing period of 15 years. However, we feel that this period is too long to be practical; therefore, a two-year testing period is chosen here.[25] The three-year training and three-year validation periods start in 1993, 1995, and 1997, with the out-of-sample test periods starting in 1999, 2001, and 2003.

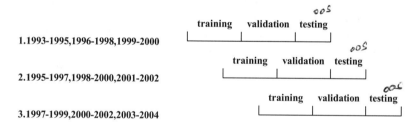

Fig. 3. The use of foreign exchange rate data: Training, validation, and testing.

5 Results

Genetic programming based on what we describe in Section 2 and Table 1 is applied to the data in the way we depicted in Figures 10–13. The results are organized into two parts. First, we present the profitability performance of GP according to the fitness function Equations (2) and (10). Second, we then conduct an analysis of the GP-discovered programs.

---

[25] Given the constantly changing nature, it is hard to accept that an underlying regularity can be sustained for a long span without being exploited. A testing period that is two years long seems to be more practical. In addition, see Pereira (2002).

5.1 Profitability

Stock Markets: GP and Buy-and-Hold

The results shown in Table 2 are the total returns as defined in Equation (3). Since there are 50 runs for each market, the results reported in Table 2 from column 2 to column 6 are the mean, the standard deviation, the medium, the maximum, and the minimum of the 50 runs. In other words, these results are based on the average performance of 50 GP trading programs. The mean return ($\bar{R}$) can therefore be regarded as the return of a uniform portfolio of these 50 programs. Alternatively, since these 50 trading programs will generally give different trading suggestions, we may follow what the majority suggests.[26] Column 7 gives the performance of this *majority rule*. The last column is the performance of the buy-and-hold strategy, which is conventionally taken as a benchmark to be compared with any potential competitors.

The results are presented separately with respect to three different test phases. As we can see from the table, the results differ by markets and by periods. The only one showing a quite consistent pattern is the Taiwan stock market. In this market, GP persistently beats the buy-and-hold strategy, regardless of whether we follow the portfolio rule or the majority rule, while the latter gives an even better result. In particular, in the first period, Taiwan's stock index experiences a negative total return of 23.66%.[27] Nevertheless, GP performs quite superbly, with a positive return of 18.46% (majority rule) or 16.20% (portfolio rule). The loss of B&H is sustained in the second test period with a return of -10.91%, but the return for the majority rule of GP remains positive, i.e., 2.41%. In the last period, when the Taiwan stock market comes back from a low, and B&H earns a return of 34.34%, the portfolio rule and the majority rule each earn an even higher return, 36.31% and 36.67%, respectively. In addition, the statistical test of the null that the total return of GP is no greater than the total return of B&H is significantly rejected in all three test periods.

The general superiority of GP is, however, not shared by other markets. If we follow the portfolio rule of GP, then only in 8 out of the remaining 21 cases can GP outperform B&H, and only in four of these eight does the dominance reach a significant level. On the other hand, if we follow the majority rule, GP wins 3 and loses 6 of the 21 cases and is tied with B&H in the remaining 12 cases. Therefore, in general, GP does not perform better than B&H. This result is consistent with the earlier findings (Allen and Karjalainen, 1999; Wang, 2000).

---

[26] Since GP is a population-based search algorithm, one can also think about the majority rule by using all the trading programs in the last generation. In doing so, we do not have to exclusively concentrate on only the best rule for each run. Given the size of this paper, we have to leave this interesting experiment to future research.

[27] This means that an initial capital of 1 dollar at the beginning of the year 1999 will result in only 77 cents by the end of the year 2000.

Table 2. The Total Return R of GP: Case Without Short Sale

Country	Mean	Stdev	Median	Max	Min	Majority	B&H
Test Period: 1999–2000							
USA	0.0655	0.0342	0.0644	0.1171	−0.1294	0.0644	0.0644
UK	0.0459	0.0908	0.0478	0.2368	−0.2850	0.0478	0.0478
Canada	0.3660	0.1030	0.3601	0.5837	0.0982	0.4136	0.3495
Germany	0.1490	0.1114	0.1491	0.4233	−0.1865	0.2473	0.2127
Spain	−0.0666	0.0904	−0.0511	0.0805	−0.2560	−0.0523	−0.0511
Japan	0.0024	0.0540	0.0016	0.1767	−0.1444	0.0016	0.0173
Taiwan	0.1620**	0.1353	0.1416	0.4675	−0.1893	0.1846	−0.2366
Singapore	0.1461	0.1866	0.1903	0.6178	−0.2801	0.2122	0.3625
Test Period: 2001–2002							
USA	−0.3171	0.0498	−0.3212	−0.0461	−0.3486	−0.3228	−0.3228
UK	−0.3625**	0.0284	−0.3700	−0.2988	−0.3700	−0.3700	−0.3700
Canada	−0.1761**	0.1065	−0.2251	0.0994	−0.2718	−0.2610	−0.2407
Germany	−0.4772**	0.2242	−0.5169	−0.0392	−0.6384	−0.5474	−0.5474
Spain	−0.2780	0.0910	−0.2870	0.0763	−0.3821	−0.2870	−0.2870
Japan	−0.0722**	0.1520	−0.0480	0.1900	−0.3990	−0.0582	−0.3796
Taiwan	0.0376**	0.1899	0.0249	0.6807	−0.3267	0.0241	−0.1091
Singapore	−0.3123	0.0333	−0.3013	−0.2436	−0.3490	−0.3013	−0.3013
Test Period: 2003–2004							
USA	0.3065	0.0334	0.3199	0.3199	0.1173	0.3199	0.3199
UK	0.1797	0.0300	0.1888	0.2402	0.0944	0.1888	0.1888
Canada	0.3109	0.0585	0.3625	0.4657	0.0974	0.3625	0.3625
Germany	0.3318	0.0544	0.3571	0.4614	0.1009	0.3571	0.3571
Spain	0.3355	0.1292	0.4454	0.4454	−0.0581	0.4454	0.4454
Japan	0.0212	0.0843	0.0002	0.2695	−0.1476	0.0540	0.3054
Taiwan	0.3631**	0.0665	0.3685	0.5787	0.1665	0.3667	0.3434
Singapore	0.2512	0.0735	0.2524	0.5072	0.0686	0.2685	0.5311

The "**" refers to the rejection of the null that the total return of GP is no greater than the total return of B&H at a significance level of 0.05.

However, why can GP perform so well in the Taiwan market but not other markets? One possible clue is that *they have quite different time series patterns*. As we mentioned earlier, the United States, UK, Germany, Canada, and Spain all share a very similar pattern, i.e., they exhibit an upward tendency from the beginning to the year 2000, and after 2000 they decline for the next three years before rising again in the year 2003. This changing pattern gives rise to a sharp difference between the general property of the training set and the testing set. The smooth upward tendency of the stock index appearing in the training set will easily drive GP to pick up B&H, and if B&H strongly dominates in the training phase, then no matter what patterns appear in the validation set, it will not make any difference because B&H is the only survivor there. As a result, the validation mechanism may fail to function in the way we expect.[28] This explains why GP in so many markets strongly recommends B&H. When using the majority rule, 12 out of 24 cases are essentially using B&H.[29] This also explains why B&H has never been recommended by GP in both Taiwan's and Japan's markets: The former has a fluctuating pattern, whereas the latter is largely characterized by a declining tendency.

Table 3 gives the performance of GP when short sales are possible. The original purpose of making GP perform short-sale operations was to make it able to trade and make profits in both directions, i.e., downturns and upturns. However, when what GP learns is only B&H, then introducing such a new function will not make anything different. This is exactly what is revealed by Table 3, which indicates the identical performance between B&H and GP in many markets. This is even true for the second test period when the markets experienced a big slump, and short sales can be extremely profitable, but GP behaves very quietly: When using the majority rule, five out of the eight markets strongly recommend B&H. As a total, 16 out of 24 cases essentially use B&H, a frequency that is even bigger than that for the case without a short sale. With such a high intensity of using B&H, it is therefore not expected that there will be any significant difference in the return between B&H and GP, when the majority rule is used.[30]

The only two exceptions are the Taiwan and Japan stock markets: the two markets which do not have a long initial upward trend. The Taiwan market fluctuates, whereas the Japan market declines. Both situations may allow GP to learn to do short selling and to take advantage of market downturns. Indeed, the majority rule of GP does very well in the first test period (1999–2000), when the TAIEX drops quite dramatically (see Figure 10). It earns a return of 60.99%, which is almost four times higher than the case when short sales are not infeasible. In the case of Japan, during the second period, when the stock index keeps on declining, the majority rule of GP earns a return of 38.08%, which is not only higher than the corresponding case of no short sales (−0.05%), but is also greater than B&H (−37.96%). However, we also notice that when the stock price moves upward, GP programs with the short-selling

---

[28] If this is the case, then we can see the early termination of the evolutionary cycle. The actual number of generations tends to much less than 100, and is not far away from 50. See Section 5.2 and Figure 8 for details.

[29] Since the majority rule follows the outcome of votes from 50 programs, one cannot tell directly from that outcome whether the rule learned is actually B&H. Nonetheless, we can confirm this by checking the medium statistics.

[30] Here, we don't have to go through the portfolio rule of GP except by pointing out that the portfolio rule of GP performs a little worse than the case without a short sale. Only in two 2 of 24 cases is the null of no excess return rejected.

Table 3. The Total Return R of GP: Case with Short Sale

Country	Mean	Stdev	Median	Max	Min	Majority	B&H
Test Period: 1999–2000							
USA	0.0685	0.0243	0.0644	0.2362	0.0647	0.0644	0.0644
UK	0.0444	0.0258	0.0478	0.0481	−0.1347	0.0478	0.0478
Canada	0.3414	0.2025	0.3495	0.8211	−0.1775	0.3495	0.3495
Germany	0.2113	0.1063	0.2127	0.6397	0.0078	0.2127	0.2127
Spain	−0.0874	0.1297	−0.0511	0.2173	−0.5844	−0.0511	−0.0511
Japan	0.0317	0.1834	−0.0377	0.3893	−0.3437	−0.0377	0.0173
Taiwan	0.5265**	0.3726	0.4658	1.4440	−0.1096	0.6099	−0.2366
Singapore	0.1620	0.3711	0.1342	0.9163	−0.8403	0.1342	0.3625
Test Period: 2001–2002							
USA	−0.3231	0.0018	−0.3228	−0.3228	−0.3353	−0.3228	−0.3228
UK	−0.3658	0.0329	−0.3700	−0.1382	−0.3859	−0.3700	−0.3700
Canada	−0.2367	0.0307	−0.2407	−0.0502	−0.3175	−0.2407	−0.2407
Germany	−0.5512	0.0439	−0.5474	−0.3352	−0.7010	−0.5474	−0.5474
Spain	−0.2852	0.0504	−0.2870	0.0231	−0.4509	−0.2870	−0.2870
Japan	0.1745**	0.3016	0.3653	0.5259	−0.5557	0.3808	−0.3796
Taiwan	−0.0470	0.3578	−0.0098	0.8076	−0.7441	−0.0788	−0.1091
Singapore	−0.2939	0.0370	−0.3013	−0.0927	−0.3013	−0.3013	−0.3013
Test Period: 2003–2004							
USA	0.3109	0.0460	0.3199	0.3215	0.0483	0.3199	0.3199
UK	0.1817	0.0275	0.1888	0.1897	0.0506	0.1888	0.1888
Canada	0.3389	0.1131	0.3625	0.5505	−0.0423	0.3625	0.3625
Germany	0.3457	0.0794	0.3571	0.5701	−0.0147	0.3571	0.3571
Sapin	0.2729	0.2754	0.4454	0.4641	−0.5535	0.4454	0.4454
Japan	−0.1792	0.2682	−0.1819	0.2970	−0.7490	−0.0613	0.3054
Taiwan	0.2740	0.2417	0.3073	0.7756	−0.5718	0.2248	0.3434
Singapore	−0.0183	0.1917	−0.0384	0.3859	−0.4422	−0.0930	0.5311

The "**" refers to the null that the total return of GP is no greater than the total return of B&H is rejected at a significance level of 0.05.

devices may not work well. This is clearly reflected by the relatively inferior performance of GP in the last testing period of both the Taiwan and Japan stock markets.

In sum, our finding is consistent with Potvin et al. (2004). "The results show that the trading rules generated by GP are generally beneficial when the market falls or when it is stable. On the other hand, these rules do not match B&H when the market is rising," (Ibid. p. 1046).

Stock Market: GP and Practical Technical Trading Strategies

While B&H is frequently used as the benchmark in academic research, financial practitioners are seldom interested in such an inactive strategy. In this paper, we extend our test to 21 technical trading strategies that have actually been used by investors in financial securities firms. These strategies are basically composed of the historical data of prices and trading volumes. In addition to the closing price, they also involve the opening price, and the daily highest and lowest.[31] Since the data on trading volume are not available from some markets, our testing is inevitably limited to only those markets whose data are sufficient, which includes the United States, the UK, Canada, Taiwan and Singapore.[32]

Tables 4 to 6 present the performances of the technical trading strategies. Notice that, unlike GP, the 21 technical trading strategies are deterministic in the sense that their performance is fixed once the data are fixed. Therefore, there is no need to conduct statistical tests for these 21 trading strategies, and the appearance of a "*" in these figures simply means that the total return of the corresponding trading strategy is greater than that of B&H. In these tables, we also include the total returns of B&H, and the portfolio rules of GP without short sales (GP 1) and with short sales (GP 2).

We have already noticed that the market behaves quite differently within these three periods. It is therefore interesting to observe how these differences can impact the performance of the 21 trading strategies. First of all, in the years 2003–2004, the years of a bull market, all these 21 technical trading strategies perform uniformly worse than B&H and are also inferior to GP 1. This is probably the worst period for the 21 technical trading strategies. This result, however, is in a sharp contrast to that for the years 2001–2002, the years of a bear market. During this bear market, when both B&H and GP are earning a negative return, many technical trading strategies can still have a positive return. A number of technical trading strategies can outperform GP and B&H in all markets, such as strategies #6, #13, #15, #16, #18, #19, and #21. This shows that some of these technical trading strategies are very good at dealing with the bear market.[33]

---

[31] Unfortunately, due to the business contract, we are unable to spell out exactly what these strategies are except for pointing out that these strategies are developed from many familiar technical indices, such as KD line, MACD, etc.

[32] For the case of Japan, since trading volume data are only available in the last testing period, we therefore include the results for that period for Japan only (see Table 6).

[33] These technical trading strategies are fixed and are independent of the history of the data. Hence, their performance is not path-dependent. This property is very different from GP, whose performance depends on what it learns from the past. So, the switch from a bull (bear) market to a bear (bull) market may cause GP to perform badly.

Table 4. Performances of 21 Practical Trading Strategies: 1999–2000

Rule	USA	UK	Canada	Taiwan	Singapore
B&H	0.0636	0.0478	0.3495	−0.2366	0.3625
GP 1	0.0655	0.0459	0.3660	0.1620**	0.1461
GP 2	0.0685	0.0444	0.3414	0.5265**	0.1620
1	−1.1173	−1.2855	−1.8943	−1.5102	−1.0679
2	0.0292	−0.5265	−0.9935	−0.8737	−0.8182
3	−0.1640	−0.6941	−0.2494	−0.3338	−0.7028
4	−0.9865	−0.8252	−0.1182	−0.7371	−0.5123
5	−0.0896	−0.3062	−0.9872	−0.2571	−0.6288
6	−0.7176	−0.6335	−0.0440	0.0048*	−0.7599
7	−1.1736	−1.7050	−2.1544	−1.1646	−1.9132
8	−1.2402	−1.3594	−2.1444	−0.7130	−0.8391
9	−1.3883	−1.0738	−1.6657	−1.0748	−0.7450
10	−1.6532	−1.4603	−1.5322	−1.0678	−0.4226
11	−1.0941	−0.5934	−1.4946	−0.3628	−0.9329
12	−1.4735	−1.2046	−2.6474	−1.5254	−1.6464
13	−0.9116	−0.7762	−0.1522	−0.6863	−0.3210
14	−0.2477	−0.2666	−0.9692	−0.2258*	−0.5817
15	−0.6658	−0.5571	0.0019	0.0218*	−0.7405
16	−0.7576	−0.9016	−0.1671	−0.4350	−0.0302
17	−0.1607	0.0126	−1.0631	0.3375*	−0.5044
18	−0.4397	−0.6185	−0.0055	0.1213*	−0.4336
19	−0.4240	−0.7951	−0.0942	−0.1480*	−0.1412
20	0.1419*	−0.0474	−1.0680	−0.5793	−0.5628
21	−0.4195	−0.6143	0.0827	0.2087*	−0.5644

The 21 practical technical trading strategies are each coded by a number, from 1 to 21. GP1 refers to the mean total return of the 50 runs without a short sale, whereas GP2 refers to that with a short sale. "**" refers to the rejection of the null that the total return of GP is no greater than the total return of B&H at a significance level of 0.05. "*" indicates the case where the associated trading strategy beats B&H.

Foreign Exchange Markets

The performance of GP in the foreign exchange markets is shown in Table 7. The table presents the total return of GP in three testing phases. As in the stock markets, we have run 50 trials for each foreign exchange market, so the results presented here are the sample statistics, which are the same those in Table 2. We also test the null hypothesis that the trading rules discovered by GP cannot earn positive profits, and we fail to reject the null hypothesis in 19 out of the 24 markets. Therefore, the hypothesis that GP can generate profitable trading strategies in the foreign exchange markets does not win strong support. This finding contradicts Neely and Weller (1999) and Neely et al. (1997).

Table 5. Performances of 21 Practical Trading Strategies: 2001–2002

Rule	USA	UK	Canada	Taiwan	Singapore
B&H	−0.3212	−0.3682	−0.2395	−0.1091	−0.2998
GP 1	−0.3171	−0.3625	−0.1761**	0.0376**	−0.3123
GP 2	−0.3231	−0.3658	−0.2367	−0.0470	−0.2939
1	−0.6764	−1.2137	−0.6493	−0.7024	−0.6708
2	−0.2521*	−0.7671	−0.8770	−0.3661	−0.4094
3	−0.3141*	−0.7511	−0.7826	−0.2876	−0.8150
4	0.3264*	−0.1745	0.2165*	0.1807*	0.0164*
5	−0.4818	−0.3331	−0.4180	−1.2470	−0.5208
6	0.0366*	−0.2513	0.0547*	0.3708*	0.3771*
7	−0.5690	−1.1215	−1.3767	−0.7911	−0.7403
8	−0.8376	−1.3764	−0.7633	−0.5823	−0.5115
9	−0.5734	−0.4859	−0.3960	−1.3472	−0.3864
10	−0.5684	−0.9808	−0.4071	−1.1382	−0.4474
11	−0.2926*	−0.4236	−0.1715*	−1.0765	−0.3287
12	−1.1004	−1.5990	−0.6958	−0.9736	−0.6115
13	0.2933*	−0.1761	0.1881*	0.3810*	0.0690*
14	−0.5135	−0.3186	−0.4824	−1.3545	−0.4006
15	0.0280*	−0.2329*	0.1605*	0.4168*	0.4207*
16	−0.1480*	0.0733*	0.2524*	0.3992*	0.0091*
17	0.2375*	−0.1076*	−0.5579	−0.6997	0.0303*
18	0.1103*	−0.0283*	0.1836*	0.4934*	0.4155*
19	0.3294*	−0.1682*	0.1894*	0.5084*	0.2565*
20	−0.4114	−0.4669	−0.3571	−0.7458	−0.2711
21	0.0026*	−0.1460*	0.2175*	0.5652*	0.4014*

The 21 practical technical trading strategies are each coded by a number, from 1 to 21. GP1 refers to the mean total return of the 50 runs without a short sale, whereas GP2 refers to that with a short sale. "**" refers to the rejection of the null that the total return of GP is no greater than the total return of B&H at a significance level of 0.05. "*" indicates the case where the associated trading strategy beats B&H.

Of course, what we have done here is not directly comparable to Neely et al. (1997), which used a different sample as well as a different testing scheme.[34] However, since their study is probably one of the few that has documented the profitability of GP in financial markets, we attempt to see whether we can replicate their result and, if so, whether the result can be extended to the years after their study. We therefore closely follow what Neely et al. (1997) did before.

We choose the three currencies from the six studied by them, namely, USD/DEM, USD/JPY, and USD/DBP. We also extend our data set to the year 1974 from the original 1992, and we then use the same training,

---

[34] They used 3 years of data (1975–1977) to train, the next 3 years of data (1978–1980) to validate, but tested the performance with the following 15 years of data (1981–1995).

Table 6. Performances of 21 Practical Trading Strategies: 2003–2004

Rule	USA	UK	Canada	Taiwan	Singapore	Japan
B&H	0.3199	0.1888	0.3625	0.3434	0.5311	0.3054
GP 1	0.3065	0.1797	0.3109	0.3631**	0.2512	0.0212
GP 2	0.3109	0.1817	0.3389	0.2740	−0.0183	−0.1792
1	−1.2650	−1.2230	−1.2332	−1.2960	−0.8698	−1.1913
2	−0.4914	−0.7151	−0.8140	−0.3149	−1.1061	−1.0995
3	−0.8475	−0.6523	−0.5763	−1.2569	−0.8838	−0.6142
4	−0.3036	−0.4743	0.1376	−0.2014	−0.1493	−0.3520
5	−0.9854	−0.9161	−0.9404	−0.6707	−0.9288	−0.2907
6	−0.0345	−0.2429	0.1413	−0.2529	0.0763	−0.2944
7	−1.1375	−1.5570	−1.3133	−1.8864	−1.0732	−1.7092
8	−1.5310	−1.2684	−1.4079	−1.8406	−0.7647	−1.3282
9	−1.0423	−0.5544	−1.0284	−0.2498	−0.4816	−1.2326
10	−1.1487	−1.4300	−1.0682	−1.1927	−1.1635	−1.1701
11	−0.8146	−0.5441	−0.9877	−0.3081	−0.4581	−0.6512
12	−1.6917	−1.3152	−1.6029	−1.4781	−0.8880	−1.5899
13	−0.2964	−0.5038	0.1403	−0.1907	−0.1253	−0.4830
14	−0.8910	−0.7512	−0.9259	−0.7126	−1.0833	−0.6836
15	−0.0581	−0.2264	0.0989	−0.2692	0.0976	−0.2194
16	−0.3900	−0.6218	0.1317	−0.3082	−0.4440	−0.5625
17	−0.9896	−0.8966	−0.9200	−0.6355	−0.9069	−0.4680
18	−0.1835	−0.3169	0.1129	−0.4500	0.0552	−0.5544
19	−0.3939	−0.4714	0.1534	−0.3059	−0.3021	−0.5453
20	−0.8376	−0.9450	−0.7904	−0.8147	−0.9371	−0.2721
21	−0.1286	−0.2824	0.1333	−0.3110	0.1037	−0.4935

The 21 practical technical trading strategies are coded by numbers, from 1 to 21. GP1 refers to the mean total return of the 50 runs without short sales, whereas GP2 refers to that with short sales. "**" refers to the null that the total return of GP is no greater than the total return of B&H being rejected at a significance level of 0.05. "*" indicates the case where the associated trading strategy beats B&H.

validation, and test periods as they did. However, we do not stop in the year 1995. Instead, we continue to carry out the test further, in a sliding-window manner, to the periods 1984–1998, 1997–2001, and 1990–2004. To avoid the problem of over-learning, they apply a higher transaction cost, $c = 0.2\%$, to the training and validation period, and a lower transaction cost, $c = 0.05\%$, to the test period. This setting remains unchanged; nonetheless, in order to gain more insights into the role of the parameter c, we add two more combinations, i.e., a c of 0.05% uniformly for the training, validation, and test periods, and a c of 0.2% uniformly for the same three periods. The results are shown in Table 8.

In Table 8, the first column shows the results obtained by Neely et al. (1997) in the three currencies. The second column presents our respective results. Despite slight differences, qualitatively speaking, they are largely the

Table 7. The Total Return of GP Using Short Testing Periods

FX	Mean	Stdev	Median	Max	Min	Majority
Test Period: 1999–2000						
USD/GBP	−0.0692	0.0528	−0.0629	0.0537	−0.2448	−0.0406
USD/CAD	−0.0225	0.0269	−0.0218	0.0610	−0.0727	−0.0059
USD/DEM	−0.2041	0.1390	−0.2575	0.1382	−0.3926	−0.2194
USD/ESP	−0.0583	0.1753	−0.0721	0.3540	−0.3574	−0.0016
USD/ITL	−0.1175	0.1275	−0.0795	0.1374	−0.4890	−0.1103
USD/JPY	−0.0033	0.1459	−0.0523	0.2500	−0.3448	0.0510
USD/TWD	0.0024	0.0608	0.0146	0.1277	−0.1372	0.0231
USD/SGD	−0.0864	0.0692	−0.0957	0.0493	−0.2076	−0.0531
Test Period: 2001–2002						
USD/GBP	−0.0246	0.0717	−0.0510	0.2007	−0.1376	−0.0275
USD/CAD	0.0292**	0.0633	0.0375	0.1491	−0.1477	0.0470
USD/DEM	−0.0136	0.1085	−0.0125	0.2834	−0.2026	0.0973
USD/ESP	−0.0069	0.0946	−0.0321	0.2285	−0.1576	0.0374
USD/ITL	−0.0156	0.1043	−0.0204	0.1793	−0.2585	0.1133
USD/JPY	0.0950**	0.0629	0.0866	0.2841	−0.0103	0.0647
USD/TWD	0.0731**	0.0285	0.0710	0.1788	0.0296	0.0325
USD/SGD	0.0307**	0.0278	0.0249	0.1173	−0.0692	0.0249
Test Period: 2003–2004						
USD/GBP	0.0406**	0.1006	0.0358	0.2183	−0.2151	0.0364
USD/CAD	−0.2259	0.1104	−0.3021	0.0904	−0.3021	−0.3021
USD/DEM	−0.2263	0.1503	−0.2923	0.2850	−0.3458	−0.2923
USD/ESP	−0.1728	0.1393	−0.2797	0.1634	−0.2797	−0.2797
USD/ITL	−0.2557	0.0798	−0.2786	0.0390	−0.3453	−0.2786
USD/JPY	−0.0118	0.0937	−0.0647	0.2455	−0.2053	−0.0718
USD/TWD	−0.0101	0.0440	−0.0197	0.0784	−0.1485	−0.0734
USD/SGD	−0.0539	0.0082	−0.0528	−0.0528	−0.1106	−0.0528

The result reported here is the mean of 50 runs. The "**" refers to the null that the annual return of GP that is no greater than 0 is rejected at a significance level of 0.05.

same. Indeed, with this training and testing style, GP does perform rather well in all three markets. Our extended tests also show that GP can discover profitable trading rules in 1984–1998 and 1990–2004. In sum, out of a total of 12 test periods (4 in each of the 3 currencies), GP can earn statistically significant profits in 10 periods. The significance of the transaction cost c is also reflected in the table. Remember that a higher c was intentionally chosen by Neely et al. (1997) so as to avoid overlearning. However, from the second block of Table 8, we fail to see the significance of this design: When c is decreased to 0.05% in the training set, GP performs even better

Table 8. The Mean Annual Return of GP Using Long Testing Periods

Test Period	USD/DEM	USD/JPY	USD/DBP
NWD: 1981–1995	0.0605**	0.0234**	0.0228**
$c = 0.2\%$ (training), 0.05% (testing)			
1981–1995	0.0418**	0.0327**	0.0407**
1984–1998	0.0033**	0.0512**	0.0075**
1997–2001	−0.0070	0.0200**	−0.0225
1990–2004	0.0239**	0.0103**	0.0149**
$c = 0.05\%$ (training), 0.05% (testing)			
1981–1995	0.0350**	0.0545**	0.0390**
1984–1998	0.0170**	0.0505**	0.0123**
1987–2001	−0.0005	0.0239**	−0.0312
1990–2004	0.0206**	0.0064	0.0125**
$c = 0.2\%$ (training), 0.2% (testing)			
1981–1995	0.0298**	0.0241**	0.0281**
1984–1998	−0.0030	0.0389**	−0.0002
1987–2001	−0.0132	0.0010	−0.0267
1990–2004	0.0124**	−0.0002	0.0056**

The result reported here is the mean of 50 runs. NWD refers to Neely et al. (1997). The "**" refers to the null that the annual return of GP that is no greater than 0 is rejected at a significance level of 0.05.

in 7 out of 12 markets than in the case of a higher c.[35] There is no evidence to show the relation between overlearning and direct punishment (transaction costs).[36] Nevertheless, the third block of Table 8 does indicate what we expect: When the transaction cost is uniformly leveled up from 0.05% to 0.2%, the return performance gets worse. Now, GP can only earn statistically significant positive profits in 6 of the 12 markets, down from the original 9.[37]

This replication of Neely et al. (1997) highlights the significance of the division of the data into a training set and a test set, an issue that has seldom been formally addressed in the literature. A maintained hypothesis is not to use too long a training set, since financial patterns, if they exist, may not survive too long. Consequently, based on the hypothesis, retraining is needed

---

[35] However, a statistical test shows that GP with $c = 0.05$ performs significantly better only on USD/DEM in the periods 1984–1998 and 1987–2001, and a GP with $c = 0.02$ performs better only on USD/DBP in the period 1987–2001.

[36] In addition, see Chen and Kuo (2003a).

[37] Wang (2000) questioned whether the positive result found in Neely et al. (1997) is due to the use of low transaction cost. The evidence that we have here lends partial support to this possibility. The role of the transaction cost will be further examined in Appendix B.

before one can test long enough. This hypothesis motivates the division of the data specified in Figures 2 and 3. However, our replication of Neely et al. (1997) suggests giving a second thought to this hypothesis. In fact, a short test period has its problems, too. Take Figure 3 as an example. If the pattern to learn is a cycle that will repeat itself within three years, i.e., the length of our training set, then using a two-year test period may invalidate this pattern since the cycle has not been completed yet in two years. We believe that this is an issue deserving more attention in the future (see also footnote 24).

A Different Measure: The Equity Curve

It is always questionable whether one should judge the performance of a trading strategy by a fixed duration with a fixed destination, for example, the end of a year, since sudden changes, i.e., sudden drops or sudden rises, in the financial market may make the result very sensitive to the destination that we choose. This is particularly true for B&H. Alternatively, one may like to evaluate the investment performance *dynamically* over the entire investment horizon, such as using the idea of the *equity curve* (Chen and Kuo, 2001).

Figures 14 and 15 show the equity curves of the stock markets. In each panel, there are two equity curves: one for GP, and one for B&H. For the former, since 50 trials generate 50 trading programs, and, associated with each trading program, there is an equity curve, what is presented is, therefore, the average of the corresponding fifty equity curves. The average equity curve of GP is then drawn in contrast to the equity curve of B&H for comparison reasons. The two equity curves, drawn together, help us keep good track of the superiority of the two over the entire investment horizon, rather than just a single point in time, and that may give us second thoughts regarding the conclusions made earlier. For example, even though GP outperforms B&H in the Taiwan stock market in all three test periods, GP falls behind B&H most of the time and only catches up and moves ahead at the very end of the period. Table 9 gives the percentage of time that GP outperforms B&H.

Table 9. Proportion of the Winning Time for GP

Stock Market	USA	UK	Canada	Germany	Spain	Japan	Taiwan	Singapore
1999–2000	0.710	0.179	0.191	0.237	0.118	0.020	0.193	0.034
2001–2002	0.940	0.988	0.944	0.932	0.390	0.947	0.266	0.004
2003–2004	0.135	0.249	0.203	0.292	0.044	0.212	0.434	0.183

5.2 Rule Analysis

"Hidden Knowledge"

Analyzing trading rules is always a very tempting task for GP researchers. It is particularly true when GP is applied as a tool for discovering hidden knowledge, and we are interested in knowing what we may be missing (Chen and Kuo, 2003a). Pioneers in this research area, such as Neely et al. (1996) and Allen and Karjalainen (1999), are the first group of people who are trying to disclose what these GP-discovered trading programs say to us. This work so far is still very arduous, and only limited progress has been made. There are, however, two general results. First, when the rules found by using GP are simple,[38] most likely they are familiar types of rules, such as the moving-average rules, filter rules, trading range break rules, or even B&H. What GP discovers is the possible combination of these rules, and the respective unknown parameters. Second, when the rules found are complex, the situation becomes more difficult. Three possibilities can arise: redundancy, noise, and structure. Only the last one can be considered to be rule discovery, and when that happens, what GP discovers, to some extent, is a nonlinear alteration or finer modification of the simple rules. These two findings or two types of hidden knowledge generally apply to us.

For the first type of "hidden knowledge," let us take the test period 2003–2004 for the stock markets as an illustration. B&H seems to be the most frequently discovered strategy using GP.[39] For example, it appears 38 times out of a total of 50 trials for the United States, 20 times for Canada, 7 times for the UK, 32 times for Spain, and 6 times for Germany. This result is a little boring, but not too surprising given the strong upward tendency of the training period. In the foreign exchange market, we see a similar thing.[40]

When that upward tendency disappears, B&H is no longer the most dominant trading program. In fact, it did not show up at all in the Taiwan, Japan, and Singapore markets.[41] The rules discovered are the familiar simple technical trading rules. Take Taiwan as an example. The best-performing rule out

---

[38] As to the definition and measure of complexity, see Section 5.2.

[39] The simplest case is that the entire GP parse tree has just one single terminal, namely, "True," which means to buy and to hold the stock until the end of the entire investment horizon. Of course, redundancy and noise can make a seemingly complex trading program nothing more than B&H.

[40] In fact, the frequent appearance of the B&H strategy also helps us to see that GP may fail to perform well. For example, in the last test period, the years 2003–2004, GP suffers a loss in almost all foreign exchange markets. The trading rules applied to this period are those learned from the data for the years from 1996 to 1998 and validated by the data for the years from 1999 to 2001. It can be seen from Figures 12 and 13 that, during this period, a few currencies, such as CAD, DEM, ITL, and ESP, experienced a tendency to depreciate. For currencies like these, it would not be too surprising if what comes up from GP is simply the B&H strategy. Unfortunately, when coming to the years 2003–2004, we see a major change in the trend, with almost all currencies experiencing a tendency to appreciate, which inevitably caused GP to perform quite badly.

[41] The prevalence of the B&H in GP search was argued as an evidence of market efficiency in Wang (2000). This assertion is, however, not exact. As we demonstrate here, whether GP would "rediscover" B&H depends on whether both the training and the validation set have a strong upward tendency. So, the prevalence

of the 50 trials in the test period 2003–2004, in the LISP format, is

$$(or \ (< \ min(189) \ min(148)) \ (> index \ lag(201) \ )), \quad (12)$$

whose parse-tree representation is shown in Figure 4, which says that "if the minimum price of the past 189 days is less than the minimum price of the past 148 days or the current price is larger than the the price lagged by 201 days, then take a long position; otherwise stay out of the market."[42] Rule (12) is a Boolean combination of a trading range break-out rule and a backward-looking rule.[43] These rules alone are not novel. It is the unknown parameters of these rules and the unknown possible combination of them that define the "hidden knowledge."

In addition to the recombination of primitive trading rules, the second type of "hidden knowledge" that the GP program can discover for us consists

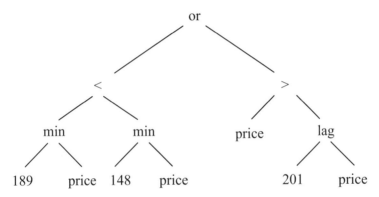

Fig. 4. GP-discovered trading program: Taiwan stock market.

This trading program is found by GP using the data of the Taiwan stock index for the years 1993–1997 and is validated by the data for the years 1998–2002. Its performance is tested using the data for the years 2003–2004.

---

of discovering B&H should not be used as strong evidence to support the efficient market hypothesis.

[42] Due to data preprocessing as detailed in Section 4, the price is referred to as the *normalized* price, and not the original price.

[43] The trading range break-out rule, the first part of (12),

$$(< \ min(189) \ min(148)),$$

is not redundant, since it is not necessarily true as might be seen at first sight. In fact, in the test period, 2003–2004, only 126 times out of 500 trading days does this inequality hold; for the rest of the time, the two are the same. If the price experiences a long upward trend, then this inequality should always be satisfied. So, this signal indicates some degree of fluctuation for the normalized price, and possibly for the price as well. This rule, therefore, tries to identify some changes in the trend.

of some kinds of nonlinear alterations to the primitive trading rules Allen and Karjalainen (1999, pages 263–264) has documented such an example. In their example, the rule found is something similar to a moving-average rule, but the window size is *time-variant*. This alteration provides us with the flexibility to search for both short-term trends and long-term trends, depending on whether the market is bullish or bearish. We also find trading programs of this kind. Figure 5 is a trading program found to be comparable to B&H in the 2003–2004 test period for the Taiwan stock market. A careful analysis would show that this rule is essentially an alteration of the fixed backward-looking rule:

$$(> index \ lag(205)). \tag{13}$$

What the program in Figure 5 does is to add a condition to this simple rule. So, in normal times, the trading program just checks whether the inequality of the backward-looking rule is satisfied when deciding to be in or out of the market. Nonetheless, when the price experiences an upward trend or a downward trend, this trading program signals caution regarding using this simple rule alone and suggests looking at other finer conditions as defined by its subtrees. The caution seems to be necessary because it reminds the trader that the ongoing trend may change at any point in time when the upward or downward tendency has gone too wild. As a result, what GP finds here is a simple backward-looking rule associated with a reversal protection mechanism, and its operation becomes much more complex than that of the original version.

Figure 6 plots the equity curves of the three trading rules, namely B&H, rule (13), and the GP-discovered trading program as shown in Figure 5. The figure does show that the equity curve of the GP-discovered program is 64.8% of the time higher than that of B&H. However, with this test data set, the modification of the original simple rule (13) does not help much. In fact, the equity curve of the simple backward-looking rule dominates that of the modification made by GP 58.8% of the time. What is particularly interesting is that both of these rules correctly recommend that the trader stay out of the market on the same day, just right at the beginning of a big drop happening around the end of April 2004. After that they never signal to buy again, and neither does the price go back to the same high again.[44]

Rule Complexity

The complexity of trading rules seems to be another issue concerning many researchers. The parse-tree structure of GP gives us a very intuitive way of measuring the complexity of a trading program, namely, the number of nodes

---

[44] We do notice that some GP-discovered trading programs share this feature, namely, *less greedy and more rational*. Whether this property can generally hold to a larger extent of the program requires more systematic study.

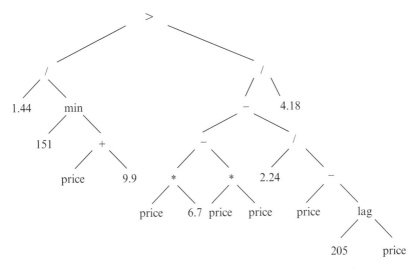

Fig. 5. GP-discovered trading program: Taiwan stock market.

This trading program is found by GP and using the data of Taiwan stock index from years 1993–1997, and is validated by the data from years 1998–2002. Its performance is tested with the data from years 2003–2004.

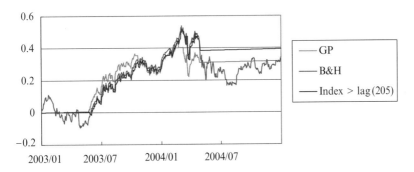

Fig. 6. Equity curves of the three trading programs.

appearing in a parse tree, which is also known as the *node complexity*. However, as Allen and Karjalainen (1999) correctly point out: "measuring the complexity of the trading rules cannot in general be done by inspection of the tree structures," (Ibid. p. 262). This is mainly due to the fact that redundancy and noise, instead of real patterns, may inflate the size of the trading programs. However, supposing that the chance of introducing redundancy and noise is equally likely for each market, then the overall complexity measure

126   S.-H. Chen et al.

can still be used as a way of examining whether rule complexity is different among markets or is different over time. It is for this purpose that we present the mean complexity of the trading program for all markets in Table 10.

Table 10. The Node Complexity of GP-Discovered Programs

Stock Market	USA	UK	Canada	Germany	Spain	Japan	Taiwan	Singap.	Mean
Without Short Sale									
1999–2000	14.1	23.2	17.3	25.5	20.2	23.4	28.8	16.2	21.1
2001–2002	10.3	14.1	10.8	23.9	19.2	19.9	26.9	18.7	18.0
2003–2004	12.6	21.8	19.2	23.8	16.6	24.4	22.3	24.5	20.6
Mean	12.3	19.7	15.8	24.4	18.6	22.5	26.0	19.8	19.9
With Short Sale									
1999–2000	7.7	19.8	19.0	17.2	21.8	19.2	26.5	15.9	18.4
2001–2002	10.2	12.9	15.5	18.7	20.4	21.7	28.4	15.8	18.0
2003–2004	9.6	12.5	10.0	25.8	15.0	22.9	24.0	23.7	18.0
Mean	9.2	15.1	14.8	20.6	19.1	21.3	26.3	18.5	18.1
FX Market	ITL	GBP	CAD	DEM	ESP	JPY	TWD	SGD	Mean
1999–2000	24.4	23.6	23.2	24.2	28.6	22.3	24.1	17.3	23.5
2001–2002	21.0	29.4	24.0	26.0	21.9	30.7	16.8	17.5	23.4
2003–2004	22.8	22.2	29.9	26.8	24.0	21.6	20.5	13.6	22.7
Mean	22.8	25.1	25.7	25.6	24.8	24.9	20.5	16.1	23.2

The result here is the mean of 50 runs of GP.

Table 10 is the mean complexity of the GP-discovered trading programs in the eight stock markets (both with and without short sales) and in the eight foreign exchange markets. The program complexity of the foreign exchange market (with a mean of 23.2) is greater than that of the stock market (with a mean of 19.9). Within the stock markets (the case without short sales), program complexity changes from a low of 12.3 (United States) to a high of 26.0 (Taiwan) and a low of 18.0 in the years 2001–2002 to a high of 21.1 in the years 1999–2000, while within the foreign exchange markets, it changes from a low of 16.1 (Singapore) to a high of 25.7 (CAD) and a low of 22.7 in the years 2003–2004 to 23.5 in the years 1999–2000. Whether or not we can relate market complexity to program complexity is an issue beyond the scope of this paper, but the statistics shown here do provoke some thoughts on the complexity of different markets at different times.

In addition to the mean, by pooling all three scenarios together, we also present the empirical distribution of the complexity in Figures 18 and 19. By pooling the three sections of the tests together, each empirical histogram gives the distribution of the number of nodes of the 150 GP-discovered trading programs. These two figures show some common patterns across all markets.

First, most GP-discovered programs are not complex. In particular, programs with a node complexity of less than 10 dominate all markets. In the U.S. market, they account for even more than 50% of all 150 programs. Second, the more complex the program, the less likely it will be discovered. This is well reflected by declining frequencies in the histogram. This may surprise those who expect to learn complex trading programs from GP.

A systematic study of the complexity is not available in the literature. Usually, authors of earlier studies only provided some partial observations. For example, in the stock market, Allen and Karjalainen (1999) noted "the structure and complexity of the rules found by the algorithm vary across different trials. The size of the rules varies from nine to 94 nodes, with a depth between five and ten levels," (Ibid. p. 261). Also, in the foreign exchange market, Neely et al. (1997) found that, out of the 100 programs, the mean number of nodes for the USD/DEM was 45.58, and only two had fewer than 10 nodes (Ibid. p. 419). Bhattacharyya et al. (2002) are probably the only ones who documented the statistics of node complexity. They found that the node complexity of the USD/DEM hourly data ranges from 18.65 to 29.65, depending on the specific design of GP used.

Another way to reflect upon the complexity of a GP-discovered trading program is to examine whether its complexity has actually contributed to the profit performance. This indirectly helps us to estimate how seriously that noise and redundancy have complicated a trading program. Table 11 gives the correlation between the complexity and the total return (R) in the stock market as well as the correlation between complexity and the total rate of return (Π) in the foreign exchange market.

The results are somewhat mixed. There are a total of 26 cases showing positive correlation, but also 22 cases showing negative correlation. Pooling all the cases together, in the stock market, the correlation coefficient is only about 0.09, and in the foreign exchange market, it is almost nil. There is thus no clear evidence indicating that complexity can contribute to the profit performance. This may be partially due to the dominance of simple strategies, such as B&H or simple technical trading rules. One typical example would be what we see in Figure 6, where the complex version of the trading program (13) appears to be no better than the original simple version.

Despite the general pattern described above, it is also worth noting some special cases. The Taiwan stock market provides such an interesting case. We notice two concurrent patterns in the Taiwan stock market. First, the node complexity is generally higher than in other markets. Second, the node complexity has a positive relation with the return in all three test periods. These two patterns associated with the performance of GP-discovered trading programs in this market, as shown in Section 5.1, provide us with one of a few ideal cases indicating that GP is working.

The rather simple structure of the evolved trading programs motivates us to ask a question regarding the *search intensity* required to find these programs, in particular, the *number of generations*. It is suspected that a simple

Table 11. The Correlation Between Program Complexity and Return

Stock Market	USA	UK	Canada	Germany	Spain	Japan	Taiwan	Singapore
1999–2000	0.18	0.08	0.25	−0.01	0.34	−0.20	0.23	0.08
2001–2002	0.08	0.36	−0.05	0.07	0.25	−0.28	0.10	−0.03
2003–2004	−0.14	−0.07	−0.38	−0.01	0.03	0.35	0.19	−0.13
Pooling = 0.09								
Exchange Market	ITL	GBP	CAD	DEM	ESP	JPY	TWD	SGD
1999–2000	0.22	−0.27	−0.22	0.00	−0.11	0.32	0.01	−0.07
2001–2002	0.18	−0.40	−0.26	0.15	−0.06	0.07	−0.10	0.19
2003–2004	0.13	−0.13	0.52	0.15	0.17	−0.11	−0.32	−0.29
Pooling = 0.00								

structure may not need a long evolution and can emerge quite quickly, so that the evolutionary cycle may terminate much earlier than the upper limit $Gen = 100$ (Table 1). We test this conjecture by first running the correlation between the number of generations effectively used and the resultant node complexity. Figure 7 gives the X-Y plot of the two. We find that this coefficient is 0.319 in the stock market and 0.575 in the foreign exchange market. This result supports our hypothesis that complex programs need a longer time to evolve than simple programs. Nonetheless, from their profitability performance (Table 11), these complex programs seem to introduce more redundancy than structure, and hence a longer evolution time does not really help.

Figure 8 gives the histogram of the actual numbers of generations used to evolve before the evolutionary cycle terminates. As we have specified in Table 1, if there have been 50 consecutive generations where we have been unable to find a better program to replace the incumbent in the winner list, the evolutionary cycle will terminate. The histogram is made by pooling $50 \times 8 \times 3 = 1,200$ trials. Of these 1,200 trials, it is interesting to notice that a very high proportion of evolution does not take too much time. In the stock market, 38% of the trials do not actually experience a real search. The best program found in the initial generation proves to be unbeatable in the validation; hence, the evolutionary cycle stops right away at 50 generations. Twenty-two percent of the trials experienced less than 10 generations of search, and terminated before 60 generations. With so little time to use, it is well expected that most programs coming out of this evolutionary cycle will not be too complex. GP in the foreign exchange markets did search more intensively than the stock markets, but still 40% of the trials terminated before 60 generations. In both the stock and foreign exchange markets, there are only about 10% to 20% of trials that actually evolved about 90 to 100 generations.

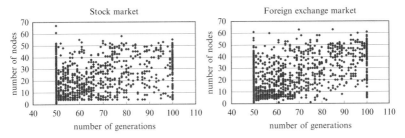

Fig. 7. Evolution time and program complexity.

The histogram is made by pooling all trials of different markets and different periods together. The correlation between evolution time and program complexity is 0.319 in the stock market and 0.575 in the foreign exchange market.

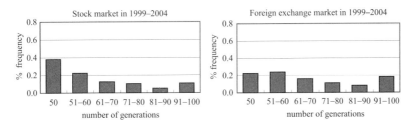

Fig. 8. Number of evolved generations.

The histogram is made by pooling all trials of different markets and different periods together.

So, generally speaking, the search intensity is lower than what we expected at first sight. Given this result, the question as to whether we have actually searched *sufficiently* arises. However, the main problem lies in the fact that it does not pay well to conduct a more intensive search. Table 12 and Figure 20 show that longer search does not bring a higher return.

Trading Frequency

Despite what the rule says or how complex it is, as long as its criterion for market timing is always satisfied, it makes no essential difference to B&H.[45]

---

[45] In this case, the trading program always outputs "True." Similarly, if the criterion is never satisfied, which is equivalent to always "False," then the investor will never invest in the stock market, and the trading frequency can even be as low as zero.

Table 12. The Correlation Between Evolving Time and Returns

Stock Market	USA	UK	Canada	Germany	Spain	Japan	Taiwan	Sing.
1999–2000	−0.02	−0.20	0.32	−0.02	−0.09	−0.30	0.38	0.07
2001–2002	0.11	0.16	0.16	−0.16	0.36	−0.11	0.03	−0.07
2003–3004	−0.29	−0.26	−0.40	−0.14	−0.29	0.36	−0.13	0.01
Pooling = 0.069								
Exchange Market	ITL	GBP	CAD	DEM	ESP	JPY	TWD	SGD
1999–2000	0.34	−0.12	−0.33	−0.07	−0.01	0.26	−0.29	−0.25
2001–2002	0.41	−0.49	−0.27	0.34	0.03	0.11	0.01	0.12
2003–3004	0.08	−0.17	0.65	0.14	0.11	0.13	−0.39	−0.44
Pooling = 0.166								

Therefore, an alternative way to analyze the GP trading rules is to examine how often they send buy or sell signals, or how sensitive they are to the price dynamics. Table 13 gives the average of the number of trades for each market, where the average is taken over the 50 trials of running GP.

Generally speaking, the trading frequency is quite low for all markets. On a two-year basis, the trading frequency ranges from a low of one to a high of 9.24. This result is comparable to Allen and Karjalainen (1999)[46] but is in a striking contrast to that of the 21 human-written trading programs. In Table 13 the row starting with "21 rules" gives the average of the trading frequency of the 21 above-mentioned technical trading rules. Using these business trading programs, one expects to finish about 25 round trips every two years.

To understand the difference, it is important to be aware that what is captured by these 21 programs comprises only patterns, but not the transaction cost. As a result, their development and use seem to be independent of the transaction cost. However, this is not the case with GP. Since the transaction cost has been taken into account in the fitness function, different transaction costs will generally result in different trading programs with different trading behavior, and this is confirmed by Table 13 when we reduce the transaction cost from 0.5% to zero. Needless to say, when c is reduced to zero, the GP-discovered trading programs in all markets become more active. This result is also consistent with Allen and Karjalainen (1999).[47] In Canada and Singapore, the GP-discovered trading programs are even more active than the 21 rules. Obviously, changing the transaction cost does drive GP to look for different trading programs.[48]

---

[46] Allen and Karjalainen (1999) found that when $c = 0.5\%$, the trading frequency drops to an average of 1.4 trades per year, which is 2.8 trades biennially (Ibid. p. 260).

[47] Allen and Karjalainen (1999) found that when the transaction cost is as low as 0.1%, the trading frequency is high, with an average of 18 trades per year or 36 trades biennially (Ibid. p. 260).

[48] Allen and Karjalainen (1999) are the first who tried to examine how the transaction cost may actually induce different trading programs.

Table 13. Trading Frequency: Stock Markets

Rule	USA	UK	Canada	Taiwan	Singapore	Japan
Test Period: 1999–2000						
GP($c = 0.5\%$)	1.94	3.56	4.12	2.72	9.24	1.26
GP($c = 0$)	13.40	25.44	84.64	10.98	77.54	3.76
21 rules($c = 0.5\%$)	25.86	27.7	27.75	24.64	25.43	NA
Test Period: 2001–2002						
GP($c = 0.5\%$)	1.08	1.00	2.00	3.40	4.00	2.92
GP($c = 0$)	3.10	4.54	97.74	14.26	75.28	9.38
21 rules($c = 0.5\%$)	24.90	27.67	25.81	24.05	23.74	NA
Test Period: 2003–2004						
GP($c = 0.5\%$)	1.28	1.50	2.04	3.34	3.56	3.90
GP($c = 0$)	13.18	19.16	85.64	9.40	54.68	9.48
21 rules($c = 0.5\%$)	26.21	27.86	25.81	24.07	23.36	25.76

"GP" refers to the average trading frequency of the 50 runs without short sales, and the "21 rules" refers to the average trading frequency of 21 practical trading strategies.

Table 14 gives the trading frequencies of the GP-discovered program in the foreign exchange market. Neely et al. (1997) earlier found that the mean number of trades for USD/DEM is 14 biennially, which is higher than our finding here (see Table 14). This may be partially due to their different design of GP. The impact of the transaction cost on the number of traders is also shown in Table 14. By increasing the transaction cost from 0.05% to 0.2%, the GP-discovered trading programs, as expected, become much less active in many markets.

Consistency of Trading Rules

GP is a stochastic search algorithm. Different trials will not necessary give the same recommendation; generally, they do not. It is therefore interesting to know how consistent these different trading rules can be. Figure 16 presents the proportion of rules indicating a long position for each stock index during the three test periods. A *consistency statistic*, $H(p)$, is used to give a summary of this picture. The statistic is defined as

$$H(p) = p(1-p), \qquad (14)$$

where p is the proportion of rules indication a long position. It is clear that $H(p)$ is minimized at 0 when p is close to the two extremes, namely, 0 and 1, and it is maximized at 0.25 when $p = 0.5$. Hence, $0 \leq H(p) \leq 0.25$. On the top of each plot in Figure 16, we also report the average of $H(p)$ over the entire three test periods followed by the average of each test period. The consistency index ranges from the lowest 0.02 (United States) to the highest 0.14 (Canada) ≈ 0.83 × 0.17, which indicates that the consistency of all GP programs is high. Even for the individual test period with the highest H, i.e., the UK 1999–2000, where H is 0.21 (= 0.7 × 0.3), we can still have 70% of the programs seeing eye to eye with each other. Similarly, Figure 17 gives the consistency statistic of the foreign exchange market. The consistency index is a little higher, ranging from 0.08 to 0.18, with the highest one being 0.21 (USD/CAD 2001–2002, USD/ESP 1999–2000, USD/TWD 1999–2000).[49]

Table 14. Trading Frequency: Foreign Exchange Markets

	ITL	GBP	CAD	DEM	ESP	JPY	TWD	SGD
Test Period: 1999–2000								
$c = 0.05\%$	7.42	10.92	8.10	8.06	9.30	11.40	5.42	32.66
$c = 0.2\%$	2.58	2.72	2.88	4.82	6.26	10.00	2.68	8.38
Test Period: 2001–2002								
$c = 0.05\%$	9.48	9.40	13.08	8.12	10.18	14.85	28.12	1.67
$c = 0.2\%$	6.60	2.94	2.08	6.04	8.02	9.46	3.12	1.00
Test Period: 2003–3004								
$c = 0.05\%$	2.28	13.16	4.72	4.02	6.24	11.94	12.46	1.42
$c = 0.2\%$	2.78	6.36	1.26	2.28	2.64	5.04	6.62	1.00

6 Concluding Remarks

The question pertaining to whether GP can discover profitable trading strategies is much harder to answer based on the limited publications we have so far. The current extensive test does, however, make a few points clearer.

6.1 Issues Learned from the Extensive Test

Data-Division Schemes

First, the performance of GP crucially depends on the experimental design. In Section 5.1, we have replicated and extended part of the empirical work done by Neely et al. (1997), and from there we are able to show the significance of the division of data into training, validation, and testing. If we do not start with a "right" division, then we may end up with a less desirable result, e.g., comparing Table 7 with Table 8. A blind division may cause great dissimilarity among the three sets, particularly, the training set and the testing set, and that may cause the GP to malfunction. The general failure of the foreign

---

[49] Neely et al. (1997) have evidence that USD/DEM has a higher degree of census than USD/Yen. This result is also obtained here.

exchange investment in the period 2003–2004 serves well to demonstrate this problem (also see footnote 40).

Evaluation Scheme

Second, in addition to the data-division scheme, the evaluation scheme can be a problem as well. Specifically, the use of B&H as a benchmark and the use of the fixed-period return, such as Equation (2) or (3), as the fitness function can cause some unintended biases, when the market has a strong upward or downward tendency, as many examples have shown in this paper. As we have see earlier in footnote 44, the GP-discovered programs quite often are less greedy and are well altered. It will not chase the upward trend (a bullish market) and will leave the market earlier before the reversal happens. With such an operation, an evaluation made before the end of the bullish market will be very unfavorable for GP as opposed to B&H. On the other hand, since the GP-discovered programs tend to stop loss, it will leave the bearish market before it is too late. In this case, an evaluation made before the end of the bullish market will be more advantageous to GP than to B&H. Briefly, bulls tend to be biased toward B&H, whereas bears tend to favor GP.

Searching Efficiency

Third, from the analysis of GP-discovered programs, we evidence that GP is able to discover the "hidden knowledge" by either combining or refining the existing simple trading rules (Figure 5). It is observed that the GP-discovered trading programs perform no worse than many human-written business programs. However, using quite intensive computing resources, GP does not seem to search very efficiently. Consider the following three hypotheses held by some financial users of GP:

- The hidden financial or hidden knowledge, if it exists, must be complex.
- Once the complex financial pattern is discovered, we may expect excess returns.
- Since it takes GP many generations to evolve (grow) or discover complex patterns, the evolution time, complexity, and return are positively correlated with each other.

In Section 5.2, we have found that two of these three hypotheses are largely false. The general findings are summarized as follows:

- Evolution time can help grow complex trading programs (Figure 7).
- However, complex trading programs need not have structured patterns as their essence. They may introduce redundant or noisy elements, which contribute little to making successful trading decisions. As a result, the correlation between complexity and return is weak, if it even exists (Table 11).

- After the removal of redundancy and noise, some complex trading programs can be simplified as just a recombination of simple rules. Real complex, well-structured programs are rarely seen (Figures 18 and 19). Therefore, either complex patterns are not prevalent, or the simple GP (the current version) is not powerful enough to capture them.
- Putting them together, we find only a weak correlation between complexity and return, and hence a weak correlation between evolving time and return (Figure 17), despite the positive correlation between complexity and evolving time.

6.2 Possible Solutions and Directions for Further Studies

In light of the conclusion reached above, we would also like to revise the issues list above by pointing out directions for further studies. First of all, one has to notice that the first two issues mentioned are not unique to GP. They largely exist for all computational intelligence techniques. As a result, their solutions can also be found from or shared with other computational intelligence techniques. The last one also has its generality, but GP does have unique causes that other techniques do not share. Solutions to the third issues are, therefore, tailor-made and exploit more advanced GP.

Reacting to the Issue of Data-Division Schemes

Regarding the issue of the data-division scheme, it has been proposed to alternate training sets and the test sets, instead of putting them in chronicle order, e.g., Bhattacharyya et al. (2002). Nonetheless, alternating training sets and test sets assumes that one can use the distant future to forecast the immediate future. From the forecasting viewpoint, this is a not practical solution to the problem.

Instead, one does not have to work with an exogeneously given division scheme, and *active learning* is an idea to leave the choice of the training sets to be endogenously determined (Lanquillon, 1999). Active learning is a design to relearn from the past if the current (on-line) performance is worse to a degree defined by a threshold. For example, if the equity curve of GP has been continuously below an ideal equity curve for a consecutive number of periods, then a relearning mechanism will be automatically triggered. This idea is not new in computational intelligence literature, and in econometrics it is well-known as the CUSUM test. In the literature of GP, Chen (1998) was probably the first to apply this idea. Nevertheless, how to design a relearning scheme may not be a simple task. Hopefully, the pile of the literature on active learning or incremental learning may provide some help.

Reacting to the Issue of Evaluation Schemes

Regarding the issue of the possibly biased evaluation scheme, the equity curse frequently used in investment literature would help. It provides a better vision of how the trading program behaves over the investment horizon and will avoid the above-mentioned evaluation bias. Furthermore, the equity curve itself can be used as the fitness function. One may as well also consider various kinds of fitness functions addressed in the literature (see footnote 12).

One common cause of the first and the second issue, or, generally, the issue of luck, is mainly due to the use of only one ensemble (single time series), and one ensemble may not represent the population well. A solution to this problem is to conduct a rigorous statistical test using the booth-trapping method or the Monte Carlo simulation (Tsao and Chen, 2004), which can help us to give a more rigorous evaluation to see why GP works or fails to work. One can generate artificial data by using relevant stochastic models, particularly, those financial econometric models. It would therefore, be nice to test whether or not GP can survive all types of stationary time series. If indeed GP can survive well with the stationary time series, then we can add disturbances with structural changes and test how quickly GP can recognize the changes and start to relearn under a given active learning scheme. This may further help us to evaluate the performance of different active learning schemes and enable us to see whether active learning is, after all, a mean to avoid the path-dependent problem.[50]

Reacting to Enhancing Search Efficiency

The last issue is most challenging. Generally, there are two directions to enhance search efficiency. First, enhancing *representations*. The idea of using strongly typed GP to solve some redundancy issues has been attempted by several researchers. Bhattacharyya et al. (2002) added *semantic restrictions*, O'Neill et al. (2002) adopted *gene expression programming*, and Yu et al. (2004) used the λ-abstraction approach, to restrict the types of trading programs to be generated.[51] Second, enriching *building blocks*. Wang (2000) used *automatic define functions* can be a case in point. As we have mentioned in Section 2, it would be interesting if we could redesign our terminal set and function set by including some advanced functions and terminals, instead of starting everything from scratch. Of course, this would need to work with securities and investment firms by incorporating their knowledge base into GP.

---

[50] One problem revealed in this paper is the *path dependence* of GP. What GP will perform in the future depends on what it learns from the past. It is generally well argued in this paper that the switch between the bull and bear market may cause GP to learn something unsuitable for the future.

[51] Of course, this is not an exhaustive list. There is newer research involving novel GP-based approaches in financial decision making (e.g., newer strongly typed GP approaches, grammar-guided GP, evolutionary neural logic networks, evolving-fuzzy systems, or other hybrid intelligent approaches, etc.). See, for example, Tsakonas et al. (2006).

Appendices

A. Significance of Validation

In Section 4, we have mentioned that using part of the data to validate what was learned from the training set, and based on the validation result decide a termination point, has become a popular device to avoid overfitting. However, the effectiveness of this device has not been well addressed in the literature. Does it really work? Chen and Kuo (2003a) cast doubt on the foundation of this device. Motivated by earlier studies, we also examine the effectiveness of the validation device.

Here, let us be more specific regarding what we mean by training without validation. Look back at Figure 2, where, according to the original procedure, the data are divided into three sections, the training set and the testing set, and in the middle the validation set. Now, when the procedure does not have validation, we simply replace the training set by the validation set, i.e., to move the training set immediately before the testing set. So, going back to Figure 2, the three training sets, 1989–1993, 1991–1995, and 1993–1997, are now completely replaced by the original three validation sets, 1994–1998, 1996–2000, and 1998–2002. This way of carrying out the procedure without validation is mainly due to the recent nature of the financial time-series data. When there is no need for validation, then the immediate past should be used for training.

Without validation, the evolutionary cycle will not terminate until it runs to 100 generations. The result is shown in Table 15. The statistical test seems to be in favor of the procedure with validation. In 11 out of a total of 24 cases, the return of the procedure with validation outperforms that without validation, while only in four cases do we have the opposite situation.

B. Significance of the Transaction Cost

As mentioned in Section 3, the transaction cost can have some impacts on the results we have. In the main text, we already presented the results associated with a c of 0.5%. To make us see how significantly the results will be altered, we choose a rather low value, $c = 0$. Table 16 presents the profitability result. By comparing this table with Table 2, we can see that the total return earned by GP improves over all the cases (in terms of the majority rule), and the number of times GP beats B&H increases from the original 7 cases to the current 13 cases.

The change to the transaction cost does not impact the profit performance directly. Its impact is channeled through the new resultant GP-discovered trading programs associated with different essence and different behavior. Table 13 has already shown that the GP-discovered trading programs can be more active in trading when the transaction cost is reduced. However, a more fundamental change starts from the trading program itself. To see

Table 15. Effectiveness of Validation

	Validation		No Validation	
Country	Mean	Stdev	Mean	Stdev
Testing Period: 1999–2000				
USA	0.0655	0.0342	0.0689	0.0599
UK	0.0459	0.0908	0.0401	0.0406
Canada	0.3660	0.1030	0.3278	0.1570
Germany	0.1490	0.1114	0.2198**	0.0822
Spain	−0.0666	0.0904	−0.0454	0.0353
Japan	0.0024	0.0540	−0.0132	0.1871
Taiwan	0.1620**	0.1353	0.0715	0.1806
Singapore	0.1461**	0.1866	0.0419	0.0897
Testing Period: 2001–2002				
USA	−0.3171	0.0498	−0.1616**	0.2209
UK	−0.3625	0.0284	−0.2725**	0.2311
Canada	−0.1761	0.1065	−0.1917	0.1296
Germany	−0.4772	0.2242	−0.5319	0.1123
Spain	−0.2780	0.0910	−0.2578	0.1285
Japan	−0.0722**	0.1520	−0.1261	0.1531
Taiwan	0.0376**	0.1899	−0.1068	0.1598
Singapore	−0.3123	0.0333	−0.0685**	0.1161
Testing Period: 2003–2004				
USA	0.3065**	0.0334	0.0722	0.0772
UK	0.1797**	0.0300	0.0777	0.0658
Canada	0.3109**	0.0585	0.1421	0.0786
Germany	0.3318**	0.0544	0.1590	0.1377
Spain	0.3355**	0.1292	0.0819	0.0803
Japan	0.0212	0.0843	0.0300	0.0847
Taiwan	0.3631**	0.0665	0.0598	0.1329
Singapore	0.2512**	0.0735	0.1409	0.1294

The "**", depending on where it appears, refers to the rejection of the null, at a significance level of 0.05, that the total return of the procedures with validation is no greater than that without it or the null that the total return of the procedure without validation is no greater than that with validation.

this, Table 17 presents the node complexity of the GP-discovered programs in the stock market when c is reduced to zero. By comparing this table with Table 10, we can see that the trading programs not only become more active, but, underlying their activeness, they also become more complex. The node complexity increases from the original 19.9 (the case without sale) to the current 24.4, and for each single market, the node complexity also increases pairwisely.

Table 16. The Total Return R of GP Without Short Sales: $c = 0$

Country	Mean	Stdev	Median	Max	Min	Majority	B&H
Test Period: 1999–2000							
USA	0.1703**	0.0886	0.1998	0.3678	−0.1464	0.2788	0.0751
UK	0.0477	0.1031	0.0564	0.2732	−0.1989	−0.1171	0.0584
Canada	0.5830**	0.2169	0.6118	0.9115	0.0295	0.5964	0.3630
Germany	0.1967	0.1249	0.1767	0.5684	−0.1475	0.2515	0.2249
Spain	0.0156**	0.1166	0.0019	0.2620	−0.1609	0.1428	−0.0416
Japan	0.0066	0.0562	0.0016	0.1523	−0.1372	0.0016	0.0276
Taiwan	0.2825**	0.1321	0.3119	0.5808	−0.1341	0.5158	−0.2289
Singapore	0.5959**	0.1319	0.6938	0.8327	0.0719	0.5050	0.3762
Test Period: 2001–2002							
USA	−0.3099	0.0232	−0.3144	−0.2238	−0.3146	−0.3144	−0.3144
UK	−0.3408**	0.0881	−0.3618	−0.0613	−0.3658	−0.3618	−0.3618
Canada	−0.1311**	0.0404	−0.1389	0.0357	−0.1821	−0.1389	−0.2319
Germany	−0.4486**	0.2401	−0.5058	0.0555	−0.5685	−0.5065	−0.5401
Spain	−0.2458**	0.1018	−0.2784	0.1087	−0.2998	−0.2784	−0.2784
Japan	−0.0386**	0.2053	−0.0133	0.3099	−0.3801	0.0089	−0.3734
Taiwan	0.1588**	0.2022	0.1509	0.8019	−0.1607	0.2446	−0.1001
Singapore	−0.1592**	0.1125	−0.1926	0.1559	−0.2928	−0.2426	−0.2928
Test Period: 2003–2004							
USA	0.3247	0.0638	0.3351	0.4546	0.0681	0.3794	0.3332
UK	0.1607	0.0714	0.1444	0.3646	0.0146	0.1714	0.2007
Canada	0.3862**	0.0351	0.4118	0.4668	0.2549	0.4118	0.3762
Germany	0.3677	0.0666	0.3707	0.5699	−0.0011	0.3707	0.3707
Spain	0.3278	0.0779	0.3372	0.4546	0.0215	0.3838	0.4599
Japan	0.1159	0.1043	0.0949	0.3926	−0.0356	0.1047	0.3185
Taiwan	0.3662	0.1006	0.3644	0.6195	0.0657	0.5122	0.3569
Singapore	0.4398	0.0888	0.4406	0.7534	0.2327	0.3812	0.5465

The "**" refers to the null that the total return of GP is no greater than the total return of B&H is rejected at a significance level of 0.05.

Earlier, in Section 5.2, we have seen that it takes time to develop complex programs. Here, we find that this is still the case. The simple correlation coefficient, by pooling together all trials, between evolution time and program complexity is 0.52 (0.32 when $c = 0.5\%$). As shown in Figure 9, we can see that the peak of the stopping time (the terminal generation) moves from the original 50 to the interval "91–100." Now, more than 20% of the trials run to or almost to the last generation. This result highlights a possibility that the GP design may not be completely independent of the market to which it is applied. The market with low transaction costs may deserve more running time or a higher search intensity than markets with high transaction costs.

Table 17. The Node Complexity of GP-Discovered Programs: Stock Markets, $c = 0$

Test period	USA	UK	Canada	Germany	Spain	Japan	Taiwan	Singap.	Mean
1999–2000	19.9	27.9	22.4	22.0	19.7	22.3	27.1	21.4	22.9
2001–2002	23.2	24.2	17.9	26.0	26.6	33.0	28.0	19.8	24.9
2003–2004	24.3	20.6	25.3	29.6	21.5	28.8	24.8	28.8	25.5
Mean	22.5	24.3	21.9	25.9	22.6	28.1	26.7	23.4	24.4

The result here is the mean of 50 runs of GP.

Furthermore, we are interested in knowing whether the increases in evolution time and program complexity can contribute to the return. In other words, we would like to know whether or not the increase in search intensity, encouraged by a lower transaction cost, has actually led to the discovery of some hidden patterns that will be neglected when the transaction cost is high. However, we do not see much change in the two properties established in Section 5.2. Both the program complexity and evolution time have little to do with the return. The correlation coefficient, by pooling all trials together, between program complexity and the return is only 0.02 (0.09, when $c = 0.5\%$), and the correlation coefficient between evolution time and the return is 0.11 as shown in Table 19 (0.07, when $c = 0.5\%$).

However, the feature that program complexity has little to do with the return remains unchanged even though the transaction cost is removed (Table 18).

Table 18. The Correlation Between Program Complexity and Returns: Stock Markets, $c = 0$

Test Period	USA	UK	Canada	Germany	Spain	Japan	Taiwan	Singap.
1999–2000	0.17	−0.11	0.06	0.04	0.12	0.24	0.04	0.12
2001–2002	0.21	0.34	−0.24	−0.13	0.01	−0.05	−0.09	−0.19
2003–2004	−0.22	0.25	0.24	0.10	−0.08	0.62	−0.11	−0.05
Pooling = 0.02								

Table 19. The Correlation Between Evolution Time and Return: Stock Markets, $c = 0$

Test Period	USA	UK	Canada	Germany	Spain	Japan	Taiwan	Singap.
1999–2000	0.20	−0.39	0.28	−0.05	0.42	−0.05	0.20	−0.08
2001–2002	0.30	0.34	0.05	−0.16	0.25	−0.27	0.00	−0.27
2003–3004	−0.20	0.13	0.35	−0.08	0.07	0.49	0.07	0.26
Pooling = 0.11								

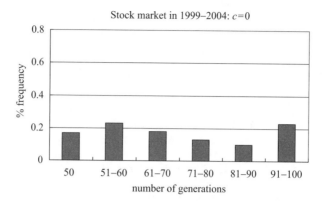

Fig. 9. Number of evolved generations.
The histogram is made by pooling all trials of different markets and different periods together.

C. Significance of Data Transformation

The hypothesis that the original data without transformation to take care of the non smoothness may be inappropriate is widely accepted among researchers. Most researchers do consider different ways of smoothing the data before it is applied to any computational intelligence tool. This is known as data preprocessing. However, absent is a real test to show how bad it can be if the original data are directly used, i.e., the price series in our example. Here, we shall fill the gap by providing some concrete evidence.

Table 20 shows the results of using the transformed data and using the original data. The transformation taken here is to divide the original price series by the 250-day moving average as we do in Section 4. The second and third columns are directly copied from the respective columns of Table 2, and the next two columns are the counterparts of the previous two using the original series.

To see the contribution of using data preprocessing, we test the differences between the return from using the transformed data and the return from using the original data. Out of a total of 24 markets under examination, there are 8 markets, and the return from using the transformed data is statistically significant compared to the return from not using the original data. Only in three cases do we see the opposite result, and in the rest of the 13 cases, they are tied. This result shows the advantages of using the transformed data. Nonetheless, the original series is not bad to a prohibitive degree, as some may expect.

Table 20. GP Using Transformed Time Series and Original Time Series

	$P_t/\text{MA}(250)$		P_t	
Country	Mean	Stdev	Mean	Stdev
Testing Period: 1999–2000				
USA	0.0655**	0.0342	0.0428	0.0599
UK	0.0459	0.0908	0.0669	0.0624
Canada	0.3660**	0.1030	0.2747	0.0645
Germany	0.1489	0.1114	0.2085**	0.0230
Spain	−0.0666	0.0904	−0.0919	0.0822
Japan	0.0024	0.0540	−0.0067	0.0497
Taiwan	0.1620**	0.1353	0.0750	0.0915
Singapore	0.1461	0.1866	0.1297	0.1599
Testing Period: 2001–2002				
USA	−0.3171	0.0498	−0.3173	0.0271
UK	−0.3625	0.0284	−0.3526	0.0689
Canada	−0.1761**	0.1065	−0.2364	0.0633
Germany	−0.4772	0.2242	−0.4283	0.2231
Spain	−0.2780**	0.0910	−0.3098	0.0714
Japan	−0.0722	0.1520	−0.0037**	0.0624
Taiwan	0.0376	0.1899	0.0063	0.2343
Singapore	−0.3123**	0.0333	−0.3709	0.0740
Testing Period: 2003–2004				
USA	0.3065	0.0334	0.3119	0.0309
UK	0.1797**	0.0300	0.1607	0.0511
Canada	0.3109	0.0585	0.3558**	0.0237
Germany	0.3318	0.0544	0.3323	0.0669
Spain	0.3355	0.1292	0.3184	0.0656
Japan	0.0212	0.0843	0.0199	0.0679
Taiwan	0.3631**	0.0665	0.1692	0.1334
Singapore	0.2512	0.0735	0.2536	0.1455

The "**", depending on where it appears, refers to the rejection of the null, at a significance level of 0.05, that the total return of using the transformed time series is no greater than that of using the original one or the null that the total return of using the original time series is no greater than that of using the transformed one.

D. Various Figures

Figures 10 and 11 show the time-series plot of the stock indexes. Figures 12 and 13 show the time-series plot of the foreign exchange rates. Figures 14 and 15 depict the equity curves of the GP-discovered trading programs in the stock markets. Figures 16 and 17 plot the time series of the proportion of rules indicating a long position in the stock markets and the foreign exchange markets, respectively. Figures 18 and 19 give the distribution of the node complexity of the GP-discovered programs for the stock markets and the foreign exchange markets. Figure 20 plots the relation between the evolution time of GP and the return in the 2003–2004 period of the stock markets.

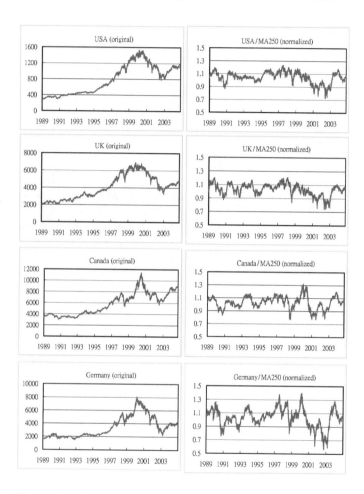

Fig. 10. The stock indexes for the United States, the UK, Canada, and Germany. In the left panel are the original series of the S&P 500 (US), FTSE 100 (UK), TSE 300 (Canada), and Frankfurt Commerzbank (Germany). In the right panel are the corresponding series after normalization by the 250-day moving average.

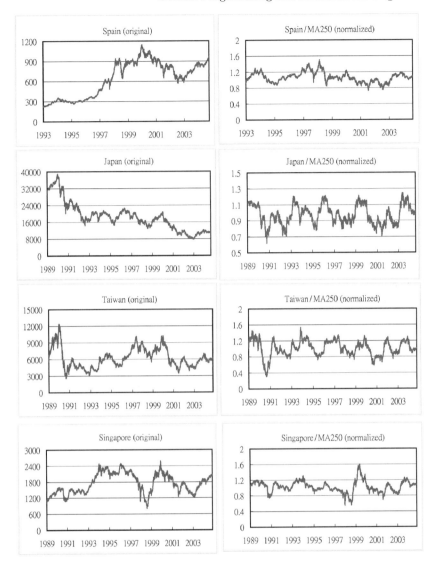

Fig. 11. The stock indexes of Spain, Japan, Taiwan, and Singapore. In the left panel are the original series of Madrid-SE (Spain), Nikkei Dow Jones (Japan), Straits Times (Singapore), and the Capitalization Weighted Stock index (Taiwan). In the right panel are the corresponding series after normalization by the 250-day moving average.

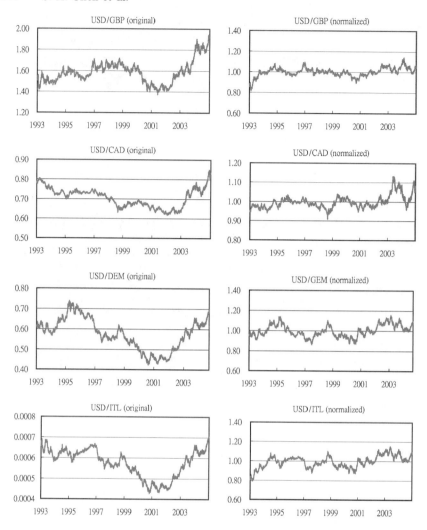

Fig. 12. Time series of foreign exchange rates: The rates between the United States and the UK, Canada, Germany, and Italy.

In the left panel are the original series of the USD per GBP, CAD, DEM, and ITL. In the right panel are the corresponding series after normalization by the 250-day moving average.

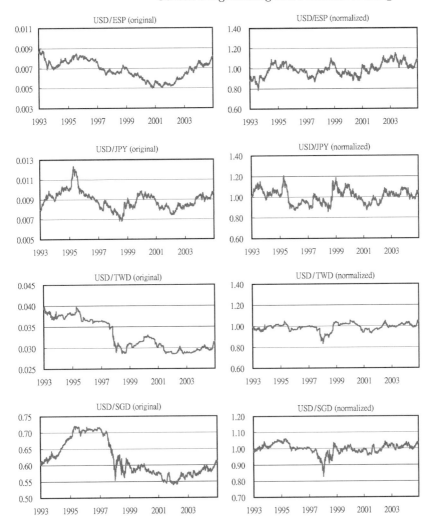

Fig. 13. Time series of foreign exchange rates: The rates between the United States and Spain, Japan, Taiwan, and Singapore.

In the left panel are the original series of the USD per ESP, JPY, TWD, and SGD. In the right panel are the corresponding series after normalization by the 250-day moving average.

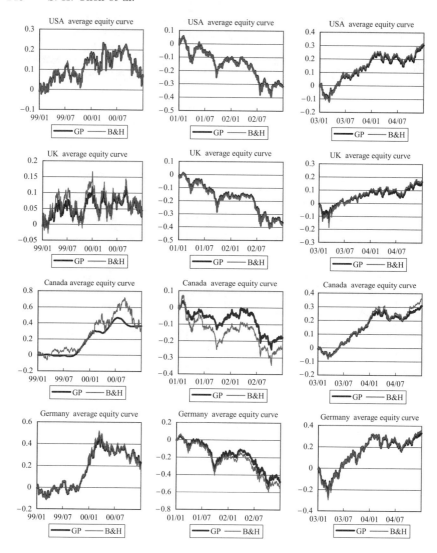

Fig. 14. Equity curves: Stock markets of the United States, the UK, Canada, and Germany.

Each equity curve drawn here is the average taken over the 50 equity curves, each of which is derived by following a single GP-discovered trading program in a single trial.

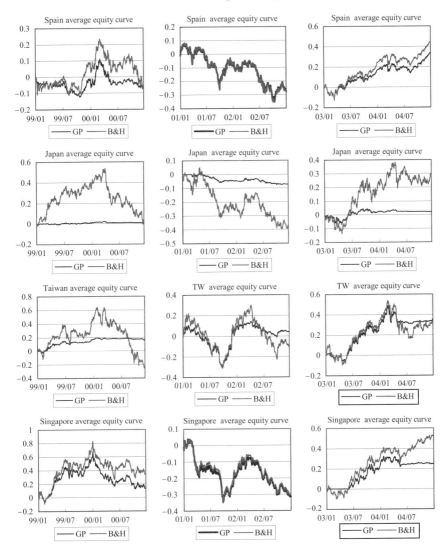

Fig. 15. Equity curves: Stock markets of Spain, Japan, Taiwan, and Singapore. Each equity curve drawn here is the average taken over the 50 equity curves, each of which is derived by following a single GP-discovered trading program in a single trial.

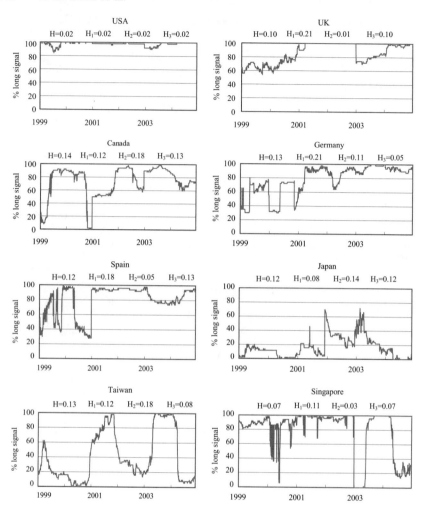

Fig. 16. Proportion of rules indicating a long position: Stock markets.

At the top of each plot is the entropy statistic H defined in Equation (14). H refers to the average of three periods, whereas H_1, H_2, H_3 refer to the average of each of the three respective periods.

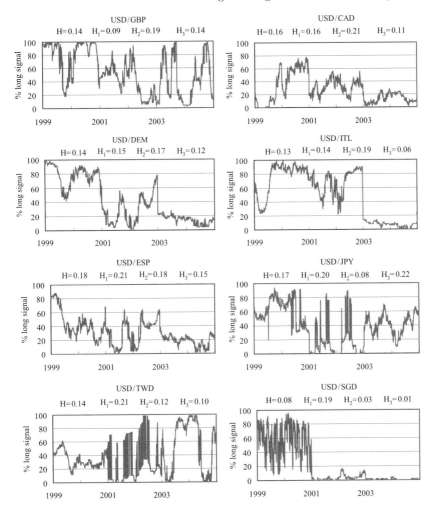

Fig. 17. Proportion of rules indicating a long position: Foreign exchange markets.
At the top of each plot is the entropy statistic H defined in Equation (14). H refers to the average of three periods, whereas H_1, H_2, H_3 refer to the average of each of the three respective periods.

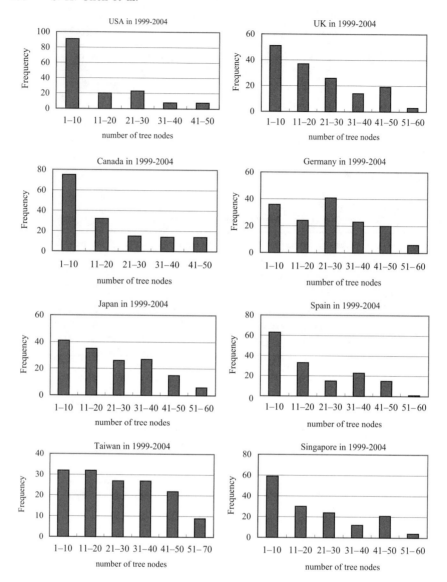

Fig. 18. Node complexity of the GP-discovered programs: Stock markets.

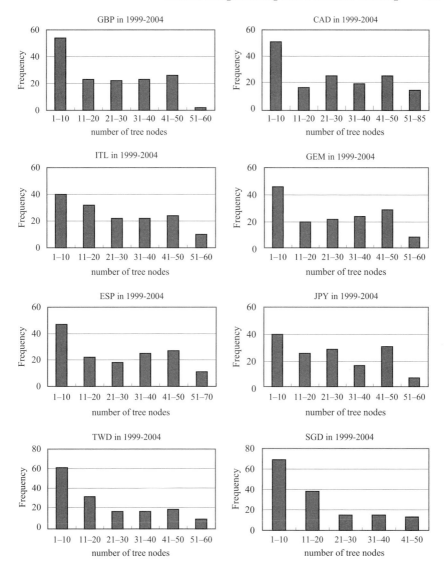

Fig. 19. Node complexity of the GP-discovered programs: Foreign exchange markets.

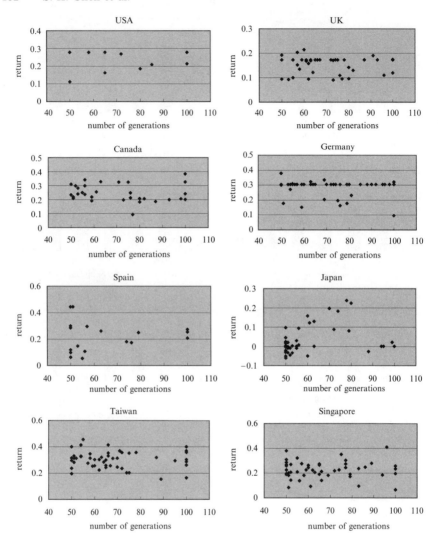

Fig. 20. Evolution time and the return: Stock markets, 2003–2004.

Acknowledgments

An earlier version of the paper was presented as a plenary speech at the *First International Symposium on Advanced Computation in Financial Markets* (ACFM'2005), Istanbul, Turkey, December 15–17, 2005. The authors greatly benefited from the interaction with conference participants, particularly from the chair Uzay Kaymak. The authors are also grateful to one anonymous referee for the very helpful suggestions. The research support from NSC 94-2415-H-004-003 is greatly acknowledged.

References

1. Allen, F., and Karjalainen, R. Using genetic algorithms to find technical trading rules. *Journal of Financial Economics*, 51(2):245–271, 1999.
2. Bhattacharyya, S., Pictet, O. V., and Zumbach, G. Knowledge-intensive genetic discovery in foreign exchange markets. *IEEE Transactions on Evolutionary Computation*, 6(2):169-181, 2002.
3. Brock, W., Lakonishok, J., and LeBaron, B. Simple technical trading rules and the stochastic properties of stock returns. *Journal of Finance*, 47:1731–1764, 1992.
4. Chen, S.-H. Modeling volatility with genetic programming: A first report. *Neural Network World*, 8(2):181–190, 1998.
5. Chen, S.-H. On the relevance of genetic programming in evolutionary economics. In K. Aruka, Editor, *Evolutionary Controversy in Economics towards a New Method in Preference of Trans Discipline*. Springer-Verlag, Tokyo, 2001, pages 135–150.
6. Chen, S.-H., and Chie, B.-T. A functional modularity approach to agent-based modeling of the evolution of technology. In A. Namatame, T. Kaizouji, and Y. Aruka, Editors, *Economics and Heterogeneous Interacting Agents*. Springer, New York, forthcoming, 2005.
7. Chen, S.-H., and Kuo, T.-W. Trading strategies on trial: A comprehensive review of 21 practical trading strategies over 56 listed stocks. In *Proceedings of the Fourth International Conference on Computational Intelligence and Multimedia Applications (ICCIMA 2001)*, IEEE Computer Society Press, 2001, pages 66–71.
8. Chen, S.-H., and Kuo, T.-W. Discovering hidden patterns with genetic programming. In S.-H. Chen and P. P. Wang, Editors, *Computational Intelligence in Economics and Finance*. Springer-Verlag, New York, 2003a, pages 329–347.
9. Chen, S.-H., and Kuo, T.-W. Modeling international short-term capital flow with genetic programming. In K. Chen et al., Editors, *Proceedings of 7th Information Sciences* (**JCIS 2003**), September 26-30, 2003, Cary, NC, 2003b, pages 1140–1144.
10. Chen, S.-H., and Kuo, T.-W. Overfitting or poor learning: A critique of current financial applications of GP. In C. Ryan, T. Soule, M. Keijzer, E. Tsang, R. Poli, and E. Costa, Editors, *Genetic Programming*. Lecture Notes in Computer Science 2610, Springer, New York, 2003c, pages 34–46.
11. Chen, S.-H., Kuo, T.-W., and Shien, Y.-P. Genetic programming: A tutorial with the software simple GP. In S.-H. Chen, Editor, *Genetic Algorithms and Genetic Programming in Computational Finance*. Kluwer, New York, 2002, pages 55–77.
12. Granville, J. E. *New Strategy of Daily Stock Market Timing for Maximum Profit*. Prentice-Hall, Englewood Cliffs, NJ, 1976.
13. Koza, J. *Genetic Programming: On the Prorgamming of Computers by Means of Natural Selection*. MIT Press, Cambridge, MA, 1992.
14. Koza, J. *Genetic Programming II: Automatic Discovery of Reusable Programs*. MIT Press, Cambridge, MA, 1994.
15. Koza, J., Bennett III, F., Andre, D., and Keane, M. *Genetic Programming III: Darwinian Invention and Problem Solving*, Morgan Kaufmann, San Francisco, 1999.

16. Koza, J., Keane, M., Streeter, M., and Mydlowec, W. *Genetic Programming IV: Routine Human-Competitive Machine Intelligence*. Springer, New York, 2005.
17. Lanquillon, C. Dynamic aspects in neural classification. *Journal of Intelligent Systems in Accounting, Finance and Management*, 8(4):281–296, 1999.
18. Neely, C., Weller, P., and Dittmar, R. Is technical analysis in the foreign exchange market profitable? A genetic programming approach. *Journal of Financial and Quantitative Analysis*, 32(4):405–426, 1997.
19. Neely, C., and Weller, P. Technical trading rules in the European monetary system. *Journal of International Money and Finance*, 18(3):429–458, 1999.
20. O'Neill, M., Brabazon, A., and Ryan, C. Forecasting market indices using evolutionary automatic programming. In S.-H. Chen, Editor, *Genetic Algorithms and Genetic Programming in Computational Finance*, Kluwer, New York, 2002, pages 175–195.
21. Pereira, R. Forecasting ability but no profitability: An empirical evalaution of genetic algorithm-optimised technical trading rules. In S.-H. Chen, Editor, *Evolutionary Computation in Economics and Finance*. Physica-Verlag, Berlin, 2002, pages 287–310.
22. Potvin, J. Y., Soriano, P., and Vallee, M. Generating trading rules on the stock markets with genetic programming. *Computers & Operations Research*, 31:1033–1047, 2004.
23. Tsakonas, A., Dounias, G., Doumpos, M., and Zopounidis, C. Bankruptcy prediction with neural logic networks by means of grammar-guided genetic programming. *Expert Systems with Applications*, 30:449–461, 2006.
24. Tsao, C.-Y., and Chen, S.-H. Statistical analysis of genetic algorithms in discovering technical trading strategies. *Advances in Econometrics*, 17:1–43, 2004.
25. Wang, J. Trading and hedging in S&P 500 spot and futures markets using genetic programming. *Journal of Futures Markets*, 20(10):911–942, 2000.
26. Yu, T., Chen, S.-H., and Kuo, T.-W. Discovering financial technical trading rules using genetic programming with lambda abstraction. In U.-M., O'Reilly, T. Yu, R. Riolo, and B. Worzel, Editors, *Genetic Programming Theory and Practice II*. Kluwer Academic Publishers, New York, 2004, pages 11–30.

Part II

Risk Management

Part II

Indoeuropean Dialects

Interest Rate Models: A Review

Christos Ioannidis, Rong Hui Miao, and Julian M. Williams

School of Management, University of Bath, Bath, BA2 7AY, UK
C.Ioannidis@bath.ac.uk

1 Introduction

The pricing of rate-sensitive instruments is a complex and demanding task. From a financial perspective a popular approach to modeling the term structure of interest rates is to use an approach driven by one or more stochastic processes. This review will address three general approaches to interest rate modeling: single and multifactor models of the short rate, models of forward rates, and finally LIBOR models. This review will focus on key results and pertinent pricing formulas and demonstrate several practical approaches to implementing short-rate models. For extended reviews on interest rate modeling, the interested reader is directed to Musiela and Rutkowski (2004) and Brigo and Mercurio (2006), who offered comprehensive coverage of each of the models reviewed here.

The notation conventions that are used are shown in Table 1.

2 Continuous-Time Models of Interest Rates

The success of the Vasicek (1977) and Cox–Ingersoll–Ross (1985) models has been mainly attributed to the analytical tractability of the price of a default-free, fixed-income bond. Representations of the evolution of interest rates using continuous-time stochastic processes require, in general, three desirable features: first, that the resultant process is always positive; second, a tractable solution to bond and option prices; and finally, nonexplosive properties to the money market account. Most of the popular interest rate models result in log-normal distributions to the interest rate. Unfortunately, a byproduct of these models is often that the third desirable property, nonexplosive money market accounts, does not hold. This peculiar property of some log-normal interest rate models was studied in Sandmann and Sondermann (1997). The issue is generally overcome in a practical sense, when using numerical techniques to evaluate the interest rate pathway. For example, when using multinomial

Table 1. Notation Conventions

Notation	Definition	Notes
$r(\cdot)$	Functional of interest process	$r(t)$, is the function of the rate w.r.t. time
r_t	Evolution of the rate process	
$r(t,T)$	The rate of change of account between t and T	
X	n length column vector of state processes	$X = [X^1, X^2, \ldots, X^n]'$
y, x	Arbitrary state/response variables	
$f(\cdot)$	Arbitrary function, usually of forward rates	
$g(\cdot), a(\cdot), b(\cdot)$	Arbitrary functions $k \to d$-dimensional functions	$a : \mathbb{R}^k \to \mathbb{R}^d$
$W(\cdot)$	Functional form of an n-length Weiner process	$W(\cdot) = [W^1(\cdot), \ldots, W^n(\cdot)]'$
W_t	Evolution of an n-length Weiner process	$W_{t+h} - W_t \sim N(0, h\mathbf{I}_{n \times n})$
$\underset{d \times d}{\mathbf{I}}$	d-dimensional identity matrix	
$\nabla f(\cdot)$	Vector of first-order partial derivatives of the multivariate function $f(\cdot)$	$\nabla f(x, y) = [f_x, f_y]'$
$\nabla^2 f(\cdot)$	Matrix of second-order partial derivatives of $f(\cdot)$	$\nabla^2 f(x, y) = \begin{bmatrix} f_{xx} & f_{xy} \\ f_{yx} & f_{yy} \end{bmatrix}$
$P(\cdot)$	Pricing function of a zero coupon bond	$P(t,T)$
P_t	Price evolution of a zero coupon bond	
$\mathbb{Q}$	$\mathbb{Q}$ martingale measure	

trees, the finite number of states available and the number of periods generally negate the explosive properties intrinsic to the model; see Brigo and Mercurio (2006) for extended discussion.

3 Simple Short-Rate Models

Before we begin to look at specific models, we first need to understand the basic relationship between the value of a rate-dependent security and the stochastic process that describes the evolution of the interest rate. The basic model for fixed-income markets utilizes an instantaneous interest rate, $r(t)$, whose evolution is described by some stochastic differential equation,

$$dr(t) = \mu(t, r(t)) dt + \sigma(t, r(t)) dW(t), \tag{1}$$

where $W(t)$ is a Brownian motion under some risk-neutral probability measure, $\mathbb{Q}$, $\mu(\cdot)$, and $\sigma(\cdot)$ are generally affine functions with parameter vector, θ, and recall that $W(t+h) - W(t) \sim N(0, h)$. It is often suggested that the diffusion process, $r(t)$, is the shortest rate available (e.g., the overnight borrowing rate). The following section looks at the pricing of a simple fixed-income contract in the presence of a general stochastic short-rate diffusion model.

Consider a zero coupon bond, with price at time, t, designated, $P(t,T)$, i.e., a contract that promises to pay a certain "*face*" amount at some future time, T. In order to price this contract, a function $f(\cdot)$ needs to be defined, which is of the following form,

$$P(t,T) = f(t, r(t)). \qquad (2)$$

More specifically, this is a generic pricing function, with substitutable terms $f(t,r)$, and the sets of first and second partial derivatives of $f(t,r)$ are given as

$$\nabla f(t,r) = \left\{ \frac{\partial f(t,r)}{\partial t} = f_t, \frac{\partial f(t,r)}{\partial r} = f_r \right\},$$

$$\nabla^2 f(t,r) = \left\{ \frac{\partial^2 f(t,r)}{\partial r \partial t} = \frac{\partial^2 f(t,r)}{\partial t \partial r} = f_{rt}, \frac{\partial^2 f(t,r)}{\partial r^2} = f_{rr}, \frac{\partial^2 f(t,r)}{\partial t^2} = f_{tt} \right\}. \qquad (3)$$

In addition, we define a continuous-time discount process $D(t)$, as an integral process over t_0 to t, as

$$D(t) = \exp\left(-\int_0^t r(s)\, ds\right). \qquad (4)$$

The intertemporal change of a single unit currency in a money market account is simply

$$\frac{1}{D(t)} = \exp\left(\int_0^t r(s)\, ds\right). \qquad (5)$$

The evolution of the discount process is therefore

$$dD(t) = -r(t) D(t)\, dt. \qquad (6)$$

Now consider the current time t value of a zero coupon bond paying one unit of currency at time T; this is in effect the expected cumulative discount that one unit of currency undergoes between t and T. Therefore,

$$D(t) P(t) = E(D(T)|\mathcal{F}(t)),$$
$$P(t,T) = E\left(\exp\left(-\int_t^T r(s)\, ds\right)\Big|\mathcal{F}(t)\right), \qquad (7)$$

where $\mathcal{F}(t)$ is some filtration up to time t. The next task now is to compute the yield between time t and T, which we shall define as $Y(t,T)$. This is defined as the compounding rate of return that is consistent with the observed process of a zero coupon bond over the period t and T,

$$Y(t,T) = -(T-t)^{-1} \log(P(t,T)). \tag{8}$$

Alternatively, we can rewrite the bond price in terms of the yield:

$$P(t,T) = \exp(-Y(t,T)(T-t)), \tag{9}$$

where $T - t$ is a proportion of the base time unit, e.g., 6 months is 0.5 years; therefore, $T - t = 0.5$.

To find the partial differential equation that generates the pricing formula, from (2), we simply need to substitute the interest rate and discount rate differential equations into the bond pricing formula, (7), and differentiate using the results from (3).

$$\begin{aligned} d(D(t)P(t,T)) &= f(t,r(t)\,dD(t) + D(t)\,df(t,r(t))) \\ &= D(t)\left(-rf\,dt + f_t\,dt + f_r\,dR + \tfrac{1}{2}f_{rr}d^2r\right) \\ &= D(t)\left(-rf + f_t + \mu f_r + \tfrac{1}{2}\sigma f_{rr}\right)dt + D(t)\,\sigma f_r\,dW(t). \end{aligned} \tag{10}$$

Setting $dt = (t+h) - t$ and $h = 0$, the pricing formula has the following solution:

$$f_t(t,r) + \mu(t,r)f_r(t,r) + \tfrac{1}{2}\sigma^2(t,r)f_{rr}(t,r) = rf(t,r), \tag{11}$$

as the final payoff of the zero coupon bond is set to 1; therefore, the terminal condition is

$$f(T,r) = 1, \forall r. \tag{12}$$

From these conditions given a one-dimensional stochastic differential equation that determines the spot rate model with tractable terminal solutions and deterministic drift and volatility functions that are twice differentiable, a pricing formula for a zero coupon bond may be found using (11).

We review nine of the classical short-rate models illustrating the underlying stochastic differential equation and the subsequent analytical bond pricing formulation, if one is available.

1. **Vasicek (1977)**:
   - SDE: $dr_t = k(\theta - r_t)dt + \sigma dW_t$
   - Feasible domain of r_t: $r_t \in \mathbb{R}$
   - Available tractable analytical solution for the price of bond:

$$P(t,T) = A(t,T)\exp(-B(t,T)r(t))$$

$$A(t,T) = \exp\left(\left(\theta - \frac{\sigma^2}{2k^2}\right)B(t,T) - \frac{\sigma^2}{2k^2}B(t,T)^2\right)$$

$$B(t,T) = \frac{1}{k}(1 - \exp(-k(T-t)))$$

- *Parameters and restrictions*:
$$r_0 \in \mathbb{R}_+, k \in \mathbb{R}_+$$
$$\theta \in \mathbb{R}_+, \sigma \in \mathbb{R}_+$$

- *Comments*: Classical model. The instantaneous spot rate is not necessarily positive.

2. **Cox, Ingersoll, and Ross, (1985):**
   - *SDE*: $dr_t = k(\theta - r_t) dt + \sigma \sqrt{r_t} dW_t$
   - *Feasible domain of r_t*: $r_t \in \mathbb{R}_+$
   - *Available tractable analytical solution for the price of bond*:
$$P(t,T) = A(t,T) \exp(-B(t,T) r(t))$$
$$A(t,T) = \frac{2h \exp\left(\frac{1}{2}(k+h)(T+t)\right)}{2h + (k+h)(\exp(h(T-t)) - 1)}$$
$$B(t,T) = \sqrt{k^2 - 2\sigma^2}$$

   - *Parameters and restrictions*:
$$r_0 \in \mathbb{R}_+, k \in \mathbb{R}_+$$
$$\theta \in \mathbb{R}_+, \sigma \in \mathbb{R}_+$$
$$2k\theta > \sigma^2$$

   - *Comments*: Square root diffusion model, yields a log-normal rate. Explosive money account issue.

3. **Dothan (1978):**
   - *SDE*: $dr_t = \alpha r_t dt + \sigma r_t dW_t$
   - *Feasible domain of r_t*: $r_t \in \mathbb{R}_+$
   - *Available tractable analytical solution for the price of bond*: No closed form zero coupon bond formula available.
   - *Parameters and restrictions*: -
   - *Comments*: Only approach to have a tractable solution for pure discount bonds, using a modified Bessel function approach.

4. **Exponential Vasicek; see Brigo and Mercurio (2006):**
   - *SDE*: $dr_t = r_t (\eta - \alpha \ln(r_t)) dt + \sigma r_t dW_t$
   - *Feasible domain of r_t*: $r_t \in \mathbb{R}_+$
   - *Available tractable analytical solution for the price of bond*: No analytical closed-form zero coupon bond formula available
   - *Parameters and restrictions*: -
   - *Comments*: Not an affine term structure model and suffers from the explosive money account properties, present in several log-normal models.

5. **Hull and White (1990):**
   - *SDE*: $dr_t = k(\theta - r_t) dt + \sigma dW_t$
   - *Feasible domain of r_t*: $r_t \in \mathbb{R}$

- *Available tractable analytical solution for the price of bond*:

$$P(t,T) = A(t,T)\exp(-B(t,T)r(t))$$
$$A(t,T) = \frac{P^M(0,T)}{P^M(0,t)}\exp\left(A(t,T)f^M(0,t) - \frac{\sigma^2}{4a}(1-\exp(-2at))A(t,T)^2\right)$$
$$B(t,T) = \frac{1}{a}(1-\exp(-a(T-t)))$$

- *Parameters and restrictions*:

$$r_0 \in \mathbb{R}_+, k \in \mathbb{R}_+$$
$$\theta \in \mathbb{R}_+, \sigma \in \mathbb{R}_+$$
$$a \in \mathbb{R}_+$$

- *Comments*: Hull and White (1990) extend the Vasicek model and improve the fit by endogenizing the mean revision parameter.

6. **Black and Karasinski (1991):**
    - *SDE*: $dr_t = r_t(\eta_t - \alpha\ln(r_t))dt + \sigma r_t dW_t$
    - *Feasible domain of r_t*: $r_t \in \mathbb{R}_+$
    - *Available tractable analytical solution for the price of bond*: No analytical closed-form zero coupon bond formula available
    - *Parameters and restrictions*: -
    - *Comments*: One of the most commonly implemented models, the Black and Krasanski (1991) model offers a guaranteed positive rate but yields infinite money accounts under all maturities.

7. **Mercurio and Moraleda (2000):**
    - *SDE*: $dr_t = r_t\left(\eta_t - \left(\lambda - \frac{\gamma}{1+\gamma t}\right)\ln(r_t)\right)dt + \sigma r_t dW_t$
    - *Feasible domain of r_t*: $r_t \in \mathbb{R}_+$
    - *Available tractable analytical solution for the price of bond*: No analytical closed-form zero coupon bond formula available
    - *Parameters and restrictions*: -
    - *Comments*: Extended version of the Hull and White model and is in many respects a bridge to the Heath, Jarrow, and Morton approach.

8. **Extended Cox, Ingersoll, and Ross; see Brigo and Mercurio (2006):**
    - *SDE*:

$$r_t = x_t + \varphi_t$$
$$dx_t = k(\theta - x_t)dt + \sigma\sqrt{x_t}dW_t$$

    - *Feasible domain of r_t*: $r_t \in \mathbb{R}_+$
    - *Available tractable analytical solution for the price of bond*:

$$P(t,T) = \frac{P^M(0,T) A(0,T) \exp(-B(0,t)x_0)}{P^M(0,T) A(0,t) \exp(-B(0,T)x_0)} A(t,T) \exp(-B(t,T)r(t))$$

$$A(t,T) = \left(\frac{2h \exp\left(\frac{1}{2}(k+h)(T+t)\right)}{2h + (k+h)(\exp(h(T-t))-1)} \right)$$

$$B(t,T) = \sqrt{k^2 - 2\sigma^2}$$

- *Parameters and restrictions*:

$$r_0 \in \mathbb{R}_+, k \in \mathbb{R}_+$$
$$\theta \in \mathbb{R}_+, \sigma \in \mathbb{R}_+$$
$$h = \sqrt{k^2 + 2\sigma^2}$$

- *Comments*: Extension of the Cox–Ingersoll–Ross approach, which includes a deterministic shift, φ_t, to improve the model fit.

9. **Extended Vasicek Model; see Brigo and Mercurio (2006)**:
   - *SDE*:

$$r_t = x_t + \varphi_t$$
$$dx_t = x_t(\eta - \alpha \ln(x_t)) dt + \sigma x_t dW_t$$

   - *Feasible domain of r_t*: $r_t \in \mathbb{R}_+$
   - *Available tractable analytical solution for the price of bond*: No analytical closed-form zero coupon bond formula available
   - *Parameters and restrictions*: -
   - *Comments*: The extension to the Vasicek model again incorporates a deterministic shift in the rate in order to improve the model fit over the basic specification.

4 Estimating Interest Rate Models

This section will demonstrate the discretization and empirical estimation of a single-factor Vasicek model from first principles. This is a mean-reverting process and yields an interest rate process with a general Gaussian distribution and tractable bond and option pricing formulas. To capture mean reversion, the Vasicek (1977) model assumes that the short rate follows an Ornstein–Uhlenbeck process:

$$dr_t = \kappa(\theta - r_t) dt + \sigma dW_t, \tag{13}$$

where κ measures the speed of mean reversing, θ is the unconditional mean (long-term level of short rate), and σ is the instantaneous volatility of the short rate.

With this Gaussian model under true probability measure, the normally distributed short rate has the mean and variance

$$E_{r,t}(r_T) = \theta + (r - \theta)e^{-\kappa(T-t)}, \qquad (14)$$

$$\text{var}_{r,t}(r_T) = \sigma^2 (2\kappa)^{-1}\left(1 - e^{-2\kappa(T-t)}\right). \qquad (15)$$

Assuming the market price of risk is a constant, that is, $\lambda(r,t) = \lambda$, the dynamic of the short rate under risk-neutral measure $\mathbb{Q}$ will be

$$dr_t = \kappa\left(\hat{\theta} - r_t\right)dt + \sigma dW_t^{\mathbb{Q}}, \hat{\theta} = \theta - \lambda\theta/\kappa \qquad (16)$$

where we can see from (13) and (16) that the short-rate dynamic under both probability measures has the same qualitative properties when the market price of risk is defined as $\lambda(r,t) = \lambda$.

4.1 Bond Pricing in Continuous Time

Assume that bond prices follow a geometric Brownian motion:

$$dP(t,T) = \mu_P(t,T)P(t,T)dt + \sigma_P(t,T)P(t,T)dW. \qquad (17)$$

Using Ito's lemma, we get the PDE of bond prices:

$$\frac{1}{2}\frac{\partial^2 P}{\partial r_t^2}\sigma^2 + \frac{\partial P}{\partial r_t}[\kappa(\theta - r_t) - \lambda\sigma] + \frac{\partial P}{\partial t_t} - r_t P = 0. \qquad (18)$$

For $dr_t = \kappa(\theta - r_t)dt + \sigma dW_t$ and $\lambda(r,t) = \lambda$, with the boundary condition $P(T,T) = 1$ and assuming exponential-affine form of the bond price, $P(t,\tau) = \exp[A(\tau) + B(\tau)r_t]$ with $\tau = T - t$, it is possible to solve $A(\tau)$ and $B(\tau)$ analytically.

Under a Euler scheme, the continuous factor diffusion is rewritten as a discrete process. Campbell et al. (1997) and Backus et al. (1998) gave a discrete-time version of the Vasicek single-factor model. Under discrete time, the single factor x_t is assumed to follow an AR(1) process:

$$\begin{aligned}x_{t+1} &= \varphi x_t + (1-\varphi)\theta + \sigma\varepsilon_{t+1},\\ \varepsilon_{t+1} &\sim N(0,1).\end{aligned} \qquad (19)$$

θ is the mean of this factor, and the unconditional variance is $\sigma^2/(1-\varphi^2)$, where φ is the persistence parameter, i.e., the factor is expected to revert toward its long-term mean, θ, at a rate of $(1-\varphi)$.

4.2 Bond Pricing in Discrete Time

Following Backus et al. (1998) closely, bond pricing theory predicts that in an arbitrage-free market there is a positive stochastic factor denoted by M_t (also known as the pricing kernel) that determines the price at time t of a bond with nominal cash flows P_t in terms of its discounted future payoff,

$$P_t = E_t \left(P_{t+1} M_{t+1} \right). \tag{20}$$

Assuming the distribution of the stochastic discount factor M_{t+1} is conditionally log-normal, taking logs of 20, we have

$$\log \left(E_t \left(P_{t+1} M_{t+1} \right) \right) \sim N \left(\mu, \sigma^2 \right), \tag{21}$$

where

$$\mu = E_t \left(m_{t+1} + p_{t+1} \right),$$
$$\sigma^2 = var_t \left(m_{t+1} + p_{t+1} \right) \tag{22}$$
$$\Rightarrow p_t = E_t \left(m_{t+1} + p_{t+1} \right) + \frac{1}{2} var_t \left(m_{t+1} + p_{t+1} \right),$$

where p_t and m_t represent the natural log of P and M, respectively.

From the assumption that the log of the pricing kernel satisfies a linear relationship with factor x_t, then

$$-m_{t+1} = \xi + x_t + \lambda \varepsilon_{t+1}, \tag{23}$$

where the parameter λ determines the covariance between the factor and shocks to pricing kernel, finally we set $\xi = \lambda^2 / 2$, with the boundary condition, $P(T,T) = 1$. From (21) and (23) the one-period bond price will satisfy

$$p_{t,1} = -\xi - x_t + \lambda^2 / 2, \tag{24}$$

and the resulting short rate is therefore

$$r_t = -p_{t,1} = x_t. \tag{25}$$

The price of long bonds will take the form of

$$-p_{t,n} = A_n + B_n x_t, \tag{26}$$

with $x_{t+1} = \varphi x_t + (1 - \varphi) \theta + \sigma \varepsilon_{t+1}$, and starting with the condition that $A_0 = B_0 = 0$, (24) implied $A_1 = 0$ and $B_1 = -1$. Given the bond maturity n, we could evaluate from (21) that

$$E_t \left(m_{t+1} + p_{t+1,n} \right) = - \left[A_n + B_n (1 - \varphi) \theta \right] - (1 + B_n \varphi) x_t \tag{27}$$

and

$$var_t \left(m_{t+1} + p_{t+1,n} \right) = (\lambda + B_n \sigma)^2; \tag{28}$$

therefore, with (26), we have the recursion

$$A_n = A_{n-1} + \xi + B_{n-1} (1 - \varphi) \theta - \tfrac{1}{2} (\lambda + B_n \sigma)^2 \tag{29}$$

and

$$B_n = 1 + B_{n-1} \varphi. \tag{30}$$

5 Multifactor Models of Interest Rates

The basic single-factor models of interest rates assume that the spot rate is driven by a one-dimensional Markov process; therefore, they are driven by a single unique state variable. Another approach to modeling interest rates is to derive the term structure assuming that the driving variable is a vector process containing multiple degrees of uncertainty. In order to look at multifactor models from a practical point of view, we need to specify an n-dimensional vector diffusion process and then use this process to identify the spot rate model. Consider a vector diffusion process, $X = [X_1, X_2, ..., X_n]'$, whose diffusion is described by a stochastic differential equation as follows:

$$dX_t = \mu(X_t, t)\,dt + \sigma(X_t, t)\,dW_t, \tag{31}$$

where $\mu(.)$ and $\sigma(.)$ are coefficients with domain in $\mathbb{R}^n$ and $\mathbb{R}^{n \times n}$, respectively. The simplest multifactor models assume that there are n sources of variation in the system. In general, we restrict our attention to two-factor models, primarily as empirical research suggests that two factors account for the majority of the variation in the system. These state variables have a variety of economic interpretations beyond the scope of this chapter; however, certain models assign these state variables as the yield of bonds of certain finite maturities; see Duffie and Singleton (1993).

5.1 Affine Multifactor Models

A simple class of multifactor models is the affine models. From the previous section we defined that a model is affine if it permits the following representation:

$$P(t, T) = \exp(m(t, T) - g(t, T)\,r_t), \forall t \in [t, T] \tag{32}$$

for some functions $m(.)$ and $g(.)$. If we assume that the state vector evolves via Equation (31), i.e., that X is a time homogenous Markov process with some state space $X \in \mathbb{R}^n$, then we can define the evolution of the spot rate as being some function $f : \mathbb{R}^n \to \mathbb{R}$, which in the simplest case is simply the summation of the factors. In the simplest sense,

$$r_t = f(X_t). \tag{33}$$

In a discrete-time model, the log of the pricing kernel now is a linear combination of several factors $x'_t = [x_{1,t} \cdots x_{j,t}]$ and has the form

$$-m_{t+1} = \xi + \gamma' x_t + \lambda' V(x_t)^{1/2} \varepsilon_{t+1}, \tag{34}$$

where $V(x_t)$ is the variance-covariance matrix of the error term and is defined as a diagonal matrix with the ith diagonal element given by $v_i(x_i) = \alpha_i + \beta'_i x_t$. It is important to note that β_i and as such the parameters of the volatility function are constrained such that v_i is always positive. The error term, ε_t, is therefore independently normally distributed noise, $\varepsilon_t \sim N(0, I)$. λ' is the market price of risks, which governs the covariance between the pricing kernels and the factors.

The j-dimensional vector of factors (state variables) x is given by

$$x_{t+1} = (I - \Phi)\theta + \Phi x_t + V(x_t)^{1/2} \varepsilon_{t+1}. \qquad (35)$$

Φ has typical positive diagonal elements to keep the state variables stationary, and θ is specified as the long-run mean of the state variable x. The bond price in terms of log-linear functions of the state variables takes the form

$$-p_{n,t} = A_n + B'_n x_t; \qquad (36)$$

considering the restrictions on the end-maturity log bond price, that must equal zero, and the corresponding restrictions on A_n and B_n, to the common normalization of $A_0 = B_0 = 0$, the recursion is therefore

$$A_n = A_{n-1} + \xi + B'_{n-1}(I - \Phi)\theta - \frac{1}{2}\sum_{i=1}^{j}(\lambda_i + B_{i,n-1})^2 \alpha_i. \qquad (37)$$

Subsequently,

$$B'_n = (\gamma' + B'_{n-1}\Phi) - \frac{1}{2}\sum_{i=1}^{j}(\lambda_i + B_{i,n-1})^2 \beta'_i. \qquad (38)$$

Duffie and Kan (1996) demonstrated a closed-form expression for the spot, forward, volatility, and term structure curves. Following Beckus et al. (1996, 1998) and Cassola and Luis (2001), in discrete time the one-period interest rate is defined as

$$y_{1,t} = \xi - \frac{1}{2}\sum_{i=1}^{j}\alpha_i \lambda_i^2 + \left(\gamma' - \frac{1}{2}\sum_{i=1}^{j}\beta'_i \lambda_i^2\right) x_t; \qquad (39)$$

the expected short rate is therefore

$$E_t(y_{1,t+n}) = E_t \left(\xi - \frac{1}{2}\sum_{i=1}^{j}\alpha_i \lambda_i^2 + \left(\gamma' - \frac{1}{2}\sum_{i=1}^{j}\beta'_i \lambda_i^2\right) x_{t+n}\right)$$

$$= \xi - \frac{1}{2}\sum_{i=1}^{j}\alpha_i \lambda_i^2 + \left(\gamma' - \frac{1}{2}\sum_{i=1}^{j}\beta'_i \lambda_i^2\right) E(x_{t+n})$$

$$= \xi - \frac{1}{2}\sum_{i=1}^{j}\alpha_i \lambda_i^2 + \left(\gamma' - \frac{1}{2}\sum_{i=1}^{j}\beta'_i \lambda_i^2\right)((I - \Phi^n)\theta + \Phi^n x_t). \qquad (40)$$

The volatility curve derived from the variance-covariance matrix is given by

$$\text{var}_t(y_{n,t+1}) = n^{-2} B'_n V(x_t) B_n, \tag{41}$$

and the one-period forward curve that is derived by using the recursive restrictions in (38) takes the form

$$f_{n,t} = p_{n,t} - p_{n+1,t}. \tag{42}$$

Finally, the term premium is computed as the log one-period excess return of the n maturity bond over the one-period short interest rate as

$$\Lambda_{n,t} = E_t(p_{n,t+1}) - p_{n+1,t} - y_{1,t}$$
$$= -\sum_{i=1}^{j}\left(\alpha_i \lambda_i B_{i,n} - \frac{1}{2}\alpha_i B_{i,n}^2\right) - \sum_{i=1}^{j}\left(\lambda_i B_{i,n} + B_{i,n}^2\right)\beta'_i x_t. \tag{43}$$

From the above formulation of forward rate, short-term interest rate, and the term premium, previously we demonstrated that the forward rate equals the sum of the expected future short-term interest rate, the term premium, and a constant.

5.2 Vasicek Model in Discrete Time

The Vasicek multifactor model is a special case of Duffie–Kan affine models and implies that some form of the expectation hypothesis theory holds as it assumes constant risk premium.

As specified in Duffie and Kan (1996), the factors are assumed to be first-order autoregressive series with zero mean:[1]

$$x_{i,t+1} = \varphi_i x_{i,t} + \sigma_i \varepsilon_{i,t+1}. \tag{44}$$

The relevant characterization of parameters for this model as referenced in Backus et al. (1998) is

$$\begin{aligned}
\theta_i &= 0, \\
\Phi &= \text{diag}(\varphi_1 \cdots \varphi_k), \\
\alpha_i &= \sigma_i^2, \\
\beta_i &= 0, \\
\xi &= \delta + \sum_{i=1}^{j} \frac{1}{2}\lambda_i^2 \sigma_i^2 \\
\gamma_i &= 1,
\end{aligned} \tag{45}$$

---

[1] Zero mean specification as referenced in Cassola and Luis (2001) corresponds to the consideration of the differences between the true factors and their means.

and then the corresponding recursive restrictions are derived as

$$A_n = A_{n-1} + \delta + \frac{1}{2}\sum_{i=1}^{j}\left(\lambda_i^2 \sigma_i^2 - (\lambda_i \sigma_i + B_{i,n-1}\sigma_i)^2\right), \quad (46)$$

$$B_{i,n} = 1 + B_{i,n-1}\varphi_i \quad \text{or} \quad B_{i,n} = (1 - \varphi_i^n)/(1 - \varphi_i). \quad (47)$$

For stationarity, the AR(1) parameter is constrained as $-1 < \varphi_i < 1$.

This model is also called the *constant volatility model*, so the terms on the right-hand side of (34) relating to the risk should be zero. The implication of this result is that there should be no interactions between the risk and the state variables (state variables do not enter the volatility curve); therefore, the term premium will be a constant in this Gaussian model.

The short-term interest rate is given by (39), and the Gaussian model general restrictions of (45) result in

$$y_{1,t} = \delta + \sum_{i=1}^{j} x_{i,t}; \quad (48)$$

by applying the same set of restrictions to the forward curve, the new one-period forward rate is

$$f_{n,t} = \delta + \frac{1}{2}\sum_{i=1}^{j}\left(\lambda_i^2 \sigma_i^2 - \left(\lambda_i \sigma_i + \frac{1 - \varphi_i^n}{1 - \varphi_i}\sigma_i\right)^2\right) + \sum_{i=1}^{j}(\varphi_i^n x_{i,t}). \quad (49)$$

This forward curve would accommodate very different shapes. Given the constraint that $\varphi_i < 1$, the limiting value will not depend on the state variables; as such the following limit exists:

$$\lim_{n\to\infty} f_{n,t} = \delta + \sum_{i=1}^{j}\left(-\frac{\lambda_i \sigma_i^2}{1 - \varphi_i} - \frac{\sigma_i^2}{2(1 - \varphi_i)^2}\right), \quad (50)$$

but if $\varphi_i = 1$, the interest rates will be nonstationary, in this case the limiting value of the one-period forward rate will have infinite values but will be time-varying.[2] From the constraints demonstrated in (45), the following definitions are derived:

$$\text{var}_t(y_{n,t+1}) = n^{-2}\sum_{i=1}^{j}\left(B_{i,n}^2 \sigma_i^2\right), \quad (51)$$

$$\Lambda_{n,t} = E_t(p_{n,t+1}) - p_{n+1,t} - y_{1,t}$$

---

[2] A linear combination of nonstationary state variables will also be a nonstationary series.

$$= \frac{1}{2} \sum_{i=1}^{j} \left(\lambda_i^2 \sigma_i^2 - \left(\lambda_i \sigma_i + \frac{1-\varphi_i^n}{1-\varphi_i} \right)^2 \right) \qquad (52)$$

$$= \sum_{i=1}^{j} \left(-\lambda_i \sigma_i^2 B_{i,n} - \frac{1}{2} B_{i,n}^2 \sigma_i^2 \right).$$

Since the state variables have constant volatility, the yield curve volatility will not depend on the level of the state variables in (51). For the term premium in (52), the one-period forward rate in this Gaussian model is constructed by summing the term premium with a constant and with the state variables weighted by its autoregressive parameters. Taking the limiting case into consideration that when $\varphi_i < 1$, the finite limiting value of the risk premium differs only by δ from the forward rate. So we can conclude that for a constant volatility model, the expected instantaneous rate is the average short-term interest rate δ.

Correspondingly, the expected future short-term rate with restrictions from (45) will be defied as

$$E_t(y_{1,t+n}) = \delta + \sum_{i=1}^{j} (\varphi_i^n x_{i,t}). \qquad (53)$$

Combining the results of (49), (52) and (53), the expectation hypothesis of the term structure holds and with a constant term premium will be as follows:

$$f_{n,t} = E_t(y_{1,t+n}) + \Lambda_n. \qquad (54)$$

Therefore, it is equivalent to saying that testing the adequacy of this Gaussian model is the same as testing the validity of the expectation hypothesis of the term structure with constant term premiums. From (54), this conclusion implies the expected one-period forward curve should be flat when the risk premium equals zero.

6 Estimating a Two-Factor Model for German Interest Rates

The data set utilized is the zero coupon bond yields published by the Bank of Germany.[3] This data set comprises month-end spot rates that are generated by interpolating German listed federal securities for maturities of 1 to 10 years between January 1973 and July 2006.[4]

---

[3] http://www.bundesbank.de/statistik/statistik_zeitreihen.en.php (Version 2006).
[4] The zero coupon bond estimation method is explained in
http://www.bundesbank.de/download/volkswirtschaft/kapitalmarktstatistik/2006/capitalmarketstatistics072006.pdf (page 24, Version 2006).

6.1 Descriptive Analysis on Yield Data

From the descriptive analysis in Table 2 and the average yield curve plot in Figure 1, we notice first, the average yield curve over this period was upward sloping, with a 11year yield of 5.63% and a 10-year yield of 6.91%. In relating to the Gaussian model, we estimate that this average upward sloping indicated the expected excess returns on bonds are constant and positive. The associated standard deviation of these rates deceased from around 2.6% to 1.8% at a constant rate. The yields are very persistent, with first-order autocorrelation coefficients all above 0.98 for the entire maturities, suggesting a possible nonstationary structure. Finally, the correlation matrix shows high cross-correlation of yields along the curve. High cross-correlations suggest that a small number of common factors drive the co-movement of the bond yields across different maturities. However, the correlation coefficients differ from unity, lending evidence to the importance of nonparallel shifts. This is equivalent to suggesting that a one-factor model of the term structure of interest rates would be insufficient, and this is confirmed by the findings in the principal component analysis undertaken.

6.2 Principal Component Analysis

Principal component (PC hereafter) analysis is based on the idea that it is possible to describe the interrelationships between a large number of correlated random variables in terms of a smaller number of uncorrelated random variables (principal components). This allows us to determine the main factors that drive the behavior of the original correlated bond yield.[5] Suppose we have a vector $Y = (Y_1, Y_2, \ldots, Y_N)'$ with covariance matrix Σ. The covariance matrix will be positively defined for all nonlinear combinations of one Y_i on the others. When it is positively defined with dimension N, it will have a complete set of N distinct positive eigenvalues such that $A'\Sigma A = D$, where A is the orthogonal matrix and D is the diagonal matrix with the eigenvalues of Σ along the diagonal. Now consider a vector W defined as $W = A'Y$; the covariance matrix of W is given by $E(WW') = A'E(YY')A = A'\Sigma A = D$, from which we can construct a set of uncorrelated variables as $W*_i = b'_i Y$, where b_i is the ith eigenvector of Σ, and the corresponding eigenvalue is λ_i. Given the property of orthogonality in A, we can show that the following holds:

$$\Sigma = \lambda_i b_i b'_i \quad \text{for } i = 1, \ldots, N. \tag{55}$$

In PC analysis, we use only $n < N$ linear combinations (principal components), and the first n linear combinations are adequate to describe the

---

[5] Principal component analysis in this chapter followed the theoretical procedures described in Bolder et a. (2004), An empirical analysis of the Canadian term structure of zero-coupon interest rates. Working paper 2004-48, Bank of Canada.

Table 2. Descriptive Analysis of German Zero Coupon Bond Yields (01/1973–07/2006)

Maturity	1y	2y	3y	4y	5y	6y	7y	8y	9y	10y
Mean (%)	5.636	5.892	6.131	6.326	6.483	6.608	6.710	6.792	6.860	6.917
Std.Dev.	2.463	2.300	2.192	2.107	2.035	1.975	1.922	1.876	1.835	1.800
Skewness	0.634	0.410	0.265	0.170	0.105	0.060	0.029	0.008	−0.005	−0.013
Kurtosis	−0.382	−0.665	−0.773	−0.813	−0.820	−0.808	−0.779	−0.738	−0.686	−0.627
AutoCorr.	0.988	0.988	0.988	0.988	0.988	0.988	0.987	0.987	0.986	0.986
Normality test:										
$\chi^2(2)$	70.299	39.255	25.834	19.954	16.9940	15.0420	13.2850	11.4700	9.5750	7.7050
$P(X > x)$	0.000	0.000	0.000	0.000	0.0002	0.0005	0.0013	0.0032	0.0083	0.0212

Correlation matrix:										
1y	1.000	0.989	0.972	0.954	0.936	0.92	0.905	0.89	0.877	0.864
2y		1.000	0.995	0.984	0.972	0.959	0.946	0.934	0.922	0.911
3y			1.000	0.997	0.99	0.981	0.972	0.962	0.953	0.943
4y				1.000	0.998	0.993	0.987	0.98	0.972	0.965
5y					1.000	0.999	0.995	0.99	0.985	0.979
6y						1.000	0.999	0.996	0.992	0.988
7y							1.000	0.999	0.997	0.994
8y								1.000	0.999	0.998
9y									1.000	0.999
10y										1.000

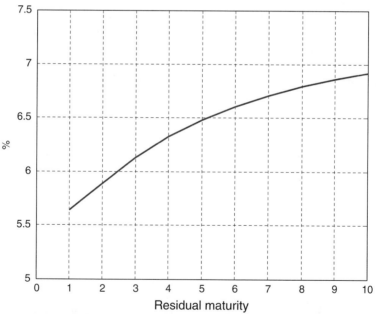

Fig. 1. Average yield curve of German zero coupon bond yields (01/1973–07/2006).

correlation of the original variables Y_i. These n principal components are regarded as the driving forces behind the co-movement of Y_i (yields in our case).

Litterman and Scheinkman (1991) were the first to use PC analysis. They found that over 98% of the variation in the returns on government fixed-income securities can be explained by three factors, labeled as *level*, *slope*, and *curvature*. From the plot in Figure 2, we see that the sensitivities of the rates to the first factor are roughly constant across maturities. Thus, if this PC increased by a given amount, we would observe a (approximately) parallel shift in the zero coupon term structure. This PC corresponds to the level factor, and in our context we find the most important first principal component captured 96.8% of the variation in yield, as shown in Table 3 and Figure 3. The second PC tends to have an effect on short-term rates that is opposite to its effect on long-term rates. An increase in this PC causes the short end of the yield curve to fall and the long end of the yield curve to rise. This is the slope factor — a change in this factor will cause the yield curve to steepen (positive change) or flatten (negative change). In our context, the second PC extended the proportion of variation explained to 99.8%. The third PC corresponds to the curvature factor, because it causes the short and long ends to increase, while decreasing medium-term rates. This gives the shape of the zero coupon bond yield curve more or less curvature. This PC seems to be the least significant of the three, accounting for an average of less than 0.5% of the total variation in term-structure movements.

Table 3. Percent Variation in Yield Levels Explained by the First k PCs

	PC1	PC2	PC3	PC4	PC5
Std. dev.	6.403176	1.126706	0.270761	0.099705	0.030346
% variation explained	0.968038	0.029973	0.001731	0.000235	2.17E-05
% cumulative explained	0.968038	0.998011	0.999742	0.999976	0.999999

The interpretation of the principal components in terms of level, slope, and curvature turns out to be useful in thinking about the driving forces behind the yield curve variation. The latent factors implied by the affine term structure models typically behaved like these principal components. As Piazzesi (2003) reported, this empirical finding applies to different data set and model specifications. For example, the Chen and Scott (1993) square-root process, the Gong and Remolona (1996) Gaussian process, the mixture model of Balduzzi et al. (1996), and the Dai and Singleton (2002) model find the coefficients estimated from $y_{n,t} = n^{-1}\left(A(n) + B(n)' x_t\right)$ exhibited similar general pattern as plotted in Figure 2 for the case of three state variables. However, there is no exact mapping between "stochastic mean"/"stochastic volatility" and the Litterman-Scheinkman labels. For instance, some studies find that "stochastic volatility" behaves like a curvature factor, yet others find it was as persistent as the level factor. But low-dimensional models with the number of state variables less than 3 find that the yield coefficients correspond to the first N ($N < 3$) principal components. These models include the $N = 2$

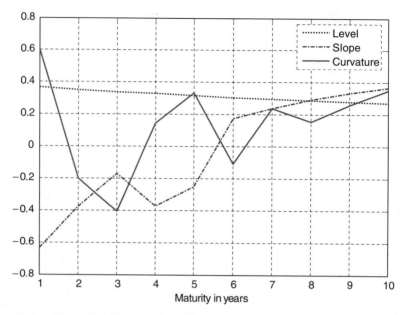

Fig. 2. Loadings of yields on principal components of German zero coupon bond yields (01/1973–07/2006).

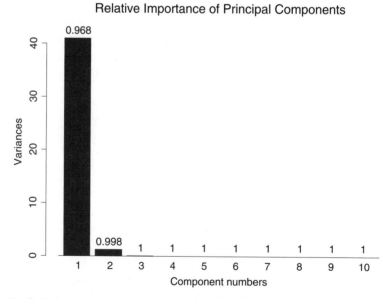

Fig. 3. Relative importance of principal components on yield data (01/1973–07/2006).

square-root model as in Chen and Scott (1993) and the Gaussian model as in Balduzzi et al. (1998). Therefore, in this chapter, according to the evidence found above, we shall adopt to use a two-factor model.

6.3 Test on the Expectation Hypothesis

The model utilized here belongs to the Gaussian (two-factor Vasicek) model; as a result, the yield data used in this paper should satisfy the Gaussian model's specification of the validity of expectation hypothesis.

The common statement of the expectations hypothesis is that forward rates are conditional expectations of future short rates, which implies that forward rates are martingales. As this is easily rejected by the data, which show average forward rates change systematically across maturities, we interpret the expectations hypothesis as including a constant term premium as in (54). According to Backus et al. (2001), by the definition of martingale and the law of iterated expectations, (54) can be rewritten as

$$f_{n,t} = E_t\left(E_{t+1}\left(y_{1,t+n}\right) + \Lambda_n\right) = E_t\left(f_{n-1,t+1}\right) + (\Lambda_n - \Lambda_{n-1}), \qquad (56)$$

and by subtracting the short-term interest rate on both sides of (56) and adding an error term yields

$$f_{n-1,t+1} - y_{1,t} = \kappa + \eta_n\left(f_{n,t} - y_{1,t}\right) + \varepsilon, \qquad (57)$$

where κ is the constant in regression, and the ε is error term. From (57), the expectation hypothesis holds if the forward regression slope of $\eta_n = 1$ in the regression, and a rejection of this hypothesis may be seen as evidence of changing term premiums over time. If we define η_n as $(\mathrm{var}\left(f_{n,t} - f_{0,t}\right))^{-1} \mathrm{cov}\left(f_{n-1,t+1} - f_0, f_{n,t} - f_{0,t}\right)$ and given that $y_{1,t} = f_{0,t}$ where

$$\begin{aligned}\mathrm{cov}\left(f_{n-1,t+1} - f_{0,t}, f_{n,t} - f_{0,t}\right) &= (B_1 + B_n - B_{n+1})'\Gamma_0 \\ &\quad (B_1 - \Phi'(B_n - B_{n-1})),\end{aligned} \qquad (58)$$

$$\mathrm{var}\left(f_{n,t} - f_{0,t}\right) = (B_1 + B_n - B_{n+1})'\Gamma_0\left(B_1 + B_n - B_{n+1}\right), \qquad (59)$$

Γ_0 is the unconditional variance of the state vectors x with solution

$$\Gamma_0 = \Phi\Gamma_0\Phi' + V. \qquad (60)$$

V is a diagonal matrix, and the elements along the diagonal represent the variance, so the solution is therefore

$$\mathrm{vec}\left(\Gamma_0\right) = \left(I - \Phi \otimes \Phi'\right)^{-1}\mathrm{vec}\left(V\right). \qquad (61)$$

Thus, the slope of the regression takes the form

$$\eta_n = \frac{(B_1 + B_n - B_{n+1})' \Gamma_0 (B_1 - \Phi' (B_n - B_{n-1}))}{(B_1 + B_n - B_{n+1})' \Gamma_0 (B_1 + B_n - B_{n+1})}. \tag{62}$$

For $n \to \infty$, B_n converges to

$$\lim_{n \to \infty} \eta_n = \frac{B_1' \Gamma_0 B_1}{B_1' \Gamma_0 B_1} = 1. \tag{63}$$

The expectation hypothesis testing forward rate regression result corresponding to (57) is represented in Table 4 and plotted in Figure 4. There we see that the results differ from those anticipated by the expectations hypothesis, and the largest deviations from the expectations hypothesis come at short maturities of 1 year to 3 years, while the regression slopes are close to 1 for maturities from 4 years afterwards. As stated in Backus et al. (2001), in a broad class of stationary models, the theoretical regression slope approaches 1 as n approaches infinity.

This feature of our results is plausible and fits nicely into the stationary bond pricing theory. The estimated coefficients are about 0.96 at long maturities, less than unity as suggested by theory, with standard errors of 0.02 or smaller. The evidence suggests, then, that changes in forward rates at short maturities consist of changes in both short rates and term premiums. It also suggests that the behavior of long forward rates make this constant term premium Gaussian model a possible approximation.

Table 4. Expectation Hypothesis Testing Forward Rate Regression (01/1986–04/2006)

Maturity	1	2	3	4	5	6	7	8	9
Slope coeff.	0.899	0.936	0.951	0.954	0.965	0.965	0.969	0.966	0.971
Std. error	0.031	0.020	0.014	0.012	0.010	0.010	0.012	0.012	0.012

6.4 Econometric Methodology

As stated in Duffee (2002), filtering is a natural approach when the state factors are not observable. In our model, the state factors that determine the dynamics of the yield curve are nonobservable but are defined as Gaussian with the errors that are normally distributed.[6] The Kalman filter and a maximum likelihood procedure would be chosen for the estimation and since x_t has affine dynamics, analytic expressions of the first two moments of the conditional density are available.

---

[6] According to Duffee (2002), "Outside the Gaussian class of term structure models, parameter estimates obtained directly from Kalman filter estimation are inconsistent. There is Monte Carlo evidence that when the underlying model is linear but heteroskedastic, the inconsistency may be of limited importance in practice."

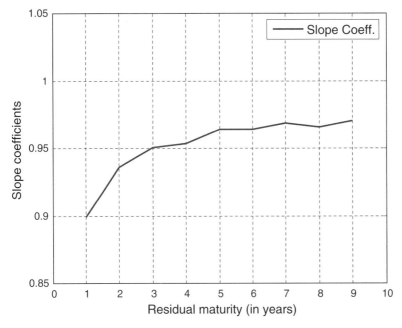

Fig. 4. Expectation hypothesis testing of German zero coupon bond yields (01/1986–04/2006).

The starting point for the derivation of the Kalman filter is to write the model in linear state-space form. Following Dunis et al. (2003), this consists of two equations: One is the observation equation and the other is the state equation:

$$Y_t \underset{r\times 1}{} = \underset{r\times n}{A} \cdot \underset{n\times 1}{X_t} + \underset{r\times k}{H} \cdot \underset{k\times 1}{S_t} + \underset{r\times 1}{w_t}, \tag{64}$$

$$\underset{k\times 1}{S_t} = \underset{k\times 1}{C} + \underset{k\times k}{F} \cdot \underset{k\times 1}{S_{t-1}} + \underset{k\times k}{G} \cdot \underset{k\times 1}{v_t}. \tag{65}$$

r equals the number of variables to estimate, and n is the number of observable exogenous variables. Since in our model the state variables are all latent and no macro factors are identified, the matrix X_t should be a unit matrix, C is a column vector of zeros, and F is a diagonal matrix given that there is no cross-correlation between the state variables in our defined model. k represents the number of latent state variables and w_t and v_t are the identically independently distributed errors as $w_t \sim N(0, R)$ and $v_t \sim N(0, Q)$, where the R and Q are defined as:

$$\underset{r\times r}{R} = E(w_t w_t') \quad \text{and} \quad \underset{k\times k}{Q} = E(v_{t+1} v_{t+1}'). \tag{66}$$

For $y_{n,t} = n^{-1}(A_n + B_n' x_t)$, the observation equation in a Gaussian two-factor model could be written as

$$\begin{bmatrix} y_{1,t} \\ \vdots \\ y_{r,t} \end{bmatrix} = \begin{bmatrix} a_{1,t} \\ \vdots \\ a_{r,t} \end{bmatrix} + \begin{bmatrix} b_{1,1} & b_{2,2} \\ \vdots & \vdots \\ b_{1,r} & b_{2,r} \end{bmatrix} \begin{bmatrix} s_{1,t} \\ s_{2,t} \end{bmatrix} + \begin{bmatrix} w_{1,t} \\ \vdots \\ w_{r,t} \end{bmatrix}, \quad (67)$$

where the dependent column vector is the r zero coupon yields at time t with maturities $n = 12, 24, \ldots, 120$ months in our data set, with $a_n = A_n/n$, $b_n = B_n/n$. The elements $s_{1,t}$ and $s_{2,t}$ consist of two latent variables in consideration at time t, and $w_{1,t}, \ldots, w_{r,t}$ are the normally distributed independent errors with zero mean and variances equal to σ_n^2.

According to (44), the state equation is defined as follows:

$$\begin{bmatrix} s_{1,t+1} \\ s_{2,t+1} \end{bmatrix} = \begin{bmatrix} \varphi_1 & 0 \\ 0 & \varphi_2 \end{bmatrix} \begin{bmatrix} s_{1,t} \\ s_{2,t} \end{bmatrix} + \begin{bmatrix} \sigma_1 & 0 \\ 0 & \sigma_2 \end{bmatrix} \begin{bmatrix} v_{1,t+1} \\ v_{2,t+1} \end{bmatrix}, \quad (68)$$

where $s_{1,t+1}$ and $s_{2,t+1}$ are the two state variables at time $t+1$ and are related to their one-period lagged values by a stable matrix of φ's. A diagonal matrix of σ's is taken out from the error terms to make $v_{1,t+1}$ and $v_{2,t+1}$ with mean zero and variances of ones.

In order to avoid implausible estimates for the volatility curve, as explained in Cassola and Luis (2001), we shall attach the volatility curve estimation equation below the yield curve equation. Therefore, (67) is modified to:

$$\begin{bmatrix} y_{1,t} \\ \vdots \\ y_{r,t} \\ \text{var}_t(y_{1,t+1}) \\ \vdots \\ \text{var}_t(y_{r,t+1}) \end{bmatrix} = \begin{bmatrix} a_{1,t} \\ \vdots \\ a_{r,t} \\ a_{r+1,t} \\ \vdots \\ a_{2r} \end{bmatrix} + \begin{bmatrix} b_{1,1} & b_{2,1} \\ \vdots & \vdots \\ b_{1,r} & b_{2,r} \\ 0 & 0 \\ \vdots & \vdots \\ 0 & 0 \end{bmatrix} \begin{bmatrix} s_{1,t} \\ s_{2,t} \end{bmatrix} + \begin{bmatrix} w_{1,t} \\ \vdots \\ w_{r,t} \\ w_{r+1,t} \\ \vdots \\ w_{2r,t} \end{bmatrix}. \quad (69)$$

The values of the conditional variance on one-period forward lagged yield curve are estimated from the specification in (51), and in this example the number of variables to estimate increases to $2r$ [previously r in (67)].

6.5 Estimation Results

The results of the Kalman filter estimation are presented in Figure 5, which displays the comparison between the observed and estimated yield curves. The estimated yield curve exhibits more curvature in the moderate maturities and showed a slightly higher yield than the observed, but it generally underestimates the true yield at both ends (shorter maturity and longer maturities).

Comparisons between the estimated and observed volatility curves are plotted in Figure 6. Clearly, the model estimated volatility generally overestimated the shorter ends but underestimated the longer ends. This may result

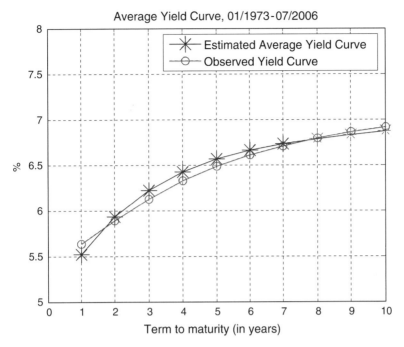

Fig. 5. Observed and estimated average yield curves.

from the constant volatility assumption embedded in this Gaussian model, as the yield volatility does not depend on the level of factor, and the factor loadings are the main contribution to the evolution of yield volatility. Nevertheless, the factors' loadings (as will be shown later in Figure 11) seemed to have larger values at the shorter ends of maturities and smaller ones at the longer ends; thus, the estimated volatility curve diverges from the observed one at both ends.

In Figure 7, the estimated term premium curve is plotted, which can be seen as a plausible result if we linked it with the results in Figure 8 (average one-period forward curve) and Figure 9 (average expected short-term interest rate). Under this model specification, the short-term interest rate should remain constant in the steady state, which leads a flat one-period forward curve under the condition of a zero term premium. Nevertheless, in our model, the term premium is predicted to increase as the maturity of bonds increases (therefore, the average yield curve is upward sloping), so the average one-period forward curve should also have an upward sloping pattern [combination effect of both the term premium and the expected short-term interest rate as shown in (54)].

In Figure 10, we plotted the time-series yield across the range of maturities and indicate the divergence between the estimated and the observed time series. The best fits come from the moderate maturities of 3-year and

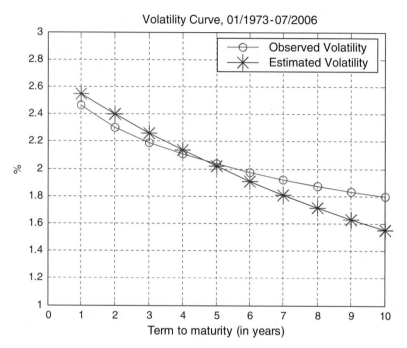

Fig. 6. Observed and estimated volatility curves.

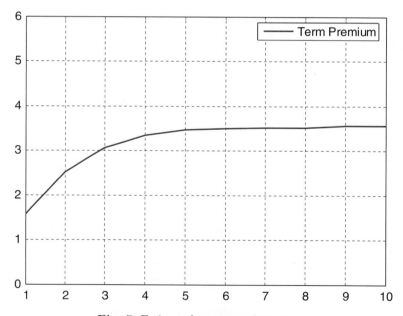

Fig. 7. Estimated term premium curve.

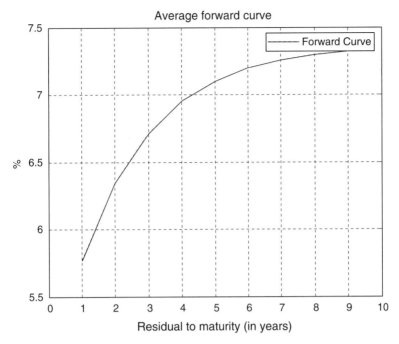

Fig. 8. Estimated average one-period forward rate.

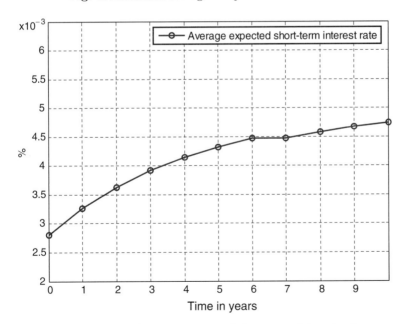

Fig. 9. Estimated average expected short-term interest rate.

4-year bond yields; neither the shorter maturities nor the longer maturities showed a sufficiently close fit between the estimated and observed series. Estimated yield evolution displayed as underestimation in shorter maturities and overestimation in longer maturities.

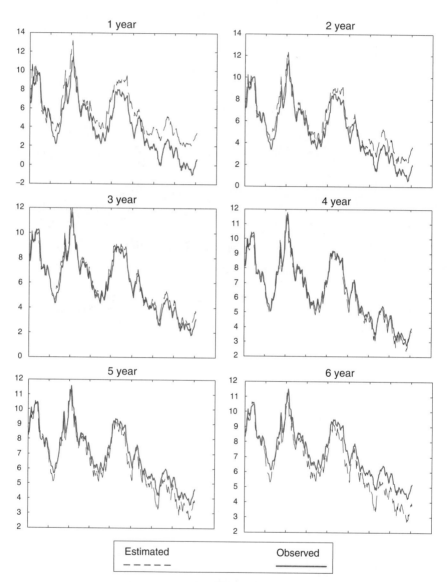

Fig. 10. Time-series yield (01/1973–07/2006).

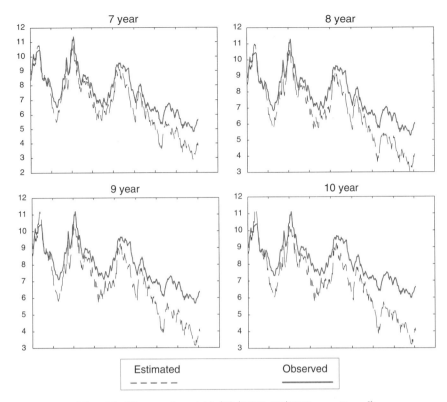

Fig. 10. Time-series yield (01/1973–07/2006, *continued*).

The reproduced average yield curve and volatility curve are plausibly consistent with the findings in Cassola and Luis (2001), yet the time-series evolution of the yields across the whole spectrum in this paper is somewhat different from the findings in their paper. The possible explanations for this would be: first, the use of a different data set; second, in contrast to Gong and Remolona (1996)'s findings of the quality of fit for the yields (U.S. term structure of yields estimated in their paper), at two different two-factor heteroskedastic models to fit for the shorter ends and the longer ends of the yield curve and the volatility curve. Compared to the Gaussian two-factor model in this chapter, the absence of state factors' dynamic in the volatility modeling would not be able to capture the volatility evolution at the whole of the maturity spectrum; therefore, the time-series yields curves showed divergence between predictions and data; third, Cassola and Luis (2001) gave a possible explanation for the observed divergence from the time-series plot starting at the beginning of 1998 as the consequence of the Russian and Asian crisis, events that would violate the stability of the estimates.

Table 5 lists the estimation results of the parameters; they are similar to the parameter estimation results in Cassola and Luis (2001). φ_1 and φ_2

in both papers are found close to 1, indicating the presence of strongly in persistent factors. Later the unit root test on the factors confirmed this point, and this has been referenced in many papers that the driving forces behind the yield curve are best described as random walks.[7]

Table 5. Kalman Filtering Estimation of Parameters of German Zero Coupon Bond Yields (01/1973–07/2006)

	φ_1	φ_2	σ_1	σ_2	$\lambda_1\sigma_1$	$\lambda_2\sigma_2$	δ
Parameters	0.9440	0.9810	0.0009	0.0025	0.2210	−0.1010	0.0050

Figures 11 and 12 present the factors' loadings and the factors time series evolution, respectively. The second factor is clearly the dominant one in explaining the dynamic of yield curve and showed higher persistency, while the first factor is relatively less persistent and volatile.

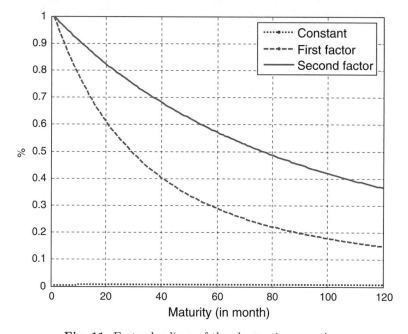

Fig. 11. Factor loadings of the observation equation.

---

[7] Result from the unit root test is not included in this chapter.

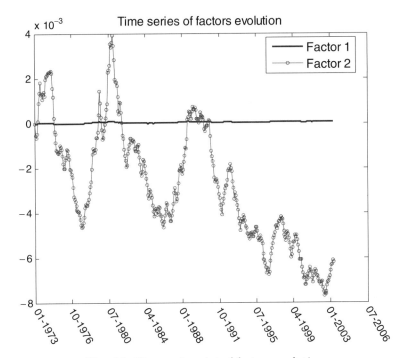

Fig. 12. Time-series plot of factors evolution.

7 Stochastic Term Structure Models

In reality there is no single basic interest rate model that can be used to capture the yield of a specific instrument as such a zero coupon bond. Essentially, there are a mixture of bonds with varying coupons and maturities and terminal payoffs, and these need to be included in the modeling process. Two approaches to the bond pricing problem maybe utilized; first, using the term structure and yield curves to estimate the parameters in a simple spot rate diffusion model, short-rate models, or driving the pricing equation of the bond from the dynamic evolution of the yield curve, and as such we can view bond pricing in terms of the forward rate curve $f(t,s)$ as

$$P(t,T) = P_T e^{-\int_t^T f(t,s)ds}. \tag{70}$$

Calculation of bond prices requires that we know the function $f(t,T)$ at time t for all maturities T. This in turn requires that the forward rate must adequately represent the market view of the future interest rates that are unknown at time t. A modeling device to overcome this problem is to develop techniques to model the evolution of the forward curve from time t, thus allowing the formation of expectations regarding future bond prices.

Markets for such securities are deep and therefore liquid; by utilizing the observed prices of currently traded bond we can extract values of discrete points of the forward curve. Using these results, dynamic models can be postulated and their success will be based on their ability to "forecast" the "observed" discrete values of $f(t,T)$. The selected short-rate model is then benchmarked against the market data for the short-term rate. The link of the forward rate curve with the short-term rate is given as follows:

$$r(t) = f(t,t). \tag{71}$$

The choice of the short-term rate model plays is very important in modeling forward curves. Technically, all short-rate models postulate the short-rate as a Markovian random process; subsequently, a partial differential equation is associated with each model.

Depending upon the choice of the short-rate model, one can derive the whole forward yield curve by considering a short-term rate model with time-dependent coefficients.

The model proposed by Heath et al. (1990) offers an encompassing framework that very well-known short-rate models can be embedded (Soloviev, 2002). They modeled the whole forward curve (which is just another name for the yield curve) by introducing a different stochastic process for the forward rate. In its most general form, the model is non-Markovian and cannot be described by a differential equation. The HJM model is considered the most appropriate device for the correct pricing of interest rate derivatives.

For any given set of bond prices, the forward rate is given as

$$f(t,T) = \frac{\partial \ln P(t,T)}{\partial T}. \tag{72}$$

Thus, from the market prices of bonds at maturities T, we compute the forward rate at the present time t, and vice versa.

This dual relationship between bond prices and forward rates can be used to develop the HJM model in terms of 1) the stochastic process of bond prices and then derive the SDE for the forward rate, or 2) develop the forward rate model and then compute from the appropriate SDE the equilibrium bond prices. Consider the equation describing the dynamic evolution of the bond price as the usual geometric Brownian motion

$$dP(t,T) = \mu(t,T)P(t,T)dt + \sigma(t,T)P(t,T)dW(t). \tag{73}$$

At maturity we expect that $\sigma(T,T) = 0$; from Ito's lemma, the stochastic differential equation for the forward rate is

$$df(t,T) = \frac{\partial}{\partial T}\left(\tfrac{1}{2}\sigma^2(t,T) - \beta(t,T)\right)dt - \frac{\partial}{\partial T}\sigma(t,T)dW(t). \tag{74}$$

This equation provides estimates of the forward rate that are consistent with the absence of arbitrage.

The replication approach to pricing derivatives allows the use of a risk-neutral random process such that $\beta(t,T) = r(t)$, where $r(t)$ is the risk-free rate. Substituting in the above, we obtain

$$df(t,T) = \sigma(t,T)\frac{\partial}{\partial T}\sigma(t,T)\,dt - \frac{\partial}{\partial T}\sigma(t,T)\,dW(t), \tag{75}$$

or more compactly,

$$df(t,T) = \alpha(t,T)dt + \phi(t,T)dW(t), \tag{76}$$

where the drift parameter is a function of the forward rate volatility $\phi(\cdot)$:

$$\begin{aligned}\alpha(t,T) &= \phi(t,T)\int_t^T \phi(t,s)ds,\\ \phi(t,s) &= \frac{\partial \sigma(t,T)}{\partial T}.\end{aligned} \tag{77}$$

The use of risk-neutral valuation requires that the model should be calibrated to reproduce the observed bond market data, thus avoiding the possibility of arbitrage. Alternatively, we can derive the same relationship between the drift and the forward rate volatility starting from the dynamic equation of the forward rate and substituting into the bond pricing equation and its dynamic evolution.

What is of interest here is that the forward rate dynamics and thus bond price evolution depend upon the forward rate volatility, the "driving" variable. The model does not provide the answer of how to compute it; it takes it as given, and it is up to the researchers to determine the most appropriate measure from market data/guess/estimate.

We can now use the analytical framework developed above to encompass the most popular short-rate models.

The solution to the forward rate dynamic equation is

$$f(t,T) = f(0,T) + \int_0^t \alpha(s,T)ds + \int_0^T \phi(s,T)dW(s), \tag{78}$$

and the resulting short rate is

$$r(t) = f(0,t) + \int_0^t \alpha(s,t)ds + \int_0^t \phi(s,t)dW(s). \tag{79}$$

The risk-neutral valuation requires that the computed short rate is calibrated to the market. To derive the Ho–Lee model, consider the following version of the HJM model:

$$df(t,T) = \sigma^2(T-t)dt + \sigma dW(t). \tag{80}$$

The short-rate model is therefore

$$r(t) = f(t,t) = f(0,t) + \sigma^2\int_0^t (t-s)ds + \sigma\int_0^t dW(s). \tag{81}$$

Taking the time derivative, we obtain

$$dr(t) = \phi(t)dt + \sigma dW(t), \tag{82}$$

which is the arbitrage-free version on the Ho–Lee model that can be calibrated to market by choosing the drift parameter $\phi(t)$, with $\phi(T) = df(0,T)/dT + \sigma^2 T$, with the forward yield curve $f(0,T)$ determined from market information (bond prices).

Consider the forward rate volatility:

$$\phi(t) = \sigma e^{-\kappa(T-t)}. \tag{83}$$

The resulting forward curve is given by

$$f(0,T) = \frac{\phi}{\kappa} + e^{-\kappa T}\left(r_0 - \frac{\phi}{\kappa}\right) - \frac{\sigma^2}{2k^2}(1 - e^{-\kappa T})^2, \tag{84}$$

and the corresponding SDE describing the dynamics of the short rate is equivalent to the Hull–White model of the short rate:

$$dr(t) = (\phi - \kappa r(t)) + \sigma dW(t) \tag{85}$$

(this is a general representation of the widely used Vasicek model that is discussed in the previous section).

To obtain the CIR model, consider the following representation of the short-rate volatility:

$$\sigma(t,T) = \sigma(t)\sqrt{r(t)}\frac{\partial P(t,T)}{\partial T}. \tag{86}$$

A bond price equation can be approximated by

$$P(t,T) = \frac{1}{\alpha}(1 - e^{-k(T-t)}). \tag{87}$$

The derivative of bond prices with respect to different maturities is

$$\frac{\partial P(t,T)}{\partial T} \approx e^{-k(T-t)} = \Delta(t,T). \tag{88}$$

The forward rate equation can be expressed in terms of the above as

$$f(0,T) = r(0)\Delta(0,T) + \int_0^T \phi_s \Delta(s,T)ds, \tag{89}$$

and the corresponding short-rate dynamics obey the CIR dynamic equation:

$$dr(t) = (\phi(t) - \alpha(t)r(t)) + \sigma(t)\sqrt{r(t)}dW(t). \tag{90}$$

8 Multifactor HJM Models

The stochastic interest rate models we have considered so far all have one stochastic process in their dynamic equations. For this reason, these models are called one-factor models. The major limitation of the single-factor model is that they predict that all the segments of the forward curve are perfectly correlated. One-factor models allow only coherent changes of the entire forward curve. This prediction is not supported by the data, as it contradicts market observations of the forward curve movements. In practice, short-term rates of interest are not perfectly correlated with long-term rates. Therefore, one-factor models cannot provide an adequate description of the time evolution of the entire forward curve.

Thus, in order to model the dynamic of the entire yield curve, we need a model with more than one stochastic factor. The simplest modification of a one-factor model is to consider two-factor models where one factor is responsible for the movements of the short end and the other is responsible for the evolution of the long end. Many two-factor models are formulated directly in terms of two interest rates: the *short rate* r and some *long rate*. A simple two-factor model allows for two different stochastic processes to enter the rates at the short and long end of the yield curve:

$$dr_s(t) = (\pi_1 - \pi_2(r_L(t) - r_S(t)))dt + \sigma_S r_S(t)dW_1(t), \tag{91}$$
$$dr_L(t) = r_L(t)(\pi_3 - \pi_4(r_S(t) - r_L(t)))dt + \sigma_L r_L(t)dW_2(t).$$

In this version of the model the short-rate equation exhibits mean reversion, implying that in the long run the short rate tends to the long rate. The stochastic terms in the model are of log-normal form. By and large, closed-form model solutions are very difficult to come by, and so numerical methods are employed for computation of the equilibrium rates.

Within the context of the HJM model, we can introduce more than one factor by adopting the following specification:

$$df(t,T) = \alpha(t,T)dt + \sum_{i=1}^{n} \phi_i(t,T)dW_i(t), \tag{92}$$

where n uncorrelated stochastic factors are included in the description of the forward curve. As in multifactor pricing models, there is no theory regarding the choice of factors. A convenient technique to incorporate the factors into the forward rate equation is principal components analysis. The technique can be applied in terms of the following sequence of calculations.

First, we determine the functional form of the volatility coefficients:

$$\phi_i(t,T) = \hat{\phi}_i(T-t). \tag{93}$$

By assuming that the number of factors equals the number of the maturities of the traded bonds, we can write the volatility coefficients for all market available maturities as the following matrix:

$$\Phi_{i,j} = \hat{\phi}_i(T_i - t). \tag{94}$$

The covariance matrix of the forward rate increments can then be calculated using time-series observations for each maturity:

$$\Psi_{i,j} = \text{cov}(df(T_i,t)df(T_j - t)) = \{\Phi\Phi'\}_{i,j}. \tag{95}$$

We can then compute the eigenvalues and eigenvectors of $\Psi_{i,j}$ and express the volatility coefficients as

$$\Phi_{i,j} = \sqrt{\lambda_i}(\gamma_i)_j, \tag{96}$$

where $(\gamma_i)_j$ denotes the jth element of the vector γ_i. The eigenvector that corresponds to the largest eigenvalue is the first principal component and denotes the most important "driving" variable in the determination of the forward curve.

9 Market Models: The LIBOR Approach

A relatively new and exciting approach to modeling interest rates is the LIBOR market model family. This is essentially arbitrage free modeling of market rates, in particular attempting to capture the properties of the evolution of the *prime* or *base* rate. LIBOR stands for the *London Inter-Bank Offered Rate*, which is the interest rate offered by banks in Eurocurrency markets to other banks. This is the floating rate commonly used in floating swap agreements and is commonly used as the basis of spreads for loans of varying risk. In general, a Heath–Jarrow–Morton approach is always used as the fundamental starting block for the majority of LIBOR models. This section will review the basic LIBOR approaches and suggest strategies for implementing LIBOR models. The interested reader is directed to Musiela and Rutkowski (2004), Rebonato (2002, 2004), and Brigo and Mercurio (2006) for extended reading on market models. In general, this section will summarize the key results of the LIBOR model and outline implementation strategies for this popular family of interest rate models.

9.1 The LIBOR Market

The basic Heath–Jarrow–Morton methodology approaches the modeling of term structure via instantaneous, continuously compounded forward rates. In practice, this does not always provide the closest fit to reality. Before the current popularity of the market model approach, the pricing of interest rate derivatives was almost exclusively tasked to the short-rate model genre, in particular those models with simple closed-form analytical pricing solutions. The following section follows the fundamental work of Musiela and Sondermann (1993) and Brace et al. (1997). Consider the jth discrete rate from n possible rates; this is related to an instantaneous forward rate as follows:

$$1 + j(t, T) = \exp(f(t, T)). \tag{97}$$

For an annual compounding system with rate $q(t, T)$, with δ compounding periods, the nominal rate will be

$$(1 + \delta q(t, T))^{\frac{1}{\delta}} = \exp(f(t, T)). \tag{98}$$

Happily, if we set $\gamma(t, T)$ as the deterministic volatility of each nominal annual rate, $q(t, T)$, we can see that

$$\sigma(t, T) = \delta^{-1} (1 - \exp(-\delta f(t, T))) \gamma(t, T). \tag{99}$$

As we do not have a money account of the form $E\left(\exp\left(\exp\left(\bar{f}(t, T)\right)\right)\right)$, the rates and the value of the money account are nonexplosive. Unfortunately, there is no closed-form solution to the price of a zero coupon bond, just an infinite sum of disappearing forward rates as such. Musiela (1994), Brace et al. (1997), and Miltersen et al. (1997) all effectively concluded that the most appropriate methodological aim should be to focus on the effective forward rate, defined by

$$(1 + \delta f_s(t, T, T + \delta))^{\delta} = \exp\left(\int_T^{T+\delta} f(t, u) \, du\right), \tag{100}$$

where $(1 + \delta f_s(t, T, T + \delta))^{\delta}$ is the effective discrete compounding forward rate over the time interval δ. When interpolating the yield curve, we effectively look at the evolution of the *tenor* structure over prespecified collection of settlement dates; ordering these dates $0 \leq T_0 < T_1 < \cdots < T_n$, with some collection of time differences, $\delta_j = T_j - T_{j-1}$, using an indexing approach $j \in [1, ..., n]$, gives us our market modeling approach to the term structure of interest rates. We can now define the jth LIBOR rate as

$$1 + \delta_{j+1} L(t, T_j) \triangleq \frac{P(t, T_j)}{P(t, T_{j+1})}. \tag{101}$$

Modeling the evolution of the jth LIBOR rate as a log-normal process then,

$$dL(t, T_j) = L(t, T_j) \lambda(t, T_j) dW_t^{T_{j+1}}, \tag{102}$$

and with this representation we can define the following expectation:

$$E(L(t, T_j)) = E\left(\frac{1 - P(T_j, T_{j+1})}{\delta_{j+1} P(T_j, T_{j+1})} \bigg| \mathcal{F}_t\right). \tag{103}$$

Having defined a possible model for the evolution of each forward rate, now we can use these diffusions to prices a variety of rate-dependent instruments.

9.2 Basic LIBOR Models: LLM/LFM

The major advantage of LIBOR market models is their ability to capture and utilize information on the forward yield curve structure of interest rates. In this sense we start by reviewing two approaches to deriving the *forward* LIBOR curve. The initial work of Miltersen et al. (1997) suggested the following stochastic differential equation:

$$dL(t,T) = \mu(t,T)\,dt + L(t,T)\,\lambda(t,T)\,dW_t^*. \tag{104}$$

This forms the basis of most common LIBOR model, the Log-Normal-LIBOR Market Model (LLM), or the Log-Normal Forward LIBOR Market Model, (LFM).

9.3 The Brace–Gatarek–Musiela Log-Normal LIBOR Market Model

This section follows closely the derivation of Brigo and Mercurio (2006), Chapter 6 and Musiela and Rutkowski (2000), Chapter 12, and describes the derivation of the Brace–Gatarek–Musiela model proposed in Brace et al. (1997). Consider a collection of Band prices and a corresponding collection of forward processes:

$$\begin{array}{ll} \text{Bond} & \text{Forward Processes} \\ P(t,T_1), P(t,T_2) & f_P(t,T_2,T_2) = \dfrac{P(t,T_1)}{(P(t,T_2))} \\ P(t,T_i), P(t,T_j) & f_P(t,T_i,T_j) = \dfrac{P(t,T_i)}{(P(t,T_j))} \end{array} \tag{105}$$

By definition from the basic setup, if the basic δ-LIBOR forward rate is defined by a simple zero drift stochastic differential equation,

$$dL(t,T) = \delta^{-1}(1+\delta L(t,T))\,\lambda(t,T,T+\delta)\,dW_t^{T+\delta}, \tag{106}$$

where for any maturity there is a real bounded process, $\lambda(t,T)$ representing the volatility for the forward LIBOR process $L(t,T)$, then the initial term structure (Musiela, 1997) is as follows:

$$L(0,T) = \frac{P(0,T) - P(0,T+\delta)}{\delta P(0,T+\delta)}. \tag{107}$$

If $\lambda(t,T)$ is deterministic, then the model collapses to a standard log-normal representation, over some fixed accrual period. If $\lambda(t,T)$ is a stochastic process, then the LIBOR model is effectively a stochastic volatility process; as such, log-normality no longer necessarily holds. The Brace–Gatarek–Musiela approach is to set the feasible LIBOR rates in a Heath–Jarrow–Morton framework and then set out the LIBOR rates as the instantaneous forward rates. Therefore, if the evolution of the forward rate is log-normal,

$$df(t,T) = \mu(t,T)\,dt + \sigma(t,T)\,dW_t^*, \qquad (108)$$

where

$$\mu(t,T) = \sigma(t,T_1)\sigma(t,T_2), \quad T_1 < T_2 + \delta, \qquad (109)$$

then the LIBOR rate is defined as

$$\sigma(t,T+\delta) - \sigma(t,T) = \int_T^{T+\delta} \sigma(t,u)\,du = \frac{\delta L(t,T)}{1+\delta L(t,T)}\lambda(t,T). \qquad (110)$$

The general Brace–Gatarek–Musiela approach sets $\sigma(t,T) = 0$ and yields a pricing model for a zero coupon bond:

$$P(t,T) = -\sum_{k=1}^{n=\frac{T}{\delta}} \frac{\delta L(t,T-k\delta)}{1+\delta L(t,T-k\delta)}\lambda(t,T-k\delta). \qquad (111)$$

This neat pricing formula is only available when $\lambda(t,T)$ is deterministic, and this model is one of the most commonly used market models.

9.4 The Musiela–Rutkowski and Jamshidian Approach

The previous pricing model uses forward induction to form an analytical pricing model for a zero coupon bond. Musiela and Rutkowski (2004) took the original concept and inverted the induction process. This backward induction approach was then modified by Jamishidian (1997), who put forward a pricing model that described the whole family of LIBOR rates in a systematic framework. Recall that the maturity of a family of bonds and the relationship to a forward LIBOR rate

$$1 + \delta_{j+1}L(t,T_j) = \frac{P(t,T_j)}{P(t,T_{j+1})}.$$

The major assumption is as follows: Assume that the evolution of all bond prices characterized by some stochastic process of the form

$$dP(t,T_j) = P(t,T_j)(a(t,T_j))\,dt + b(t,T_j)\,dW_t,$$

where $W_t \in \mathbb{R}^d$ and $a(\cdot)$ and $b(\cdot)$ are arbitrary affine functions; as such, all bond prices will in effect be determined via some collection of LIBORs. Jamishidian (1997) demonstrated that under this assumption, we can characterize the evolution of LIBOR rate as the following SDE:

$$dL(t,T_j) = \mu(t,T_j)\,dt + \zeta(t,T_j)\,dW_t^j,$$
$$\mu(t,T_j) = \frac{P(t,T_j)}{\delta_{j+1}P(t,T_{j+1})}(a(t,T_j) - a(t,T_{j+1})) - \zeta(t,T_j)b(t,T_{j+1}),$$
$$\zeta(t,T_j) = \frac{P(t,T_j)}{\delta_{j+1}P(t,T_{j+1})}(b(t,T_j) - b(t,T_{j+1})).$$
$$(112)$$

Furthermore, he showed that

$$b(t, T_{j_0}) - b(t, T_{j+1}) = \sum_{k=j_0}^{j} \frac{\delta_{k+1} \zeta(t, T_k)}{1 + \delta_{k+1} L(t, T_k)}. \tag{113}$$

From this setup we can arrive at the three major results from Jamshidian (1997): the general solution

$$dL(t, T_j) = \sum_{k=j_0}^{j} \frac{\delta_{k+1} \zeta(t, T_k) \zeta(t, T_j)}{1 + \delta_{k+1} L(t, T_k)} dt + \zeta(t, T_j) dW_t^*,$$

where dW_t^* is a d-dimensional Weiner–Brownian motion. If we set the volatility component $\zeta(t, T_j)$ to be some deterministic function of the individual volatilities of each of the available LIBOR rates,

$$\zeta(t, T_j) = \lambda_j \left(t, L(t, T_{j_0}), L(t, T_{j_0+1}), ..., L(t, T_j), ..., L\left(t, T_{\frac{T}{\delta}}\right) \right), \tag{114}$$

and simplifying further to impose the log-normal conditions, we find the following SDE representation of the individual LIBOR processes:

$$\frac{dL(t, T_j)}{L(t, T_j)} = \sum_{k=j_0}^{j} \frac{\delta_{k+1} L(t, T_k) \lambda(t, T_k) \lambda(t, T_j)}{1 + \delta_{k+1} L(t, T_k)} dt + \lambda(t, T_j) dW_t^*. \tag{115}$$

The interested reader is directed to Jamshidian (1997) for the full derivation. However, the implication of this SDE is very visible, as it sets out the evolution of an individual rate as being some function of all the other rates. In particular, the volatilities of each forward rate will in some way be correlated.

10 Volatility and Correlation in Forward Rates

The previous section suggested a fundamental set of interrelationships between LIBOR rates. In this section, we shall review the Rebonato (2004) approach to specifying the LIBOR dynamics and rerelate this to the principal component analysis suggested previously. At this point we begin to set some fundamental structure to the LIBOR rates and evolve them contemporaneously as a vector process, instead of taking each rate in isolation. Consider an n-length vector of LIBOR rates with k factors, representing k underlying stochastic processes driving the evolution of these forward rates. We can specify the evolution of this system as follows:

$$dL_t = \begin{bmatrix} dL(t, T_1) \\ dL(t, T_2) \\ \vdots \\ dL(t, T_n) \end{bmatrix}. \tag{116}$$

Simplifying our notation from the previous section, we set $dL(t, T_j) \in dL_t$; then we can define the instantaneous volatility of the jth rate as $\sigma(t, T_j)$ and the correlation between the ith and jth rates as $\rho(t, T_i, T_j)$.

The simplest specification sets a deterministic drift measure $\mu(L_t, t) \in \mathbb{R}^n$ and a single covariance matrix that describes the volatility and correlation of the Weiner processes:

$$dL_t = \mu(L_t, T) + \Sigma(L_t, T) dW_t^*. \tag{117}$$

Again, $dW_t^* \in \mathbb{R}^d$. In this case we decompose $\Sigma(L_t, t) = H(L_t, t) \circ R(L_t, t)$, where $\circ$ is the element by element multiplication of two identical arrays, i.e., the Hadamard product. Setting

$$\begin{aligned} H(L_t, t) &= [\sigma(t, T_i) \sigma(t, T_j)], \\ R(L_t, t) &= [\rho(t, T, T_j)], \end{aligned} \tag{118}$$

we implicitly assume that $n = k$, i.e., for each forward rate there is a factor underlined by an independent Wiener–Brownian motion and that the volatility/dependence structure of this variation is fully described by $\Sigma(L_t, t)$, the instantaneous covariance matrix. It is unlikely that the number of state variables exactly matches the number of available forward rates. Brigo and Mercurio (2006) suggested that between 85 and 97% of all variation may be accounted for by as few as three factors.

Let us now consider the case when $k < n$. In this circumstance we can use principal component analysis to identify the common structure. Knowing that

$$W_{t+h} - W_t \sim N\left(\underset{k\times 1}{\mathbf{0}}, \underset{k\times k}{\mathbf{I}}\right),$$

we can respecify (117) as follows

$$dL_t = \mu(L_t, T) + \mathbf{\Gamma}(L_t, T)^{'} (\mathbf{\Omega}(L_t, T) dW_t^*) \mathbf{\Gamma}(L_t, T), \tag{119}$$

where $\mathbf{\Omega}(L_t, t) \in \mathbb{R}^{k\times k}$ is the instantaneous factor covariance matrix and $\mathbf{\Gamma}(L_t, t) \in \mathbb{R}^{n\times k}$ are the instantaneous factor loading matrices. If we define

$$\Theta(L_t, T) = \mathbf{\Gamma}(L_t, T)^{'} (\mathbf{\Omega}(L_t, T)) \mathbf{\Gamma}(L_t, T) \tag{120}$$

and set ς_j as the jth eigenvalue of $\Theta(L_t, T)$, then the number of factors, k, will be the number of nonzero eigenvalues of $\Theta(L_t, T)$. This identification procedure can be used to define the factor loadings specified in the previous section.

11 Concluding Remarks

Free from the optimizing dynamic framework of speculative agents but based on arbitrage-free pricing theory, there is a wealth of models of the term structure of interest rates that has produced truly innovative models and that are of great interest to both academics and practitioners.

Government bonds are unlike other asset classes such as equities for which the most popular of analytical devices the log-normal Black–Scholes framework is universally accepted. The reason is that, unlike such securities, the simultaneous "random" fluctuation in the whole of the yield curve, a locus of the association between bond yield and maturity, is a complex phenomenon, and a single movement in the share price or exchange rate fluctuation that we may analyze using the BS model cannot approximate satisfactorily. While for a single stock or exchange we are dealing with scalar dynamics, in the case of the yield curve the model attempts to account for the dynamics of each element of a vector, the elements being correlated.

The contribution of all the interest rate models reviewed in this chapter provided a rigorous and convincing (no-free lunch) framework that allows to price interest rate-sensitive securities and their derivatives.

All the models are mathematically sophisticated, and yet they allow us freedom to specify key parameters such as the relevant dynamic stricture, the nature of volatility, and the numbers of factors. The main conceptual difference between them is the bond pricing framework that is implicitly assumed in their construction.

This assumption has allowed this review to characterize this set of models into three major categories that, although they are not mutually exclusive, allow for a more systematic study of the theoretical and applied approaches to interest rate modeling and subsequent bond pricing.

The three major families of models are spot rate, forward rate, and market models. Although all three of these prescriptions are mathematically consistent (by definition of a term structure model), each approach leads to distinct development, implementation, and calibration issues. Moreover, the resulting intuition gained from using the models, especially important for relating the pricing and hedging of products based on the model, is different in each case. Before the characteristic of each framework is elaborated and comparisons are made between them, it is useful to first discuss what is required in any term structure model.

All models are based on the arbitrage-free principle, thus ensuring that by the appropriate choice of measure the market price of risk is removed. Technically, this is achieved by formulating the drift parameter in the diffusion equation that describes the dynamics of the spot rate and/or the forward rates.

Spot Rate Models

The first generation of models developed were generally spot rate-based. This choice was due to a combination of mathematical convenience and tractability, or numerical ease of implementation. Furthermore, the most widely used of these models is one-factor models, in which the entire yield curve is specified by a single stochastic state variable, in this case the spot or short-term rate. Examples of these include the models of Vasicek (1977), Ho and Lee (1986), Hull and White (1990), Black et al. (1990), and Black and Karasinski (1991), to name but the most well known.

These models are distinguished by the exact specification of the spot rate dynamics through time, in particular the form of the diffusion process, and hence the underlying distribution of the spot rate.

Forward Rate Models

An alternative approach to modeling the term structure was offered by the Heath et al. (1990) structure. In contrast to the spot rate approach, they model the entire yield curve as a state variable, providing conditions in a general framework that incorporates all the principles of arbitrage-free pricing and discount bond dynamics. The HJM methodology uses as the driving stochastic variable the instantaneous forward rates, the evolution of which is dependent on a specific (usually deterministic) volatility function.

Because of the relationship between the spot rate and the forward rate, any spot rate model is also an HJM model. In fact, any interest rate model that satisfies the principles of arbitrage-free bond dynamics must be within the HJM framework.

Market Models

The motivation for the development of market models arose from the fact that, although the HJM framework is appealing theoretically, its standard formulation is based on continuously compounded rates and is therefore fundamentally different from actual forward LIBOR and swap rates as traded in the market. The log-normal HJM model was also well known to exhibit unbounded behavior (producing infinite values) in contrast to the use of log-normal LIBOR distribution in Black's formula for caplets. The construction of a mathematically consistent theory of a term structure with discrete LIBOR rates being log-normal was achieved by Miltersen et al. (1997) and developed by Brace et al. (1997). Jamishidian (1997) developed an equivalent market model based on log-normal swap rates.

The main conclusion of the extensive body of literature regarding the performance of these models is that there is no clear "winner" as such. The model performance relates strongly to the type of instrument one wishes to price.

Empirically, the estimation and calibration of multifactor models in an arbitrage-free framework, such as the example we have explored in this review, are computationally very intensive and require the application of very sophisticated econometric techniques. The use of affine models for interest rate modeling constitutes the dominant econometric methodology, although the choice and contribution of factors provide for a lively debate. Of crucial importance is the imposition of parameter restrictions on the affine structure; without such an imposition, the reliability of the econometric results will be questionable.

There is no doubt that this approach will be extended to more complicated functional forms and more importantly to ever-increasing dimensions as techniques from physics and other natural sciences that deal with multi-dimensional problems percolate into the empirical finance literature.

References

1. Ait-Sahalia, Y. Testing continuous-time models of the spot interest rate. *Review of Financial Studies* 9:427–470, 1996.
2. Backus, D. K., Foresi, S., and Telmer, C. I. Affine models of currency pricing. NBER Working paper, 5623, 1996.
3. Backus, D. K., Foresi, S., and Telmer, C. I. Discrete-time models of bond pricing. NBER Working paper, 6736, 1998.
4. Backus, D. K., Foresi, S., Mozumdar, A., and Wu, L. Predictable changes in yields and forward rates. NBER Working paper, 6379, 2001.
5. Balduzzi, P., Das, S. R., and Foresi, S. The central tendency: A second factor in bond yields. *Review of Economics and Statistics*, 80:62–72, 1998.
6. Balduzzi, P., Das, S. R., Foresi, S., and Sundaram, R. A simple approach to three factor affine term structure models. *Journal of Fixed Income*, 6:43–53, 1996.
7. Barr, D. G., and Campbell, J. Y. Inflation, real interest rates, and the bond market: A study of UK nominal and index-linked government bond prices. NBER Working paper, 5821, 1996.
8. Black, F., Derman, E., and Toy, W. A one factor model of interest rates and it's Application to treasury bond options. *Financial Analysis Journal*, 46:33–39, 1990.
9. Black, F. and Karasinski, P. Bond and option pricing when short rates are log-normal. *Financial Analysts Journal*, 47(4):52–59, 1991.
10. Bolder, D. J. Affine term-structure models: Theory and implementation. Bank of Canada, Working paper 15, 2001.
11. Bolder, D. J., Johnson, G., and Mezler, A. An empirical analysis of the candian term structure of zero coupon interest rates. Bank of Canada, Working paper 2004-48, 2004.
12. Brace, A., Gatarek, D., and Musiela, M. The market model of interest rate dynamics. *Mathematical Finance*, 7:127–154, 1997.
13. Brigo, D., and Mercurio, F. *Interest Rate Model-Theory and Practice: With Smile, Inflation and Credit*. Springer, Berlin, 2006.
14. Campbell, J. Y., Lo, A. W., and MacKinlay, A. C. *The Econometrics of Financial Markets*. Princeton University Press, Princeton, NJ, 1997, pp. 427–464.
15. Cassola, N., and Luis, J. B. A two-factor model of the German term structure of interest rates. European Central Bank, Working paper, 46, 2001.
16. Chen, R., and Scott, L. Multi-factor Cox-Ingersoll-Ross models of the term structure: Estimates and test from a Kalman filter. University of Georgia, Working paper, 1993.
17. Cox, J., Ingersoll, J., and Ross, S. A Theory of the term structure of interest rates. *Econometrica*, 53:385–407, 1985.
18. Dai, Q., and Singleton, K. J. Specification analysis of affine term structure models. *Journal of Finance*, 55:1943–1978, 2000.
19. Dai, Q., and Singleton, K. J. Expectation puzzles, time-varying risk premia, and affine models of the term structure. *Journal of financial Economics*, 63:415–441, 2002.

20. David, J. B., Johnson, G., and Metzler, A. An empirical analysis of the Canadian term structure of zero-coupon interest rates. Bank of Canada, Working paper, 18, 2004.
21. Dothan, U. On the term structure of interest rates. *Journal of Financial Economics*, 6:59–69, 1978.
22. Duarte, J. Evaluating an alternative risk preference in affine term structure models. *The Review of Financial Studies*, 17:379–404, 2004.
23. Duffee, G. Term premia and interest rate forecasts in affine models. *Journal of Finance*, 57:527–552, 2002.
24. Duffie, D., and Kan, R. A yield factor model of interest rates. *Mathematical Finance*, 6:379–406, 1996.
25. Duffie, D., and Singleton, K. Simulated moments estimation of Markov models of asset prices. *Econometrica*, 61:929–952, 1993.
26. Dunis, C. L., Laws, J., and Naim, P. *Applied Quantitative Methods for Trading and Investment.* John Wiley & Sons, New York, 2003, pp. 71–128.
27. Gallant, A. R., and Tauchen, G. Simulated score methods and indirect inference for continuous-time models. Department of Economics, Duke University, Working paper, 02-09, 2002.
28. Gerlach, S. The information content of the term structure: Evidence for Germany. BIS Working paper, 29, 1995.
29. Glassermann P. *Monte Carlo Methods in Financial Engineering.* Springer, Berlin, 2003.
30. Gong, F. F., and Remolona, E. M. A three-factor econometric Model of the U.S. term structure. Federal Reserve Bank of New York Working paper, 1996.
31. Hamilton, J. D. *Time Series Analysis.* Princeton University Press, Princeton, NJ, 1994, pp. 372–408.
32. Heath, D., Jarrow, R., and Morton,A. Contingent claim valuation with a random evolution of interest rates. *Review of Futures Markets*, 9:54–76, 1990.
33. Ho, T. S. Y., and Lee, S. B. Term structure movements and the pricing of interest rate contingent claims. *Journal of Finance*, 41:1011–1029, 1986.
34. Hull J. C., and White, A. Pricing interest rate derivative securities. *Review of Financial Studies*, 3:573–592, 1990.
35. Jamishidian, F. LIBOR and swap market models and measures. *Finance and Stochastics*, 1:293–330, 1997.
36. Jorion, P. and Mishkin, F. A multicountry comparison of term-structure forecasts at long horizons. *Journal of Financial Economics*, 29:59–80, 1991.
37. Karatzas, I., and Shreve, S. *Brownian motion and stochastic calculus*, 2nd edition. Springer, Berlin, 1998.
38. Litterman, R., and Scheinkman, J. Common factors affecting bond returns. *Journal of Fixed Income*, 1:49–53, 1991.
39. Mercurio, F. and Morelada, J. M. An analytically tractable interest rate model with humped volatility. *European Journal of Operational Research*, 120:205–214, 2000.
40. Miltersen, K., Sandmann, K., and Sondermann, D. Closed form solutions for term structure derivatives with log-normal interest rates. *Journal of Finance*, 52:409–430, 1997.
41. Musiela M. Nominal rates and lognormal volatility structure. University of New South Wales, Working paper, 1994.
42. Musiela, M., and Rutkowski, M. *Martingale Methods in Financial Modelling.* Springer, Berlin Germany, 2004.

43. Musiela, M., and Sondermann, D. Different dynamical specifications of the term structure of interest rates and their implications. University of Bonn, Working paper, 1993.
44. Pelsser A. *Efficient Methods for Valuing Interest Rate Derivatives.* Springer, New York, 2000.
45. Piazzesi, M. *Affine Term Structure Models.* University of Chicago, Preprint, 2003.
46. Rebonato R. *Modern Pricing of Interest Rate Derivtives: The LIBOR Market Model and Beyond.* Princeton University Press, Princeton, NJ, 2002.
47. Rebonato R. *Volatility and Correlation: The Perfect Hedger and the Fox.* John Wiley & Sons, New York, 2004.
48. Ross, S. A. The arbitrage theory of capital asset pricing. *Journal of Economic Theory*, 13:341–360, 1976.
49. Shreve, S. *Stochastic Calculus for Finance II: Continuous Time Models.* Springer, New York, 2004.
50. Sandmann, K. and Sondermann, D. A note on the stability of lognormal interest rate models and the pricing of Eurodollar futures. *Mathematical Finance*, 7:119–125, 1997.
51. Soloviev, O. *Mathematical Models in Economics.* Queen Mary University of London, Department of Physics, 2002.
52. Vasicek, O. A. An equilibrium characterization of the term structure. *Journal of Financial Economics*, 5:177–88, 1977.

Engineering a Generalized Neural Network Mapping of Volatility Spillovers in European Government Bond Markets

Gordon H. Dash, Jr.[1] and Nina Kajiji[2]

[1] 777 Smith Street, Providence, RI, 02908, USA GHDash@uri.edu
[2] National Center on Public Education & Social Policy, University of Rhode Island, 80 Washington Street, Providence, RI 02903, USA nina@uri.edu

1 Introduction

The recent convergence of two allogeneous disciplines, financial and computational engineering, has created a new mode of scientific inquiry. Both disciplines rely upon theoretical analysis and physical experimentation as tools for discovering new knowledge. With these interrelated roots, it is not surprising that the two disciplines have come together to provide researchers with new and exciting tools by which to extend findings borne of more traditional approaches to scientific inquiry. Despite its relatively young age, the discipline of financial engineering has already provided a unique and substantial body of knowledge that is best described as contemporary financial innovation. The discipline's innovation process is best characterized by its ability to enumerate engineering solutions to vexing financial problems by drawing upon the multidisciplinary use of stochastic processes, optimization methods, numerical techniques, Monte Carlo simulation, and data analytic heuristics. It is within this multidisciplinary characterization that the study of European government bond markets is styled. In this chapter, we seek to utilize the innovation inherent in financial engineering to provide a new and expanded window into the structure of volatility spillovers in European government bond markets. Within this context, we rely upon the kindred field of computer engineering to provide the tools needed to design and implement a radial basis function artificial neural network: a new network that includes an optimal derivation of the regularization parameter. Simulation-based heuristics encapsulate the optimizing artificial network to help improve the art of supervising the training phase of the artificial network. And, lastly, by connecting traditional parametric-based conditional volatility estimation and a classification-directed ANN, the conjoint aim of this research is brought to fruition – to design a generalized financially engineered method of scientific inquiry that will extract new information from the distribution of residual risk that characterizes European government bond returns.

The motivation for the current inquiry can be traced to the launch of the Euro and the resultant growth that defined the securities markets of Europe. Today, the European government securities market has reached a size that rivals its U.S. counterpart. Similarity of size notwithstanding, the question for policymakers is whether the unique and diverse structure of the Euro-government bond market is one that will nurture ongoing advances in overall efficiency. Conventional wisdom agrees that achieving improved liquidity, marketability, and other efficiencies will require consistent monitoring and, importantly, new policy adjustments that reflect an increased understanding of the information flow from global participants. Such thinking is evidenced by the creation of industry partnerships between various governments and the leading primary dealers in Euro government securities. For example, the European Primary Dealers Association (EPDA) is one such organization. It has convened to focus on voluntary industry initiatives that are designed to promote best practices on primary market policies and procedures. The search for setting effective bond market policy will depend in large part on how well market participants understand the manner, persistence, and magnitude of efficiency disturbing shocks that, from time to time, propagate across these markets. The complexity of multicountry volatility linkages augurs for the continued development of explanatory investigations.

The specific aim of this research is to reexamine reported findings that describe the effects of volatility spillovers from the United States (US) and aggregate European government bond markets into the government bond markets of two EMU countries (Germany and Spain) and one non-EMU country (Sweden). The analytical examination is developed in two stages. The initial stage focuses on the process of engineering a complex nonlinear artificial neural network (ANN) mapping of government bond excess returns. To accomplish this step, the current research exploits the Kajiji-4 radial basis function (RBF) ANN. The algorithm has proven to be a fast and efficient topology for mapping financial instrument volatility across various time intervals [for example, Dash and Kajiji (2003) used the algorithm to successfully model daily volatility asymmetries of FX futures contracts and, similarly, Dash, et al. (2003) employ the method to forecast hourly futures options ticks]. The second stage establishes the overall effectiveness of the ANN to control for the known conditional volatility properties that define transmission linkages among government bond excess returns. Stated differently, if the Kajiji-4 RBF ANN method achieves its appointed task, we expect the modeled residuals to be devoid of any meaningful latent economic information. A finding to this end will expand the window of understanding into how information flows from aggregate markets lead to identifiable volatility asymmetries in individual Euro zone government bond market.

The testable volatility spillover model preferred for the research inquiry can be traced to a panoptic review on market contagion by Christiansen

(2003). The ANN models formulated here are further influenced by the linear regression-based methods of Bekaert and Harvey (1995, 1997), the VAR methods of Clare and Lekkos (2000), and associated extensions offered by Ng (2000). We also take into consideration associated refinements detailed in Bekaert et al. (2005) and in Baele (2002). Clearly, however, it is the Christiansen approach that sets a foundation for the two-stage modeling experiment that defines this research. The initial step of our inquiry is divided between modeling the excess returns of aggregate bond indices – one for the US and another for European – and the modeling of the mean return generating process for individual country government bond indices. In the latter step, that of modeling spillover effects for individual country excess returns, the function mapping equation is augmented to infer GARCH effects by including the one-period lagged returns associated with the two aggregate bond indices. As with prior specifications of country-specific volatility spillover models, the conditional volatility of the unexpected return includes a recognized dependence on the variance of both the U.S. and European idiosyncratic shocks as well as own-country idiosyncratic shocks.

The remainder of the chapter is organized as follows. Section 2 revisits the Christiansen volatility spillover model by providing a modeling extension in the form of the nonparametric Kajiji-4 RBF ANN. Section 3 describes the government bond market data available for the study. Section 4 engineers a second-stage nonlinear ANN analysis, which is applied to the excess bond returns. The objective in this section is to control for latent conditional volatility in the nonlinearities of the residual bond returns. A summary and conclusion are presented in Section 5.

2 The Volatility Spillover Model

The issues of time-varying volatility of financial time series are well documented. The ARCH model process of Engle (1982) exploited this autoregressive property where historical events leave patterns behind for a certain time after some initial action. The GARCH model of Bollerslev (1986) introduced the ability to examine volatility in terms of conditional heteroscedasticity in that the variance, which is now conditional on the available information, varies and also depends on old values of the process. To this end, the symmetric GARCH model has gained rapid acceptance for its ability to generate a reasonably good fit to actual data over a wide range of sample sizes. Unlike asymmetric counterparts, the symmetric GARCH technique models the conditional variance that depends solely on the magnitude but not on the sign of the underlying asset.

The existence of volatility leverage effects in financial time series, the observation that bad news has a larger impact on volatility than does good news, is also a well-known phenomenon (see, for example, Koutmos and Booth, 1995, and Booth et al., 1997). The EGARCH model of Nelson (1991) and Nelson and Cao (1992) has proven to be nearly ideal for capturing the leverage effects that define the overall market behavior of financial instruments. Research findings by Andersen et al. (2001) provide a useful division of stock return volatility into the following three dimensions: volatility clustering, asymmetric volatility, log-normality and long memory volatility. Within the dimension of volatility clustering, the effort of Skintzi and Refenes (2004) is of direct importance to the objectives set forth for this research. These researchers employ a dynamic correlation structure bivariate EGARCH model to yield evidence of price and volatility spillovers from the U.S. bond market and the aggregate Euro-area bond market to 12 individual European bond markets. These same results also provided new evidence that own bond market effects are significant and exhibit asymmetric impacts in the volatility-generating process. The generation of this new knowledge was inextricably linked to the innovative use of new modeling methodology

2.1 An ANN Volatility Spillover Model for the U.S. Government Bond Market

While there are many plausible modeling structures available to investigators of bond volatility spillover effects, the approach used here continues the authoring of innovative and parsimonious expression. Within the context of the Christiansen methodology, we engineer a nonlinear ANN approach that is not unwieldy, overparameterized, nor difficult to test with the temporal and spatial dynamics of bond volatility. We begin by defining the conditional return on the U.S. government bond index as an AR(1) process:

$$R_{US,t} = b_{0,US} + b_{1,US} R_{US,t-1} + \varepsilon_{US,t}. \tag{1}$$

In this model, the idiosyncratic shocks ($\varepsilon_{US,t}$) are normally distributed with a mean zero ($E|\varepsilon_{i,t}| = 0$), are uncorrelated among them ($E|\varepsilon_{i,t}\varepsilon_{j,t}| = 0; \forall i \neq j$), and the conditional variance follows an asymmetric EGARCH(1,1) specification:

$$\sigma^2_{US,t} = \omega_{US} + \alpha_{US} e^2_{US,t-1} + \gamma_{US} \sigma^2_{US,t-1}. \tag{2}$$

The constraints $\omega_{US} > 0$, σ_{US}, $\gamma_{US} \geq 0$, and $\alpha_{US} + \sigma_{US} \leq 1$ of the somewhat restrictive GARCH framework are relaxed by not imposing non-negative constraints on parameters α_k and γ_j; a characteristic that allows for the capture of directional shock effects of differing magnitudes on the volatility distribution.

Specifically, the generalized EGARCH model implemented for the all excess bond return generating models presented in this section is represented by

$$\ln(\varepsilon^2_{US,t}) = \varpi + \sum_{k=1}^{q} \alpha_k g(z_{t-1}) + \sum_{j=1}^{p} \gamma_j \ln(h^2_{t-j}), \tag{3}$$

where
$$g(z_t) = \theta z_t + \gamma \left[|z_t| - E|z_t|\right] \quad (4)$$
and
$$z_t = \frac{\varepsilon_t}{\sqrt{h_t}} \quad (5)$$
are defined as
$$E|z_t| = \left(\frac{2}{\pi}\right)^{0.5}; \quad z_t \sim N(0,1) \quad (6)$$

In our formulation, the parameter γ is set to 1. Next, we focus our attention on the return-generating process across European markets.

2.2 ANN Volatility Spillover Models for European Government Bond Markets

The degree to which economic shocks in the aggregate European government bond market influence the return-generating process of individual European countries is reestimated by application of the ANN econometric model in this research. The existence of own-market volatility spillover effects is another postulate that is reexamined under the engineering approach developed here. The reexamination of these two effects requires the proposed specification to be consistent across normative models – one for the aggregate European government bond market and another that is suitable to describe both EMU and non-EMU countries. The process begins with the former. The conditional excess return on the European total return government bond index is assumed to be a multifactor AR(1). The model is specified as:

$$R_{E,t} = b_{0,E} + b_{1,E} R_{E,t-1} + \gamma_{E,t-1} R_{US,t-1} + \phi_{E,t-1} \varepsilon_{US,t} + \varepsilon_{E,t}. \quad (7)$$

In this system, the conditional mean of the European bond excess return depends on its own lagged return as well as the spillover effects introduced by the lagged U.S. excess return, $R_{US,t-1}$, and the U.S. idiosyncratic risk shock, $\varepsilon_{US,t}$. Following the previous assumption, the conditional variance of the idiosyncratic risk shock ($\varepsilon_{E,t}$) is assumed to follow an asymmetric EGARCH(1,1) specification:

$$\ln(\varepsilon_{E,t}^2) = \varpi + \sum_{k=1}^{q} \alpha_k g(z_{t-1}) + \sum_{j=1}^{p} \gamma_j \ln(\varepsilon_{t-j}^2), \quad (8)$$

subject to the usual restrictions [e.g., see Equations (4)–(6)].

The second ANN econometric specification described here is a model that is capable of describing the conditional return-generating process for the ith individual European country government bond market from among the N markets included in the study. That is, for country i, the conditional excess return is determined by

$$R_{i,t} = 0b_{0,i} + b_{1,i}R_{i,t-1} + \gamma_{i,t-1}R_{US,t-1} + \delta_{i,t-1}R_{E,t-1} + \phi_{i,t-1}\varepsilon_{US,t} + \psi_{i,t-1}\varepsilon_{E,t} + \varepsilon_{i,t} \tag{9}$$

Within this model statement the conditional excess return depends upon the lagged performance of own-country return as well as that of the U.S. and aggregate European bond markets. More specifically, the U.S. and European spillover to the ith country is captured by the lagged returns $R_{US,t-1}$ and $R_{E,t-1}$, while volatility spillover effects are captured by $\varepsilon_{US,t}$ and $\varepsilon_{E,t}$, idiosyncratic shocks from the regional conditional return estimations, respectively. Finally, and for completeness, again we note that the idiosyncratic shocks for all N country models are subject to the same EGARCH(1,1) distributional assumptions [Equation (10)] and associated constraints [Equations (4)–(6)] as previously defined for the expected behavior of the regional return index:

$$\ln(\varepsilon_{i,t}^2) = \varpi + \sum_{k=1}^{q} \alpha_k g(z_{t-1}) + \sum_{j=1}^{p} \gamma_j \ln(\varepsilon_{t-j}^2). \tag{10}$$

With the return-generating process identified for all bond indices associated with expected volatility spillover effects, the emphasis shifts directly to the nonlinear modeling features provided by the application a neural network methodology.

2.3 The ANN Normative Extension

In a manner that is reminiscent of how the human brain processes information, ANNs follow an engineering process that relies upon a large number of highly interconnected neurons working in parallel to solve a specific problem based on approximations.[1] Of course, for artificial networks information processing must be achieved by using software to simulate the components of the brain. While a number of interesting computational topologies are currently in use for simulating interconnected neurons, the focus here is on research dimensions that have compared ANN to nonparametric statistical inference (Amari, 1990), provides an interesting review and discussion. Within this direction, Weigend et al. (1991) have shown that Bayesian estimation theory is a useful way to build parsimonious models that permit for a subsequent comparison to an analogous nonparametric regression. Like Refenes and Bolland (1996), we advance the domain of applicability of artificial networks in financial research. In this specific application the research exploits the topology's design to estimate structural models where the properties of the estimators are readily ascertained. Finally, we buttress our rational to use the proposed ANN to

---

[1] For the decision-making principles of functional segregation (not all functions of the brain are performed by the brain as a whole) and functional integration (different regions of the brain are activated for different functions, with region overlaps being available for use in different networks) under ANN approximation, see Rustichini et. al. (2002).

reexamine volatility spillover effects in European government bond markets based on the successful applications of ANNs to uncover bond features as chronicled in McNelis (2005).[2,3]

What is left unresolved by extant literature is an assessment of how efficiently Bayesian-based RBF ANN topology – a network with a documented ability to represent any Borel-measureable function – actually approximates the volatility features of government bond excess returns at local, regional, and global levels (for additional discussion on function approximation, see Polak, 1971, Saarinen et al., 1993, Schraudolph and Sejnowski, 1996, and Smagt, 1994). Of course, while examples of RBF ANN applications populate the literature, we expect the recently developed Kajiji-4 algorithm to add even greater efficiency to the function mapping process. To that end, in the next section, we introduce the data used in the normative ANN estimation of excess government bond returns. It is important to summarize the data before developing the mathematical characteristics of the Kajiji-4 algorithm since the ANN topology requires rescaling the actual data before the supervised learning process is initiated. After data transformation issues are presented, Section 4 of the chapter provides the details of the enhanced RBF ANN algorithm and its relevance to the financial engineering study at hand.

3 The Data

Weekly data for all government total return bond indices under study are obtained from Global Financial Data for the period May 2003 to January 2005 inclusive (a total of 90 observations).[4] Non-synchronous data issues are partially reduced by the use of weekly data. The two EMU-member countries, Germany (REX government bond performance index) and Spain (Spain 10-year government bond total return index), and one non-EMU country, Sweden (government bond return index w/GFD extension), defines the European local market. The U.S. effect is sampled by the inclusion of the Merrill Lynch U.S. government bond return index. Lastly, the JP Morgan European total return government bond index samples the aggregate European government bond market. Total return indices are preferred, as they are derived under the assumption that all received coupons are invested back into the bond index. All weekly returns are stated in U.S. dollar terms. Excess log returns

---

[2] Feature extraction encompasses the process where raw input data are transformed and mapped to the target variable that is being modeled. It is necessary for the mapping process of inputs onto a target to contain a sufficient amount of information to make it possible for the ANN to learn a generalized behavior that will produce accurate predictions.

[3] See Zirilli (1997) for a detailed discussion on strategies for trend prediction using ANNs as well as various ANN configurations for optimal use of networks in financial trading systems.

[4] www.globalfindata.com.

are obtained by removing (retaining) the risk-free rate from log-differenced government bond index levels. Where needed, the 90-day secondary market Treasury bill yield is used as the U.S. risk-free proxy and the 1-month LIBOR rate is used as a proxy for the European aggregate market.

By definition, the European aggregate government total return index is a weighted average of the individual local bond markets. This suggests the potential for a spurious correlation between local market returns and those of the aggregate European index. This issue is explored by Christiansen (2003), Bekaert et al. (2005), and Diamond et al. (2002). In the latter analysis, the focus is on generating an artificial aggregate European bond index by removing the local market effect. The research presented deviates from this approach for the following two reasons. First, prior research reports a correlation of 0.95 between the artificial aggregate index and each unadjusted aggregate bond index. Given the significantly high correlation measure, the gains associated with creating an artificial aggregate European government bond index appear to be marginal. Second, the Kajiji-4 RBF ANN is constructed upon a closed-form derivation of the regularization parameter, a structural approach that is designed to explicitly and directly minimize correlated variables error. In summary, because the gain from removing correlated variables appears to offer a small benefit and given the explicit objective of multicollinearity reduction embraced within the design of the Kajiji-4 RBF ANN, this research does not implement an artificially constructed aggregate European government bond index.

Figure 1 provides a comparative view of the five excess government bond returns for the sample period. Table 1 complements Figure 1 with a presentation of descriptive statistics for the five government bond indices. The range of the average weekly returns is very tight. The returns are moderately negative for each of the five series over the sample period. This finding coincides with the comparative graphical view as provided by Figure 1. Except for the United States, the variability of returns is remarkably similar across all other countries. Return volatility for non-U.S. countries ranges between 0.0002 and 0.0003. The U.S. return volatility presents a minor exception with measurements approaching zero. Further, for all European-based government bond indices, the returns (and their associated volatility) show a significant right-skew and excess kurtosis. The absence of normality for all measured bond indices is confirmed by the Shapiro–Wilk (1965) test for normality. Once again noting the exception for U.S. bond returns, the Ljung and Box (1978) test for first- to fourth-order autocorrelation indicates that all other bond returns and their squared terms exhibit significant autocorrelation. The clear presence of heteroscedasticity in the bond returns motivates us to diligently examine the ability of the engineered RBF ANN to uniformly control for this (and other) identified distributional properties. Specifically, we use this statistical summary to argumentatively examine the EGARCH properties of the estimated residuals.

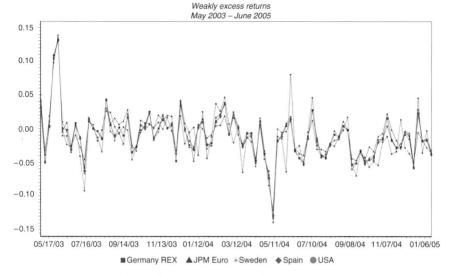

Fig. 1. Comparative view of excess returns.

Table 1. Descriptive Statistics

	Mean	Std.Dev	Skewness	Kurtosis*	Normal	AC(1)	AC(2)	AC(3)	AC(4)
Germany									
Weekly return	−0.0108	0.0339	0.7307	4.726	0.777 (0.0001)	0.328 (0.001)	0.039 (0.006)	−0.015 (0.016)	0.135 (0.017)
Weekly volatility	0.0003	0.0009	5.7485	35.549	0.432 (0.0001)	0.307 (0.003)	−0.060 (0.01)	−0.015 (0.027)	−0.021 (0.056)
Sweden									
Weekly return	−0.0105	0.0341	0.6914	4.635	0.777 (0.0001)	0.296 (0.004)	0.064 (0.013)	0.016 (0.034)	0.069 (0.025)
Weekly volatility	0.0003	0.0007	4.6452	23.103	0.420 (0.0001)	0.276 (0.008)	−0.059 (0.024)	−0.022 (0.058)	0.000 (0.112)
Spain									
Weekly return	−0.0106	0.0340	0.6307	4.406	0.788 (0.0001)	0.319 (0.002)	0.045 (0.008)	−0.018 (0.021)	0.144 (0.019)
Weekly volatility	0.0003	0.0007	4.3329	20.447	0.433 (0.0001)	0.291 (0.005)	−0.059 (0.017)	−0.017 (0.042)	−0.025 (0.083)
Euro									
Weekly return	−0.0092	0.0326	0.8472	5.348	0.751 (0.0001)	0.355 (0.001)	0.041 (0.002)	0.031 (0.007)	0.172 (0.005)
Weekly volatility	0.0002	0.0008	6.4010	39.871	0.401 (0.0001)	0.340 (0.001)	−0.069 (0.004)	−0.001 (0.011)	0.023 (0.024)
USA									
Weekly return	−0.0122	0.0388	0.5690	2.824	0.815 (0.0001)	0.224 (0.031)	0.092 (0.065)	0.094 (0.097)	0.070 (0.148)
Weekly volatility	0.0000	0.0001	0.0000	−3.000	0.502 (0.0001)	0.185 (0.074)	−0.085 (0.144)	0.036 (0.262)	0.119 (0.252)

p-values in parentheses represent the significance level for the Ljung–Box statistic.

4 ANN Estimation of Volatility Spillover

This section of the chapter estimates bond market spillover effects by applying the Kajiji-4 RBF ANN to the aggregate European bond model [Equation (6)] and to the individual country model [Equation (8)]. However, before the exact estimation of spillover effects is presented, we provide algorithmic detail as it pertains to the ANN modeling methodology. Section 4.1 sets forth the algorithmic properties of the Kajiji-4 RBF ANN in a context that advances the efficient modeling of a near-chaotic target variable. This section differentiates the enhanced dual objective Kajiji-4 algorithm from the more traditional uni-objective RBF ANN. Section 4.2 is devoted to the specifics of generalizing the application of the Kajiji-4 RBF ANN to modeling of European government bond returns and their associated volatility. The first subsection of 4.2 is devoted specifically to data transformation and scaling.

The second subsection of 4.2 continues the engineering of the generalized network. As a supervised learning method there is a need to infer a functional mapping based on a set of training examples. Hence, an important research consideration is lodged in the determination of how many observations are required to effectively train the ANN. To achieve a meaningful generalization, we invoke a simulation analysis that is designed to find the ideal minimum bias-variance (modeling error, or MSE) over a range of training values such that it is possible to visualize a global optimum. By identifying an ideal training size (or, training range), it is possible to reduce the negative impacts of overfitting the mapping function.

Section 4.3 presents the empirical results generated from the actual mapping of spillover effects onto the excess bond market returns for each individual European country. Within this section we also report the effectiveness of the ANN topology by examining the distributional properties of the squared residuals obtained by application of the Kajiji-4 RBF ANN to the country-specific return-generating models as well as the model for the aggregate European bond market. In addition to the nonlinear ANN examination of the squared residuals, this section is augmented to include a linear diagnostic of the same squared residuals. Section 4.4 closes the fourth part of the chapter with an interpretive deduction of the policy implications generated by the engineered neural network for mapping volatility spillovers in European government bond markets.

4.1 The Kajiji-4 RBF ANN: Modeling the Conditional Return-Generating Process

The process of engineering a nonparametric RBF ANN bond spillover model(s) is an extension of the parametric-based three-step approach proffered by Christiansen (2003). To estimate the coefficients of her model, Christiansen employed the Quasi Maximum Likelihood method with Gaussian likelihood functions using a combination of the Berndt et al. (1974) and the Newton–Raphson numerical optimization algorithm. Our efforts deviate from the parametric approach by introducing the Gaussian-based Kajiji-4 RBF ANN to extract relevant model coefficients.

The Uni-Objective RBF ANN Architecture

The appointed task of an ANN is to produce an analytic approximation for the input–output mappings as described by the following noisy data stream:

$$\{[x(t), y(t)] : [\mathbb{R}^n, \mathbb{R}]\}_{t=1}^{\infty}, \qquad (11)$$

where $x(t)$ is the input vector at time t, $y(t)$ is the output at time t, and n is the dimension of the input space. The data are drawn for the set

$$\{[x(t), y(t) = f(x(t) + \varepsilon(t)]\}_{t=1}^{\infty}. \qquad (12)$$

A radial basis function artificial neural network is a topology defined by three associated computational layers. The input layer has no particular calculating power; its primary function is to distribute the information to the hidden layer. The neurons of the hidden middle layer embrace a nonlinear radial basis transfer function – the most popular being the Gaussian. Radial basis transfer functions are well regarded for their ability to interpolate multidimensional space. The estimated weights of the third and final layer, the output layer, are described by a linear transfer function. The linearity property infers that if one fixes the basis functions (i.e., picks the centers and widths of the Gaussians), then the predictions are linear in the coefficients (weights). Researchers rely upon the linearity of output weights to assist in the identification of main input effects (for the foundations of RBF ANNs, see Broomhead and Lowe, 1988; Lohninger, 1993; Parthasarathy and Narendra, 1991, as well as Sanner and Slotine, 1992).[5]

The Kajiji-4 RBF ANN is an enhanced Bayesian network that incorporates a closed-form determination of the regularization parameter.[6] Before we explore the extensions proffered by Kajiji to enhance the mapping efficiency of the single-objective RBF ANN, it is useful to describe the basic underlying architecture of the uni-objective algorithm. Within the underlying architecture the optimal output weighting values are estimated by applying a supervised least-squares method to the training set. The term "supervised" describes a technique for creating an output function by training the network using pairs

---

[5] The RBF ANN is a type of nonlinear regression model where it is possible to estimate the output layer weights by any of the methods normally used in nonlinear least squares or maximum likelihood. However, we note that if every observation was used as an RBF center, this would result in a vastly overparameterized model.

[6] The Kajiji-4 algorithm is one of several artificial network methods supported within version 3.15 (and above) of the WinORSe-AI software system produced by The NKD-Group, Inc. (www.nkd-group.com).

of training input data with corresponding desired outputs. For the general topology of radial basis function neural networks, the learning function is stated as

$$y = f(x), \qquad (13)$$

where y, the output vector, is a function of x, the input vector, with n the number of inputs. The supervised learning function architecture is

$$f(x_i) = \sum_{j=1}^{m} w_j h_j(x), \qquad (14)$$

where m is the number of basis functions (centers), h is the hidden units, w is the weight vector, and $i = 1 \ldots k$ are the output vectors (target outputs). The flexibility of $f(x)$ and its ability to model many different functions are inherited from the freedom to choose different values for the weights. Within the RBF architecture, the weights are found through optimization of an objective function. The most common objective function employs the least-squares criterion. This is equivalent to minimizing the sum of squared errors (SSE) as measured by

$$SSE = \sum_{i=1}^{p} (\hat{y}_i - f(x_i))^2. \qquad (15)$$

The Multiple-Objective RBF ANN Architecture

Kajiji (2001) reasoned that some modeling problems are best examined by considering at least two objectives: smoothness and accuracy. To achieve these dual objectives, Kajiji augmented the generalized RBF to include a modified Tikhonov regularization equation (1977). Tikhonov regularization adds a weight decay parameter to the error function to penalize mappings that are not smooth. By adding a weight penalty term to the SSE optimization objective, the modified SSE is restated as the following cost function:

$$C = \sum_{i=1}^{p} (\hat{y}_i - f(x_i))^2 + \sum_{j=1}^{m} k_j w_j^2, \qquad (16)$$

where the k_j are regularization parameters or weight decay parameters. Under this specification the function to be minimized is stated as

$$C = \frac{\mathrm{argmin}}{k} \left(\varsigma \sum_{i=1}^{p} (y_i - f(x_i \,|\, \overline{k}))^2 + \sum_{j=1}^{m} k_j w_j^2 \right). \qquad (17)$$

Iterative techniques are commonly employed to compute the weight decay vector $\overline{k}$. With the introduction of weight decay methods by Hoerl and Kennard (1970) and Hemmerle (1975), iterative techniques were found to lack specificity and to be computationally burdensome (see, Orr, 1996, 1997). Furthermore, computational experience has established that iteration-based methods often stop at local minima or produce inflated residual sums of squares whenever the weight decay parameter approaches infinity. The Kajiji-4 RBF algorithm was designed to overcome these inefficiencies through its incorporation of a globally optimized regularization parameter based on Crouse et al.'s (1995) Bayesian enhancement to optimal ridge regression. Taken together, the aforementioned extensions embraced by the Kajiji-4 RBF ANN allow the dual-objective algorithm to directly attack the twin evils that deter efficient ANN modeling: the *curse* of dimensionality (multicollinearity or overparameterization) and inflated residual sum of squares (inefficient weight decay).

4.2 Engineering an Efficient ANN Mapping of Market Volatility

A primary reason for a decision maker to utilize a neural network is to generate an efficient mapping of a complex target variable. To achieve this objective it is necessary for the modeler to execute several well-defined steps. The ANN modeling progression begins with an assessment of the input data. At this point in the analysis it is always advisable to transform the data to a scale that contributes to efficient supervised learning. That is, preprocessing the input data by rescaling generally leads to improved performance as the data are shaped to a range that is more appropriate for the network and the distribution of the target variable. In the next step, the parameters that control the algorithm's learning phase must be stated and calibrated. This includes the identification of the network's transfer function and the associated bias-error measure. The transfer function accepts input data in any range and responds nonlinearly to the distance of points from the center of the radial unit. Contrastingly, the generated output occurs in a strictly limited range, a fact that augers for preprocessing as defined in the prior step. Prior to initiating the final step – executing the modeling algorithm – one must determine just how many data cases are needed to supervise efficient training of the network so that it learns to recognize patterns without learning idiosyncratic noise. The following subsections explore this three-step approach as we engineer an efficient ANN model to examine bond volatility spillover across European government bond markets.

Data Scaling by Transformation

Neural networks learn more efficiently with lower prediction errors when the distribution of input variables is modified to better match outputs. For example, if one input variable ranges between 1 and 10,000, and another ranges between 0.001 and 0.100, the accuracy of the estimated network weights would be impacted substantially by these different magnitudes. To estimate an efficient network model, it is desirable to help the network cope with these different ranges by scaling the data in such a way that all the weights can remain in small, predictable ranges. Shi (2000) provided a more robust discussion on the comparative merits of three popular transformation methods to rescale input data (linear transformation, statistical standardization, and mathematical functions). More importantly, the Shi analysis adds to this discussion by proposing a new method for efficient data scaling as a preprocessor to neural network modeling – it is a method that relies upon the use of cumulative distribution functions (distribution transformation). The point to emphasize here is the importance of data scaling by one of any number of recognized transformations as an integral part of the neural network engineering process.

The Kajiji-4 RBF ANN supports several alternative data transformation algorithms. For the bond volatility modeling application described herein, we choose to scale the data by the *Normalized Method 1* technique as defined in Table 2 and Equations (18)–(20).[7] We note that this transformation technique scales the data to $[0, 1.01]$ when $S_L = 0\%$ and $S_U = 1\%$. That is, the post-transformation minimum data value is 0 and the maximum data value is 1 plus the S_U value. Setting the headroom parameters S_L and S_U to any other positive value produces a scaled data series that is bounded between two positive numbers based on the actual minimum and maximum data values. For example, if the actual data minimum is -1 and the actual data maximum is 0; the scaled data after transformation will lie in the range $[-0.01, 1.0]$.

Table 2. Notation for RBF ANN Scaling

Term	Definition
D_i	Actual data point
D_L	Lower scaling value
D_U	Upper scaling value
$D_{\min}$	Minimum data value
$D_{\max}$	Maximum data value
D_V	Normalized data value
S_L	Lower headroom (%)
S_U	Upper headroom (%)

The algorithm proceeds by computing D_L and D_U as defined by Equations (18) and (19):

$$D_L = D_{\min} - \frac{(D_{\max} - D_{\min})S_L}{100}, \tag{18}$$

$$D_U = D_{\max} + \frac{(D_{\max} - D_{\min})S_U}{100}. \tag{19}$$

---

[7] In addition to the Normalized Method 1, we found interesting results by application of *Normalized Method 2*: $D_V = N_L + (N_U - N_L)\frac{(D_i - D_{\min})}{(D_{\max} - D_{\min})}$, where D_V, D_i, $D_{\min}$, and $D_{\max}$ are as previously defined. N_L and N_U specify the lower and upper bounds of the scaled data, respectively (e.g., to scale the data to $[-1, +1]$, set $N_L = -1$ and $N_U = +1$).

The scaled data values, D_V, replace the original data series by Equation (20):

$$D_V = \frac{D_i - D_L}{D_U - D_L}. \tag{20}$$

Algorithmic Parameterization

Like all supervised learning networks, the RBF network approximates an unknown mapping, F^*, between pairs $(\mathbf{x}_i, y_i)$, for $i = 1, \ldots, n$ for m observations. Mapping to the target response surface is by the use of a radial function, a function that decreases (increases) monotonically with the distance from a central point. The parameterization of the RBF network begins with the judicious choice of a transfer function. Neural network researchers understand that sigmoidal functions may be better estimates for some data, while Gaussian functions may be better approximators for other kinds of data.[8] For the purpose of mapping bond volatility, we implement a radial Gaussian transfer function. The central point, or *radius*, for the Gaussian transfer function is set to 1.0. To avoid over- and underfitting (inefficient training) of the network, it is important to 1) identify the appropriate number of data cases to include in the training session and 2) choose a *bias-variance* control that matches the complexity of the ANN to the complexity of the data. The former issue is discussed in more detail in the next section. As for the choice of *bias-variance* control, using the options available within the Kajiji-4 RBF ANN, we estimate the parameters of the bond volatility experiment by invoking the generalized cross-validation (GCV) error minimization method.[9]

Efficient Supervised Learning

RBF ANN algorithms are trained to predict the target variable by supervising the use of an increasing number of cases (observations) on the input variables up to the point where modeling improvements become redundant. Overfitting occurs at the point where the ANN ceases to learn about the underlying process, but, instead, it begins to memorize the peculiarities of the training cases. By way of example, using a financial training model, Mehta and Bhattacharyya (2004) demonstrated how an overfit network effectively obscured important short-term patterns of profitable trading. ANN researchers agree that prior to initiating the economic modeling process, it is important to determine the efficient number of cases over which to train the network. To achieve

---

[8] Alternative radial functions supported by the Kajiji-ANN framework include (1) Cauchy, (2) multiquadric, and (3) inverse multiquadric.
[9] Support for alternative bias-error minimization methods include (1) unbiased estimate of variance (UEV), (2) final prediction error (FPE), and (3) Bayesian information criterion (BIC).

the goal of locating an ideal minimum of training cases, the Kajiji-4 ANN framework supports a stepwise minimization of the ANN cost function:[10]

$$i^* = \min \sum_i^{Q_{\max}} C_i. \qquad (21)$$

In this equation, C_i is the cost function C as defined in equation (16) computed for the ith data observation and $Q_{\max}$ is the maximum number of cases in the training set. For the comparative modeling process invoked in this chapter, we step the network over $Q_{\max}$ vectors to find i^*, the case associated with a global (ideal) minimization of the network cost function (C^*). Table 3 summarizes the simulation results over the 80 total observation points available to each spillover model. Figure 2 presents a graphical summary of the Kajiji-4 training simulation when applied to the return-generating models. The headroom setting (scaling control) for data transformation is set to [0, 1]. The column $MinN$ reports the base number of cases required to obtain a global minimum bias-error measurement under supervised training. The column labeled i^* reports the result of the country-specific simulations that identified the observation associated with the smallest MSE within the simulation range. By way of example, the results of the simulation suggest that when applied to a specific volatility model, like the German government bond volatility index, the Kajiji-4 RBF model should be trained using 55 observations, which by contrast, would be increased to 72 observations when modeling the Swedish government bond volatility index.

Table 3. RBF ANN Simulation Settings

Model	Min N	i^*
Germany	11	55
Sweden	11	72
Spain	11	71
Euro	11	74
US	11	52

4.3 The Estimated Kajiji-4 RBF ANN Spillover Model

Before presenting the policy inferences drawn from an interpretation of the optimal RBF ANN weights, it is useful to investigate the overall efficiency of the RBF ANN function approximation. The Kajiji-4 algorithm provided an accurate mapping of excess government bond returns for the aggregate European

---

[10] Also supported within the ANN framework are the following methods as differentiated by their treatment of the regularization parameter: (1) Kajiji-1 (Hemmerle's calculation, 1975), (2) Kajiji-2 (Crouse et al. closed-form estimation, 1995), and (3) Kajiji-3 (Kajiji-1 with prior information).

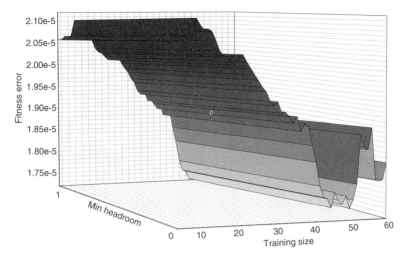

Fig. 2. Simulation results.

government bond market and the individual country bond markets. Table 4 reports output measures generated by an application of the Kajiji-4 RBF ANN for the individual European country government bond spillover models. We note the value of R-square is evenly reported across countries, ranging from a low of 86.27% (Germany) to a high of 87.37% (Sweden). Additionally, the two ANN performance measures defined as Direction and *Modified Direction* each report accuracy and consistency across all countries.[11]

Figure 3 presents a visual reference using the German Government bond market to support the reported function mapping accuracy for all countries (the spillover modeling results are similar for all other bond markets). For both the training and validation subsamples, the full range of the predictability chart reflects a close and effective function mapping.

Before proceeding with an interpretation of policy implications, we turn our attention to the distributional properties of the residual error terms obtained by solving the Kajiji-4 excess return government bond spillover model(s). The basic intuition applied here is related to the nonlinear nature of the network. If the nonlinear mapping ability of the Kajiji-4 algorithm met its appointed task, then we expect the residual error terms to be de-

---

[11] The *Direction* measure captures the number of times the target prediction followed the up-and-down movement of the actual target variable. The *Modified Direction* measure augments the *Direction* measure to consider a ratio of number of correct up-tick predictions to number of down-tick predictions. Both measures are scaled between 0 and 1. TDPM (not reviewed in this analysis) is a correction weight that compensates for incorrect directional forecasts by overall magnitude of the movement. The smaller the weight, the more accurate the training phase. Large weights are indicative of a missed direction, an incorrect magnitude adjustment, or some combination of the two.

Table 4. RBF ANN Government Bond Spillover Models

	Germany	Sweden	Spain	Euro
Computed Measures				
Lambda	0.018761	0.000013	0.000005	0.323282
Actual error	0.032489	0.028351	0.028568	0.024464
Training error	0.000439	0.000332	0.000335	0.000196
Validation error	0.000102	0.000116	0.000166	0.000031
Fitness error	0.000310	0.000291	0.000301	0.000168
Performance Measures				
Direction	0.733333	0.777778	0.777778	0.755556
Modified direction	0.647321	0.764368	0.689304	0.680460
TDPM	0.000048	0.000044	0.000043	0.000026
R-Square	86.27%	87.37%	86.98%	92.70%
Model Characteristics				
Training (N)	55	72	71	74
Training (%)	61.80%	80.90%	79.78%	83.15%
Transformation	Norm:1	Norm:1	Norm:1	Norm:1
Min/Max/SD	0% / 1%	0% / 1%	0% / 1%	0% / 1%
Radius	1	1	1	1
Algorithmic Settings				
Method	Kajiji-4	Kajiji-4	Kajiji-4	Kajiji-4
Error min. rule	GCV	GCV	GCV	GCV
Transfer function	Gaussian	Gaussian	Gaussian	Gaussian

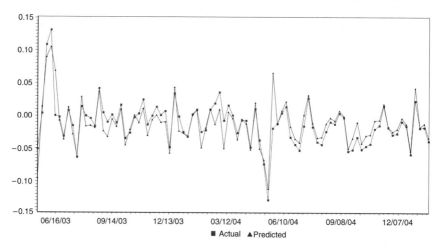

Fig. 3. Kajiji-4 RBF-ANN actual and predicted values (Germany).

void of any latent economic information [but not necessarily $N(0,1)$ i.i.d.]. By our earlier-stated hypothesis, we expect an absence of the heteroscedastic EGARCH effects, which were reported to exist by prior studies conducted on both bond and equity returns.

In the next section we describe the statistical properties of the ANN residuals. After the presentation of descriptive statistics, the investigation proceeds by testing for latent ARCH-framework effects within the Kajiji-4 generated residuals. This is followed by both a linear and nonlinear components analysis of the European sector return residuals. The linear investigation is accomplished by extracting principal components. The nonlinear analysis rests upon the application of an alternative ANN topology: the Kohonen self-organizing map (K-SOM). The K-SOM method is an effective counterpart to nonlinear principal components.

RBF Residuals – Descriptive Statistics

Table 5 presents the descriptive statistics for country-level idiosyncratic residual returns obtained by the application of the Kajiji-4 spillover model. At four significant digits, the results show a mean of zero and are relatively small to near-zero variance measurement. For both Sweden and Spain, the skewness measure is small and negative. The skewness coefficient for Germany is also negative but at a much higher level (-1.3266). Excess kurtosis is also evident for all residual return time series, with the range among the countries spanning from 6.776 (Germany) to a relatively small coefficient of 2.983 (Spain). The Shapiro–Wilk's test for normality (W-statistic) confirms the expectation of nonnormality.

Table 5. Idiosyncratic Term Descriptive Statistics

	Germany	Sweden	Spain	Euro
Mean	0.0000	0.0000	0.0000	0.0000
Variance	0.0003	0.0003	0.0003	0.0002
Skewness	−1.3266	−0.5513	−0.2867	−3.7772
Kurtosis	6.7757	4.1138	2.9828	17.8946
W-Statistic	0.8108	0.8237	0.8556	0.6330
	(0.0001)	(0.0001)	(0.0001)	(0.0001)
LB(12)	33.4180	14.5250	17.2830	10.9250
	(0.0010)	(0.2680)	(0.1390)	(0.5350)
LB2 (4)	0.8500	8.4470	10.1040	0.3800
	(0.9320)	(0.0770)	(0.0390)	(0.9840)
LB2 (8)	1.4860	9.4690	11.3200	0.5600
	(0.9930)	(0.3040)	(0.1840)	(1.0000)
LB2(12)	5.0860	10.4130	11.7030	0.6000
	(0.9550)	(0.5800)	(0.4700)	(1.0000)

LB(.) and LB^2(.) are the Ljung–Box statistics for heteroscedasticity in the residual and squared residual series, respectively; W-statistic is the Shapiro–Wilk's statistic for normality. p-values are shown in parentheses.

The autocorrelation patterns among the residuals are somewhat mixed. Germany is the only country that exhibits a significant 12-period (12-week) lag in the first-order autocorrelation of the excess return residual. The analysis is moderately stronger upon the examination for nonlinear dependence as captured by the squared residuals. At the 5% and 10% levels of significance, the Ljung–Box statistic reports evidence of a four period nonlinear dependence for both Spain and Sweden. In summary, we find that applying the Kajiji-4 RBF ANN to the government bond spillover model for European markets results in a solution where the idiosyncratic residual returns exhibit moderate nonlinear dependence, an absence of skewness at the individual country level, and for one country (Germany) moderately heavy tails.

RBF Residuals – ARCH Effects

This section of the research is directed toward obtaining a better understanding of what, if any, latent economic information remains in the idiosyncratic residuals after solving the Kajiji-4 excess return-generating models. Nonlinear dependence and heavy-tailed unconditional distributions are characteristic of conditionally heteroscedastic data. The premise behind the use of the RBF ANN is that its penchance for resolving nonlinear structure will result in the extraction of all relevant economic variability patterns. We take the additional step of applying the EGARCH(1,1) framework to the squared idiosyncratic residuals. While it is not the only method within the ARCH framework by which it is possible to capture some of the most important stylized features of return volatility, we have previously reported that the EGARCH(1,1) model is well suited to capture time-series clustering, asymmetric correlation, log-normality, and, with proper specification, long memory (for a comprehensive review, see Andersen, 2001). Table 6 presents the results of applying the EGARCH(1,1) model.

Table 6. EGARCH Estimates of the Idiosyncratic RBF ANN Volatility

Index	μ	α_0	α_1	β	θ	R^2	Log Lik.	$\Pr > \xi^2$
Germany	0.0004	−13.9817	−1.5427	0.0499	−1.1194	1.76%	530.25	0.0001
	(0.9966)	(0.0001)	(0.0001)	(0.6177)	(0.0001)			
Sweden	0.0004	21.6825	0.0399	−0.4648	38.0063	3.10%	532.85	0.0001
	(0.0001)	(0.0001)	(0.0764)	(0.0001)	(0.0391)			
Spain	0.0002	−14.5430	1.4386	0.0009	−0.2956	7.76%	542.44	0.0001
	(0.0001)	(0.0001)	(0.0001)	(0.9757)	(0.2278)			
Euro	0.0002	−14.3619	−1.7946	0.0648	−0.8626	1.09%	550.34	0.0001
	(0.6769)	(0.0001)	(0.0001)	(0.5171)	(0.0001)			
US	0.0000	−18.8703	0.0000	0.0000	0.0000	2.09%	714.44	0.0001
	(0.9935)	(0.0001)	(0.2827)	(1.0000)	(1.0000)			

p-values in parentheses.

The reported EGARCH results fail to identify significant residual return dependence and leverage effects. Except for Sweden (a non-EMU country), none of the explanatory variables is jointly significant for any of the bond indices.

RBF Residuals – Linear Diagnostics

The results of the prior section suggest that the country-specific residuals generated by the Kajiji-4 RBF ANN appear to be statistically well ordered, as they do not exhibit any significant ARCH effects. In this section we extend the examination of the residuals to test for statistical independence. Specifically, we apply a principal components analysis (PCA) to the European residuals in order to uncover whether the observed distributional properties are caused by some hidden factor(s). The PCA method is widely used for finding the most important directions in the data in mean-square space. In this application there are K common factors F_t through the N observations of the $N \times K$ factor loading matrix Γ, where

$$\varepsilon_{i,t} \equiv \Gamma F_t \qquad (22)$$

The unobservable factors F_t are assumed to be i.i.d. distributed with

$$E[\varepsilon_t] = 0, \quad \text{Var}[\varepsilon_t] = \sigma^2 I_N, \qquad (23)$$
$$E[F_t] = 0, \quad \text{Var}[F_t] = I_K, \qquad (24)$$
$$\text{Cov}[F_t, \varepsilon_t], \qquad (25)$$

where I_N is an $N \times N$ identify matrix and the zeros indicate null vectors and matrices of the appropriate size. The results of the PCA are presented in Table 7.

Table 7. PCA of RBF ANN Spillover Residuals

	Factor 1	Factor 2	Factor 3	Factor 4
Sweden	0.954	-	-	-
Spain	0.936	-	-	-
Germany	0.926	-	-	-
Euro	0.847	0.487	-	-
Eigenvalue	3.360	0.467	0.161	0.012
Cumulative percentage of variance explained	0.840	0.957	0.997	1.000

If the Kajiji-4 RBF ANN has achieved its appointed task, then the linear factor-analytic solution should not be able to discern any latent independent factor structure. That is, the simple structure of the data should be reduced to a single variance-maximizing factor. The results derived from this analysis provide strong evidence that one latent factor dominates the analysis. Factor 1 alone explains 84% of the variability in the idiosyncratic risk terms. The remaining factors add decreasing contributions to the decomposition of variability among the residuals. For example, factors 2 and 3 add 11.7% and 4.0%, respectively, to the decomposition of total variance. The one-factor dominance of this solution provides additional evidence that the RBF ANN spillover model removed a significant amount of structured economic volatility from the excess returns of European government bonds.

RBF Residuals – Nonlinear Diagnostics

A self-organizing map (SOM) is a type of neural network (a vector quantization method) that implements a k-means cluster algorithm. The Kohonen SOM (K-SOM) has the supplementary property that it maps the distribution of any input space to which it is exposed (see Kohonen, 1990, for theory and examples). Not only is the method computationally fast to solve, but the output from this method presents the latent features of the data in an ordered set of clusters that contributes to their data-specific interpretation. SOMs have proven to be a useful classification tool, when applied to large high-dimensional data sets (Craven and Shavlik, 1997; Lu et al., 1996; Kaski and Kohonen, 1996; Kaski, 1997).

In this section of the study, we apply the K-SOM to the standardized residuals produced by application of the RBF spillover model. Unlike the PCA solved above, the K-SOM method does not impose any distributional assumptions on the components. Our objective is to employ this nonparametric visualization method to locate any latent clusters that may remain in the residuals of the associated spillover models. As with the application of factor-analytic methods, we do not expect the SOM application to locate more than one dominant cluster when applied to the RBF-generated standardized residuals. A low-dimensional topology in residual error space is consistent with a model that has accounted for all relevant variability.

Figure 4 provides a visualization of the K-SOM solution. At a value of approximately 2.0 on the independent axis and 1.0 on the dependent axis, a large significant cluster is visible. Although we observe several smaller clusters, the ad hoc review argues for a single dominant source of similarity within the residual error measurements. This observation is reinforced by the ordered parameter identities as summarized in Table 8. The desired outcome of this application was to obtain a single dominant cluster. Based on the combined review of Table 8 and Figure 4, it is immediately apparent that the Kajiji-4 RBF ANN spillover model achieved its appointed task – to model the time-dependent sources of volatility spillover effects in European government bond markets.

Table 8. K-SOM Clusters Identity

Cluster 1/1	Cluster 2/1
Week of June 16, 2003	All other
Week of May 24, 2004	observations

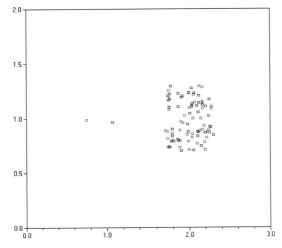

Fig. 4. K-SOM clusters within RBF ANN residuals.

4.4 Policy Implications

Table 9 reports the computed network weights obtained as a result of estimating the individual country spillover models [Equations (7) and (9)]. To assess which European policy effects for individual government bond markets were extracted by application of the spillover model, we focus on the sign and magnitude of the weights produced by applying the Kajiji-4 RBF ANN. For the individual country models, the algorithmic solution weights assist in the determination of the mean- and volatility-spillover effects to each country from both the U.S. and aggregate European government bond markets. We note, however, that the estimated weights are not unit-specific with regard to the data (like regression parameters). That is, the RBF weights do reflect direction and magnitude. Hence, it is possible to infer an economic interpretation that follows along the usual interpretation of estimated parameters from a parametric estimation method.

For the most part, the empirical results for individual government bond markets corroborate extant findings for post Euro introduction (Christiansen, 2003). Except for Germany, the sign of each RBF weight is positive. That is, the mean spillover effect from the European aggregate bond market to individual countries is large and positive for all countries except Germany (-0.1692). However, the magnitude of the RBF weights suggests that the spillover effect is weaker (by half) for the non-Euro zone country (Sweden) than it is for EMU

Table 9. RBF ANN Spillover Model Weights

Return-Generating Model	Lagged Country $b_{1,t}$	Lagged Euro $\delta_{i,t-1}$	Lagged U.S. $\gamma_{i,t-1}$	Euro Residual $\psi_{i,t-1}$	U.S. Residual $\phi_{i,t-1}$
U.S.	n/a	n/a	0.4137	n/a	n/a
Euro	n/a	0.1470	0.1328	n/a	0.1385
Germany	0.267	−0.1692	0.2520	−0.0760	0.1629
Sweden	2.033	6.2023	0.2249	1.8278	−9.7189
Spain	3.554	15.4974	−0.1560	3.6120	−21.8662

member Spain. The negative and smaller weight reported for Germany suggests a minor spillover effect from the German bond market to the aggregate Euro government bond market. Skintzi and Refenes (2004) reported contradictory results for the study period 1991 to 2002. After including additional EMU countries, this study reports that spillovers were to Germany from the aggregate Euro area bond index and that the effect became stronger after the introduction of the Euro.

The inferences from the signs and relative magnitudes of the weights derived from the U.S. mean spillover effect corroborate prior research findings. For example, we note that the RBF model is consistent in finding a larger U.S. mean-spillover effect to the German market than that provided by the aggregate European bond market (0.2520 and −0.1692, respectively). By contrast, for the other Euro member, Spain, the U.S. mean spillover effect is 100 times smaller than that reported for the aggregate European bond markets (−0.1560 and 15.4974, respectively).

The evidence of volatility spillover effects uncovered by the application of the Kajiji-4 ANN supports prior findings with one important caveat. At the outset, each country exhibits strong ARCH effects in excess government bond returns as each country displays a positive weight for its own index's past values. The German weight is almost 10 times less contributory than that reported for either Spain or Sweden. The ANN results for the German market produce additional findings that challenge contemporary wisdom when compared to parametric-based analytics. By way of example, we note that the U.S. volatility spillover effect is just over 2.2 times as important as the European volatility spillover effect and, when compared to the other countries, the signs for the U.S. spillover are reversed. Thus, while increased U.S. volatility leads to increased volatility in German bond market, just the opposite outcome is realized from increased aggregate European bond market volatility in excess return. That is, the Kajiji-4 findings provide strong evidence that German bond volatility spills into the aggregate European markets. Not only does this finding contradict to prior reports, but the contradiction is amplified when we note that the magnitude of the U.S. and Euro bond market volatility effects is reversed for the other two countries. As shown in Table 9, for both Spain and Sweden the U.S. volatility effect is negative

(−21.8662 and −9.7189), while that for the Euro zone is positive (3.6120 and 1.8278). This finding suggests that as U.S. excess bond volatility increases (decreases), country-specific excess bond volatility in both Spain and Sweden will decrease (increase). For Spain and Sweden the volatility effects are 6.1 and 5.3 times larger from U.S. sources than from European sources, respectively. In succinct terms, the policy implications derived from the application of the Kajiji-4 RBF ANN to the defined sample period support the existence of volatility spillover effects from U.S. and aggregate European government bond markets, but the findings stand in contradiction to the extant literature that reports aggregate European spillover effects dominate those generated by U.S. bond markets. The results presented here also raise a question about the length of the time required to properly train and develop a generalized RBF neural network. While the length of the data series proved to be sufficient for adequate network training, left unanswered by this research is any assessment of how unique economic peculiarities may impact the ANN learning process as it relates to the volume and direction of volatility transmission linkages.

5 Summary and Conclusions

The objective of this chapter centered on the use of a two-step econometric analysis to gain a more in-depth understanding of the volatility transmission linkages among U.S., aggregate European, and individual country government bond markets. The intended outcome was motivated by global policymakers who, collectively, desired to achieve increased efficiency in the overall operation of the European government bond market. This chapter introduced an innovative financial engineering approach to investigate the historical volatility of European government bond markets. With a focus on transmission linkages among U.S., aggregate European and individual European government bond markets, the analytical phase of the chapter set out to generalize the engineering of a nonlinear neural network mapping of European volatility spillover. The effectiveness of the engineered model was tested based on its ability to control for the ARCH effects that are known to characterize the distributional properties of bond excess returns. The Kajiji-4 RBF ANN algorithm was chosen for the modeling task based primarily on its use of a closed-form solution to derive the regularization parameter, a design consideration that is used to mitigate the ill effects of within-model collinearity. The network proved to be extremely efficient in the separation of global, regional, and local volatility effects. The econometric effectiveness of the findings enumerated from Kajiji-4-based volatility spillover models was interrogated by conducting both a parametric and nonparametric examination of its estimated time-series residuals. In all post-application analyses, the structural independence of modeled equations was more characteristic of initial assumptions than not.

The econometric modeling experiment produced several important findings for researchers who engineer nonlinear mappings of economic time series by intelligent networks. First, as with all ANN applications, calibration of a supervised network is a necessary step along the way to producing an efficient nonlinear mapping of a target variable. In this regard, it was found that among the choices for data transformation, the *normalize method 1* approach produced the smallest fitness MSE. Second, for the Kajiji-4 method, the demarcation between supervised training and model validation is best accomplished by a simulation-based application for each individual country as well as for all broad market bond indices. Default settings for all remaining algorithmic control parameters (e.g., radius, transfer functions, etc.) produced efficient and accurate mapping for all financial time series. Third, the two-stage analytical approach preferred in this research demonstrated the effective use of a classification directed neural network topology to confirm the efficient engineering of time-series mapping neural networks. Specifically, for the intent of this study, in the second stage we were able to successfully invoke a Kohonen self-organizing map to document the absence of latent economic effects in the squared residuals produced by the first-stage time-series mapping of volatility effects by the Kajiji-4 RBF ANN. For comparative analytics, orthogonal PCA was also applied to the squared residuals, a process that produced one well-organized principal component with no latent economic meaning.

Lastly, an examination of the weights produced by ANN algorithmic procedure yielded policy inferences that supported for prior findings, but with a degree of ambiguity. The engineered network model reported in this research confirmed the volatility linkages reported by Christiansen (2003) of mean and volatility spillover effects from both the U.S. and aggregate European bond markets to individual European country government bond markets. The results also confirmed the importance of U.S. government bond market volatility as a determinant of local government bond market volatility (see Skintzi and Refenes, 2004). The ambiguity arises when interpreting the reversed signs attached to the parameter estimates. Our research suggests that U.S. volatility operates inversely with volatility measurements in local markets. Moreover, the modeling effect here finds that German bond market volatility spills into the aggregate European market to affect both EMU and non-EMU countries. Due to the targeted time span implemented in this research, one may conjecture that the reported evidence may be time domain-dependent. The question of time domain length as it relates to unique economic shocks is deserving of increased scrutiny given the inverted signs of extracted parameters. Additionally, when compared to extant literature, the efficient findings extracted by engineering a nonlinear mapping network call for an extension of the current analysis to include the emerging bond markets of Europe. Such an extension would lead to a more tractable generalization of generated policy inferences.

References

1. Amari, S.-I. Mathematical foundations of neurocomputing. *IEEE Transactions on Neural Networks*, 78(9):1443–1463, 1990.
2. Andersen, T., Bollerslev, T., Diebold, F., and Ebens, H. The distribution of realized stock return volatility. *Journal of Financial Economics*, 61(1):43–76, 2001.
3. Baele, L. Volatility spillover effects in European equity markets: Evidence from a regime switching model. Ghent University, Working paper, 2002.
4. Bekaert, G., and Harvey, C. R. Time-varying world market integration. *Journal of Finance*, 50(2):403–444, 1995.
5. Bekaert, G., and Harvey, C. R. Emerging equity market volatility. *Journal of Financial Economics*, 43:29–77, 1997.
6. Bekaert, G., Harvey, C. R., and Ng, A. Market integration and contagion. *Journal of Business*, 78:39–69, 2005.
7. Berndt, E. K., Hall, B. H., Hall, R. E., and Hausman, J. A. Estimation and inference in nonlinear structural models. *Annals of Economic and Social Measurement*, 3:653–665, 1974.
8. Bollerslev, T. Generalized autoregressive conditional heteroskedasticity. *Journal of Econometrics*, 31:307–327, 1986.
9. Booth, G. G., Martikainen, T., and Tse, T. Price and volatility spillovers in Scandinavian stock markets. *Journal of Banking and Finance*, 21:811–823, 1997.
10. Broomhead, D. S., and Lowe, D. Multivariate functional interpolation and adaptive networks. *Complex Systems*, 2:321–355, 1988.
11. Christiansen, C. Volatility-spillover effects in European bond markets. University of Aarhus, Denmark, Working paper series no. 162, 2003.
12. Clare, A., and Lekkos, I. An analysis of the relationship between international bond markets. Bank of England, Working paper, 2000.
13. Craven, M., and Shavlik, J. Using neural networks for data mining. Future Generation Computer Systems, 1997.
14. Crouse, R. H., Jin, C., and Hanumara, R. C. Unbiased ridge estimation with prior information and ridge trace. *Communication in Statistics*, 24(9):2341–2354, 1995.
15. Dash, G., Hanumara, C., and Kajiji, N. Neural network architectures for efficient modeling of FX futures options volatility. *Operational Research: An International Journal*, 3(1):3–23, 2003.
16. Dash, G., and Kajiji, N. New evidence on the predictability of South African FX volatility in heterogenous bilateral markets. *The African Finance Journal*, 5(1):1–15, 2003.
17. Diamond, F., Simons, J. et al. J.P. Morgan government bond indices. J.P. Morgan, Portfolio Research Report, 2002.
18. Engle, R. F. Autoregressive conditional heteroskedasticity with estimates of the variance of U.K. inflation. *Econometrica*, 50:987–1008, 1982.
19. Hemmerle, W. J. An explicit solution for generalized ridge regression. *Technometrics*, 17(3):309–314. 1975.
20. Hoerl, A. E., and Kennard, R. W. Ridge regression: Biased estimation for nonorthogonal problems. *Technometrics*, 12(3):55–67, 1970.
21. Kajiji, N. Adaptation of alternative closed form regularization parameters with prior information to the radial basis function neural network for high frequency financial rime series. University of Rhode Island, 2001.

22. Kaski, S. Data exploration using self-organizing maps. Neural Networks Research Centre, Helsinki University of Technology, 1997.
23. Kaski, S., and Kohonen, T. Exploratory data analysis by the self-organizing map: Structures of welfare and poverty in the world. In A. Refenes, Y. Abu-Mostafa, J. Moody, and A. Weigend, Editors, *Neural Networks in Financial Engineering*, World Scientific, Singapore, 1996, pages 498–507.
24. Kohonen, T. The self-organizing map. *Proceedings of the IEEE*, 78(9):1464–1480 1990.
25. Koutmos, G., and Booth, G. G. Asymmetric volatility transmission in international stock markets. *Journal of International Money and Finance*, 14:747–762, 1995.
26. Ljung, G., and Box, G. On a measure of lack of fit in time series models. *Biometrika*, 67:297–303, 1978.
27. Lohninger, H. Evaluation of neural networks based on radial basis functions and their application to the prediction of boiling points from structural parameters. *Journal of Chemical Information and Computer Sciences*, 33:736–744, 1993.
28. Lu, H., Setiono, R., and Liu, H. Effective data mining using neural networks. *IEEE Transactions on Knowledge and Data Engineering*, 8(6):957–961, 1996.
29. McNelis, P. D. *Neural Networks in Finance: Gaining Predictive Edge in the Market*. Elsevier Academic Press, Burlington, MA, 2005.
30. Mehta, K., and Bhattacharyya, S. Adequacy of training data for evolutionary mining of trading rules. *Decision Support Systems*, 37(4):461–474, 2004.
31. Nelson, D. B. Conditional heteroskedasticity in asset returns: A new approach. *Econometrica*, 59:347–370, 1991.
32. Nelson, D. B., and Cao, C. Q. Inequality constraints in the univariate GARCH model. *Journal of Business and Economic Statistics*, 10:229–235, 1992.
33. Ng, A. Volatility spillover effects from Japan and the US to the pacific-basin. *Journal of International Money and Finance*, 19:207–233, 2000.
34. Orr, M. J. L. *Introduction to Radial Basis Function Networks*. Center for Cognitive Science, Scotland, UK, 1996.
35. Orr, M. J. L. *MATLAB Routines for Subset Selection and Ridge Regression in Linear Neural Networks*. Center for Cognitive Science, Scotland, UK, 1997.
36. Parthasarathy, K., and Narendra, K. Stable adaptive control of a class of discrete-time nonlinear systems using radial basis neural networks. Yale University, Report No. 9103, 1991.
37. Polak, E. *Computational Methods in Optimization*. Academic Press, New York, 1971.
38. Refenes, A. N., and Bolland, P. Modeling quarterly returns on the FTSE: A comparative study with regression and neural networks. In C. H. Chen, Editor, *Fuzzy Logic and Neural Network Handbook*. McGraw-Hill, New York, 1996.
39. Rustichini, A., Dickhaut, J., Ghirardato, P., Smith, K., and Pardo, J. V. A brain imaging study of procedural choice. University of Minnesota, Working paper, 2002.
40. Saarinen, S., Bramley, R., and Cybenko, G. Ill-conditioning in neural network training problems. *SIAM Journal of Scientific Computing*, 14(3):693–714, 1993.
41. Sanner, R. M., and Slotine, J.-J. E. Gaussian networks for direct adaptive control. *IEEE Transactions on Neural Networks*, 3:837–863, 1992.
42. Schraudolph, N. N., and Sejnowski, T. J. Tempering backpropagation networks: Not all weights are created equal. In D. S. Touretzky, M. C. Moser, and

M. E. Hasselmo, Editors, *Advances in Neural Information Processing Systems*, MIT Press, Cambridge, 1996, pages 563–569.
43. Shapiro, S. S., and Wilk, M. B. An analysis of variance test for normality (complete samples). *Biometrika*, 52(3/4):597–611, 1965.
44. Shi, J. J. Reducing prediction error by transforming input data for neural networks. *Journal of Computing in Civil Engineering*, 14(2):109–116, 2000.
45. Skintzi, V. D., and Refenes, A.-P. N. Volatility spillovers and dynamic correlation in European bond markets. Financial Engineering Research Center, Athens University of Economics and Business, Athens, 2004.
46. Smagt, P. Minimization methods for training feed-forward networks. *Neural Networks*, 7(1):1–11, 1994.
47. Tikhonov, A., and Arsenin, V. *Solutions of Ill-Posed Problems*. Wiley, New York, 1977.
48. Weigend, A. S., and Rumelhart, D. Generalization of by weight-elimination applied to currency exchange rate prediction. *Proceedings of IJCNN*, New York, 1991.
49. Zirilli, J. S. *Financial Prediction Using Neural Networks*. International Thompson Computer Press, London, 1997.

Estimating Parameters in a Pricing Model with State-Dependent Shocks

Leonard MacLean[1], Yonggan Zhao[2], Giorgio Consigli[3], and William Ziemba[4]

[1] School of Business Administration, Dalhousie University, Halifax, Nova Scotia B3H 3J5, Canada `lmaclean@mgmt.dal.ca`
[2] RBC Centre for Risk Management, Faculty of Management, Dalhousie University, Halifax, Nova Scotia B3H 3J5, Canada `Yonggan.Zhao@dal.ca`
[3] University of Bergamo, Via Salvecchio 19 24129, Bergamo, Italy
[4] Sauder School of Business, University of British Columbia, Vancouver, BC V6T 1Z2, Canada `ziemba@interchange.ubc.ca`

1 Introduction

The expansion of international liquidity and free capital mobility across global financial markets in the last two decades has increased the likelihood and scope of financial instability events induced by sudden changes of market agent expectations. These include unprecedented exchange rates devaluations in developed and emerging markets, Eurobond spreads sudden tightening (Russia 1998, Brazil 1999, Argentina 2001, Venezuela 2001, etc.), and stock markets turmoil due to prevailing speculative conditions (Wall Street, 1987, 1989, 1997, 2002; Japan 1990; Nasdaq dot com 2000–2003). A particular form of volatility is the price bubble, with a steady increase in stock prices followed by a collapse.

A rationale for speculative bubbles, common among practitioners and equity market research units in large financial institutions, is based on the so-called irrational exuberance model, whose origin dates back to 1996 after a speech given by the Fed Chairman Alan Greenspan. He postulated that a long-term equilibrium level for the equity market benchmark implied yield, the 10-year Treasury yield. Price earnings, according to this approach, are thus expected to fluctuate over time around the inverse of the 10-year Treasury rate: Risk premiums will accordingly reduce to negligible values or in some periods even become negative (as observed, for instance, in the early part of this decade). In the Fed model, the correction to stock prices comes in the form of a shock whose size and probability depend on the overvaluation.

In this chapter, we define a model of market returns that under certain conditions determines the emergence of a speculative bubble in the economy and more generally drives bond and equity returns. The model has diffusion parameters that are random variables plus shock/jump terms with random coefficients. There are a number of existing models with jumps (Chernov et al., 2002), but the jumps are modeled by a homogeneous Poisson process. The theory behind the bubble implies that the jump is dependent on the excess returns on stocks. That is, the jump process is state-dependent and therefore is nonhomogeneous. In Section 2, a model is defined with jump terms having a probability and size depending on the differential between the returns on stocks and the long-run equilibrium return. The model incorporates both over- and undervaluation of stocks. A methodology for estimating parameters in the model from actual returns is presented in Section 3. The basis of the approach is the excess volatility in stock returns. Excess returns are assumed to have a shock component, so the location of shocks can be inferred from returns. An algorithm for locating shocks and maximum likelihood estimates for parameters conditional on the shocks are presented. In Section 4, the methodology is tested on returns from the U.S. securities and bond markets. The results support the dependence of shocks on the yield differential.

2 The Pricing Model

The accumulation of wealth is achieved through investment in risky assets in a dynamic securities market. In this section, a general model for asset prices is developed, which accommodates the generation of speculative bubbles and subsequent crashes as prescribed by the Fed model. However, the model is general in that normal and undervalued periods are possible. The presentation will consider three assets: stock, bond, and cash, but the equations and methods can be extended to multiple stocks and bonds. A discrete-time model is presented, since it provides the framework for observation of prices at regular intervals such as days, and the estimation of parameters from observations.

The manifest variables in the pricing model are

$$S(t) = \text{stock price at time } t,$$
$$P(t) = \text{bond price at time } t,$$
$$B(t) = \text{cash return at time } t.$$

The stock and bond prices are random variables defined on a probability space (Ω, B, P) representing the uncertain dynamics of the market.

The prices at regular intervals in times are the result of accumulated changes between observations, so continuous-time differential equations are integrated to get difference equations for changes over an interval.

Consider then the prices of the stock and bond in log form at regular intervals in time $s = 1, \ldots, t$. Let $Y_{1s} = \ln(P_s)$ and $Y_{2s} = \ln(S_s)$, with the initial prices $y_{10} = \ln(p_0)$ and $y_{20} = \ln(s_0)$. So

$$Y_{is} = y_{i,s-1} + \Delta Y_{is}, \quad i = 1, 2, \tag{1}$$

where ΔY_{is} is the change in log-price between times $s-1$ and s. It is assumed that

$$\Delta Y_{1s} = \alpha_{1s} + \delta_1 Z_{1s}, \qquad (2)$$
$$\Delta Y_{2s} = \alpha_{2s} + \delta_2 Z_{2s} + I_s \vartheta_s, \qquad (3)$$

where $Z_{is} \propto N(0,1), i = 1, 2$. ϑ_s is the shock size and I_s is the indicator for a shock in period s, where $I_s = 1$ with probability λ_s, for $s = 1, \ldots, t$. For t days we have $(I_1, \ldots, I_t)$ specifying the days for which there was a shock. Then the time since last shock τ is

$$\tau_s = \tau_{s-1}(1 - I_{s-1}) + 1. \qquad (4)$$

The shock parameter ϑ_s is the aggregate effect of shocks and the drift parameter $\alpha_{is}, i = 1, 2$, are the aggregate effects of drift in the sth time interval. The drift and shock parameters are random variables, defined for $i = 1, 2$, by the equations

$$\alpha_{is} = \mu_i + \gamma_i F, \qquad (5)$$
$$\vartheta_s = \tau_s \theta + \eta Z, \qquad (6)$$

where $F \propto N(0,1)$ is a common factor to stocks and bonds, and $Z \propto N(0,1)$ defines the variation in shock size. So α_{1s} and α_{2s} are correlated. The parameter θ captures the size per period of the differential from the long-run rate. In (6) the expected shock size grows linearly with the time since the last shock.

It is assumed that the intensity λ_s follows the power law, with shape parameter β and size parameter ϕ. The intensity is

$$\lambda_s = \left(\frac{\beta}{\phi}\right)\left(\frac{\tau_s}{\phi}\right)^{\beta-1}. \qquad (7)$$

The power law specification for the intensity fits into the type of formulation where indicator variables are used to define a shock on a given day. So the probability of a shock increases with the time since the last shock, or equivalently with the differential. There are theoretical foundations for the power law with growth models, and so it is the preferred form for the intensity (Gabaix et al., 2003).

The drift parameters $\alpha_{is}, i = 1, 2$, are not dynamic. It is anticipated that a moving window of data will be used to estimate parameters, and within a window a constant prior is justified. The shock size ϑ_s is dynamic, since it depends on the time since the last shock and occurs with probability

$$q_s = \frac{\lambda_{s-\tau_s}}{\lambda_s} \prod_{j=s-\tau_s+1}^{s}(1 - \lambda_j). \qquad (8)$$

Putting the equations together yields the formulas for prices on bonds and stocks at time t:

$$Y_{1t} = y_{10} + \mu_1 t + \gamma_1 \sum_{s=1}^{t} F_s + \delta_1 \sum_{s=1}^{t} Z_{1s}, \tag{9}$$

$$Y_{2t} = y_{20} + \mu_2 t + \gamma_2 \sum_{s=1}^{t} F_s + \delta_2 \sum_{s=1}^{t} Z_{2s} + \vartheta_{\{t\}}, \tag{10}$$

where

$$\vartheta_{\{t\}} = \theta \sum_{s=1}^{t} I_s \tau_s + \eta \sum_{s=1}^{t} I_s Z_s. \tag{11}$$

The model for price dynamics defined by the above equations has similar components to existing models (Chernov et al., 2002). The distinguishing feature of the model here is the time dependence of the point process $\vartheta_{\{t\}}$. This feature is the mechanism for irrational exhuberance and bubbles, but the distribution of prices is considerably complicated by the time dependence. An approach that makes the problem tractable is to separate the price distributions into the conditional distributions, given a sequence of shocks, and the identification of an optimal sequence of shocks.

With the parameters represented as $\Theta = (\mu_1, \mu_2, \theta, \gamma_1, \gamma_2, \delta_1, \delta_2, \eta)$ and the shock sequence $I = (I_1, \ldots, I_t)$, the conditional distribution of log-prices at time t, given the shock sequence and parameters, is Gaussian:

$$(Y_{1t} \mid \Theta, I) \propto N(\mu_{1t}(\Theta), \sigma_{1t}^2(\Theta)), \tag{12}$$
$$(Y_{2t} \mid \Theta, I) \propto N(\mu_{2t}(\Theta, I), \sigma_{2t}^2(\Theta, I)), \tag{13}$$

where

$$\mu_{1t}(\Theta) = y_{10} + \mu_1 t, \tag{14}$$
$$\mu_{2t}(\Theta, I) = y_{20} + \mu_2 t + \theta K_t(I), \tag{15}$$
$$\sigma_{1t}^2(\Theta) = \gamma_1^2 t + \delta_1^2 t, \tag{16}$$
$$\sigma_{2t}^2(\Theta, I) = \gamma_2^2 t + \delta_2^2 t + \eta^2 N_t(I), \tag{17}$$
$$K_t(I) = \sum_{s=1}^{t} I_s \tau_s, \tag{18}$$
$$N_t(I) = \sum_{s=1}^{t} I_s. \tag{19}$$

Based on the independence of F_s, Z_{1s}, Z_{2s}, and Z_s, the covariance between log-prices on stocks and bonds is

$$\sigma_{12t}(\Theta, I) = \gamma_1 \gamma_2 t. \tag{20}$$

3 Parameter Estimation

Consider the set of observations on daily closing prices

$$\{y_1, \ldots, y_t\},$$

where

$$y'_s = (y_{1s}, y_{2s}), s = 1, \ldots, t.$$

Corresponding to the observed prices on stocks is an *unobserved* sequence of shocks $I = (I_1, \ldots, I_t)$. Fitting the model to the data implies estimating the parameters and inferring the sequence of shocks. The changes in prices: $\Delta Y_s = Y_s - y_{s-1}$, are defined by (10) and (11). Consider the observed daily changes in log-prices

$$e_s = y_s - y_{s-1}, \quad s = 1, \ldots, t. \tag{21}$$

The conditional distribution for e_s given I and (Θ, Ξ) is a bivariate normal distribution with density

$$f_s(e_s \mid \Theta, I) = (2\pi)^{-1} |\Sigma_s(I)|^{-\frac{1}{2}} \exp\left[-\tfrac{1}{2}(e_s - \xi_s(I))' \Sigma_s^{-1}(I)(e_s - \xi_s(I))\right], \tag{22}$$

where

$$\xi_s(I) = \begin{pmatrix} \xi_{1s}(I) \\ \xi_{2s}(I) \end{pmatrix} = \begin{pmatrix} \mu_1 \\ \mu_2 + I_s \tau_s \theta \end{pmatrix}, \tag{23}$$

$$\Sigma_s(I) = \begin{pmatrix} \sigma_{1s}^2(I) & \sigma_{12}(I) \\ \sigma_{12}(I) & \sigma_{2s}^2(I) \end{pmatrix} = \begin{pmatrix} \gamma_1^2 + \delta_1^2 & \gamma_1 \gamma_2 \\ \gamma_1 \gamma_2 & \gamma_2^2 + \delta_2^2 + \eta^2 I_s \end{pmatrix}. \tag{24}$$

The structure in the covariance matrix is important for model fitting. Consider $\Gamma' = (\gamma_1, \gamma_2)$ and $\Delta_s(I_s) = \begin{bmatrix} \delta_1^2 & 0 \\ 0 & \delta_2^2 + \eta^2 I_s \end{bmatrix}$. Then $\Sigma_s(I) = \Gamma\Gamma' + \Delta_s(I_s)$. This decomposition of the covariance matrix into a matrix determined by common market factors and a matrix of specific variances will be important in the estimation of parameters.

3.1 Estimating Model Parameters with Given Shocks

For given values of the shocks indicators $I = (I_1, \ldots, I_t)$ and data $e = (e_1, \ldots, e_t)$, the data can be split into two sets based on times with shocks. Let $A = \{s|I_s = 1\}$ and $\bar{A} = \{s|I_s = 0\}$. Consider the statistics for the subsamples: $n_A = $ the number of values in A, $n_{\bar{A}} = $ the number of values in $\bar{A}$,

$$\hat{\xi}_A = \frac{1}{n_A} \sum_{s \in A} e_s = \begin{pmatrix} \bar{e}_{1A} \\ \bar{e}_{2A} \end{pmatrix}, \tag{25}$$

$$\hat{\xi}_{\bar{A}} = \frac{1}{n_{\bar{A}}} \sum_{s \in \bar{A}} e_s = \begin{pmatrix} \bar{e}_{1\bar{A}} \\ \bar{e}_{2\bar{A}} \end{pmatrix}, \tag{26}$$

$$S_A = \frac{1}{n_A} \sum_{s \in A} (e_s - \hat{\xi}_A)(e_s - \hat{\xi}_A)' = \begin{pmatrix} \hat{\sigma}_{1A}^2 & \hat{\sigma}_{12A} \\ \hat{\sigma}_{12A} & \hat{\sigma}_{2A}^2 \end{pmatrix}, \tag{27}$$

$$S_{\bar{A}} = \frac{1}{n_{\bar{A}}} \sum_{s \in \bar{A}} (e_s - \hat{\xi}_{\bar{A}})(e_s - \hat{\xi}_{\bar{A}})' = \begin{pmatrix} \hat{\sigma}_{1\bar{A}}^2 & \hat{\sigma}_{12\bar{A}} \\ \hat{\sigma}_{12\bar{A}} & \hat{\sigma}_{2\bar{A}}^2 \end{pmatrix}. \tag{28}$$

The subsample statistics are the basis of maximum likelihood estimates for parameters. Consider the notation $J = \begin{bmatrix} 0 & 0 \\ 0 & 1 \end{bmatrix}$, $T_s = [0, \tau_s]'$, $\Sigma_A = \Sigma + \eta^2 J$, and $\mu_A = \mu + \theta T_s$, where μ is the expected return and Σ is the covariance, respectively, for the returns without shocks. Assuming the shock sequence is known, the likelihood function for the parameters $\Theta = (\mu, \Sigma, \theta, \eta)$ is Gaussian, and maximizing the log-likelihood produces the estimates in Table 1. In the formulas, $\tilde{\Sigma}$ is a prior value for the covariance matrix. It is anticipated that the formulas will be used iteratively, so the prior value is the result from the previous iteration.

Table 1. Conditional Maximum-Likelihood Estimates for Θ

Parameter	Estimate
μ	$\hat{\mu} = (n_A \tilde{\Sigma}_A^{-1} + n_{\bar{A}} \tilde{\Sigma}^{-1})^{-1}[(\tilde{\Sigma}_A^{-1} \sum_{s \in A}(e_s - \theta T_s)) + (\tilde{\Sigma}^{-1} \sum_{s \in \bar{A}} e_s)]$
Σ	$\hat{\Sigma} = \frac{1}{n}[S_A \tilde{\Sigma}_A^{-1} \tilde{\Sigma} + \tilde{\Sigma}^{-1} \tilde{\Sigma}_A S_{\bar{A}}] - \frac{n_{\bar{A}}}{n} \eta^2 J$
θ	$\theta = \frac{\sum_{s \in A}(e_s - \mu)' \tilde{\Sigma}^{-1} T_s}{\sum_{s \in A} T_s' \tilde{\Sigma}^{-1} T_s}$
η^2	$\eta^2 = \frac{|\tilde{\Sigma}|}{\tilde{\sigma}_{11}} \left(\frac{tr(\frac{1}{n_A} S_A \tilde{\Sigma}^{-1} J \tilde{\Sigma}^{-1})}{tr \tilde{\Sigma}^{-1} J} - 1 \right)$

For the covariance matrices, there is a similar structure defined by the common factor in the drift. The structural solution is presented in Table 2.

With a given sequence of shocks it is straightforward to estimate $\Xi = (\beta, \phi)$ using the definition of the power law.

Table 2. Structural Solution: $\Sigma = \Gamma\Gamma' + \Delta$

Parameter	Estimate
Γ	$\hat{\Gamma}' = (\sqrt{\hat{\rho}}\hat{\sigma}_1, \sqrt{\hat{\rho}}\hat{\sigma}_2)$
Δ	$\hat{\Delta} = \mathrm{diag}((1-|\hat{\rho}|)\hat{\sigma}_1^2, (1-|\hat{\rho}|)\hat{\sigma}_2^2)$
ρ	$\hat{\rho} = \frac{\hat{\sigma}_{12}}{\hat{\sigma}_1 \hat{\sigma}_2}$

Consider the times between shocks calculated from the given sequence $I = (I_1, \ldots, I_t)$, defined as

$$x = \{x_1, \ldots, x_{n_A+1}\}.$$

Note that the first time x_1 will include the estimated time since a shock at the start of observation, which is $\hat{x}_0 = (y_{20} - y_{10})/\hat{\theta}$. Also, the process at time t is possibly truncated before a shock, so the last time x_{n+1} may be truncated. The likelihood for Ξ, given I and x, is

$$L(\Xi \mid I, x) = \left(\frac{\beta}{\phi}\right)^{n_A} \left[\prod_{i=1}^{n_A} \left(\frac{x_i}{\phi}\right)^{\beta-1}\right] \exp\left[-\left(\frac{x_{n_A+1}}{\phi}\right)^{\beta}\right]. \qquad (29)$$

With

$$\ell(\Xi \mid I, x) = \ln L(\Xi \mid I, x),$$

the maximum-likelihood estimates are given in Table 3.

Table 3. Conditional Power Law Estimates

Parameter	Estimate
$G(\beta)$	$n_A \sum_{i=1}^{n_A} x_i^{\beta} + (\sum_{i=1}^{n_A} \ln(x_i)) \sum_{i=1}^{n_A} \beta x_i^{\beta} - n_A \sum_{i=1}^{n_A} \beta^2 x_i^{\beta-1}$
ϕ	$\hat{\phi} = \frac{\sum_{i=1}^{n_A} x_i^{\hat{\beta}}}{n_A},$
β	$\hat{\beta} \ni G(\hat{\beta}) = 0.$

For the power law, the time between shocks follows a Weibull distribution, with shape parameter β. If the probability of a shock increases with the time since the last shock, then $\beta > 1$, and the estimate $\hat{\beta}$ can be used to verify the hypothesis that the shock intensity is state-dependent.

4 Fitting the Shocks: A Peaks Method

The identification of periods with shocks is a key step in implementing the maximum-likelihood methodology. It remains to determine the specification of shocks that is most compatible with the observed log-price increments. If the motivation for shocks is the distribution of returns, and in particular the extreme values, then specifying a critical size for increments and assuming that periods with increments beyond the critical size must contain a shock component is reasonable. The critical size should be linked to the actual distribution of returns. An approach to identifying shocks based on extreme value concepts is presented in this section.

Let $e = (e_1, \ldots, e_n)$ be the vector of one-period observations of log returns and $I = (I_1, \ldots, I_n)$ be the associated jump sequence where $I_i = 1$ indicates a jump. The joint likelihood is $p(e, I) = p(e|I) \times p(I)$, and then the log of the joint likelihood is $\ln(p(e, I)) = \ln(p(e|I)) + \ln(p(I))$.

The diffusion (random walk) parameters including the jump distribution parameters can be estimated using the conditional log-likelihood $\ln(p(e|I))$ for a given jump sequence, and the intensity parameters ϕ and β can be estimated via $\ln(p(I))$. Note that $\ln(p(e|I)) = \ln \prod_{i=1}^{n} p(e_i|I_i) = \sum_{i=1}^{n} \ln(p(e_i|I_i))$.

For computations, first the diffusion and the jump distribution parameters are estimated by maximizing the above log-likelihood, $\ln(p(e|I))$ for a given sequence I. To determine the "best" jump sequence for a given set of observations, maximize the above total (joint) log-likelihood, $\ln(p(e, I))$, over all "possible" jump sequences. The parameter estimates in the propositions are for a given set of shocks. To determine the "best" sequence of shocks, the conditional mle $\hat{\Theta}(I)$ is used. So the shock space is searched and for each selection the conditional likelihood value is calculated for the optimal estimates. The shock sequence with the highest conditional likelihood value is the goal. This is assumed to be close to the joint likelihood value. The number of shock sequences is very large, so a search method based on the observed trajectory of prices is proposed. It is referred to as the PEAKS METHOD since it looks for changes/increments above a specified size in identifying shock times. The method proceeds as follows:

1. Calculate the mean $\bar{e}$ and standard deviation s_1 from the observed increments on stocks.
2. Specify a grid size $\omega > 0$ and a size interval (L, U), and set $k = 0$.
3. Determine a critical increment deviation size $L \leq k\omega s_1 \leq U$, and identify times/indices $T_k = \{i \mid |e_i - \bar{e}| > k\omega s_1, 1 \leq i \leq n\}$.
4. Assume there is a shock at times $i \in T_k$, i.e. $I_i = 1, i \in T_k$. For this sequence of shocks, calculate the conditional maximum-likelihood estimates for $\Theta(k)$, and corresponding conditional likelihood value $l(k)$. Set $k = k + 1$, and return to [2], unless $k \geq (U - L)/\omega s_1$.
5. Find the best sequence of shocks from

$$k^* = \arg\max \left\{ l(k), 0 \leq k \leq (U - L)/\omega s_1 \right\}.$$

In the PEAKS METHOD fewer shocks are identified as k increases. The definition of shock is determined by the data, or more specifically the shock size that produces the best conditional likelihood for the data. A plot $\{k, l(k)\}$ would indicate the pattern of the likelihood with shock size.

5 Numerical Tests

Two issues about the estimation methodology need to be checked: (1) If prices follow the model specified by the equations, are the estimation procedures accurate? (2) If the estimation methodology is sound, what do parameter estimates from actual price trajectories tell us about the significance of shocks and the dependence of shocks on the state of prices?

5.1 A Monte Carlo Study

The soundness of the procedures is checked through simulation. Assume the parameter sets $\Theta = [\mu_1, \mu_2, \sigma_1^2, \sigma_{12}, \sigma_2^2, \theta, \eta]$ and $\Xi = (\phi, \beta)$ for generating the stock returns. θ and η are the mean and the standard deviation of the jumps. Let

$$\Theta = [0.0023, \quad 0.0089, \quad 0.0052, \quad -0.0021, \quad 0.0124, \quad -0.0070, \quad 0.0312],$$
$$\Xi = (20, 5).$$

An example of simulated trajectories of stocks and bonds is shown in Figure 1.

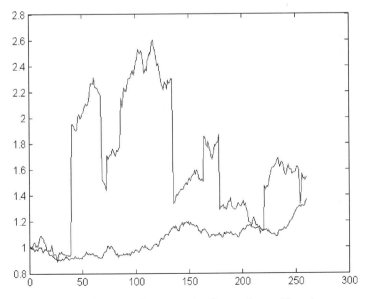

Fig. 1. Simulated trajectories for stocks and bonds.

To test the estimation methodology, 200 trajectories of daily prices were generated and parameter estimates calculated. The results are presented in Table 4.

Table 4. Monte Carlo Results

Parameters	True Value	Estimate	Standard Error	95% Interval
μ_1	0.0023	0.0023	0.0003	(0.0018, 0.0029)
μ_2	0.0089	0.0090	0.0008	(0.0076, 0.0102)
σ_{11}	0.0052	0.0051	0.0002	(0.0047, 0.0055)
σ_{12}	−0.0021	−0.0009	0.0003	(−0.0014, −0.0004)
σ_{22}	0.0124	0.0126	0.0005	(0.0118, 0.0135)
θ	−0.0070	−0.0060	0.0086	(−0.0176, 0.0079)
η	0.0132	0.0287	0.0068	(0.0179, 0.0394)
ϕ	20.0000	20.3636	1.1418	(18.2563, 22.1391)
β	5.0000	5.7206	1.2343	(4.2370, 8.1255)

The estimation results are very reasonable. All estimates are statistically significant and are within the 95% confidence intervals for the true values. The procedures are internally consistent. If the model is an accurate description of price behavior, then the statistical methods are valid for estimating parameters. It is possible that there is model error, but the model presented by Equations (9)–(11) is an enhancement of existing models. The new feature is the power law for the intensity in the Poisson process, and that accomodates constant intensity (existing models) and time-/state-dependent intensity.

5.2 Shock Intensity and Price Differentials

The setup in the pricing model emphasizes the possible dependence of the shock intensity on the time since the last shock and therefore on the gap in prices between stocks and bonds. The shock is a correction, and in the absence of a correction, the stock price can become over- or undervalued, depending on investor sentiment. As the stock price becomes increasingly misvalued, the chances for a correction grow.

To test the intensity for dependence, the parameters in the model for asset price dynamics with (possibly) time-dependent shocks are estimated from actual trajectories of daily prices for stocks and bonds in the U.S. market between 1998 and 2002. The data are from Datastream, with the stock price representing the total market price index, and the bond price representing the yield on 10-year government benchmark bonds. For the computational experiment the 5-year period was divided into 10 consecutive half-year intervals with 130 trading days. A plot of the stock returns in the 10 intervals is shown in Figure 2.

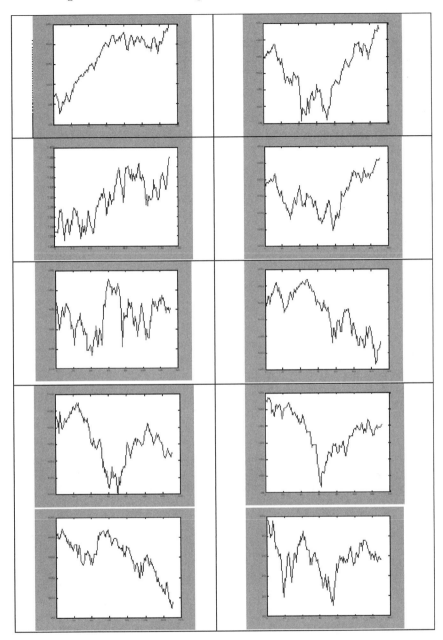

Fig. 2. Daily stock returns for half-year periods.

In each period, the PEAKS METHOD for shock selection was implemented and conditional maximum-likelihood estimates for model parameters were determined.

The final results for shock selection are presented in Table 5. The "best" shock size in terms of the number of standard deviations is very consistent across periods.

Table 5. Peaks Method

Period i	Log Likelihood $l(k^*)$	Critical Size $k^*\omega$	Number of Shocks n_A
1	1023.14	2.36	3
2	871.52	2.40	4
3	960.63	2.14	1
4	885.52	2.14	4
5	913.45	2.16	3
6	990.97	2.10	4
7	855.77	2.42	4
8	913.77	1.96	3
9	971.12	2.10	4
10	889.79	2.12	3

The parameter estimates for the best selection of shocks are given in Table 6. Estimates for the shape parameter β are consistently above 1.0, indicating that the probability of a shock depends on the time since the last shock.

Table 6. Parameter Estimates

Period i	Bond μ_1	Stock μ_2	Bond γ_1	Stock γ_2	Bond δ_1^2	Stock δ_2^2	Scale ϕ	Shape β
1	0.00020	0.00166	−0.00086	0.00113	0.00571	0.00753	45.25	1.53
2	−0.00048	0.00070	−0.00469	0.00607	0.00886	0.01148	19.95	0.99
3	−0.00103	0.00090	0.00271	0.00413	0.00728	0.01111	–	–
4	0.00028	0.00019	0.00823	0.00542	0.01153	0.00759	25.20	1.49
5	0.00053	0.00027	−0.00178	0.00314	0.00741	0.01311	30.06	1.50
6	0.00045	−0.00123	−0.00139	0.00283	0.00495	0.01011	32.40	1.01
7	−0.00139	−0.00133	−0.00337	0.00324	0.01207	0.01160	23.36	1.43
8	0.00031	−0.00007	−0.00188	0.00211	0.00933	0.01047	16.15	0.65
9	0.00038	−0.00161	−0.00310	0.00483	0.00577	0.00900	36.00	2.40
10	0.00055	−0.00153	−0.00577	0.01285	0.00574	0.01277	28.35	1.07

The statistics on the mean and standard deviation of shocks are presented in Table 7. These statistics characterize the set of shocks in each period. The total effect of shocks in a period is small, indicating that market corrections almost cancel out. However, the variance is large, indicating that the individual shocks are large. This supports the use of Poisson processes to fit extreme values (Chernov et al. 2002).

The key finding in this analysis is the placement of shocks. *In the data, the probability of a shock depends on the state of prices for stocks and bonds,*

Table 7. Shock Statistics

Period i	Mean θ_2	Variance η_2^2
1	−0.00038	0.01557
2	−0.00020	0.04968
3	−	−
4	0.00030	0.02610
5	0.00026	0.04680
6	0.00050	0.02520
7	−0.00016	0.04242
8	−0.00081	0.02644
9	0.00042	0.02478
10	0.00114	0.03053

and in particular on the gap in prices. In the usual model with Poisson terms, the intensity is constant and the shocks occur randomly in time. From the perspective of the distribution of returns, the timing of shocks may not appear important since the excess volatility will be captured. From the perspective of market dynamics, it is very important when shocks occur.

6 Conclusion

The addition of shock or jump terms in models for the dynamics of stock prices is useful in fitting the excess volatility of returns. The usual models have shocks with constant intensity. However, there are strong economic reasons to support the dependence of the shock intensity on the recent history of returns. In particular, over- or undervaluation of returns for an extended period should be a precursor of a market correction in the form of a shock.

In this chapter, a model for stock price dynamics is proposed that accomodates state-dependent shocks. Parameter estimation in this more complex model requires a predetermination of the size of shocks. An algorithm that iterates through shock sizes is used to fix a best sequence of shocks over the time periods in an estimation interval.

The test results on the model and estimation methods indicate that the procedures are able to accurately estimate parameters and that the dependence of shock intensity on the state of returns is supported by actual data.

References

1. Chernov, M., Ghysels, E., Gallant, A. R., and Tauchen, G. Alternative models for stock price dynamics. *Journal of Econometrics*, 116(1):225–258, 2002.
2. Consigli, G. Tail estimation and mean-variance portfolio selection in markets subject to financial instability. *Journal of Banking and Finance*, 26(7):1355–1382, 2002.

3. Cox, D. R., and Miller, H. D. *The Theory of Stochastic Processes*. Methuen & Co. Ltd., London, 1970
4. Bjorn, E., Michael, J., and Polson, N. The impact of jumps in volatility and returns. *Journal of Finance*, 53:1269–1300, 2003
5. Gabaix, X., Gopikrishan, P., Pevou, V., and Stanley, H. E. A theory of power-law distributions in financial markets. *Nature*, 423:267–270, 2003.
6. Jorion, P. On jump processes in the foreign exchange and stock markets. *Review of Financial Studies*, 1:427–445, 1988.
7. Lawley, D. N., and Maxwell, A. E., *Factor Analysis as a Statistical Method*. Butterworths, London, 1971.
8. Merton, R. C. *Continuous Time Finance*. Blackwell Publishers Inc., Malden, MA, 1990.

Controlling Currency Risk with Options or Forwards[*]

Nikolas Topaloglou, Hercules Vladimirou, and Stavros A. Zenios

HERMES European Center of Excellence on Computational Finance and Economics, School of Economics and Management, University of Cyprus, P.O.Box 20537, CY-1678 Nicosia, Cyprus hercules@ucy.ac.cy

1 Introduction

International investment portfolios are of particular interest to multinational firms, institutional investors, financial intermediaries, and high-net-worth individuals. Investments in financial assets denominated in multiple currencies provide a wider scope for diversification than investments localized in any market and mitigate the risk exposure to any specific market. However, internationally diversified portfolios are inevitably exposed to currency risk due to uncertain fluctuations of exchange rates.

Currency risk is an important aspect of international investments. With the abandonment of the Bretton Woods system in 1973, exchange rates were set free to float independently. Since then, exchange rates have exhibited periods of high volatility; correlations between exchange rates, as well as between asset returns and exchange rates, have also changed substantially. Stochastic fluctuations of exchange rates constitute an important source of risk that needs to be properly considered (see, e.g., Eun and Resnick, 1988). Thus, it is important to investigate the relative effectiveness of alternative means for controlling currency risk.

Surprisingly, in practice, and usually in the literature as well, international portfolio management problems are addressed in a piecemeal manner. First, an aggregate allocation of funds across various markets is decided at the strategic level. These allocations are then managed pretty much independently, typically by market analysts who select investment securities in each market and manage their respective portfolio. Performance assessment is usually based on comparisons against preselected benchmarks. Currency hedging is often viewed as a subordinate decision; it is usually taken last so as to cover exposures of foreign investments that were decided previously. Changes in the

---

[*] Research partially supported by the HERMES Center of Excellence on Computational Finance and Economics, which is funded by the European Commission, and by a research grant from the University of Cyprus.

overall portfolio composition are not always coordinated with corresponding adjustments to the currency hedging positions. Important and interrelated decisions are considered separately and sequentially. This approach neglects possible cross-hedging effects among portfolio positions and cannot produce a portfolio that jointly coordinates asset and currency holdings so as to yield an optimal risk-return profile. Jorion (1994) criticized this overlay approach; he showed that it is suboptimal to a holistic view that considers all the interrelated decisions in a unified manner.

We consider models that jointly address the international diversification, asset selection, and currency hedging decisions. An important part of this study is the comparison of alternative instruments and tactics for controlling currency risk in international financial portfolios. In dynamic portfolio management settings, this is a challenging problem. We employ the stochastic programming paradigm to empirically assess the relative performance of alternative strategies that use either currency forward contracts or currency options as a means of controlling currency risk.

A currency forward contract constitutes an obligation to sell (or buy) a certain amount of a foreign currency at a specific future date, at a predetermined exchange rate. A forward contract eliminates the downside risk for the amount of the transaction, but at the same time it forgoes the upside potential in the event of a favorable movement in the exchange rate. By contrast, a currency put option provides insurance against downside risk, while retaining upside potential as the option is simply not exercised if the exchange rate appreciates. So, currency forward contracts can be considered as more rigid hedge tools in comparison to currency options.

Few empirical studies on the use of currency options are reported in the literature. Eun and Resnick (1997) examined the use of forward contracts and protective put options for handling currency risk. In ex ante tests, they found that forward contracts generally provide better performance in hedging currency risk than single protective put options. Albuquerque (2007) analyzed hedging tactics and showed that forward contracts dominate the use of single put options as hedges of transaction exposures. The reason is that forward contracts pay more than single options on the downside; hence, less currency needs to be sold forward to achieve the same degree of hedging; a smaller hedge ratio is required and the cost for hedging is less. Maurer and Valiani (2003) compared the effectiveness of currency options versus forward contracts for hedging currency risk. They found that both ex-post, as well as in out-of-sample tests, forwards contracts dominate the use of single-currency put options. Only put in-the-money options produce comparable results with optimally hedged portfolios with forwards. Their results indicate more active use of put in-the-money options than at-the-money or out-of-the money put options, revealing the dependence of a hedging strategy based on put options on the level of the strike price.

Conover and Dubofsky (1995) considered American options. They empirically examined portfolio insurance strategies employing currency spot and future options. They found that protective puts using future options are generally dominated by both protective puts that use options on spot currencies and by fiduciary calls on futures contracts. Lien and Tse (2001) compared the hedging effectiveness of currency options versus futures on the basis of lower partial moments (LPM). They concluded that currency futures provide a better hedging instrument than currency options; the only situation in which options outperform futures occurs when the decision maker is optimistic (with a large target return) and not too concerned about large losses.

Steil (1993) applied an expected utility analysis to determine optimal contingent claims for hedging foreign transaction exposure as well as optimal forward and option hedge alternatives. Using quadratic, negative exponential, and positive exponential utility functions, Steil concluded that currency options play a limited useful role in hedging contingent foreign exchange transaction exposures.

There is no consensus in the literature regarding a universally preferable strategy to hedge currency risk, although the majority of results indicates that currency forwards generally yield better performance than single protective put options. Earlier studies did not jointly consider the optimal selection of internationally diversified portfolios. Our study addresses this aspect of the portfolio management problem in connection with the associated problem of controlling currency risk and contributes to the aforementioned debate. We empirically examine whether forward contracts are effective hedging instruments, or whether superior performance can be achieved by using currency options–either individual protective puts or combinations of options with appropriate payoffs.

To this end, we extend the multistage stochastic programming model for international portfolio management that was developed in Topaloglou et al. (forthcoming) by introducing positions in currency options to the decision set at each stage. The model accounts for the effects of portfolio (re)structuring decisions over multiple periods, including positions in currency options among its permissible decisions. The incorporation of currency options in a practical portfolio optimization model is a novel development. A number of issues are addressed in the adaptation of the model. Currency options are suitably priced at each decision stage of the stochastic program in a manner consistent with the scenario set of exchange rates. The scenario-contingent portfolio rebalancing decisions account for the discretionary exercise of expiring options at each decision point.

The dynamic nature of portfolio management problems motivated our development of flexible multistage stochastic programming models that capture in a holistic manner the interrelated decisions faced in international portfolio management. Multistage models help decision makers adopt more effective decisions; their decisions consider longer-term potential benefits and avoid myopic reactions to short-term movements that may lead to losses.

We use the stochastic programming model as a testbed to empirically assess the relative effectiveness of currency options and forward contracts to control the currency risk of international portfolios in a dynamic setting. We analyze the effect of alternative strategies on the performance of international portfolios of stock and bond indices in backtesting experiments over multiple time periods. Our empirical results confirm that portfolios with optimally selected forward contracts outperform those that involve a single protective put option per currency. However, we find that trading strategies involving suitable combinations of currency options have the potential to produce better performance. Moreover, we demonstrate through extensive numerical tests the viability of a multistage stochastic programming model as a decision support tool for international portfolio management. We show that the dynamic (multistage) model consistently outperforms its single-stage (myopic) counterpart.

The chapter is organized as follows. In Section 2, we present the formulation of the optimization models for international portfolio selection. In Section 3, we discuss the hedging strategies employed in the empirical tests. In Section 4, we describe the computational tests and we discuss the empirical results. Section 5 concludes. Finally, in the Appendix we describe the procedure for pricing European currency options consistently with the discrete distribution of exchange rates on a scenario tree.

2 The International Portfolio Management Model

The international portfolio management model aims to determine the optimal portfolio that has the minimum shortfall risk at each level of expected return over the planning horizon. The problem is viewed from the perspective of a U.S. investor who may hold assets denominated in multiple currencies. The portfolio is exposed to market and currency risk. To cope with the market risk, the portfolio is diversified across multiple markets. International diversification exposes the foreign investments to currency risk. To control the currency risk, the investor may enter into currency exchange contracts in the forward market, or buy currency options–either single protective puts, or combinations of options that form a particular trading strategy.

In this section, we develop scenario-based stochastic programming models for managing investment portfolios of international stock and government bond indices. The models address the problems of optimal portfolio selection and currency risk management in an integrated manner. Their deterministic inputs are the initial asset holdings, the current prices of the stock and bond indices, the current spot exchange rates, the forward exchange rates, or the currency option prices–depending on which instruments are used to control currency risk–for a term equal to the decision interval. We also specify scenario-dependent data, together with associated probabilities, that represent the discrete process of the random variables at any decision stage in terms of a scenario tree. The prices of the indices and the exchange rates at any node of the scenario tree are generated with the moment-matching procedure of Høyland et al. (2003); these, in turn, uniquely determine the option payoffs at any node of the tree.

We explore single-stage as well as multistage stochastic programming models to manage international portfolios of financial assets. The multistage model determines a sequence of buying and selling decisions at discrete points in time (monthly intervals). The portfolio manager starts with a given portfolio and with a set of postulated scenarios about future states of the economy represented in terms of a scenario tree, as well as corresponding forward exchange rates or currency option prices depending on the postulated scenarios. This information is incorporated into a portfolio restructuring decision. The composition of the portfolio at each decision point depends on the transactions that were decided at the previous stage. The portfolio value depends on the outcomes of asset returns and exchange rates realized in the interim period and, consequently, on the discretionary exercise of currency options whose purchase was decided at the previous decision point. Another portfolio restructuring decision is then made at that node of the scenario tree based on the available portfolio and taking into account the projected outcomes of the random variables in subsequent periods.

The models employ the conditional value-at-risk (CVaR) risk metric to minimize the excess losses, beyond a prespecified percentile of the portfolio return distribution, over the planning horizon. The decision variables reflect asset purchase and sale transactions that yield a revised portfolio. Additionally, the models determine the levels of forward exchange contracts or currency option purchases to mitigate currency risk. Positions in specific combinations of currency options–corresponding to certain trading strategies–are easily enforced with suitable linear constraints. The portfolio optimization models incorporate practical considerations (no short sales for assets, transaction costs) and minimize the tail risk of final portfolio value at the end of the planning horizon for a given target of expected return. The models determine jointly the portfolio compositions (not only the allocation of funds to different markets, but also the selection of assets within each market) and the levels of currency hedging in each market via forward contracts or currency options.

To ensure the internal consistency of the models, we price the currency options at each decision node on the basis of the postulated scenario sets. To this end, we adapt a suitable option valuation procedure that accounts for higher-order moments exhibited in historical data of exchange rates, as described in the Appendix. The option prices are used as inputs to the optimization models together with the postulated scenarios of asset returns and exchange rates. We confine our attention to European currency options that may be purchased at any decision node and have a maturity of one period. At any decision node of the scenario tree, the selected options in the portfolio may be exercised and new option contracts may be purchased.

We use the following notation:

Sets:

C_0	the set of currencies (synonymously, markets, countries),
$\ell \in C_0$	the index of the base (reference) currency in the set of currencies,
$C = C_0 \setminus \{\ell\}$	the set of foreign currencies,
I_c	the set of assets denominated in currency $c \in C_0$ (these consist of one stock index, one short-term, one intermediate-term, and one long-term government bond index in each country),
N	the set of nodes of the scenario tree,
$n \in N$	a typical node of the scenario tree ($n = 0$ is the root node at $t = 0$),
$N_t \subset N$	the set of distinct nodes at time period $t = 0, 1, \ldots, T$,
$N_T \subset N$	the set of leaf (terminal) nodes at the last period T, that uniquely identify the scenarios,
$S_n \subset N$	the set of immediate successor nodes of node $n \in N \setminus N_T$. This set of nodes represents the discrete distribution of the random variables at the respective time period, conditional on the state of node n.
$p(n) \in N$	the unique predecessor node of node $n \in N \setminus \{0\}$,
J_c	the set of available currency options for foreign currency $c \in C$ (differing in terms of exercise price).

Input Data:

(a) <u>Deterministic parameters:</u>

T	length of the time horizon (number of decision periods),
b_{ic}	initial position in asset $i \in I_c$ of currency $c \in C_0$ (in units of face value),
h_c^0	initially available cash in currency $c \in C_0$ (surplus if +ve, shortage if -ve),
δ	proportional transaction cost for sales and purchases of assets,
d	proportional transaction cost for currency transactions in the spot market,
μ	prespecified target expected portfolio return over the planning horizon,
α	prespecified percentile for the CVaR risk measure,
π_{ic}^0	current market price (in units of the respective currency) per unit of face value of asset $i \in I_c$ in currency $c \in C_0$,
e_c^0	current spot exchange rate for foreign currency $c \in C$,
f_c^0	currently quoted one-month forward exchange rate for foreign currency $c \in C$,
K_j	the strike price of an option $j \in J_c$, on the spot exchange rate of foreign currency $c \in C$.

(b) Scenario-dependent parameters:

p_n probability of occurrence of node $n \in \mathbf{N}$,
e_c^n spot exchange rate of currency $c \in \mathbf{C}$ at node $n \in \mathbf{N}$,
f_c^n one-month forward exchange rate for foreign currency $c \in \mathbf{C}$ at node $n \in \mathbf{N} \setminus \mathbf{N_T}$,
π_{ic}^n price of asset $i \in \mathbf{I_c}$, $c \in \mathbf{C_0}$ on node $n \in \mathbf{N}$ (in units of local currency),
$cc^n(e_c^n, K_j)$ price of European call currency option $j \in \mathbf{J_c}$ on the exchange rate of currency $c \in \mathbf{C}$, at node $n \in \mathbf{N} \setminus \mathbf{N_T}$, with exercise price K_j and maturity of one month,
$pc^n(e_c^n, K_j)$ price of European put currency option $j \in \mathbf{J_c}$ on the exchange rate of currency $c \in \mathbf{C}$, at node $n \in \mathbf{N} \setminus \mathbf{N_T}$, with exercise price K_j and maturity of one month.

All exchange rates $(e_c^0, f_c^0, e_c^n, f_c^n)$ are expressed in units of the base currency per one unit of the foreign currency $c \in \mathbf{C}$. Of course, the exchange rate of the base currency to itself is trivially equal to 1, $f_\ell^n = e_\ell^n \equiv 1$, $\forall n \in \mathbf{N}$. The prices cc and pc of currency call and put options, respectively, are expressed in units of the base currency ℓ.

Computed Parameters:

V_ℓ^0 total value (in units of the base currency) of the initial portfolio.

$$V_\ell^0 = h_\ell^0 + \sum_{i \in I_\ell} b_{i\ell}\, \pi_{i\ell}^0 + \sum_{c \in C} e_c^0 \left(h_c^0 + \sum_{i \in I_c} b_{ic}\, \pi_{ic}^0 \right) \quad (1)$$

Decision Variables:

Portfolio (re)structuring decisions are made at all non-terminal nodes of the scenario tree, thus $\forall n \in \mathbf{N} \setminus \mathbf{N_T}$.

(a) Asset purchase, sale, and hold quantities (in units of face value):

x_{ic}^n units of asset $i \in \mathbf{I_c}$ of currency $c \in \mathbf{C_0}$ purchased,
v_{ic}^n units of asset $i \in \mathbf{I_c}$ of currency $c \in \mathbf{C_0}$ sold,
w_{ic}^n resulting units of asset $i \in \mathbf{I_c}$ of currency $c \in \mathbf{C_0}$ in the revised portfolio.

(b) Currency transfers in the spot market:

$x_{c,e}^n$ amount of base currency exchanged in the spot market for foreign currency $c \in \mathbf{C}$,
$v_{c,e}^n$ amount of the base currency collected from a spot sale of foreign currency $c \in \mathbf{C}$.

(c) Forward currency exchange contracts:

$u_{c,f}^n$ amount of base currency collected from sale of currency $c \in \boldsymbol{C}$ in the forward market (i.e., amount of a forward contract, in units of the base currency). A negative value indicates a purchase of the foreign currency forward. This decision is taken at node $n \in \boldsymbol{N} \setminus \boldsymbol{N_T}$, but the transaction is actually executed at the end of the respective period, i.e., at the successor nodes $\boldsymbol{S_n}$.

(d) Variables related to currency options transactions:

$ncc_{c,j}^n$ purchases of European call currency option $j \in \boldsymbol{J_c}$ on the exchange rate of currency $c \in \boldsymbol{C}$, with exercise price K_j and maturity of one month,

$npc_{c,j}^n$ purchases of European put currency option $j \in \boldsymbol{J_c}$ on the exchange rate of currency $c \in \boldsymbol{C}$, with exercise price K_j and maturity of one month.

When currency options are used in the portfolio management model, only long positions in the respective trading strategies of options are allowed.

Auxiliary Variables:

y_n auxiliary variables used to linearize the piecewise linear function in the definition of the CVaR risk metric; they measure the portfolio losses at leaf node $n \in \boldsymbol{N_T}$ in excess of VaR,

z the value-at-risk (VaR) of portfolio losses over the planning horizon (i.e., the αth percentile of the loss distribution),

V_ℓ^n the total value of the portfolio at the end of the planning horizon at leaf node $n \in \boldsymbol{N_T}$ (in units of the base currency),

R_n return of the international portfolio over the planning horizon at leaf node $n \in \boldsymbol{N_T}$.

2.1 International Portfolio Management Models

We consider either forward contracts or currency options in the optimization models, but not both, as means to mitigate the currency risk of international portfolios. Hence, we formulate two different variants of the international portfolio optimization model; the models differ in the cashflow balance constraints and the computation of the final portfolio value.

Portfolio Optimization Model with Currency Options

This model minimizes the conditional value-at-risk (CVaR) of portfolio losses over the planning horizon, while also requiring that expected portfolio return meets a prespecified target, μ, (2i). Expectations are computed over the set of terminal states (leaf nodes). The objective value (2a) measures the CVaR of portfolio losses at the end of the horizon, while the corresponding VaR of portfolio losses (at percentile α) is captured by the variable z; see Rockafellar and Uryasev (2002), Topaloglou et al. (2002).

$$\min z + \frac{1}{1-\alpha} \sum_{n \in \mathbf{N_T}} p_n y_n \tag{2a}$$

$$\text{s.t.} h_\ell^0 + \sum_{i \in \mathbf{I_\ell}} v_{i\ell}^0 \pi_{i\ell}^0 (1-\delta) + \sum_{c \in \mathbf{C}} v_{c,e}^0 (1-d) = \sum_{i \in \mathbf{I_\ell}} x_{i\ell}^0 \pi_{i\ell}^0 (1+\delta) + \sum_{c \in \mathbf{C}} x_{c,e}^0 (1+d)$$
$$+ \sum_{c \in \mathbf{C}} \sum_{j \in \mathbf{J_c}} \left[npc_{c,j}^0 * pc^0(e_c^0, K_j) \right] \tag{2b}$$

$$h_c^0 + \sum_{i \in \mathbf{I_c}} v_{ic}^0 \pi_{ic}^0 (1-\delta) + \frac{1}{e_c^0} x_{c,e}^0 = \sum_{i \in \mathbf{I_c}} x_{ic}^0 \pi_{ic}^0 (1+\delta) + \frac{1}{e_c^0} v_{c,e}^0, \ \forall \ c \in \mathbf{C} \tag{2c}$$

$$h_\ell^n + \sum_{i \in \mathbf{I_\ell}} v_{i\ell}^n \pi_{i\ell}^n (1-\delta) + \sum_{c \in \mathbf{C}} v_{c,e}^n (1-d) + \sum_{c \in \mathbf{C}} \sum_{j \in \mathbf{J_c}} \left[npc_{c,j}^{p(n)} * \max(K_j - e_c^n, 0) \right]$$
$$= \sum_{i \in \mathbf{I_\ell}} x_{i\ell}^n \pi_{i\ell}^n (1+\delta) + \sum_{c \in \mathbf{C}} x_{c,e}^n (1+d) + \sum_{c \in \mathbf{C}} \sum_{j \in \mathbf{J_c}} \left[npc_{c,j}^n * pc^n(e_c^n, K_j) \right],$$
$$\forall \ n \in \mathbf{N} \setminus \{\mathbf{N_T} \cup 0\} \tag{2d}$$

$$h_c^n + \sum_{i \in \mathbf{I_c}} v_{ic}^n \pi_{ic}^n (1-\delta) + \frac{1}{e_c^n} x_{c,e}^n = \sum_{i \in \mathbf{I_c}} x_{ic}^n \pi_{ic}^n (1+\delta) + \frac{1}{e_c^n} v_{c,e}^n ,$$
$$\forall \ c \in \mathbf{C}, \ \forall \ n \in \mathbf{N} \setminus \{\mathbf{N_T} \cup 0\} \tag{2e}$$

$$V_\ell^n = \sum_{i \in \mathbf{I_\ell}} w_{i\ell}^{p(n)} \pi_{i\ell}^n + \sum_{c \in \mathbf{C}} \left\{ e_c^n \left[\sum_{i \in \mathbf{I_c}} w_{ic}^{p(n)} \pi_{ic}^n \right] \right.$$
$$\left. + \sum_{j \in \mathbf{J_c}} \left[npc_{c,j}^{p(n)} * \max(K_j - e_c^n, 0) \right] \right\}, \ \forall n \in \mathbf{N_T} \tag{2f}$$

$$\sum_{j \in \mathbf{J_c}} npc_{c,j}^n \leq \sum_{i \in \mathbf{I_c}} e_c^n \left(w_{ic}^n \pi_{ic}^n \right), \ \forall c \in \mathbf{C}, \ \forall n \in \mathbf{N} \setminus \mathbf{N_T} \tag{2g}$$

$$R_n = \frac{V_\ell^n}{V_\ell^0} - 1, \ \forall n \in \mathbf{N_T} \tag{2h}$$

$$\sum_{n \in \mathbf{N_T}} p_n R_n \geq \mu, \tag{2i}$$

$$y_n \geq L_n - z, \ \forall n \in \mathbf{N_T} \tag{2j}$$

$$y_n \geq 0, \qquad \forall n \in \mathbf{N_T} \tag{2k}$$

$$L_n = -R_n, \qquad \forall n \in \mathbf{N_T} \tag{2l}$$

$$w_{ic}^0 = b_{ic} + x_{ic}^0 - v_{ic}^0, \qquad \forall \ i \in \mathbf{I_c}, \ \forall \ c \in \mathbf{C_0} \tag{2m}$$

$$w_{ic}^n = w_{ic}^{p(n)} + x_{ic}^n - v_{ic}^n, \quad \forall \ i \in \mathbf{I_c}, \ \forall \ c \in \mathbf{C_0}, \ \forall \ n \in \mathbf{N} \setminus \{\mathbf{N_T} \cup 0\} \tag{2n}$$

$$x_{ic}^n \geq 0, \ w_{ic}^n \geq 0, \qquad \forall \ i \in \mathbf{I_c}, \ \forall \ c \in \mathbf{C_0}, \ \forall \ n \in \mathbf{N} \setminus \mathbf{N_T} \tag{2o}$$

$$0 \leq v_{ic}^0 \leq b_{ic}, \qquad \forall \ i \in \mathbf{I_c}, \ \forall \ c \in \mathbf{C_0} \tag{2p}$$

$$0 \leq v_{ic}^n \leq w_{ic}^{p(n)}, \qquad \forall \ i \in \mathbf{I_c}, \ \forall \ c \in \mathbf{C_0}, \ \forall \ n \in \mathbf{N} \setminus \{\mathbf{N_T} \cup 0\} \tag{2q}$$

We adopt the CVaR risk metric, as it is suitable for asymmetric distributions. Asymmetry in the returns of the international portfolios arises not only because of the skewed and leptokurtic distributions of exchange rates, but mainly because of the highly asymmetric payoffs of options. The choice of the CVaR metric that captures the tail risk is entirely appropriate for the purposes of this study that aims to explore the effectiveness of currency options–or forward contracts–as means to mitigate and control the currency risk so as to minimize the excess shortfall of portfolio returns over the planning horizon.

The purchase of an option entails a cost (price) that is payable at the time of purchase. The cost of option purchases is considered in the cash balance constraints of the base currency (2b) and (2d). Similarly, the conditional payoffs of the options are also accounted for in the cash balance conditions at the respective expiration dates. We consider only European options. Specifically, we use options with a single-period maturity (one month in our implementation). So, options purchased at some decision stage mature at exactly the next decision period, at which time they either are exercised, if they yield a positive payoff, or are simply left to expire.

The exercise prices of the options are specified exogenously as inputs to the model. By considering multiple options with different strike prices on the same currency, we can provide the model flexibility to choose the most appropriate options at each decision stage. The option prices at each node of the scenario tree are computed according to the valuation procedure summarized in the Appendix. The corresponding payoffs at the successor nodes on the tree are also computed and entered as inputs to the portfolio optimization program. The optimal portfolio rebalancing decisions, as well as the optimal positions in currency options, are considered in a unified manner at each decision node of the scenario tree. The model does not directly relate positions in options on different currencies, thus allowing selective hedging choices.

Model (2a)–(2q) is a stochastic linear program with recourse. Equations (2b) and (2c) impose the cash balance conditions in every currency at the first decision stage (root node), the former for the base currency, ℓ, and the latter for the foreign currencies, $c \in C$. Each constraint equates the sources and the uses of funds in the respective currency. The availability of funds stems from initially available cash reserves, revenues from the sale of initial asset holdings, and amounts received through incoming currency exchanges in the spot market. Correspondingly, the uses of funds include the expenditures for the purchase of assets, the outgoing currency exchanges in the spot market, and the costs for the purchase of currency options, the latter for the cash equation of the base currency only. All currency exchanges are made through the base currency. Linear transaction costs (i.e., proportional to the amount of a transaction) are considered for purchases and sales of assets, as well as for currency exchanges in the spot market. Note that all available funds are placed in the available assets; that is, we don't have investments in money market accounts in any currency, nor do we have borrowing. These could be simple extensions of the model.

Similarly, Equations (2d) and (2e) impose the cash balance conditions at subsequent decision states for the base currency, ℓ, and the foreign currencies, $c \in \boldsymbol{C}$, respectively. Now cash availability comes from exogenous inflows, if any, revenues from the sale of asset holdings in the portfolio at hand, incoming spot currency exchanges, and potential payoffs from the exercise of currency option contracts purchased at the predecessor node. Again, the uses of funds include the purchase of assets, outgoing currency exchanges in the spot market, and the purchase of currency options with maturity one period ahead. The cash flows associated with currency options (purchases and payoffs) enter only the cash balance equations of the base currency.

The final value of the portfolio at leaf node $n \in \boldsymbol{N_T}$ is computed in (2f). The total terminal value, in units of the base currency, reflects the proceeds from the liquidation of all final asset holdings at the corresponding market prices and the payoffs of currency put options expiring at the end of the horizon. Revenues in foreign currencies are converted to the base currency by applying the respective spot exchange rates at the end of the horizon.

Constraints (2g) limit the put options that can be purchased on each foreign currency. The total position in put options of each currency is bounded by the total value of assets that are held in the respective currency after the portfolio revision. So, currency puts are used only as protective hedges for investments in foreign currencies and can cover up to the foreign exchange rate exposure of the portfolio held at the respective decision state.

Equation (2h) defines the return of the portfolio during the planning horizon at leaf node $n \in \boldsymbol{N_T}$. Constraint (2i) imposes a minimum target bound, μ, on the expected portfolio return over the planning horizon. Constraints (2j) and (2k) are the definitional constraints for CVaR, while Equation (2l) defines portfolio losses as negative returns. Equations (2m) enforce the balance constraints for each asset, at the first decision stage, while Equations (2n) similarly impose the balance constraint for each asset, at subsequent decision states. These equations determine the resulting composition of the revised portfolio after the purchase and sale transactions of assets at the respective decision nodes. Short positions in assets are not allowed, so constraints (2o) ensure that asset purchases, as well as the resulting holdings in the rebalanced portfolio, are nonnegative. Finally, constraints (2p) and (2q) restrict the sales of each asset by the corresponding holdings in the portfolio at the time of a rebalancing decision.

Starting with an initial portfolio and using a representation of uncertainty for the asset prices and exchange rates by means of a scenario tree, as well as the prices and payoffs of the currency put options at each decision node, the multistage portfolio optimization model determines optimal decisions under the contingencies of the scenario tree. The portfolio rebalancing decisions at each node of the tree specify not only the allocation of funds across markets but also the positions in assets within each market. Moreover, positions in currency options are appropriately determined so as to mitigate the currency risk exposure of the foreign investments during the holding period (i.e., until the next portfolio rebalancing decision).

Portfolio Optimization Model with Currency Forward Contracts

Positions in currency forwards shell the value of foreign investments against potential depreciations of exchange rates. However, by fixing the exchange rate of forward transactions, these contracts forgo potential gains in the event of potential appreciations of exchange rates; this is the "penalty" for the protection against downside risk. We consider currency forward contracts with a single-period term (one month in our implementation). Hence, forward contracts decided in one period are executed in the next decision period. Positions in currency forward contracts are introduced as decision variables ($u_{c,f}^n$) at each decision state of the multistage portfolio optimization program. These decisions are determined jointly with the corresponding portfolio rebalancing decisions in an integrated manner. Forward exchange contracts in different currencies are not explicitly connected. The model can choose different coverage of the foreign exchange exposures in the different currencies (i.e., different hedge ratios across currencies), reflecting a flexible selective hedging approach.

$$\min \ z + \frac{1}{1-\alpha} \sum_{n \in N_T} p_n y_n \tag{3a}$$

$$\text{s.t.} \ h_\ell^0 + \sum_{i \in I_\ell} v_{i\ell}^0 \pi_{i\ell}^0 (1-\delta) + \sum_{c \in C} v_{c,e}^0 (1-d) = \sum_{i \in I_\ell} x_{i\ell}^0 \pi_{i\ell}^0 (1+\delta) + \sum_{c \in C} x_{c,e}^0 (1+d) \tag{3b}$$

$$h_c^0 + \sum_{i \in I_c} v_{ic}^0 \pi_{ic}^0 (1-\delta) + \frac{1}{e_c^0} x_{c,e}^0 = \sum_{i \in I_c} x_{ic}^0 \pi_{ic}^0 (1+\delta) + \frac{1}{e_c^0} v_{c,e}^0, \ \forall \ c \in C \tag{3c}$$

$$h_\ell^n + \sum_{i \in I_\ell} v_{i\ell}^n \pi_{i\ell}^n (1-\delta) + \sum_{c \in C} \left(v_{c,e}^n (1-d) + u_{c,f}^{p(n)} \right)$$
$$= \sum_{i \in I_\ell} x_{i\ell}^n \pi_{i\ell}^n (1+\delta) + \sum_{c \in C} x_{c,e}^n (1+d), \ \forall n \in N_T \setminus \{N_T \cup 0\} \tag{3d}$$

$$h_c^n + \sum_{i \in I_c} v_{ic}^n \pi_{ic}^n (1-\delta) + \frac{1}{e_c^n} x_{c,e}^n = \sum_{i \in I_c} x_{ic}^n \pi_{ic}^n (1+\delta) + \frac{1}{e_c^n} v_{c,e}^n + \frac{1}{f_c^{p(n)}} u_{c,f}^{p(n)},$$
$$\forall \ c \in C, \ \forall \ n \in N \setminus \{N_T \cup 0\} \tag{3e}$$

$$V_\ell^n = \sum_{i \in I_\ell} w_{i\ell}^{p(n)} \pi_{i\ell}^n + \sum_{c \in C} \left[u_{c,f}^{p(n)} + e_c^n \left[\sum_{i \in I_c} w_{ic}^{p(n)} \pi_{ic}^n - \frac{1}{f_c^{p(n)}} u_{c,f}^{p(n)} \right] \right], \ \forall n \in N_T \tag{3f}$$

$$0 \le u_{c,f}^n \le \sum_{m \in S_n} \frac{p_m}{p_n} e_c^m \left(\sum_{i \in I_c} w_{ic}^n \pi_{ic}^m \right), \ \forall c \in C, \ \forall n \in N \setminus N_T \tag{3g}$$

and also constraints (2h)–(2q).

This formulation differs from the previous model, which employs currency options, in the cash balance constraints and the valuation of the portfolio at the end of the planning horizon. Equations (3b) and (3c) impose the cash balance conditions in the first stage for the base currency, ℓ, and the foreign currencies, $c \in C$, respectively. Equations (3d) and (3e) impose the cash balance conditions for every currency at subsequent decision states. These equations account for the cash flows associated with currency forward contracts that were decided at the predecessor state.

Equation (3f) computes the value of the portfolio, in units of the base currency, at leaf node $n \in \boldsymbol{N_T}$. The terminal value reflects the proceeds from the liquidation of the final asset holdings at the corresponding market prices and the proceeds of outstanding forward contracts in foreign currencies. The values in foreign currencies are converted to the base currency by applying the respective spot exchange rates at the end of the horizon, after settling the outstanding forward contracts.

Constraints (3g) limit the currency forward contracts. The amount of a forward contract in a foreign currency is restricted by the expected value of all asset holdings in the respective currency after the revision of the portfolio at that state. This ensures that forward contracts are used only for hedging, and not for speculative purposes. The right-hand side of (3g) reflects the expected value of the respective foreign positions at the end of the holding period. The conditional expectation is taken over the discrete outcomes at the successor nodes ($\boldsymbol{S_n}$) of the decision state $n \in \boldsymbol{N} \setminus \boldsymbol{N_T}$.

2.2 Scenario Generation

The scenario generation is a critical step of the modeling process. The set of scenarios must adequately depict the projected evolution of the underlying financial primitives (asset returns and exchange rates) and must be consistent with market observations and financial theory. We generate scenarios with the moment-matching method of Høyland et al. (2003). The outcomes of asset returns and exchange rates at each stage of the scenario tree are generated so that their first four marginal moments (mean, variance, skewness, and kurtosis) as well as their correlations match their respective statistics estimated from market data. Thus, the outcomes on the scenario tree reflect the empirical distribution of the random variables as implied by historical observations.

We analyze the statistical characteristics of exchange rates over the period 05/1988–11/2001 that were used in the static and dynamic tests. As Table 1 shows, the monthly variations of spot exchange rates exhibit skewed distributions. They also exhibit considerable variance in comparison to their mean, as well as excess kurtosis, implying heavier tails than the normal distribution. Jarque–Berra tests (1980) on these data indicate that normality hypotheses cannot be accepted.[1]

The skewed and leptokurtic distributions of the financial random variables (asset returns and exchange rates) motivated our choice of the moment-matching scenario generation procedure, as this approach can capture the statistical characteristics implied by historical market data. An essential requirement in modeling stochastic financial variables is that they must satisfy

---

[1] The Jarque–Berra statistic has a $\mathcal{X}^2$ distribution with two degrees of freedom. Its critical values at the 5% and 1% confidence levels are 5.99 and 9.21, respectively. The normality hypothesis is rejected when the Jarque–Berra statistic has a higher value than the corresponding critical value at the respective confidence level.

Table 1. Statistical Characteristics and Jarque–Berra Statistic of Historical Monthly Changes of Spot Exchange Rates Over the Period 05/1988–11/2001

Exchange Rate	Mean	Std. Dev.	Skewness	Kurtosis	Jarque–Berra Statistic
UKtoUS	−0.116%	2.894%	−0.755	5.672	102.39
GRtoUS	−0.124%	3.091%	−0.215	3.399	5.69
JPtoUS	0.030%	3.615%	0.942	6.213	105.27

the fundamental no-arbitrage condition. We exhaustively tested all the scenario sets used in the numerical experiments and empirically verified that the no-arbitrage requirement was always satisfied.

We note that the stochastic programming models presented above are not restricted to the moment-matching scenario generation procedure. A user may adopt an alternative approach that he finds preferable to project the evolution of the stochastic asset prices and exchange rates, as long as the scenarios effectively reflect the empirical distribution of the random variables and satisfy fundamental financial principles (i.e., must be arbitrage-free). Dupačová et al. (2000) reviewed alternative scenario generation approaches.

In order to ensure internal consistency of the model, we adapt a suitable valuation procedure to price the currency options consistently with the postulated scenarios of exchange rates. This option pricing approach–summarized in the Appendix–accounts for the higher moments of exchange rate fluctuations.

3 Currency Hedging Strategies

The pursuit of effective means to hedge currency risk in international financial portfolios has been a subject of active research. Over the years, there has been considerable debate on this subject. The observation that, historically, changes in exchange rates had fairly low correlations with foreign stock and bond returns had raised doubts as to the potential benefit of currency hedging. This lack of a systematic relationship could, in principle, lower portfolio risk. Another argument is that, over a long time horizon, currency movements cancel out–the mean-reversion argument. In other words, exchange rates have an expected return of zero in the long run. On the other hand, for active portfolio managers who are concerned with shorter-term horizons, it is important to account for the impact of currency movements on the risk-return characteristics of international portfolios. Moreover, currency returns tend to be episodic; exchange rates can be volatile over short horizons, and the impact of this volatility needs to be controlled through appropriate means in the context of international portfolios. Currency movements also tend to exhibit some degree of persistence (volatility clamping). For these reasons, effective hedging strategies are actively sought by researchers and practitioners to improve the performance of international portfolios.

Currency hedging decisions typically concern the choice between foreign exchange forward contracts or currency options. A forward contract is an agreement between two parties (the investor and a bank) to buy (or sell) a certain amount of foreign currency at a future date at an exchange rate specified at the time of the agreement. Foreign exchange forward contracts are sold by major commercial banks and typically have fixed short-term maturities of one, six, or nine months.

Currency forwards provide a simple and cost-effective way to alter the variability of revenues from foreign currency sources, but they may not be equally effective for all types of risk management problems. These contracts fix both the rate as well as the amount of a foreign exchange transaction and thus protect the value of a certain amount in foreign currency against a potential reduction in the exchange rate. If the amount of foreign revenues is known with certainty, then an equivalent currency forward contract will completely eliminate the currency risk. However, in the case of foreign holdings of financial assets, their final value is stochastic due to their uncertain returns during the interim period; hence, full hedging is not attainable in this case. A limitation of currency forwards is that by fixing the exchange rate they forgo the opportunity for gains in the event that the exchange rate appreciates. One could argue that mitigating risk should be the primary consideration, while potential benefits from favorable exchange rate movements should be of a secondary concern. But it is the entire risk-return trade off that usually guides portfolio management decisions.

Options provide alternative means to control risks. An exporter could "shell" the future foreign exchange receipts by purchasing a currency put. A portfolio manager could protect his foreign asset holdings by buying currency options to mitigate currency risk exposure associated with foreign investments in the portfolio. Currency put options protect from losses in the event of a significant drop in the exchange rate, but with no sacrifice of potential benefits in the event of currency appreciation, as they would simply not be exercised in such a case. However, currency options entail a cost (purchase price).

In this study we experiment with two different trading strategies involving currency options that have different payoff profiles.

Using Protective Put Options

By buying a European put currency option, the investor acquires the discretionary right to sell a certain amount of foreign currency at a specified rate (exercise price) at the option's maturity date. In the numerical tests, we consider protective put options for each foreign currency with three different strike prices, K_j ("in-the-money," ITM, "at-the-money," ATM, and "out-of-the-money," OTM). These three options constitute the set of available options, J_c, for each foreign currency $c \in C$. The options have a term (maturity) of one month that matches the duration of each decision stage in the model.

The ATM options have strike prices equal to the respective spot exchange rates at the time of issue. As decisions are considered at non-leaf nodes of the scenario tree, the strike prices of the ATM options are equal to the scenario-dependent spot exchange rates specified for the corresponding node of the scenario tree. The ITM and OTM put options have strike prices that are 5% higher, respectively 5% lower, than the corresponding spot exchange rates at the respective decision state. These levels of option strike prices have been chosen fairly arbitrarily. Obviously, a larger set of options with different strike prices and cost can easily be included in the model.

The model is allowed to take only long positions in the protective put options. Thus, we add nonnegativity constraints for the positions in options in model (2):

$$npc_{c,j}^n \geq 0, \qquad \forall j \in J_c, \ \forall c \in C, \ \forall n \in N \setminus N_T.$$

Obviously, the OTM put option has a lower price than the ATM put, which, in turn, is cheaper than the ITM put option. In the numerical tests we have observed that when the model selects options in the portfolios, these are OTM put options–ITM and ATM options are never selected in the solutions when they are considered together with OTM put options.

Using BearSpread Strategies

A BearSpread strategy is composed of two put options with the same expiration date. It involves a long position in an ITM put and a short position in an OTM put. The strike prices of the constituent options are set as described above.

Let $npc_{c,ITM}^n$ and $npc_{c,OTM}^n$ be the long position in the ITM and the short position in the OTM currency put option, respectively, constituting a BearSpread position in foreign currency $c \in C$ at decision node $n \in N \setminus N_T$. To incorporate the BearSpeard strategy in the optimization model (2), we additionally impose the following constraints:

$$npc_{c,OTM}^n + npc_{c,ITM}^n = 0, \ \forall c \in C, \ \forall n \in N \setminus N_T,$$
$$npc_{c,ITM}^n \geq 0, \ \forall c \in C, \ \forall n \in N \setminus N_T.$$

The first constraint ensures that the positions in the respective put options have the same magnitude, while the second constraint ensures that the long position is in the ITM option.

The payoff profile of the BearSpread is contrasted in Figure 1 to that of a long position in an OTM currency put option. We test both of these option tactics in the context of an international portfolio management problem. The numerical experiments aim to empirically assess the relative effectiveness of these tactics to control currency risk and enhance portfolio performance.

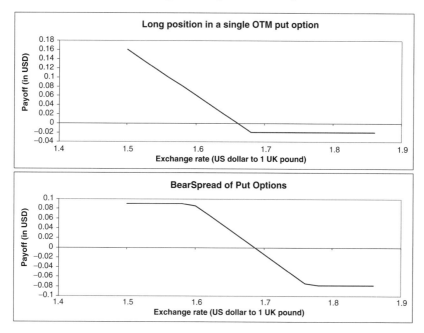

Fig. 1. Payoff patterns of currency option strategies.

We have also tested alternative option trading strategies composed of long positions in put and call options with the same expiration date. The straddle, strip, and strap strategies are composed of ATM put and call options, i.e., with the same strike price; they differ only in the proportion of their long positions in the put and the call option. The proportions of positions in the put to the call option for the straddle, strip, and strap strategies are 1:1, 2:1, and 1:2, respectively. All three strategies yield positive payoffs in the event of sufficient movement in the underlying exchange rate, either downside or upside. Their proportional payoffs in the event of currency depreciation, or appreciation, differ depending on the proportion of the put option with respect to the corresponding call option. Although these strategies yield gains in the event of even moderate volatility in exchange rates, they have a higher cost as they are composed of ATM options. The strangle strategy involves long positions (equal in magnitude) in an OTM put and an OTM call option. It provides coverage against larger movements of the underlying exchange rate, in comparison to the previous three strategies, but at a lower cost.

We do not report results with these option trading strategies because in the numerical tests the straddle, strip, and strap strategies were dominated both by the forward contracts as well as by the use of a single protective OTM put option per currency. The performance of the strangle strategy was essentially indistinguishable from that of a single protective put per currency.

4 Empirical Results

As prior research suggests, currency risk is a main aspect of the overall risk of international portfolios; controlling currency risk is important to enhance portfolio performance. We examine the effectiveness of alternative tactics to control the currency risk of international diversified portfolios of stock and bond indices. Alternative strategies, using either forward exchange contracts or currency options, are evaluated and compared in terms of their performance in empirical tests using market data.

We solved single-stage and two-stage instances of the stochastic programming models described in Section 2. The results of the numerical tests enable a comparative assessment of the following:

- forward exchange contracts versus currency options,
- alternative tactics with currency options,
- single-stage vs. two-stage stochastic programming models.

First, we compare the performance of a single-stage model with forward contracts against that of a corresponding model that uses currency options as a means to mitigate currency risk. Second, we compare the performance of alternative tactics that provide coverage against unfavorable movements in exchange rates by means of currency options. Finally, we consider the performance of two-stage variants of the stochastic programming models. The two-stage models permit rebalancing at an intermediate decision stage, at which currency options held may be exercised, and new option contracts can be purchased. We investigate the incremental improvements in performance of the international portfolios that can be achieved with the two-stage models, over their single-stage counterparts.

As explained in Section 2, the models select internationally diversified portfolios of stock and bond indices, and appropriate positions in currency hedging instruments (forward contracts or currency options), in order to minimize the excess downside risk while meeting a desirable target of expected return. Selective hedging is the norm followed in all tests.

Performance comparisons are made with static tests (in terms of risk-return efficient frontiers) as well as with dynamic tests. The dynamic tests involve backtesting experiments over a rolling horizon of 43 months: 04/1998–11/2001. At each month we use the historical data from the preceding 10 years to calibrate the scenario generation procedure: We calculate the four marginal moments and correlations of the random variables and use these estimates as the target statistics in the moment-matching scenario generation procedure. We price the respective currency options on the nodes of the scenario tree using the method described in the Appendix. The scenario tree of asset prices and exchange rates, and the option prices, are used as inputs to the portfolio optimization model. Each month we solve one instance of the optimization model (single- or multistage) and record the optimal first-stage decisions. The clock is advanced one month and the market values of the random variables are revealed. Based on these we determine the actual return using the composition of the portfolio at hand, the observed market prices for the assets and the exchange rates, and the payoffs from the discretionary exercise of currency options that are held. We update the cash holdings accordingly. Starting with the new initial portfolio composition, we repeat the same procedure for the following month. The ex post realized returns are compounded and analyzed over the entire simulation period. These reflect the portfolio returns that would have been obtained had the recommendations of the respective model been adopted during the simulation period.

We ran backtesting experiments for each investment tactic that is studied, using the CVaR metric to minimize excess shortfall in all cases.

4.1 Efficient Frontiers

We now examine the potential performance of forward and currency options, in comparison to unhedged portfolios, in terms of the risk-return profiles of their respective portfolios. Thus, we examine the potential effects of currency hedging through alternative means by comparing the efficient frontiers resulting from the alternative decision tactics. Single-stage CVaR models were used for all tests reported in this section.

Figure 2 contrasts the efficient frontiers of CVaR-optimized international portfolios on August 2001 using optimal positions in currency options or forward contracts, versus totally unhedged portfolios. We consider two different strategies with currency options: (1) a single protective put option ("at-the-money," "in-the-money," or "out-of-the-money"); (2) a BearSpread strategy of put options.

We observe that risk-return efficient frontiers of hedged portfolios (using either forwards or currency options) clearly dominate the efficient frontiers of unhedged portfolios. The efficient frontiers of unhedged portfolios extend into a range of higher risk levels. Clearly, the selectively hedged portfolios are preferable to unhedged portfolios; at any level of expected return, efficient hedged portfolios have a substantially lower level of risk–as measured by the CVaR metric–compared to efficient unhedged portfolios. We also observe that currency risk hedging (regardless of the strategy used) yields higher benefits, in terms of higher expected returns compared to the unhedged case, for the medium- and high-risk portfolios, rather than for the low-risk portfolios. The potential gain from risk reduction is increasing for more aggressive targets of expected portfolio returns.

These ex ante results indicate that forward contracts exhibit superior potential as hedging instruments compared to currency options. The results in Figure 2 show that, while the various strategies of currency options produce efficient frontiers that dominate that of the unhedged portfolios, the most dominant efficient frontier is the one produced with the optimal selection of forward contracts. For any value of target expected return, the optimal hedged portfolios with forwards exhibit a lower level of risk than the efficient

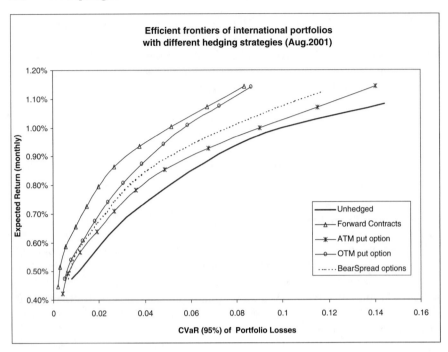

Fig. 2. Efficient frontiers of CVaR-optimized international portfolios of stock and bond indices, and currency hedging instruments.

portfolios of any other strategy. The use of currency options improves the performance of international portfolios compared to unhedged portfolios, but forward contracts exhibit the most dominant performance in the static tests.

Among the trading strategies using currency options, we observe that the efficient frontier with "out-of-the-money" options is the closest to that obtained with forward contracts, especially in the highest levels of target expected return (i.e., most aggressive investment cases). The efficient frontier of the BearSpread strategy follows next, but the differences from the first two are increasing for more aggressive targets of expected portfolio returns. Portfolios with "in-the-money" or "at-the-money" options give almost indistinguishable risk-return efficient frontiers.

4.2 Dynamic Tests: Ex-post Comparative Performance of Portfolios with Currency Options

The results of the previous section indicate that, in static tests, forward contracts demonstrated better ex ante performance potential compared to currency options. We additionally ran a number of backtesting experiments on a rolling horizon basis for a more substantive empirical assessment of alternative currency hedging strategies.

Single-Stage Models

First, we compare the ex-post realized performance of portfolios with various hedging strategies that are incorporated in single-stage stochastic programming models. The models have a holding period of one month and consider portfolio restructuring decisions at a single point during the planning horizon. The joint distribution of the random variables (asset returns and exchange rates) during the one month horizon of the models is represented by sets of 15,000 discrete scenarios.

Figure 3 contrasts the ex-post performance of portfolios with optimal forward contracts with that of portfolios using different strategies of put options. The first graph compares performance in the minimum risk case–i.e., when the models simply minimize the CVaR risk measure at the end of the planning horizon, without imposing any minimal target on expected portfolio return; for the second graph the target expected return during the one-month planning horizon is $\mu = 1\%$.

We observe that in the minimum risk case of the dynamic tests, forward contracts and the use of a single protective put per currency resulted in very similar performance, regardless of the exercise price of the options (i.e., ITM, ATM, or OTM). Forward contracts exhibited the most stable return path throughout the simulation period, indicating their effectiveness in hedging currency risk. In this minimum risk case, the models did not select a large number of currency options, resulting in low hedge ratios.

Figure 3 also presents the ex-post performance of portfolios that use a combination of currency put options comprising the BearSpread strategy. We observe a noticeable improvement in the performance of portfolios when the BearSpread strategy is employed. In the minimum risk case, the BearSpread strategy yields discernibly higher gains in the period of Sept.–Oct. 1999; this was due to its positions in Japanese bonds during this period, that allowed it to capitalize on the appreciation of the yen at that time. The remaining strategies had very limited positions in Japanese assets during that period.

In the minimum risk case, the optimal portfolios (regardless of the currency hedging strategy) were positioned almost exclusively in short-term government bonds in various currencies throughout the simulation period. These portfolios were able to weather the storm of the September 11, 2001, crisis unscathed, and actually generated profits during that period. That crisis had affected primarily the stock markets for a short period and had no material impact on the international bond markets.

The second graph in Figure 3 shows the performance of more aggressive portfolios–when a target expected return $\mu = 1\%$ is imposed over the models' one-month horizon. The differences in the performance of the currency hedging tactics are more pronounced in this case. Again, we observe that portfolios with forward contracts exhibit the most stable path of realized returns. We also observe that the BearSpread strategy of put options materially outperformed all other tactics. In this case of an aggressive target on expected return,

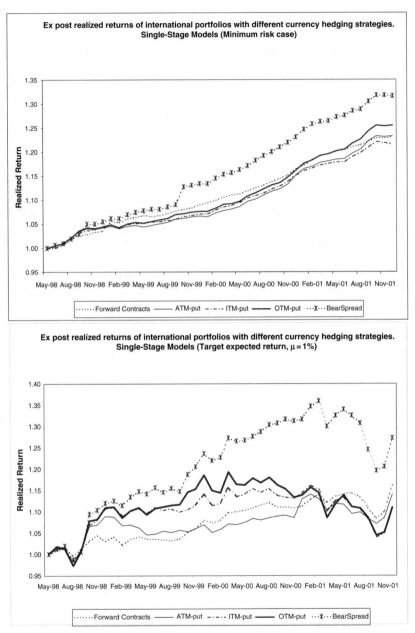

Fig. 3. Ex-post realized returns of single-stage CVaR-optimized internationally diversified portfolios of stock and bond indices, and currency hedging instruments. Backtesting simulations over period 04/1998–11/2001.

the models selected portfolios that involved sizable positions in the U.S. stock index for most of the simulation period and thus did not avoid the effects of the crisis in September 2001.

Multistage Models

We also tested two-stage instances of the portfolio management models presented in section 2. Figure 4 contrasts the ex-post performance of CVaR-optimized portfolios with alternative tactics for controlling currency risk. The first graph shows realized returns in the minimum risk case, and the second graph shows the realized returns for the aggressive investment case corresponding to a target expected return $\mu = 2\%$ during the two-month planning horizon of the models.

The comparative performance of the various currency hedging tactics remains similar, at least qualitatively, to that we had observed with the single-stage models. Again, forward contacts yield the most stable path of realized returns and the BearSpread strategy of currency put options results in the best ex-post performance. Portfolios with ITM currency options show a noticeable improvement in performance when the two-stage models are used; although these portfolios exhibit higher fluctuations in returns compared to the other strategies, they result in higher cumulative returns. In the multistage setting, the model with ITM options benefits the most from favorable exchange rate movements of the Japanese yen in Sept.–Oct. 1999 and the German mark in Nov.–Dec. 2000; the model maintained sizable positions in these currencies during the respective periods.

Overall, the results indicate that although forward contracts are generally more effective in hedging the currency risk compared to single protective put options per currency, appropriate combinations of put options lead to performance improvements.

Next, we turn to a comparative assessment of single- and two-stage models for international portfolio management in dynamic tests with real market data. The two-stage models use scenario trees composed of 150 joint outcomes of the random variables in the first month, each followed by a further 100 joint outcomes of the random variables in the subsequent month; thus, we have 15,000 scenarios over the two-month planning horizon of the models.

Figure 5 compares the performance of the models with currency options or forward contracts. The first graph presents the results of experiments minimizing the CVaR risk measure without any constraint on expected portfolio return. The second graph corresponds to more aggressive portfolios (target expected return $\mu = 1\%$ for single-stage models and $\mu = 2\%$ for two-stage models).

We observe that in the minimum risk case the models exhibit stable portfolio returns throughout the simulation period, with small losses in only very few periods. The more aggressive cases exhibit larger fluctuations in returns, reflecting riskier portfolios. In Figure 5, we observe that when currency risk is hedged with forward contracts, the two-stage model gives only slightly better results compared to the single-stage model. However, when currency options are used, the performance improvements of the two-stage models compared to the corresponding single-stage models are more evident, particularly for the cases that use ITM put options. Performance improvements with the two-stage model are also observed when the BearSpread strategy is employed. In all tests, and regardless of the trading strategy of options that is used, the two-stage models result in improved performance compared to the corresponding single-stage models.

In comparison to their single-stage counterparts, two-stage models incorporate the following advantages: (1) a longer planning horizon that permits to assess the sustained effects of investment choices; (2) increased information content as it accounts for the evolution of the random variables over the longer planning horizon; (3) the opportunity to account for the effect of portfolio rebalancing at an intermediate point during the planning horizon. The combined effects of these features lead to the selection of more effective portfolios with the two-stage models. The performance improvements are evident in higher and more stable portfolio returns that are achieved with the two-stage models in comparison to their single-stage counterparts.

Empirical comparisons of multistage stochastic programming models and single-stage models are scantly found in the literature. The results of this study demonstrate the performance improvements that are achievable with the adoption of multistage stochastic programs–that account for information and decision dynamics–in comparison to single-stage (myopic) models.

Figure 6 shows the degree of currency hedging in each country (percentage of foreign investments hedged), when using forward contracts or currency options that form the BearSpread strategy to control currency risk; these results correspond to the first-stage decisions of the two-stage optimization models. The differences are evident. When using currency options, the model consistently chooses to hedge to a high degree–hedge ratios between 90% and 100%–foreign investments in the selected portfolios (see the second graph in Figure 6). The hedge ratio is zero only when the selected portfolio does not include asset holdings in a particular foreign market. The degree of hedging is evidently different when forward contracts are used to control currency risk (first graph in Figure 6). In this case the model makes much more use of the selective hedging flexibility; hedge ratios varying both in magnitude as well as across currencies are observed during the simulation period. Moreover, we have observed that, in comparison to the single-stage models, the two-stage models select more diversified portfolios throughout the simulation period and exhibit lower portfolio turnover.

Finally, we compute some measures to compare the overall performance of the models. Specifically, we consider the following measures of the ex-post realized monthly returns over the simulation period: geometric mean, standard deviation, Sharpe ratio, and the upside potential and downside risk ratio (UP_{ratio}) proposed by Sortino and van der Meer (1991). This ratio

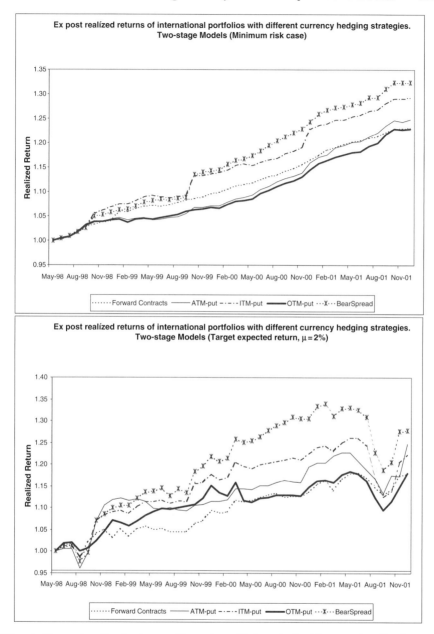

Fig. 4. Ex-post realized returns of two-stage CVaR-optimized internationally diversified portfolios of stocks and bonds indices, and currency hedging tactics. Backtesting simulations over the period 04/1998–11/2001.

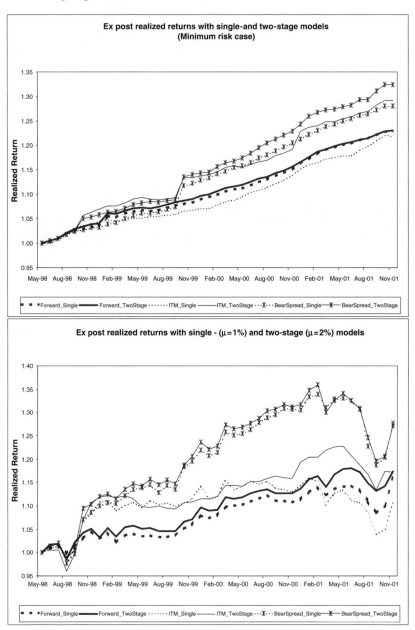

Fig. 5. Ex-post realized returns of CVaR-optimized international portfolios of stocks and bond indices, and currency hedging tactics. Comparison of single- and two-stage models.

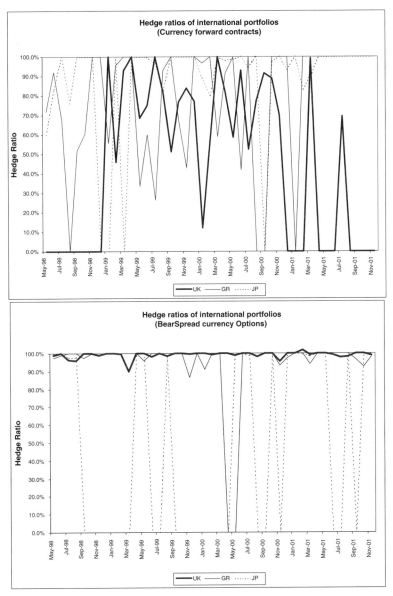

Fig. 6. Comparison of optimal hedge ratios in each currency for different decision tactics. The first graph represents the use of forward contracts; the second graph shows the hedge ratios for the BearSpread strategy of currency put options.

contrasts the upside potential against a specific benchmark with the shortfall risk against the same benchmark. We use the risk-free rate of one-month T-bills as the benchmark. This ratio is computed as follows. Let r_t be the realized return of a portfolio in month $t = 1, \ldots, k$ of the simulation, where $k = 43$ is the number of months in the simulation period 04/1998–11/2001. Let ρ_t be the return of the benchmark (riskless asset) at the same period. Then the UP_{ratio} is

$$UP_{\text{ratio}} = \frac{\frac{1}{k}\sum_{t=1}^{k} \max[0, r_t - \rho_t]}{[\frac{1}{k}\sum_{t=1}^{k}(\max[0, r_t - \rho_t])^2]^{1/2}}. \qquad (4)$$

The numerator is the average excess return compared to the benchmark, reflecting the upside potential. The denominator is a measure of downside risk, as proposed in Sortino et al. (1999), and can be thought of as the risk of failing to meet the benchmark.

Table 2. Statistics of Realized Monthly Returns with Alternative Models and Decision Tactics

Statistic	Forward Contracts	ITM Put	ATM Put	OTM Put	BearSpread Strategy
Statistics of Monthly Realized Returns, Single-Stage Model ($\mu = 1\%$)					
Geometric mean	0.0043	0.0035	0.0021	0.0022	0.0057
Stand. dev.	0.0139	0.0148	0.0193	0.0222	0.0205
Sharpe ratio	−0.0307	−0.0808	−0.1324	−0.1128	0.0468
UP_{ratio}	0.9500	0.8800	0.6910	0.7060	0.8300
Statistics of Monthly Realized Returns, Single-Stage Model (Minimum Risk)					
Geometric mean	0.0057	0.0058	0.0054	0.0063	0.0076
Stand. dev.	0.0027	0.0045	0.0036	0.0042	0.0061
Sharpe ratio	0.3796	0.2350	0.1911	0.3641	0.4725
UP_{ratio}	11.1355	5.4853	6.7168	6.7518	25.1671
Statistics of Monthly Realized Returns, Two-Stage Model ($\mu = 2\%$)					
Geometric mean	0.0046	0.0061	0.0056	0.0045	0.0065
Stand. dev.	0.0124	0.0163	0.0193	0.0153	0.0208
Sharpe ratio	−0.0120	0.0862	0.0458	−0.1023	0.0864
UP_{ratio}	0.9690	0.9470	0.7670	0.8380	0.833
Statistics of Monthly Realized Returns, Two-Stage Model (Minimum Risk)					
Geometric mean	0.0058	0.0061	0.0071	0.0057	0.0078
Stand. dev.	0.0025	0.0047	0.0078	0.0042	0.0075
Sharpe ratio	0.4132	0.3006	0.3083	0.2302	0.4108
UP_{ratio}	12.4510	5.4270	10.1260	4.6344	21.134

The statistics of the realized monthly portfolio returns over the simulation period 04/1988–11/2001 are shown in Table 2. First, we observe that in the single-stage models, the optimal portfolios with forwards exhibit better statistics compared to portfolios with protective put options. When forward contracts are used, the Sharpe ratio is higher than that for the use of protective put options (ITM, ATM, or OTM), and the standard deviation is lower. Also, the UP_{ratio} is substantially higher, indicating improved upside potential relative to downside risk. In some cases the portfolios with protective currency put options exhibit higher geometric mean than the portfolios with forward contracts; however, the rest of their statistics are worse in all simulation tests.

Next, we note that multistage models clearly outperform the corresponding single-stage models. All the statistics for the multistage models are improved compared to the results of their single-stage counterparts. These results clearly show that incremental benefits are gained over myopic models–in terms of improved performance (higher returns and lower risk)–with the adoption of two-stage (dynamic) portfolio optimization models.

Finally, we observe that a combination of currency put options that forms the BearSpread strategy outperforms the use of forward contracts. In our simulation experiments, this was the case both for the two-stage and the single-stage models, and for both the minimum risk case as well as the more aggressive case (higher expected return targets). Hence, the judicious choice of option trading strategies with suitable payoff patterns can provide the means to improve performance in international portfolio management.

5 Conclusions

This chapter investigated alternative strategies for controlling currency risk in international portfolios of financial assets. We carried out extensive numerical tests using market data to empirically assess the effectiveness of alternative means for controlling currency risk in international portfolios.

Empirical results indicate that the optimal choice of forward contracts outperforms the use of a single protective put option per currency. The results of both static as well as dynamic tests show that optimal portfolios with forward contacts achieve better performance than portfolios with protective put options. However, combinations of currency put options, like the Bear-Spread strategy, exhibit performance improvements. Yet, forward contracts consistently produced the more stable returns in all simulation experiments.

Finally, we point out the notable performance improvements of international portfolios that result from the adoption of dynamic (multistage) portfolio optimization models instead of the simpler, but more restricted myopic (single-stage) models. The two-stage models tested in this study consistently outperformed their single-stage counterparts in all cases (i.e., regardless of the decision tactics that were tested as means to control currency risk). These results strengthen the argument for the development, implementation, and use of more sophisticated dynamic stochastic programming models for portfolio management. These models are more complex, have higher information demands (i.e., complete scenario trees), and are computationally more demanding because of their significantly larger size. However, as the results of this study demonstrate, dynamic models yield superior solutions, that is, more effective portfolios that attain higher returns and lower risk. Such improvements are important in an increasingly competitive financial environment.

The next step of this work is to investigate the use of appropriate instruments and decision tactics, in the context of dynamic stochastic programming models, so as to jointly control multiple risks that are encountered in international portfolio management. For example, we can incorporate in the models options on stocks as a means to control market risk in addition to forward exchange contracts or currency options to control currency risk. We expect that such an approach to jointly manage all risk factors in the problem should yield additional benefits.

Appendix: Pricing Currency Options

In order to incorporate currency options in the stochastic programming model and maintain internal consistency, the options must be priced in accordance with the discrete distributions of the underlying exchange rate, as represented by the scenario tree. Conventional option pricing methods are not applicable, as they rely on specific distributional assumptions on the underlying exchange rate that are not satisfied in this case. In this study, the scenarios of asset returns and exchange rates are generated with a moment-matching method so as to closely reflect the empirical distributions of the random variables implied by historical market data.

We price the currency options based on empirical distributions of the exchange rates using a valuation procedure developed by Corrado and Su (1996), based on an idea of approximating the density of the underlying by a series expansion that was introduced by Jarrow and Rudd (1982). Backus et al. (1997) extended this approach to price currency options; we adapt their methodology.

We consider European currency options with maturity equaling a single period of the portfolio optimization model (i.e., one month). To price such a European option at a nonterminal node (state) $n \in \boldsymbol{N} \setminus \boldsymbol{N_T}$ of the scenario tree, the essential inputs are the price of the underlying (exchange rate) at the option's issue date (i.e., at node n) and the distribution of the underlying exchange rate at the maturity date, conditional on the state at the issue date (i.e., conditional on state n). In the context of a scenario tree, this conditional distribution is represented by the discrete outcomes of the underlying exchange rate associated with the immediate successor nodes (set $\boldsymbol{S_n}$) of node n. Hence, the option is priced on the basis of this discrete conditional distribution.

We use the following notation:

e_t the underlying spot exchange rate at node n (deterministic for the pricing problem),

$\tilde{e}_{t+1}$ the random value of the underlying exchange rate at the maturity of the option,

r_t^d the riskless rate in the base currency for the term of the option,
r_t^f the riskless rate in the foreign currency for the term of the option,
K the exercise price of the currency option (USD to one unit of the foreign currency).

The log of the underlying's appreciation during the term of the option, starting from node n, is

$$\tilde{x}_{t+1} = \ln(\tilde{e}_{t+1}) - \ln(e_t) = \ln\left(\frac{\tilde{e}_{t+1}}{e_t}\right). \tag{5}$$

Then

$$\tilde{e}_{t+1} = e_t \exp(\tilde{x}_{t+1}) \tag{6}$$

and the conditional distribution of $\tilde{e}_{t+1}$ depends on that of $\tilde{x}_{t+1}$.

In the risk-neutral setting, the price at node n of a European call option (in units of the base currency) on the exchange rate $\tilde{e}_{t+1}$ with strike price K is computed as:

$$\begin{aligned} cc_t(e_t, K) &= \exp\left(-r_t^d\right) E_t\left[(\tilde{e}_{t+1} - K)^+\right] \\ &= \exp\left(-r_t^d\right) \int_{\ln(K/e_t)}^{\infty} (e_t \exp(x) - K) f(x)\, dx, \end{aligned} \tag{7}$$

where $f(.)$ is the conditional density of $\tilde{x}_{t+1}$. Typically, the conditional density is not analytically available, and this is the case in this study where the uncertainty in exchange rates is represented in terms of an empirical distribution.

Corrado and Su (1996) applied a Gram–Charlier series expansion to approximate the empirical distribution of the underlying's logreturns in order to derive the option price. The series expansion approximates the underlying distribution with an alternate (more tractable) distribution, specifically, with the log-normal. Hence, the normal density is augmented with additional terms capturing the effects of skewness and kurtosis in the distribution of the underlying random variable. The resulting truncated series may be viewed as the normal probability density function multiplied by a polynomial that accounts for the effects of departure from normality. The coefficients in the expansion are functions of the moments of the original and the approximating distribution. The underlying theory is described by Johnson et al. (1994) and Kolassa (1994).

The series expansion represents an approximate density function for a standardized random variable that differs from the standard normal in having nonzero skewness and kurtosis. If the one-period log-change ($\tilde{x}_{t+1}$) in the spot exchange rate e has conditional mean μ and standard deviation σ, the standardized variable is

$$\tilde{\omega} = (\tilde{x}_{t+1} - \mu)/\sigma. \tag{8}$$

A truncated Gram–Charlier series expansion defines the approximate density for $\tilde{\omega}$ by

$$f(\tilde{\omega}) \approx \varphi(\tilde{\omega}) - \gamma_1 \frac{1}{3!} D^3 \varphi(\tilde{\omega}) + \gamma_2 \frac{1}{4!} D^4 \varphi(\tilde{\omega}), \tag{9}$$

where

$$\varphi(\tilde{\omega}) = \frac{1}{\sqrt{2\pi}} \exp\left(-\tilde{\omega}^2/2\right) \tag{10}$$

is the standard normal density and D^j denotes the jth derivative of what follows.

Using the Gram–Charlier expansion in (9), Backus et al. (1997) solved the pricing equation (7) and obtained the following result for the price of a European currency call option:

$$\begin{aligned}cc_t(e_t, K) = {} & e_t \exp\left(-r_t^f\right) N(d) - K \exp\left(-r_t^d\right) N(d - \sigma) \\ & + e_t \exp\left(-r_t^d\right) \varphi(d) \sigma \left[\frac{\gamma_1}{3!}(2\sigma - d)\right. \\ & \left. - \frac{\gamma_2}{4!}\left(1 - d^2 + 3d\sigma - 3\sigma^2\right)\right],\end{aligned} \tag{11}$$

where

$$d = \frac{\ln(e_t/K) - (r_t^f - r_t^d) + \sigma^2/2}{\sigma}. \tag{12}$$

$\varphi(.)$ is the standard normal density, $N(.)$ is the cumulative distribution of the standard normal, $\gamma_1 = \mu_3/\mu_2^{(3/2)}$ and $\gamma_2 = \mu_4/\mu_2^2$ are the Fisher parameters for skewness and kurtosis, and μ_i is the ith central moment.

To apply this pricing procedure at a non-leaf node $n \in \boldsymbol{N} \setminus \boldsymbol{N_T}$ of the scenario tree, we first calculate the first four moments of the underlying exchange rate over the postulated outcomes on the immediate successor nodes, $\boldsymbol{S_n}$. These estimates of the moments, and the other parameters that are deterministic (current spot exchange rate e_t, exercise price K, interest rates r_t^f and r_t^d), are used as inputs in Equation (11) to price the European currency call option. The price of a European currency put option with the same term and strike price K is determined by put-call parity:

$$pc_t(e_t, K) = cc_t(e_t, K) + K \exp(-r_t^d) - e_t. \tag{13}$$

This method of approximating the density of the underlying has been applied by Abken et al. (1996a, 1996b), Brenner and Eom (1997), Knight and Satchell (2000), Longstaff (1995), Madan and Milne (1994) and Topaloglou et al. (forthcoming).

References

1. Abken, P., Madan, D., and Ramamurtie, S. Estimation of risk-neutral and statistical densities by Hermite polynomial approximation: With an application to eurodolar futures options. Manuscript, Federal Reserve Bank of Atlanta, 1996a.

2. Abken, P., Madan, D., and Ramamurtie, S. Pricing S&P500 index options using a Hilbert space basis. Manuscript, Federal Reserve Bank of Atlanta, 1996b.
3. Ait-Sahalia, Y., and Lo, A. W. Non parametric risk management and implied risk aversion. *Journal of Econometrics*, 94(1-2):9–51, 2000.
4. Albuquerque, R. Optimal currency hedging. *Global Finance Journal*, 18(1):16–33, 2007.
5. Backus, D., Foresi, S., Li, K., and Wu, L. Accounting for biases in Black-Scholes. Working paper no. 02011, CRIF, 1997.
6. Brenner, M., and Eom, Y. No-arbitrage option pricing: New evidence on the validity of the martingale property. Manuscript, Federal Reserve Bank of New York, 1997.
7. Conover, J. A., and Dubofsky, D. A. Efficient selection of insured currency positions, protective puts versus fiduciary calls. *Journal of Financial and Quantitative Analysis*, 30(2):295–312, 1995.
8. Corrado, C. J., and Su, T. Skewness and kurtosis in S&P500 index returns implied by option prices. *The Journal of Financial Research*, XIX(2):175–192, 1996.
9. Dupačová, J., Consigli, G., and Wallace, S. W. Generating scenarios for multistage stochastic programs. *Annals of Operations Research*, 100:25–53, 2000.
10. Eun, C. S., and Resnick, B. G. Exchange rate uncertainty, forward contracts and international portfolio selection. *Journal of Finance*, 43(1):197–215, 1988.
11. Eun, C. S., and Resnick, B. G. International diversification of investment portfolios: U.S. and Japanese perspectives. *Management Science*, 40(1):140–161, 1994.
12. Fama, E. F. The behavior of stock-market prices. *Journal of Business*, 38:34–105, 1965.
13. Høyland, K., Kaut, M., and Wallace, S. W. A heuristic for generating scenario trees for multistage decision problems. *Computational Optimization and Applications*, 24(2-3):169–185, 2003.
14. Jarque, C. M., and Berra, A. K. Efficient tests for normality, homoscedasticity and serial independencce of regression residuals. *Economic Letters*, 6(3):255–259, 1980.
15. Jarrow, R., and Rudd, A. Approximate option valuation for arbitrary stochastic processes. *Journal of Financial Economics*, 10(3):347–369, 1982.
16. Johnson, N., Kotz, S., and Balakrishnan, N. *Continuous Univariate Distributions*. Wiley, New York, 1994.
17. Jorion, P. Mean-variance analysis of currency overlays. *Financial Analyst Journal*, 50(3):48–56, 1994.
18. Knight, J., and Satchell, S. *Return Distributions in Finance*. Butterworth and Heinemann, New York, 2000.
19. Kolassa, J. *Series Approximation Methods in Statistics*. Springer-Verlag, New York, 1994.
20. Lien, D., and Tse, Y. K. Hedging downside risk: Futures vs. options. *International Review of Economics and Finance*, 10:159–169, 2001.
21. Longstaff, F. Options pricing and the martingale restriction. *Review of Financial Studies*, 8(4):1091–1124, 1995.
22. Madan, D., and Milne, F. Contingent claims valued and hedged by pricing and investing in a basis. *Mathematical Finance*, 4(3):223–245, 1994.
23. Markowitz, H. Portfolio selection. *Journal of Finance*, 8:77–91, 1952.

24. Maurer, R., and Valiani, S. Hedging the exchange rate risk in international portfolio diversification: Currency forwards versus currency options. Working paper, Göethe-Universität Frankfurt am Main, 2003.
25. Rockafellar, R.T., and Uryasev, S. Conditional Value-at-Risk for general distributions. *Journal of Banking and Finance*, 26(7):1443–1471, 2002.
26. Sortino, F. A., and van der Meer, R. Downside risk–Capturing what's at stake in investment situations. *Journal of Portfolio Management*, 17(4):27–31, 1991.
27. Sortino, F. A., Plantinga, A., and van der Meer, R. The Dutch triangle–A framework to measure upside potential relative to downside risk. *Journal of Portfolio Management*, 26(1):50–58, 1999.
28. Steil, B. Currency options and the optimal hedging of contingent foreign exchange exposure. *Economica*, 60(240):413–431, 1993.
29. Topaloglou, N., Vladimirou, H., and Zenios, S. A. CVaR models with selective hedging for international asset allocation. *Journal of Banking and Finance*, 26(7):1535–1561, 2002.
30. Topaloglou, N., Vladimirou, H., and Zenios, S. A. A dynamic stochastic programming model for international portfolio management. *European Journal of Operations Research* (forthcoming).
31. Topaloglou, N., Vladimirou, H., and Zenios, S. A. Pricing options on scenario trees. *Journal of Banking and Finance* (forthcoming).

Part III

Operations Research Methods in Financial Engineering

Asset Liability Management Techniques

Kyriaki Kosmidou and Constantin Zopounidis

Technical University of Crete, Dept. of Production Engineering and Management,
Financial Engineering Laboratory, University Campus, 73100 Chania, Greece
kikikosmidou@yahoo.com kostas@dpem.tuc.gr

1 Introduction

Nowadays, because of the uncertainty and risk that exist due to the integrating financial market and technological innovations, investors often wonder how to invest their assets over time to achieve satisfactory returns subject to uncertainties, various constraints, and liability commitments. Moreover, they speculate how to develop long term strategies to hedge the uncertainties and how to eventually combine investment decisions of asset and liability in order to maximize their wealth.

Asset liability management is the domain that provides answers to all these questions and problems. More specifically, asset liability management (ALM) is an important dimension of risk management, where the exposure to various risks is minimized while maintaining the appropriate combination of asset and liability, in order to satisfy the goals of the firm or the financial institution (Kosmidou and Zopounidis, 2004).

Through the 1960s, liability management was aimless. In their majority, the banking institutions considered liabilities as exogenous factors contributing to the limitation of asset management. Indeed, for a long period the greater part of capital resources originated from savings deposits and deposits with agreed maturity.

Nevertheless, the financial system has radically changed. Competition among the banks for obtaining capital has become intense. Liability management is the main component of each bank strategy in order to ensure the cheapest possible financing. At the same time, the importance of decisions regarding the amount of capital adequacy is enforced. Indeed, the adequacy of the bank as far as equity contributes to the elimination of bankruptcy risk, a situation in which the bank cannot satisfy its debts toward clients who make deposits or others who take out loans. Moreover, the capital adequacy of banks is influenced by the changes of stock prices in relation to the amount of the capital stock portfolio. Finally, the existence of a minimum amount of equity is an obligation of commercial banks to the Central Bank for supervisory reasons. It is worth mentioning that based on the last published data (Dec. 31, 2001) the Bank of Greece assigns the coefficient for the Tier 1 capital at 8%, while the corresponding European average is 6%. This results in the configuration of the capital adequacy of the Greek banking system at higher levels than the European average rate. The high capital adequacy index denotes large margins of profitability amelioration, which reduces the risk of a systematic crisis.

Asset management in a contemporary bank cannot be distinct from liability management. The simultaneous management of assets and liabilities, in order to maximize the profits and minimize the risk, demands the analysis of a series of issues.

First is the substantive issue of strategic planning and expansion, that is, the evaluation of the total size of deposits that the bank wishes to attract and the total number of loans that it wishes to provide.

Second is the issue of determination of the "best temporal structure" of the asset liability management, in order to maximize the profits and to ensure the robustness of the bank. Deposits cannot all be liquidated in the same way. From the point of view of assets, the loans and various placements to securities constitute commitments of the bank's funds with a different duration time. The coordination of the temporal structure of the asset liability management is of major importance in order to avoid the problems of temporary liquidity reduction, which might be very injurious.

Third is the issue of risk management of assets and liabilities. The main focus is placed on the assets, where the evaluation of the quality of the loans portfolio (credit risk) and the securities portfolio (market risk) is more easily measurable.

Fourth is the issue of configuration of an integrated invoice, which refers to the entire range of bank operations. It refers mainly to the determination of interest rates for the total of loans and deposits as well as for the various commissions that the bank charges for specific mediating operations. It is obvious that in a bank market that operates in a competitive environment, there is no issue of pricing. This is true even in the case where all interest rates and commissions are set by monetary authorities, as was the situation in Greece before the liberalization of the banking system.

In reality, bank markets have the basic characteristics of monopolistic competition. Thus, the issue of planning a system of discrete pricing and product diversification is of major importance. The problem of discrete pricing, as far as the assets are concerned, is connected to the issue of risk management. It is a common fact that the banks determine the borrowing interest rate on the basis of the interest rates, which increase in proportion to the risk as they assess it in each case. The product diversification policy includes all the loan and deposit products and is based on thorough research that ensures the best possible knowledge of market conditions.

Lastly, the management of operating cost and technology constitutes an important issue. The collaboration of well-selected and fully skilled personnel, as well as contemporary computerization systems and other technological applications, constitutes an important element in creating a low-cost bank. This results in the acquisition of a significant competitive advantage against other banks, which could finally be expressed through a more aggressive policy of attracting loans and deposits with low loan interest rates and high deposit interest rates. The result of this policy is the increase of the market stake. However, the ability of a bank to absorb the input of the best strategic technological innovations depends on the human resources management.

The present research focuses on the study of bank asset liability management. Many reasons lead us to study bank asset liability management as an application of ALM. First, bank asset/liability management has always been of concern to bank managers, but in recent years and especially today its importance has increasingly grown. The development of information technology has led to such increasing public awareness that the bank's performance, its politics, and its management are closely monitored by the press and the bank's competitors, shareholders, and customers and thereby highly affect the bank's public standing.

The increasing competition in the national and international banking markets, the changeover toward the monetary union, and the new technological innovations herald major changes in the banking environment and challenge all banks to make timely preparations in order to enter into the new competitive monetary and financial environment.

All the above factors drove banks to seek out greater efficiency in the management of their assets and liabilities. Thus, the central problem of ALM revolves around the bank's balance sheet; the main question that arises is, what should be the composition of a bank's assets and liabilities on average given the corresponding returns and costs, in order to achieve certain goals, such as maximization of the bank's gross revenues?

It is well known that finding an appropriate balance among profitability, risk, and liquidity considerations is one of the main problems in ALM. The optimal balance between these factors cannot be found without considering important interactions that exist between the structure of a bank's liabilities and capital and the composition of its assets.

Bank asset/liability management is defined as the simultaneous planning of all asset and liability positions on the bank's balance sheet under consideration of the different banking and bank management objectives and legal, managerial, and market constraints. Banks are looking to maximize profit and minimize risk.

In this chapter we make a brief overview of bank ALM techniques as an application of ALM. This overview is traced by classifying the models in two main categories. Finally, the concluding remarks are discussed.

2 Bank ALM Techniques

Asset and liability management models can be deterministic or stochastic. Deterministic models use linear programming, assume particular realizations for random events, and are computationally tractable for large problems. The banking industry has accepted these models as useful normative tools (Cohen and Hammer, 1967). Stochastic models, however, including the use of chance-constrained programming, dynamic programming, sequential decision theory, and linear programming under uncertainty, present computational difficulties.

The theoretical approach of these models is outlined in the following section, whereas the mathematical programming formulation is described in the Appendix.

2.1 Deterministic Models

Looking to the past, we find the first mathematical models in the field of bank management. The deterministic linear programming model of Chambers and Charnes (1961) is the pioneer on asset and liability management. Chambers and Charnes were concerned with formulating, exploring, and interpreting the uses and constructs that may be derived from a mathematical programming model that expresses more realistically than past efforts the actual conditions of current operations. Their model corresponds to the problem of determining an optimal portfolio for an individual bank over several time periods in accordance with requirements laid down by bank examiners, which are interpreted as defining limits within which the level of risk associated with the return on the portfolio is acceptable.

Cohen and Hammer (1967), Robertson (1972), Lifson and Blackman (1973), and Fielitz and Loeffler (1979) are successful applications of Chambers and Charnes'model. Even though these models have differed in their treatment of disaggregation, uncertainty, and dynamic considerations, they all have in common the fact that they are specified to optimize a single objective profit function subject to the relevant linear constraints.

Eatman and Sealey (1979) developed a multiobjective linear programming model for commercial bank balance sheet management. Their objectives are based on profitability and solvency. The profitability of a bank is measured by its profit function. Since the primary goals of bank managers, other than profitability, are stated in terms of liquidity and risk, measures of liquidity and risk would seem to reflect the bank's solvency objective. Many measures of liquidity and risk could be employed, just as there are many measures used by different banks and regulatory authorities. Eatman and Sealey measured liquidity and risk by the capital-adequacy (CA) ratio and the risk-asset to capital (RA) ratio, respectively. The capital-adequacy ratio is a comprehensive measure of the bank's liquidity and risk because both asset and liability composition are considered when determining the value of the ratio. Since liquidity diminishes and risk increases as the CA ratio increases, banks can maximize liquidity and minimize risk by minimizing the CA ratio. The other objective reflecting the bank's solvency is the risk-asset to capital (RA) ratio. Using the RA ratio as a risk measure, the bank is assumed to incur greater risk as the RA ratio increases. Therefore, in order to minimize risk, the RA ratio is minimized. The constraints considered in the model of Eatman and Sealey are policy and managerial.

Apart from Eatman and Sealey, Giokas and Vassiloglou (1991) developed a multiobjective programming for bank assets and liabilities management. They supported that apart from attempting to maximize revenues, management tries to minimize risks involved in the allocation of the bank's capital, as well as to fulfill other goals of the bank, such as retaining its market share, increasing the size of its deposits and loans, etc. Conventional linear programming is unable to deal with this kind of problem, as it can only handle a single goal in the objective function. Goal programming is the most widely used approach in the field of multiple criteria decision making that enables the decision maker to incorporate easily numerous variations of constraints and goals.

The description of the linear goal programming model is presented at the Appendix.

2.2 Stochastic Models

Apart from the deterministic models, several stochastic models have been attempted since the 1970s. These models, in their majority, originate from the portfolio selection theory of Markowitz (1959) and are known as static mean-variance methods. According to this approach, the risk is measured by the variance in a single-period planning horizon, the returns are normally distributed, and the bank managers use risk-averse utility functions. In this case, the value of an asset depends not only on the expectation and variance of its return but also on the covariance of its return with the returns of all other existing and potential investments. Pyle (1971) applied Markowitz's theory in his static model where a bank selects the asset and liability levels it wishes to hold throughout the period. He considered only the risk of the portfolio and not other possible uncertainties. The model omits trading activity, matching assets and liabilities, transactions costs, and other similar features. A more sophisticated approach was that of Brodt (1978), who adapted Markowitz's theory and presented an efficient dynamic balance sheet management plan that maximizes profits for a given amount of risk over a multiperiod planning horizon. His two-period, linear model included uncertainty, based on Markowitz's portfolio selection theory, he tried to build an efficient frontier between the function of expected profits and the linear one of its deviations. Instead of the variance, he used the mean absolute deviation or the semi-absolute deviation that is taken by varying the value of the upper or lower bound of one of the two functions.

Charnes and Thore (1966), and Charnes and Littlechild (1968) developed chance-constrained programming models. These models express future deposits and loan repayments as joint, normally distributed random variables, and replace the capital adequacy formula by chance constraints on meeting withdrawal claims. These approaches lead to a computationally feasible scheme for realistic situations. Pogue and Bussard (1972) formulated a 12-period chance-constrained model in which the only uncertain quantity is the future cash requirement. The major weakness is that the chance-constrained procedure cannot handle a differential penalty for either varying magnitudes of constraint violations or different types of constraints.

In 1969, Wolf proposed the sequential decision-theoretic approach that employs sequential decision analysis to find an optimal solution through the use of implicit enumeration. This technique does not find an optimal solution to problems with a time horizon beyond one period, because it is necessary to enumerate all possible portfolio strategies for periods preceding the present decision point in order to guarantee optimality. In order to explain this drawback, Wolf asserts that the solution to a one-period model would be equivalent to a solution provided by solving an n-period model. This approach ignores the problem of synchronizing the maturities of assets and liabilities. Bradley and Crane (1972) developed a stochastic decision tree model that has many of the desirable features essential to an operational bank portfolio model. The Bradley–Crane model depends upon the development of economic scenarios that are intended to include the set of all possible outcomes. The scenarios may be viewed as a tree diagram for which each element (economic condition) in each path has a set of cash flows and interest rates. The problem is formulated as a linear program, whose objective is the maximization of expected terminal wealth of the firm and the constraints refer to the cash flow, the inventory balancing, the capital loss, and the class composition. To overcome computational difficulties, they reformulated the asset and liability problem and developed a general programming decomposition algorithm that minimizes the computational difficulties.

Another approach to stochastic modeling is dynamic programming. The approach dates to the work of Samuelson (1969), Merton (1969, 1990), and others. The main objective of this approach is to form a state space for the driving variables at each time period. Instead of discerning the scenarios, stochastic control perplexes the state space. Either dynamic programming algorithms or finite-element algorithms are available for solving the problem. Merton (1969) explored two classes of reasons why optimal endowment investment policy and expenditure policy can vary significantly among universities. This is done by relating the present value of the liability payments to the driving economic variables. The analysis suggests that managers and others who judge the prudence and performance of policies by comparisons across institutions should take account of differences in both the mix of activities of the institutions and the capitalized values of their no-endowment sources of cash flows. Eppen and Fama (1971) modeled asset problems. The basic idea is to set up the optimization problem under uncertainty as a stochastic control model using a popular control policy. This model reallocates the portfolio in the end of each period such that the asset proportions meet the specified targets. The continuous sample space is represented via a discrete approximation. The discrete approximation offers a wider range of application and is easy to implement. These models are dynamic and account for the inherent uncertainty of the problem.

An alternative approach in considering stochastic models is the stochastic linear programming with simple recourse (SLPSR), also called linear programming under uncertainty (LPUU). This technique explicitly characterizes each realization of the random variables by a constraint with a limited number of possible outcomes and time periods. The general description of the model is presented in the Appendix. Cohen and Thore (1970) viewed their one-period model more as a tool for sensitivity analysis than as a normative decision tool. Crane (1971) on the other hand, modulated the model to a two-period one. The computational intractability and the perceptions of the formulation precluded consideration of problems other than those that were limited both in terms of time periods and in the number of variables and realizations. Booth (1972) applied this formulation by limiting the number of possible realizations and the number of variables considered, in order to incorporate two time periods. Kallberg et al. (1982) formulated a firm's short-term financial planning problem as a stochastic linear programming with a simple recourse model where forecasted cash requirements are discrete random variables. The main goal of their paper was to minimize costs of the various sources of funds employed plus the expected penalty costs due to the constraint violations over the four-quarter horizon. They concluded that even with symmetric penalty costs and distributions, the mean model is significantly inferior to the stochastic linear programming formulation. Kusy and Ziemba (1986) employed a multiperiod stochastic linear program with simple recourse to model the management of assets and liabilities in banking while maintaining computational feasibility. Their model tends to maximize the net present value of bank profits minus the expected penalty costs for infeasibility and includes the essential institutional, legal, financial, and bank-related policy considerations and their uncertainties. It was developed for the Vancouver City Savings Credit Union for a five-year planning period. The results indicate that ALM is theoretically and operationally superior to a corresponding deterministic linear programming model and that the effort required for the implementation of ALM and its computational requirements are comparable to those of the deterministic model. Moreover, the qualitative and quantitative characteristics of the solutions are sensitive to the model's stochastic elements, such as the asymmetry of cash flow distributions. This model had (1) multiperiodicity incorporating changing yield spreads across time, transaction costs associated with selling assets prior to maturity, and the synchronization of cash flows across time by matching maturity of assets with expected cash outflows; (2) simultaneous considerations of assets and liabilities to satisfy accounting principles and match the liquidity of assets and liabilities; (3) transaction costs incorporating brokerage fees and other expenses incurred in buying and selling securities;

(4) uncertainty of cash flows incorporating the uncertainty inherent in the depositors' withdrawal claims and deposits that ensures that the asset portfolio gives the bank the capacity to meet these claims; (5) the incorporation of uncertain interest rates into the decision-making process to avoid lending and borrowing decisions that may ultimately be detrimental to the financial well-being of the bank; and 6) legal and policy constraints appropriate to the bank's operating environment. The Kusy and Ziemba model did not contain end effects, nor was it truly dynamic since it was solved two periods at a time in a rolling fashion. The scenarios were high, low, and average returns that were independent over time.

Another application of the multistage stochastic programming is the Russell–Yasuda Kasai model (Carino et al., 1994), which aims at maximizing the long-term wealth of the firm while producing high income returns. This model builds on this previous research to make a large-scale dynamic model with possibly dependent scenarios, end effects, and all the relevant institutional and policy constraints of Yasuda Kasai's business enterprise. The multistage stochastic linear program used by Carino et al. incorporates Yasuda Kasai's asset and liability mix over a five-year horizon followed by an infinite horizon, steady-state, end-effects period. The objective is to maximize expected long-run profits less expected penalty costs from constraint violations over the infinite horizon. The constraints represent the institutional, cash flow, legal, tax, and other limitations on the asset and liability mix over time.

Based on one or more decision rules, it is possible to create an ALM model for optimizing the setting of decision rules or even to create a scenario analysis. These optimization problems are relatively small, but they often result in nonconvex models and it is difficult to identify the global optimal solution. Examples of optimizing decision rules are Falcon Asset Liability Management (Mulvey et al., 1997) and Towers Perrin's Opt: Link System (Mulvey, 1996). In general, scenario analysis is defined as a single deterministic realization of all uncertainties over the planning horizon. The process constructs, mainly, scenarios that represent the universe of possible outcomes (Glynn and Iglehart, 1989; Dantzig and Infanger, 1993). The main idea is the construction of a representative set of scenarios that are both optimistic and pessimistic within a risk-analysis framework. Such an effort was undertaken by Towers Perrin, one of the largest actuarial firms in the world, which employs a capital market scenario generation system, called CAP: Link. This was done in order to help its clients to understand the risks and opportunities relating to capital market investments. The system produces a representative set of individual simulations – typically 500 to 1,000 – starting with the interest rate component. Towers Perrin employs a version of the Brennan and Schwartz (1982) two-factor interest rate model. The other submodels are driven by the interest rates and other economic factors. Towers Perrin has implemented the system in over 14 countries in Europe, Asia, and North America.

Derwa (1972), Robinson (1973), and Grubmann (1987) reported successful implementations of simulation models developed for various financial institutions. Derwa, for example, used a computer model, operating now at Société Générale de Banque, to improve management decision making in banks. The model was conceived as a form of a decision tree, which made it possible to proceed step by step and examine the factors converging on the essential objectives of the bank. Derwa concluded that the problems raised by introducing models into management are much more difficult to solve than the technical ones connected with mathematics or date processing.

Mulvey and Vladimirou (1989) used dynamic generalized network programs for financial planning problems under uncertainty. They developed a model in the framework of a multiscenario generalized network that captures the essential features of various discrete-time financial decision problems and represented the uncertainty by a set of discrete scenarios of the uncertain quantities. However, these models are small and are not able to solve practical-sized problems. Mulvey and Crowder (1979) and Dantzig and Glynn (1990) used the methods of sampling and cluster analysis, respectively, to limit the required number of scenarios to capture uncertainty and maintain computational tractability of the resulting stochastic programs.

Korhonen (2001) presented a multistage stochastic programming approach to the strategic financial management of a multicompany financial conglomerate. He created a comprehensive strategy that simultaneously covered a number of future scenarios within a multiperiod planning horizon. The strategy includes multiple conflicting goals specified for a group level, company level, or individual business area level. Moreover, the decision maker's preferences were allowed to change over time to reflect changing operating conditions and trade-off relationships between the goals.

Kouwenberg (2001) developed a scenario generation methodology for asset liability management. He proposed a multistage stochastic programming model for a Dutch pension fund. Both randomly sampled event trees and event trees fitting the mean and the covariance of the return distribution were used to generate the coefficients of the stochastic program. In order to investigate the performance of the model and the scenario generation procedures, he conducted rolling horizon simulations. The average cost and the risk of the stochastic programming policy were compared to the results of a simple fixed mix model. He compared the average switching behavior of the optimal investment policies; the results of this analysis proved that the performance of the multistage stochastic program could be improved by choosing an appropriate scenario generation method.

2.3 A Goal Programming Formulation

Kosmidou and Zopounidis (2004) developed an asset liability management (ALM) methodology into a stochastic environment of interest rates in order to select the best direction strategies to the banking financial planning. The

ALM model was developed through goal programming in terms of a one-year time horizon. The model used balance sheet and income statement information for the year $t-1$ to produce a future course of ALM strategy for the year $t+1$. As far as model variables are concerned, we used variables familiar to management and facilitated the specification of the constraints and goals. For example, goals concerning measurements such as liquidity, return, and risk have to be expressed in terms of utilized variables.

More precisely, the asset liability management model that was developed can be expressed as follows:

$$\min\ z = \sum_P p_k(d_k^+ + d_k^-), \tag{1}$$

$$\text{s.t.}\ LB_{X'} \leq X' \leq UB_{X'}, \tag{2}$$

$$LB_{Y'} \leq Y' \leq UB_{Y'}, \tag{3}$$

$$\sum_{i=1}^n X_i = \sum_{j=1}^m Y_j, \tag{4}$$

$$\sum_{j \in \Pi_{Y''}} Y_j - a \sum_{i \in E_{X''}} X_i = 0, \tag{5}$$

$$\sum_{j \in \Pi_1} Y_j - \sum_{i \in E} w_i X_i - d_s^+ + d_s^- = k_1, \tag{6}$$

$$\sum_{i \in E_x} X_i - k_2 \sum_{j \in \Pi_k} Y_j + d_l^- - d_l^+ = 0, \tag{7}$$

$$\sum_{i=1}^n R_i^X X_i - \sum_{j=1}^m R_j^Y Y_j - d_r^+ + d_r^- = k_3, \tag{8}$$

$$\sum_{i \in E_p} X_i + d_p^- - d_p^+ = l_p,\ \forall p, \tag{9}$$

$$\sum_{j \in \Pi_p} Y_j + d_p^- - d_p^+ = l_p,\ \forall p, \tag{10}$$

$$X_i, Y_j, d_k^+, d_k^- \geq 0,\ \forall i = 1,\ldots,n,\ j = 1,\ldots,m,\ k \in P. \tag{11}$$

where

- X_i: the asset element $i = 1,\ldots,n$, with n the number of asset variables,
- Y_j: the liability element $j = 1,\ldots,m$, with m the number of liability variables,
- $LB_{X'}$ ($LB_{Y'}$) is the low bound of specific asset accounts X' (liability Y'),
- $UB_{X'}$ ($UB_{Y'}$) is the upper bound of specific asset accounts X' (liability Y'),
- $E_{X''}$ are specific categories of asset accounts,

- $E_{Y''}$ are specific categories of liability accounts,
- a is the desirable value of specific asset and liability data,
- Π_1 is the liability set, which includes the equity,
- E is the set of assets,
- w_i is the degree of riskiness of the asset data,
- k_1 is the solvency ratio, as defined by the European Central Bank,
- k_2 is the liquidity ratio, as defined by the bank policy,
- E_X is the set of asset data, which includes the loans,
- Π_k is the set of liability data, which includes the deposits,
- R_i^X is the expected return of the asset i,
- R_j^Y is the expected return of the liability j,
- k_3 is the expected value for the goal of asset and liability return,
- P is the set of goals imposed from the bank,
- l_p is the desirable value goal for the goal constraint p defined by the bank,
- d_k^+ is the overachievement of the goal $k \in P$,
- d_k^- is the underachievement of the goal $k \in P$,
- p_k is the priority degree (weight) of the goal $k \in P$.

Certain constraints are imposed by the banking regulation on particular categories of accounts. Specific categories of asset accounts (X') and liability accounts (Y') are detected, and the minimum and maximum allowed limit for these categories are defined based on the strategy and policy that the bank intends to follow (constraints 2–3).

The structural constraints (4–5) include those that contribute to the structure of the balance sheet and especially to the performance of the equation Assets = Liabilities + Net Capital.

The bank management should determine specific goals, such as the desirable structure of each financial institution's assets and liabilities for the units of surplus and deficit, balancing the low cost and the high return. The structure of assets and liabilities is significant, since it swiftly affects the income and profits of the bank.

Referring to the goals of the model, the solvency goal (6) is used as a risk measure and is defined as the ratio of the bank's equity capital to its total weighted assets. The weighting of the assets reflects their respective risk, greater weights corresponding to a higher degree of risk. This hierarchy takes place according to the determination of several degrees of significance for the variables of assets and liabilities. That is, the variables with the largest degrees of significance correspond to categories of the balance sheet accounts with the highest risk stages.

Moreover, a basic policy of commercial banks is the management of their liquidity and specifically the measurement of their needs that is relative to the progress of deposits and loans. The liquidity goal (7) is defined as the ratio of liquid assets to current liabilities and indicates the liquidity risk, which indicates the possibility of the bank to respond to its current liabilities with a security margin, which allows the probable reduction of the value of some current data.

Furthermore, the bank aims to maximize its efficiency, which is the accomplishment of the largest possible profit from the best placement of its funds. Its aim is the maximization of its profitability; therefore, precise and consistent decisions should be taken into account during bank management. These decisions will guarantee the combined effect of all the variables that are included in the calculation of the profits. This decision process emphasizes several selected variables that are related to the bank management, such as to the management of the difference between the asset return and the liability cost, the expenses, the liquidity management, and the capital management. The goal (8) determines the total expected return based on the expected returns for all the assets R^X and liabilities R^Y.

Beside the goals of solvency, liquidity, and return of assets and liabilities, the bank could determine other goals that concern specific categories of assets and liabilities, in proportion to the demands and preferences of the bank managers. These goals are the deposit goal, the loan goal, and the goal of asset and liability return.

The drawing of capital, especially from the deposits, constitutes a major part of commercial bank management. All sorts of deposits constitute the major source of capital for the commercial banks, in order to proceed to the financing of the economy, through the financing of firms. Thus, special significance is given to the deposits goal.

The goal of asset and liability return defines the goal for the overall expected return of the selected asset-liability strategy over the year of the analysis.

Finally, there are goals reflecting that variables such as cash, cheques receivables, deposits to the Bank of Greece, and fixed assets should remain at the levels used in previous years. More analytically, it is known that the fixed assets are the permanent assets, which have a natural existence, such as buildings, machines, locations, and equipment. Intangible assets are the fixed assets, which have no natural existence but constitute rights and benefits. They have significant economic value, which sometimes is larger than the value of the tangible fixed assets. These data have stable character and are used productively by the bank for the regular operation and performance of its objectives. Since the fixed assets, tangible or intangible, are presented at the balance sheet at their book value, which is the initial value of cost minus the depreciation until today, it is assumed that their value does not change during the development of the present methodology.

At this point, Kosmidou and Zopounidis (2004) took into account that the banks should manage the interest rate risk, the operating risk, the credit risk, the market risk, the foreign exchange risk, the liquidity risk, and the country risk.

More specifically, the interest rate risk indicates the effect of the changes to the net profit margin between the deposit and borrowing values, which are evolved as a consequence of the deviations to the dominant interest rates of assets and liabilities. When the interest rates diminish, the banks accomplish high profits since they can refresh their liabilities to lower borrowing values. The reverse stands for high borrowing values. It is obvious that the changes to the inflation have a relevant impact on the above sorts of risk.

Considering the interest rate risk as the basic uncertainty parameter to the determination of a bank asset liability management strategy, the crucial question that arises concerns the determination of the way through which this factor of uncertainty affects the profitability of the prespecified strategy. The estimation of the expected return of the prespecified strategy and of its variance can render a satisfactory response to the above question.

The use of Monte Carlo techniques constitutes a particular widespread approach for the estimation of the above information (expected return–variance of bank asset liability management strategies). Monte Carlo simulation consists in the development of various random scenarios for the uncertain variable (interest rates) and the estimation of the essential statistical measures (expected return and variance), which describe the effect of the interest rate risk to the selected strategy. The general procedure of implementation of Monte Carlo simulation based on the above is presented in the Figure 1.

During the first stage of the procedure, the various categories of the interest rate risks are identified. The risk and the return of the various data of bank asset and liability are determined from the different forms of interest rates. For example, bank investments in government or corporate bonds are determined from the interest rates that prevail in the bond market, which are affected equally by the general economic environment as by the rules of demand and supply. Similarly, the deposits and loans of the bank are determined from the corresponding interest rates of deposits and loans, which are assigned by the bank according to the conditions that prevail to the bank market. At this stage, the categories of the interest rates, which constitute crucial uncertain variables for the analysis, are detected. The determined interest rates categories depend on the type of the bank. For example, for a decisive commercial bank, the deposit and loan interest rates have a role, whereas for an investment bank more emphasis is given to the interest rates and the returns of various investment products (repos, bonds, interest-bearing notes, etc.).

After the determination of the various categories of interest rates, which determine the total interest rate risk, at the second stage of the analysis the statistical distribution that follows each of the prespecified categories should be determined.

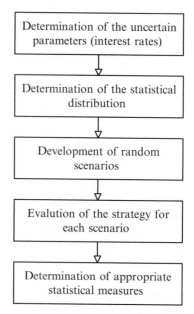

Fig. 1. General Monte Carlo simulation procedure for the evaluation of the asset liability management strategies.

Having determined the statistical distribution that describes the uncertain variables of the analysis (interest rates), a series of random independent scenarios is developed, through a random number generator. Generally, the largest the number of scenarios that are developed, the more reliable conclusions can be derived. However, the computational effort increases significantly, since for each scenario the optimal asset liability strategy should be determined and moreover its evaluation for each other scenario should take place. Thus, the determination of the number N of simulations (scenarios), which will take place should be determined, taking into account both the reliability of the results and the available computational resources.

For each scenario s_i $(i = 1, 2, \ldots, N)$ over the interest rates, the optimal asset liability management strategy Y_i is determined through the solution of the goal programming problem. It is obvious that this strategy is not expected to be optimal for each of the other scenarios s_j $(j \neq i)$. Therefore, the results obtained from the implementation of the strategy Y_i under the rest $N - 1$ possible scenarios s_j should be evaluated. The evaluation of the results can be implemented from various directions. The most usual is the one that uses the return. Representing as r_{ij} the outcome (return) of the strategy Y_i under the scenario s_j, the expected return $\bar{r}_i$ of the strategy can be easily determined based on all the other $N - 1$ scenarios s_j $(j \neq i)$ as follows:

$$\bar{r}_i = \frac{1}{N-1} \sum_{j \neq i} r_{ij}.$$

At the same time, the variance σ_i^2 of the expected return can be determined as a risk measure of the strategy Y_i as follows:

$$\sigma_i^2 = \frac{1}{N-1} \sum_{j \neq i} (r_{ij} - \bar{r}_i)^2.$$

These two statistical measures (average and variance) contribute to the extraction of useful conclusions concerning the expected efficiency of the asset liability management strategy as well as the risks that it carries. Moreover, these two basic statistical measures can be used for the expansion of the analysis of the determination of other useful statistical information, such as the determination of the confidence interval for the expected return, the quantiles, etc.

The above ALM model was applied to a commercial bank of Greece. This model provides the possibility to the financial institutions and more specifically to the banks to proceed to various scenarios of their economic progress for the future, aiming at the management of deposit, loan, and bond interest rates emerging from the changes of the market variables.

3 Conclusions

The main purpose of the present chapter was to provide a brief outline of the bank ALM techniques in the banking industry.

The banking business has recently become more sophisticated due to technological expansion, economic development, creation of financial institutions, and increased competition. Moreover, the mergers and acquisitions that have taken place the last years have created large groups of banking institutions. The success of a bank depends mainly on the quality of its asset and liability management, since the latter deals with the efficient management of sources and uses of bank funds concentrating on profitability, liquidity, capital adequacy, and risk factors.

It is obvious that in the last two decades modern finance has developed into a complex, mathematically challenging field. Various and complicated risks exist in financial markets. For banks, interest rate risk is at the core of their business; managing it successfully is crucial to whether or not they remain profitable. Therefore, it has been essential to the creation of the department of financial risk management within the banks. Asset liability management is associated with the changes of the interest rate risk. Although several models exist regarding asset liability management, most of them are focused on the general aspects and methodologies of this field and do not refer extensively to the hedging of bank interest rate risk through asset liability management. Thus, we believe that the development of a bank asset liability management model that takes into account the exogenous factors and the economic parameters of the market as well as the uncertainty of variations of the financial risks become essential. The investigation of the above will contribute to a more complete definition of the ALM of commercial banks, through an integrated information system that gives the possibility to the decision maker to proceed to various scenarios of the economic process of the bank in order to monitor its financial situation and to determine the optimal strategic implementation of the composition of assets and liabilities.

Finally, despite the approaches described here, little academic work has been done so far to develop a model for the management of assets and liabilities in the European banking industry. Based on the above, we conclude that the quality of asset liability management in the European banking system has become significant as a resource of competitive advantage. Therefore, the development of new technological approaches in bank asset liability management in Europe is worth further research.

Appendix

Linear Programming Formulation of Assets and Liabilities Management

Assuming that r_j is the unit revenue of asset i (in real terms) and c_j is the unit cost of liability j, the objective function is

$$\max z = \sum r_i x_i - \sum c_j Y_j,$$

where X_i is the mean balance of asset i, Y_j is the mean balance of liability j and z represents the difference between the bank's interest income and interest expense, i.e., its revenues ignoring operational expense. The objective function is maximized under a set of constraints.

Linear Goal Programming Model

The problem of a bank's assets and liabilities management can be formulated as the following goal programming model:

$$\min z = f(d_i^+ + d_i^-), \tag{12}$$

$$\text{s.t.} \sum_{j=1}^{n} c_{mj} x_j \leq \theta_m, \quad \forall m = 1, \ldots, M \quad \text{(rigid constraints)}, \tag{13}$$

$$\sum_{j=1}^{n} a_{ij} x_j = b_i + d_i^+ - d_i^- \quad \forall i = 1, \ldots, I \quad \text{(goals)}, \tag{14}$$

$$x_j, d_i^+, d_i^- \geq 0, \tag{15}$$

where

- x_j is the mean balance of asset or liability j (structural variables),
- Y_j is the mean balance of liability j,
- a_{ij} is the technological coefficient attached to x_j in goal i,
- θ_m is the available amount of resource m,
- c_{mj} is the consumption coefficient corresponding to x_j in constraint m,
- b_i is the target value for goal i,
- d_i^+, d_i^- are the positive and negative deviations from the target value of goal i.

The rigid constraints (13) reflect the availability limitations of resources m and correspond to the constraints in the conventional linear programming model. The goals (14) represent the objectives set by management, with the right-hand side of each goal consisting of the target value b_i and the positive/negative deviation d_i^+, d_i^- from it.

The difference in formulation between rigid constraints and goals can be handled in a number of ways. In the sequential linear goal programming model applied at a bank, these constraints are transformed to the same form as the goals. Thus, (13) becomes

$$\sum_{j=1}^{n} c_{mj} x_j = \theta_m + d_m^+ - d_m^-, \quad \forall m = 1, \ldots, M.$$

The achievement function (objective function) has the following form:

$$\min Z = \left\{ \begin{array}{l} P_1 \left[\sum_{m=1}^{M} W_{1m}(d_m^+, d_m^-) \right], P_2 \left[\sum_{i=1}^{I} W_{2i}(d_i^+, d_i^-) \right], \ldots, \\ P_\varphi \left[\sum_{i=1}^{I} W_{\varphi i}(d_i^+, d_i^-) \right] \end{array} \right\},$$

where

- P_φ are the priority levels, with $P_1 > P_2 > \cdots > P_\varphi$,
- $W_{\varphi i}$ is the linear weighting function of the deviation variables of constraint i at priority level φ,
- $\varphi \leq i+1$, i.e., the number of priority levels is less than or equal to the number of goals plus 1, since all the rigid constraints appear at the first priority level.

Stochastic Linear Programming with Simple Recourse

The general n-stage (SLPSR) model is

$$\max\ c_1 x_1 - E_{\xi_1} \left\{ \min \begin{bmatrix} q_1^+ y_1^+ + q_1^- y_1^- + \ldots + \\ \min \begin{bmatrix} c_n x_n + E_{\xi_n/\xi_{n-1}\ldots\xi_1} + \\ \{ \min\ [q_n^+ y_n^+ + q_n^- y_n^-] \} \end{bmatrix} \ldots \end{bmatrix} \right\},$$

s.t. $\sum_{j=1}^{i} T_{ij} x_i + I y_i^+ - I y_i^- = \xi_i, \quad i = 1, \ldots, n.$

The objective function implies the maximization of the net present value of monthly profits minus the expected penalty costs for constraint violations.

The approximation procedure aggregates $x_2, \ldots, x_n$ with x_1 and $\xi_2, \ldots, \xi_n$ with ξ_1. Thus, one chooses $\mathbf{x} = (x_1, \ldots, x_n)'$ in stage 1, observes $\boldsymbol{\xi} = (\xi_1, \ldots, \xi_n)'$ at the end of stage 1, and these steps together determine $(y^+, y^-) = [(y_1^+ y_1^-), \ldots, (y_n^+, y_n^-)]$ in stage 2. This approach yields a feasible procedure for the true dynamic model that is computationally feasible for large problems and incorporates partial dynamic aspects, since penalty costs for periods $2, \ldots, n$ are considered in the choice of $x_1, \ldots, x_n$. Aggregating all future period decision variables into x_1 would make the first period decision function as if all future period decisions were the same regardless of the scenario. The decision maker is primarily interested in the immediate revision of the bank's assets and liabilities.

References

1. Booth, G. G. Programming bank portfolios under uncertainty: An extension. *Journal of Bank Research*, 2:28–40, 1972.
2. Bradley, S. P., and Crane, D. B. A dynamic model for bond portfolio management. *Management Science*, 19:139–151, 1972.
3. Brennan, M. J., and Schwartz E. S. An equilibrium model of bond pricing and a test of market efficiency. *Journal of Financial and Quantitative Analysis*, 17:75–100, 1982.
4. Brodt, A. I. Dynamic balance sheet management model for a Canadian chartered bank. *Journal of Banking and Finance*, 2(3):221–241, 1978.
5. Carino, D. R., Kent, T., Muyers, D. H., Stacy, C., Sylvanus, M., Turner, A. L., Watanabe, K., and Ziemba, W. T. The Russell—Yasuda Kasai model: An asset/liability model for a Japanese insurance company using multistage stochastic programming. *Interfaces*, 24:29–49, 1994.
6. Chambers, D., and Charnes, A. Inter-temporal analysis and optimization of bank portfolios. *Management Science*, 7:393–410, 1961.
7. Charnes, A., and Littlechild, S. C. Intertemporal bank asset choice with stochastic dependence. Systems Research Memorandum no. 188, The Technological Institute, Nortwestern University, Evanston, Illinois, 1968.
8. Charnes, A., and Thore, S. Planning for liquidity in financial institution: The chance constrained method. *Journal of Finance*, 21(4):649–674, 1966.
9. Cohen, K. J., and Hammer, F. S. Linear programming and optimal bank asset management decisions. *Journal of Finance*, 22:42–61, 1967.

10. Cohen, K. J., and Thore, S. Programming bank portfolios under uncertainty. *Journal of Bank Research*, 2:28–40, 1970.
11. Crane, B. A stochastic programming model for commercial bank bond portfolio management. *Journal of Financial Quantitative Analysis*, 6:955–976, 1971.
12. Dantzig, B., and Glynn, P. Parallel processors for planning under uncertainty. *Annals of Operations Research*, 22:1–21, 1990.
13. Dantzig, G. B., and Infanger, G. Multi-stage stochastic linear programs for portfolio optimization. *Annals of Operations Research*, 45:59–76, 1993.
14. Derwa, L. Computer models: Aids to management at Societe Generale de Banque. *Journal of Bank Research*, 4(3):212–224, 1972.
15. Eatman, L., and Sealey, C. W. A multi-objective linear programming model for commercial bank balance sheet management. *Journal of Bank Research*, 9:227–236, 1979.
16. Eppen, G. D., and Fama, E. F. Three asset cash balance and dynamic portfolio problems. *Management Science*, 17:311–319, 1971.
17. Fielitz, D., and Loeffler, A. A linear programming model for commercial bank liquidity management. *Financial Management*, 8(3):44–50, 1979.
18. Giokas, D., and Vassiloglou, M. A goal programming model for bank assets and liabilities. *European Journal of Operational Research*, 50:48–60, 1991.
19. Glynn, P. W., and Iglehart, D. L. Importance sampling for stochastic simulations. *Management Science*, 35:1367–1391, 1989.
20. Grubmann, N. BESMOD: A strategic balance sheet simulation model. *European Journal of Operational Research*, 30:30–34, 1987.
21. Kallberg, J. G., White, R. W., and Ziemba, W. T. Short term financial planning under uncertainty. *Management Science*, 28:670–682, 1982.
22. Korhonen, A. Strategic financial management in a multinational financial conglomerate: A multiple goal stochastic programming approach, *European Journal of Operational Research*, 128:418–434, 2001.
23. Kosmidou K., and Zopounidis C. Combining goal programming model with simulation analysis for bank asset liability management. *Information Systems and Operational Research Journal*, 42(3):175–187, 2004.
24. Kouwenberg, R. Scenario generation and stochastic programming models for asset liability management. *European Journal of Operational Research*, 134:279–292, 2001.
25. Kusy, I. M., and Ziemba, T. W. A bank asset and liability management model. *Operations Research*, 34(3):356–376, 1986.
26. Lifson, K. A. and Blackman, B. R. Simulation and optimization models for asset deployment and funds sources balancing profit liquidity and growth. *Journal of Bank Research*, 4(3):239–255, 1973.
27. Markowitz, H. M. *Portfolio Selection: Efficient Diversification of Investments.* John Wiley and Sons, New York, 1959.
28. Merton, R. C. Lifetime portfolio selection under certainty: The continuous time case. *Review of Economics and Statistics*, 3:373–413, 1969.
29. Merton, R. C. *Continuous-Time Finance.* Blackwell Publishers, New York, 1990.
30. Mulvey, J. M. Generating scenarios for the Towers Perrin investment system. *Interfaces*, 26:1–15, 1996.
31. Mulvey, J. M., and Crowder, H. P. Cluster analysis: An application of lagrangian relaxation. *Management Science*, 25(4):329–340, 1979.
32. Mulvey, J. M., and Vladimirou, H. Stochastic network optimization models of investment planning. *Annals of Operations Research*, 2:187–217, 1989.

33. Mulvey, J. M., Correnti, S., and Lummis, J. Total integrated risk management: Insurance elements. Princeton University, Report SOR-97-2, 1997.
34. Pogue, G. A., and Bussard, R. N. A linear programming model for short term financial planning under uncertainty. *Sloan Management Review*, 13:69–98, 1972.
35. Pyle, D. H. On the theory of financial intermediation. *Journal of Finance*, 26:737–746, 1971.
36. Robertson, M. A bank asset management model. In S. Eilon and T. R. Fowkes, Editors, *Applications of Management Science in Banking and Finance*. Gower Press, Epping, Essex, 1972, pages 149–158.
37. Robinson, R. S. BANKMOD: An interactive simulation aid for bank financial planning. *Journal of Bank Research*, 4(3):212–224, 1973.
38. Samuelson, P. Lifetime portfolio selection by dynamic stochastic programming. *Review of Economics and Statistics*, 51(3):239–246, 1969.
39. Wolf, C. R. A model for selecting commercial bank government security portfolios. *Review of Economics and Statistics*, 1:40–52, 1969.

Advanced Operations Research Techniques in Capital Budgeting

Pierre L. Kunsch

Vrije Universiteit Brussel MOSI Department, Pleinlaan 2, BE-1050 Brussels, Belgium pikunsch@ulb.ac.be

1 Scope of the Chapter

Capital budgeting is a very important topic in corporate finance to make investment decisions. The purpose of the present chapter is to provide an overview in tutorial form about modern operations research (OR) techniques that can be used in such a seemingly rather traditional discipline of finance.

I have held the provided references to the useful minimum necessary for further personal studies.

Given an investment project portfolio, the problem to be solved is to select a number of projects that can create value under satisfactory conditions for the corporation. Project are adopted or deleted from the portfolio according to multiple criteria. When only monetary criteria are considered by managers, the following well-known decision methodologies are usual:

- Discounted cash flow (DCF) is based on cost-benefit analysis: the net present value (NPV) is calculated of all initial outlays (costs) and future free cash flows generated by the project after its completion (benefits). Only projects with a positive NPV will pass the DCF test.
- Payback time (PBT) is a measure of the number of years necessary to repay the initial outlay by means of the future cash flows generated by the project. Only projects with a PBT below a defined ceiling threshold will pass the PBT test.
- Internal rate of return (IRR): IRR is the discount rate at which the NPV of the initial outlays and the future cash flows will vanish. Only projects with an IRR above a defined floor threshold will pass this test.

Complications arise for many reasons, even in the case when only monetary criteria are considered. Note in particular that three main issues need to be addressed:

1. There are constraints on the available budgets (this is the very reason for the name "capital budgeting").

2. In many investment decision problems, other points of view of different stakeholders are to be considered beyond pure monetary criteria appearing in the free cash flow previsions.
3. There are many uncertainties, especially in defining future cash flows to be generated by the project, or in calculating the initial outlay.

In Section 2, I briefly describe the traditional capital budgeting toolkit used for addressing the first and third issues. The literature here is so extensive that only very basic information can be provided, mainly on the DCF technique. The readers are referred to textbooks like Mishan (1988) on cost-benefit analysis, Brealey and Myers (1991), and Besley and Brigham (2005), both on corporate finance for a more detailed introduction. Moreover, due to the limited space, many aspects will be entirely ignored. No information will be provided on issues like the way free cash flows are deduced from corporate accounting statements, the choice of a discount rate from the evaluation of the weighed average cost of capital (WACC) or from capital asset pricing model (CAPM), the use of decision trees and certainty equivalents to cope with stochastic uncertainties, etc.

The following sections describe more advanced techniques to address the three described issues.

In Section 3, the decision-making process is expanded beyond the mere monetary aspects by considering additional criteria and additional constraints, e.g., on available budgets, with the purpose of addressing the first and second issues. For illustration of the resulting multicriteria-decision aid (MCDA) approach, the well-known PROMETHEE technique is introduced: It is easy to use and quite popular among managers.

In Sections 4 and 5, advanced uncertainty treatments are presented, to address the third issue.

In Section 4, fuzzy numbers and fuzzy-inference systems are introduced to cope with uncertainties on important data appearing in the decision process, like initial outlays, free cash flows, discount rates, etc. Some elements will be given on how MCDA problems are "fuzzified," considering herewith all three issues within the same problem.

In Section 5, real options present in some investment decisions are discussed: They can improve the flexibility of the decision-making process and the NPV of the project. Examples of options are given; they are solved by means of the binomial-tree technique. Moreover, elements are given on how to introduce the real-option value as a supplementary criterion in capital budgeting using MCDA techniques.

A short conclusion is given in Section 6.

2 The Traditional Discounted Cash Flow Approach

2.1 Basics

Assume three projects ($i = 1, 2, 3$) with cash flow streams, given in Table 1, generated over a common lifetime of five years. The negative cash flow CF_0 in year $y = 0$ represents the initial outlays. Subsequent cash flows CF_y in years $y = 1, 2, \ldots, 5$ are all positive in this example: In real problems cash flows could become negative in some years (note that in this case there are well-known difficulties in obtaining just one IRR value: This is discussed in all textbooks).

Table 1. Cash Flows (CF), Net Present Values (NPV) and Profitability Indices (PIN) of Three Investment Projects with 5-Year Lifetime

Project i	Rate	Year 0	Year 1	Year 2	Year 3	Year 4	Year 5	NPV	PIN
1	7%	−1000	220	300	400	180	150	38.43	4%
2	7%	−500	100	150	200	200	200	182.91	37%
3	7%	−375	50	75	80	90	100	−57.50	na

The cost of capital is $R = 7\%$ per year. The net present values are computed with the DCF approach as being given by the following expression:

$$\text{NPV}(\text{project } i = 1,\ 2,\ 3) = CF_0(i) + \sum_{y=1}^{5} \frac{CF_y(i)}{(1+R)^y}. \tag{1}$$

Only the two first projects pass the DCF test and are considered for potential investment. It is also mentioned and demonstrated in textbooks that the NPV technique is the most suitable approach for selecting investment projects. In case the projects are exclusive, only one can be adopted.

To decide which projects are selected, a ranking is made using the NPV values. In this very simple example, the second project is ranked first on the basis of the larger NPV. This is, however, a special case: For example, it will be difficult to compare projects with different lifetimes and also with different magnitudes in investment and cash flows.

A possible attractive approach is to use a nonnegative profitability index (PIN), obtained by calculating the ratio between the NPV and the absolute value of the negative initial outlay as follows:

$$\text{PIN}(\text{project } i) = \frac{\text{NPV}(\text{project } i)}{|CF_0|}, \quad 1 \leq i \leq 3. \tag{2}$$

Note that many more approaches are described in textbooks on corporate finance, but this one is certainly easy to calculate.

From Equation (2) we obtain PIN(project 1; project 2; project 3) = (4%; 37%; na), so that indeed project 2 should be selected: But this a special case where it works both ways.

2.2 Budget Constraints – The Knapsack Approach

Of course, in general, there are constraints, either on the budgets, which are available at start for investment, or on the cash drains as the cash flows may become negative in some years; sometimes working capital requirements are imposed in addition.

This capital budgeting problem with constraints can be solved using integer linear programming (ILP) with binary variables:

A binary variable x_i with values 0 or 1 is introduced for each project i. Assuming that there are N candidate projects in the portfolio, all with a positive NPV, and that there is only one constraint on the total budget to be spent for the initial outlays for each project i, the following ILP problem must be solved:

$$\max z = \sum_{i=1}^{N} \text{PIN}_i x_i,$$
$$\text{s.t.} \sum_{i=1}^{N} x_i CF_0^i \leq \text{total budget}, \quad (3)$$
$$x_i = 0, 1, \quad 1 \leq i \leq N.$$

Of course, several additional constraints could be considered in addition. Commercial packages implementing the Branch-and-Bound algorithm, or simply the SOLVER add-in in Excel, are then used for resolution of the ILP problem.

This is exactly the well-known "knapsack problem": It has to be decided which of the more or less indispensable N objects have to be introduced in a knapsack, given some constraints on the maximum available space, total permissible weight, etc.

Figure 1 brings a simple example with one budget constraint solved by using SOLVER. Projects have to be selected among a portfolio of seven projects, all having positive NPVs. The budget constraint is 1000 AU (arbitrary accounting units). As seen in this figure, the constraint imposes that the second project, which has a PIN = 40%, cannot be kept because its budget is too large. The sixth project, which has a PIN = 21%, is chosen instead.

2.3 Stochastic Treatment of Uncertainties

Usually, input data to the DCF analysis are affected with all kinds of uncertainty, including the discount rate itself.

Many techniques are available to handle uncertainties from certainty equivalents to decision trees. Another popular technique is Monte Carlo analysis, in which important parameters are represented by probability distributions, so that for each project a probability distribution of NPVs and other useful parameters can be obtained, like the PINs defined in Equation (2). Techniques for comparing probability distributions can be used to define a ranking of

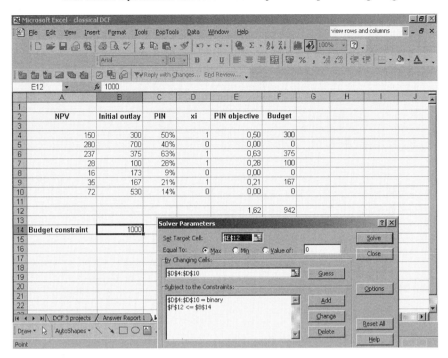

Fig. 1. Resolution with SOLVER of the knapsack problem applied to capital budgeting: Seven projects with positive PINs and one budget constraint are considered. A project is selected when $x_i = 1$.

the projects on the basis the NPV distributions, generalizing the approach presented in Section 2.1.

As an example, assume a project consisting in building a production plant for some manufactured items to be sold on a market, the uncertain yearly revenues could be estimated as follows in some year y:

$$\text{revenue}(y) = \text{price} \times \text{size of the market}(y) \times \text{market share}(y), \quad (4)$$

where the price is a known parameter, and both "size of the market (S)" and "market share (M)" in year y are random variables represented by some estimated cumulative probability distributions FS and FM. The Monte Carlo procedure would have the following steps, assuming a large enough number of random drawings, up to an upper counter limit (CL), e.g., CL = 500 drawings, and no correlation between both probability distributions:

1. set counter = 0;
2. generate random number RS and RM uniformly distributed on [0,1];
3. calculate the corresponding values of size of the market, market share by calculating, respectively, i.e., by using the inverse cumulative probability distributions;

4. calculate and record the revenue (Counter) from Equation (4) by using the obtained values and the given price;
5. increment counter:=counter+1;
6. if counter < CL, go back to step 2; otherwise, go to step 7;
7. prepare a frequency histogram of revenue; calculate mean, standard deviation, etc.

From the produced probability distribution of NPVs (or PINs) for each project in the portfolio, it is possible to estimate the probability NPV > 0, or IRR > $k\%$, given the required minimum percentage return k, etc.

A ranking can also be established, e.g., NPV1 > NPV2, with confidence level 95%, etc.

However, both difficulties discussed in Besley and Brigham (2005) should not be underestimated: (1) defining probability distributions for the difference parameters in the cash flow models; and (2) taking into account the correlations that usually exist between those parameters.

Section 4 discusses how to approach uncertainties in parameters by using fuzzy numbers rather than probability distributions.

3 Multicriteria Analysis

3.1 Basics of Multicriteria Analysis

In many decision problems other aspects than purely monetary must be considered, i.e., multiattribute or multicriteria dimensions.

This is, of course, also valid for capital budgeting problems: It is necessary to go beyond monetary values to consider additional dimensions, like the impact on available knowledge for a R&D project, environmental impact of a technological project, flexibility in a decision to invest, etc.

The treatment of these multidimensional aspects in decision problems is the object of Multicriteria Decision Aid (MCDA). An abundant literature is dedicated to the numerous available techniques. Vincke (1992) gave a good overview.

Basically, two important families of techniques can be distinguished:

1. Multiattribute techniques are based on utility theory; they have been developed in the UK and North America. A classical reference book is Keeney and Raiffa (1976). Note that the weighed sum rule is a special case of additive linear utility functions;
2. Outranking methods are based on pairwise comparisons of alternatives; they have been developed in continental Europe, and especially in France around the work of Roy (1985). A more recent reference is Roy and Bouyssou (1993).

Most techniques are valuable, but not all can be mentioned or compared with respect to their respective merits in the present chapter.

For the sake of illustrating the concepts used in many outranking methods, the following explanations will spin around the PROMETHEE approach, an easy-to-use methodology developed in Brans and Vincke (1985), and Brans and Mareschal (1994),(2002). It also has the capability of including constrained budgeting and other constraints.

3.2 A Simple Multicriteria Technique

The basic principles of the PROMETHEE technique are now presented.

Let $f_1(\cdot), f_2(\cdot), \ldots, f_K(\cdot)$ be K evaluation criteria the decision makers wish to take into account. Consider N investment strategies; suppose, in addition, that the evaluations of these projects are real numbers expressed on their particular criterion units or real number scores placed on numerical scales, such as 0 to 10, or 0 to 100, etc. The following evaluation table is set up for N investment strategies S_i, $i = 1, 2, \ldots, N$.

Table 2. Multicriteria Evaluation Table in PROMETHEE in Case N Investment Strategies $S_i, i = 1, 2, \ldots, N$ are Evaluated Against Multiple Criteria $k = 1, 2, \ldots, K$

Strategies	$f_1(\cdot)$	$f_2(\cdot)$	$\cdots$	$f_j(\cdot)$	$\cdots$	$f_K(\cdot)$
S_1	$f_1(S_1)$	$f_2(S_1)$	$\cdots$	$f_j(S_1)$	$\cdots$	$f_K(S_1)$
S_2	$f_1(S_2)$	$f_2(S_2)$	$\cdots$	$f_j(S_2)$	$\cdots$	$f_K(S_2)$
$\cdots$	$\cdots$	$\cdots$	$\cdots$	$\cdots$	$\cdots$	$\cdots$
S_i	$f_1(S_i)$	$f_2(S_i)$	$\cdots$	$f_j(S_i)$	$\cdots$	$f_K(S_i)$
$\cdots$	$\cdots$	$\cdots$	$\cdots$	$\cdots$	$\cdots$	$\cdots$
S_N	$f_1(S_N)$	$f_2(S_N)$	$\cdots$	$f_j(S_N)$	$\cdots$	$f_K(S_N)$

First, preference functions $P_k(S_n, S_m)$ are defined for the sake of comparing strategy pairs S_n, S_m for all criteria $f_k(\cdot)$, $k = 1, 2, \ldots, K$. Each function for a given k is defined as being monotonously increasing, as a function of the difference of the scores being obtained by the two strategies on each particular criterion. Different shapes of the preference function with the possibility of multiple preference thresholds can be chosen. An example of a linear V-shaped function with one threshold (p) is given in Figure 2; a linear function with two thresholds (q, p) is shown in Figure 3. Other possible shapes are Gaussian shapes, or staircase-like preference functions.

Suppose now that all pairwise comparisons have been performed, providing all preference-function values:

$$0 \leq P_k(S_n, S_m) \leq 1, \quad \text{for } k = 1, 2, \ldots, K; \ n, m = 1, 2, \ldots, N. \quad (5)$$

An aggregated preference index over all the criteria is then calculated:

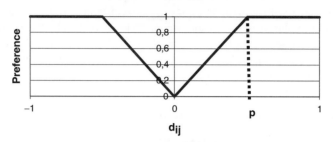

Fig. 2. A PROMETHEE V-shaped linear preference function with one preference threshold ($p = 0.5$) of the difference d_{ij} between the evaluations of a pair of actions (i, j).

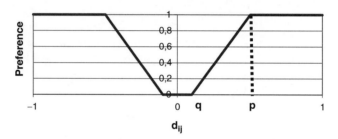

Fig. 3. A PROMETHEE linear preference function with two preference thresholds ($q = 0.1$) and ($p = 0.5$) of the difference d_{ij} between the evaluations of a pair of actions (i, j).

$$\Pi(S_n, S_m) = \sum_{j=1}^{k} P_j(S_n, S_m) w_j, \qquad (6)$$

where w_j is a weight, measuring the relative importance allocated to criterion j. The weights are supposed to be normalized, so that

$$\sum_{j=1}^{k} w_j = 1. \qquad (7)$$

The following so-called positive and negative dominance flows express the dominating and the dominated character of a strategy S_n, respectively, over all other strategies:

$$\Phi^+(S_n) = \frac{1}{n-1} \sum_{m=1}^{N} \Pi(S_n, S_m), \qquad (8)$$

$$\Phi^-(S_n) = \frac{1}{n-1} \sum_{m=1}^{N} \Pi(S_m, S_n), \qquad (9)$$

while the balance of the dominance flow gives the net flow:

$$\Phi(S_n) = \Phi^+(S_n) - \Phi^-(S_n). \qquad (10)$$

Two partial rankings give by intersection the so-called PROMETHEE I ranking. The former rankings are obtained from both the positive and the negative flows, corresponding respectively to the more or less dominating strategies (positive flows) and the more or less dominated strategies (negative flows).

Both rankings may be contradictory, so that some strategies are considered to be incomparable, i.e., they are better than other strategies for some criteria and worse for some other criteria.

Only by using the net flows in Equation (10) can a complete ranking be obtained, the so-called PROMETHEE II ranking.

The PROMCALC software in MS-DOS and the more recent Decision Lab 2000 under WINDOWS are described in Brans and Mareschal (1994, 2002). Available additional features of the software are the sensitivity analysis on the weights and graphical investigations in the so-called GAIA plane obtained by principal components analysis (PCA) (Brans and Mareschal, 1994).

An Example

Consider the following evaluation table of five projects $i = 1, 2, \ldots, 5$ for three criteria: profitability index (PIN) measured in percents, environmental index (EIN) measured on a scale from 0 (lowest value) to 10 (highest value) describing the environmental quality of each project, and the flexibility (FLEX) during the investment decision-making process, also measured on a 0–10 scale.

Table 3. Scores for Five Projects for Three Criteria

Projects	PIN (%)	EIN (0–10)	FLEX (0–10)
Proj1	50	4	6
Proj2	28	8	7
Proj3	17	7	10
Proj4	12	10	4
Proj5	34	8	2
Weights	0.4	0.25	0.35

All three criteria have to be maximized. A linear V-shaped function is adopted for the PIN criterion, with a preference threshold equal to 5%; linear functions with two thresholds are adopted for the EIN and the FLEX criteria with an indifference threshold equal to 1, and a preference threshold equal to 2.

Figure 4 shows the complete PROMETHEE II ranking established on the basis of the net flows Φs.

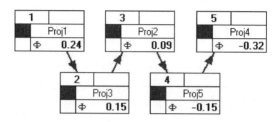

Fig. 4. The PROMETHEE II ranking with the indication of the net flows Φs.

The GAIA plane obtained by PCA is shown in Figure 5. The pi-decision axis (circle) is a representation in the GAIA plane of the weighting of the criteria. Its orientation indicates the type of compromise obtained in the PROMETHEE II ranking. The strategy is ranked first, which has the largest value of the projection in the plane on the pi-decision axis. It can easily be seen from the last figure that Proj1 has the largest projection on the pi-axis, and it is indeed ranked first.

3.3 Taking Constraints into Account

The approach has been extended to take into account constraints, common in capital budgeting, and discussed earlier in Section 2.2. Again the knapsack formulation is used. Equation (3) is directly transposed: The PIN values are replaced by coefficients derived from the net flows Φs in PROMETHEE as follows:

$$\max z = \sum_{i=1}^{N}(\Phi_i - \Phi_{\min} + \varepsilon)x_i,$$
$$\text{s.t.} \sum_{i=1}^{N} x_i CF_0^i \leq \text{total budget}, \quad (11)$$
$$x_i = 0, 1 \quad 1 \leq i \leq N,$$

where $\Phi_{\min}$ represents the (negative) smallest flow obtained in PROMETHEE,

and ε is a small positive number. In this way all coefficients of the objective function z to be maximized are positive, as requested in the knapsack problem.

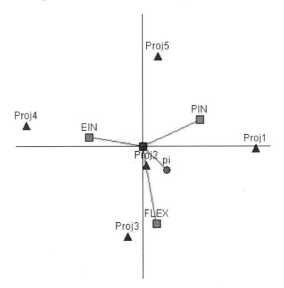

Fig. 5. The GAIA plane obtained by PCA of five projects (triangle) and three criteria (squares). The pi-axis indicates the compromise between the three criteria for the given weights.

In the original approach called PROMETHEE V described in Brans and Mareschal (2002), the net flows are directly used in the objective function z so that the formulation is no longer a knapsack problem. Because the total net flows must sum up to zero, as can be seen from analyzing Equations (5) to (10), some investments always have negative values, in particular the last ranked with the minimum flow $\Phi_{\min}$. Therefore, unless additional suitable constraints are used, investments with negative net flows will never have a chance of getting selected. It is why it is necessary to adapt the methodology as done in (11).

To illustrate this, let us use the example of Table 3 completed with budget values as given in column C of Table 4. Two constraints are considered:

- the total budget lies between 350 and 420 AU (arbitrary accounting units);
- at least three projects should be selected, so that the sum of the binary variables x_i appearing in column B of the spreadsheet is larger than or equal to 3.

The problem is solved with SOLVER, and the results are shown in Table 4: PROJ1, PROJ3, and PROJ4 are selected. Note that PROJ4 is the least-scored project, but it is still kept to satisfy the constraints.

Table 4. Multicriteria Analysis Using PROMETHEE Net Flows Submitted to Budget Constraints for a Minimum of Three Projects ($\varepsilon = 0.01$)

	x_i	Budget	Φ	$\Phi - \Phi_{\min} + \varepsilon$	$x(\Phi - \Phi_{\min} + \varepsilon)$	$x \times$ Budget
PROJ1	1	300	0.24	0.58	0.58	300
PROJ2	0	120	0.09	0.43	0.00	0
PROJ3	1	45	0.15	0.49	0.49	45
PROJ4	1	60	−0.33	0.01	0.01	60
PROJ5	0	270	−0.15	0.19	0.00	0
SUM	3	795			1.08	405
Budget $\leq$ 420						
Budget $\geq$ 350						

4 Fuzzy Treatment of Uncertainties

4.1 Basics on Fuzzy Logic

Fuzzy logic (FL) is a mathematical technique to assist decisions on the basis of rather vague statements and logical implications between variables. FL is close to the natural language, which is why some people have called it "computation with words." It was a big achievement of Bellman and Zadeh (1970) to provide this fantastic instrument to decision making. FL is indeed very useful in many technical and economic applications in which the imprecise and relatively vague judgments of experts have to be accounted for in a quantitative way, as explained for business applications in Cox (1995).

A basic ingredient of fuzzy logic is the use of "fuzzy numbers" (FNs) representing the range of possible values of a vague or imprecisely known variable or parameter. Ordinary Boolean logic would accept just one single value to represent what is then called a "crisp number."

An FN represents a possibility grade, called a *membership grade* (MG), that some parameters or variables in a model represent some semantic property or affirmation, for example, the property of a person being "middle-aged."

This MG can be represented as a function, called a *membership function* (MF), of the value of the parameter or variable, here the person's age. In this example:

MG of being middle-aged = MF "MIDDLE-AGED" (person age). (12)

MFs are often normalized, i.e., the largest possible MG is equal to 1. Common MFs are given as triangular fuzzy numbers (TFN) defined by the triplet $[a, b, c]$, or as trapezoidal fuzzy numbers (TRFN) defined by the quadruplet $[a, b, c, d]$. Other shapes are triangular-like, trapezoidal-like, Gaussian FNs, etc.

To represent an FN, the MF must have a number of properties defined in Buckley et al. (2002, pages 5–8), i.e., to be normalized and monotonically increasing or decreasing along each lateral edge.

In the quoted example, MF "MIDDLEAGED" may be a TFN, defined with MG = 0 for age values below 30 years (y), increasing to MG = 1 at 45 y, and declining to MG = 0 at 60 y.

This TFN is expressed in "years" units, and it is defined by the triplet [30, 45, 60]. Note that the membership grade (MG) 0.5 corresponds to an age of 37.5, or 52.5.

Figure 6 shows this representation and an alternate representation with the TRFN [30, 45, 52.5, 60].

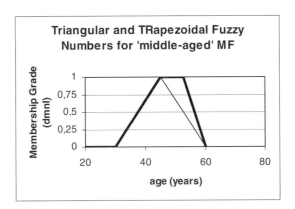

Fig. 6. Middle-aged membership function in two representations TFN and TRFN.

In the same context, other lifetimes could be represented, e.g., "CHILD," "YOUNG," "OLD." The interval of variation of the fuzzy variable is called the *universe of discourse*, in the given example for life ages, it would be in the interval [0, 100] (years).

The starting part of fuzzy logic thus consists of "fuzzification," i.e., translating imprecise variables or parameters in a model into fuzzy numbers, represented by MFs, e.g., using four MFs in describing the different ages of life.

Note that those MFs are different from probability distributions. For example, the total surface underneath any MFs is not normalized to 1, but use is made of normalized FNs; thus with a maximum at the value MG = 1. FNs are defined in the framework of possibility distributions, rather than probabilistic distributions.

In many applications a further step in fuzzification consists of developing mappings between fuzzy variables by means of fuzzy rules.

In the given example, rules connecting the life ages to the degrees of experience may be imagined:

(a) If AGE is CHILD, then EXPERIENCE is VANISHING,
(a) If AGE is YOUNG, then EXPERIENCE is LIMITED,
(c) If AGE is MIDDLE-AGED, then EXPERIENCE is APPRECIABLE,
(d) If AGE is OLD, then EXPERIENCE is IMPORTANT.

(13)

In this one-input, one-output fuzzy-inference system, the four input life ages, i.e., "CHILD," "YOUNG," "MIDDLE-AGED," and "OLD," may be represented, e.g., by triangular MFs (TFN) and the output experience levels by corresponding four trapezoidal MFs (TRFN), e.g., "VANISHING," "LIMITED," "APPRECIABLE," "IMPORTANT."

Section 4.3 discusses how such rule systems can be used in capital budgeting to link several uncertain variables.

4.2 Use of Fuzzy Numbers to Represent Uncertain Data

Fuzzy arithmetic can be developed with fuzzy numbers; thus, fuzzy DCF calculations are easy to perform. A very complete and pleasant-to-read reference is Chapter 4 of Buckley et al. (2002) on fuzzy mathematics in finance. Assume, for example, that the yearly cash flows and the discount rate are represented by FNs (provided with "hats" in the following to distinguish them from crisp data); Equation (1) can be immediately "fuzzified":

$$\overline{\text{NPV}}(\text{project } i = 1, 2, 3) = \overline{CF_0(i)} + \sum_{y=1}^{5} \frac{\overline{CF_y(i)}}{(1+\overline{R})^y}. \tag{14}$$

The fuzzy NPVs can be calculated from addition, multiplication, and division, as now explained.

To perform fuzzy arithmetic, the easiest way is to use so-called α-cuts applied on fuzzy number. For any given value of $0 \leq \alpha \leq 1$, the α-cut of any FN is the interval of values x in the universe of discourse for which $MG(x) \geq \alpha$. In the example given in Figure 6, considering the TFN "MIDDLE-AGED" [30, 45, 60], the following α-cuts in years are easily obtained:

- for $\alpha = 0$, it is the interval [30, 60],
- for $\alpha = 0.5$, it is the interval [37.5, 52.5],
- for $\alpha = 1$, it is the singleton [45, 45], etc.

Consider two TFN or TRFN $\overline{A}, \overline{B}$. For any let us note the α-cuts $\overline{A}_\alpha = [a_1, b_1]$ and $\overline{B}_\alpha = [a_2, b_2]$. Interval arithmetic operations are performed for each α-cut:

$$\text{addition: } \overline{A}_\alpha + \overline{B}_\alpha = [a_1 + a_2, b_1 + b_2], \tag{15}$$
$$\text{difference: } \overline{A}_\alpha - \overline{B}_\alpha = [a_1 - b_2, b_1 - a_2]. \tag{16}$$

The product depends on the sign of a and b. The general formula is easily established:

$$\overline{A}_\alpha * \overline{B}_\alpha = [\min(a_1a_2, a_1b_2, a_2b_1, b_1b_2), \max(a_1a_2, a_1b_2, a_2b_1, b_1b_2)]. \tag{17}$$

If we assume that zero does not belong to the two α-cuts, then $a_1b_1 \geq 0$ and $a_2b_2 \geq 0$. The following four possibilities are then considered:

$$\begin{aligned}&a_1 \geq 0 \text{ and } a_2 \geq 0: & \overline{A}_\alpha * \overline{B}_\alpha &= [a_1 a_2, b_1 b_2], \\ &b_1 < 0 \text{ and } b_2 < 0: & \overline{A}_\alpha * \overline{B}_\alpha &= [b_1 b_2, a_1 a_2], \\ &a_1 \geq 0 \text{ and } b_2 < 0: & \overline{A}_\alpha * \overline{B}_\alpha &= [a_2 b_1, a_1 b_2], \\ &b_1 < 0 \text{ and } a_2 \geq 0: & \overline{A}_\alpha * \overline{B}_\alpha &= [a_1 b_2, a_2 b_1]. \end{aligned} \qquad (18)$$

Division of A by B can be brought back to the product if zero does not belong to the interval $[a_2, b_2]$, by considering:

$$\overline{C}_\alpha = \left[\frac{1}{b_2}, \frac{1}{a_2}\right], \text{ and it happens that } \overline{A}_\alpha / \overline{B}_\alpha = \overline{A}_\alpha * \overline{C}_\alpha. \qquad (19)$$

Figure 7 shows the product of two positive TFNs performed in a rather tedious but straightforward way with EXCEL. Note that the result is still an FN, but it is no longer a TFN: Rather it is a triangular-like FN. In the same way the product/division of two TRFNs is a trapezoidal-like FN.

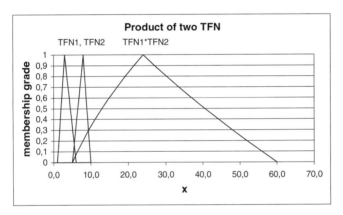

Fig. 7. Fuzzy product of two positive TFNs, TFN_1 and TFN_2, resulting in a triangular-like fuzzy number.

Example 4.2.1

Assume two projects with the same initial crisp outlay CF_0 and positive constant fuzzy cash flows (constant currency units), so that they can be interpreted as perpetuities. The values of the annual cash flows are two TRFNs ($\overline{CF}_1$ and $\overline{CF}_2$). The discount rate is also a triangular fuzzy number ($\overline{DR}$). The resulting perpetuities have to be calculated, and it must be determined which one is larger! Assume the following values:

$$\overline{CF}_1 \ [5, 6, 7, 8], \quad \overline{CF}_2 \ [6, 8, 9, 9.5], \quad \overline{DR} \ [9.5\%, 10\%, 10.5\%]. \qquad (20)$$

The two perpetuities, say P_1 and P_2, are easily calculated as the NPV of the infinite cash flow row, resulting in the following expressions, using Equation (19) to transform the fuzzy division into the fuzzy product:

$$P_i = \overline{CF_i} * \frac{1}{\overline{DR}}, \qquad i = 1, 2, \tag{21}$$

$$\frac{1}{\overline{DR}} = \left[\frac{1}{10.5\%}, \frac{1}{10\%}, \frac{1}{9.5\%} \right]. \tag{22}$$

Figure 8 shows the resulting trapezoidal-like fuzzy numbers P_1, P_2 (though they look at first sight almost like TRFNs, but the lateral edges are slightly curved, and trapezoidal-like FNs are obtained). $P_2 > P_1$ is the immediate guess, but the important overlapping between the two FNs makes it difficult to confirm this assumption without further analysis.

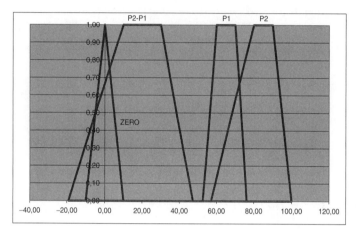

Fig. 8. Example 4.2.1; the two perpetuities P_1, P_2; the difference $P_2 - P_1$ and the fuzzy ZERO for verifying that $P_2 > P_1$.

To be sure, we may calculate the fuzzy difference $P_2 - P_1$ and compare it to zero. As we are in the fuzzy-arithmetic world, a fuzzy ZERO is needed, say the TFN $[-10, 0, 10]$, and we need to decide if the following fuzzy equality holds:

$$P_2 - P_1 > \text{ZERO}. \tag{23}$$

A more straightforward way is to check if $P_2 > P_1$ in the fuzzy sense, which has to be defined: Many approaches are described in the literature on comparing two FNs.

The methodology proposed by Buckley et al., Chapter 4 (2002), seems to be adequate for our purpose:

- when the two FNs do not overlap, the result of the comparison is immediate;
- when the two numbers overlap, the largest MG is considered at points where the lateral edges are cutting each other;

Advanced Operations Research Techniques in Capital Budgeting   317

$$\begin{array}{l} \text{if } MG(\text{edge cutting}) < \theta \text{ then} \\ \quad \text{the FN more to the right is larger} \\ \text{else} \\ \quad \text{both FN's are equal in the fuzzy sense} \end{array} \quad (24)$$

where $0 \ll \theta < 1$, e.g., $\theta = 0.8$, as recommended in Buckley et al. (2002, page 58).

In both (23) and in the direct comparison $P_2 > P_1$, it can be immediately read from the figure that the edge cutting is below $\theta = 0.8$ (through close to it from below), so that indeed it can be considered that the second project has to be preferred to the first one, from the DCF point of view.

4.3 Use of Fuzzy-Rule Systems to Infer Uncertain Data

Fuzzy logic has many more potentialities than only fuzzy arithmetic. Mostly it will be used, as in control theory, when some semantic statements can be expressed in the form of rules, as we saw in the trivial example in Equation (13) relating life ages to levels of experience.

In general, the rules result from past observations and data interpretations, or in guess work based on the experience of experts in the field. Fuzzy-rule systems like that in Equation (13) can then be used to draw conclusions and provide quantitative evaluations on data that are largely uncertain. In capital budgeting evaluations, there are, of course, many cases where this approach is relevant.

Example 4.3.1

Assume a company is developing an electric car for private use. Ignoring for the sake of simplicity all issues related to tax, capital, etc., the company analysts have estimated the annual free cash flow in year (y) as follows:

$$\begin{aligned} \text{cash flow}(y) = {} & \text{price of a car} \times \text{total demand for cars}(y) \\ & \times \text{market share of electrical cars}(y). \end{aligned} \quad (25)$$

The analysts also discovered that in a first approximation, the market share is expected to be dependent only on the price of oil: Whenever oil is cheap, the demand for electric car is expected to be "large"; conversely, in times of expensive oil, demand will be "small."

The following computations and graphics have been prepared with the Fuzzy Toolbox for Use with MATLAB (2001).

Just concentrating on the market share aspect, the analysts were able to "fuzzify" the problem. They elaborated four market share levels (MS) in percentage represented by TRFNs in the universe of discourse between 0% and 15% of the total market per year:

$$\begin{array}{ll} \text{SMALL MS} & [0,0,1,4]\%, \\ \text{REDUCED MS} & [1,4,5,8]\%, \\ \text{NOTICEABLE MS} & [5,8,10,13]\%, \\ \text{LARGE MS} & [10,13,15,15]\%. \end{array} \qquad (26)$$

Those trapezoidal MFs are shown in Figure 9.

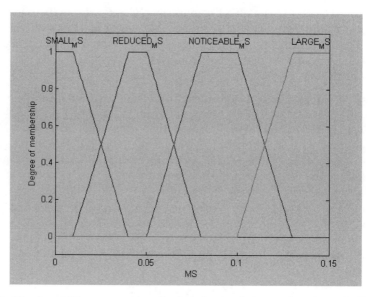

Fig. 9. The four MFs representing the four levels of the market share (MS) in the universe of discourse [0, 15%].

In this specific case, the market share was shown to depend only on the world oil price (P) in \$/barrel, for which four trapezoidal MFs are also identified in the universe of discourse [20,150]\$/barrel:

$$\begin{array}{ll} \text{SMALL P} & [20,20,27,41]\$/\text{barrel}, \\ \text{REDUCED P} & [27,41,63,77]\$/\text{barrel}, \\ \text{NOTICEABLE P} & [63,77,95,113]\$/\text{barrel}, \\ \text{LARGE P} & [95,113,120,120]\$/\text{barrel}. \end{array} \qquad (27)$$

Those four MFs are shown in Figure 10.

Further, the experts elaborated a system of four rules connecting the oil price to the market share:

Rule 1: If P is SMALL P, then MS is LARGE MS,
Rule 2: If P is REDUCED P, then MS is NOTICEABLE MS,
Rule 3: If P is NOTICEABLE P, then MS is REDUCED MS,
Rule 4: If P is LARGE P, then MS is SMALL MS.
$\qquad (28)$

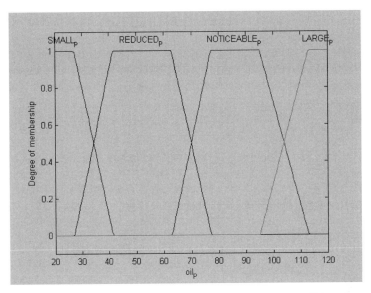

Fig. 10. The four MFs representing the four levels of the oil price (P) in the universe of discourse [20, 120]$/barrel.

This system of rules is a mapping from the universe of discourse of oil prices in [20, 120]$/barrel, which is the unique input to the fuzzy rules, to the universe of discourse of the market shares (MS) in [0, 15]%, which is the unique output, or conclusion, of the fuzzy rules.

Of course, more complicated fuzzy-rule systems may be considered. For example, the analyst could create rules with two or more inputs: An additional sensible input for determining the market share could, for example, be the price of an electric car.

The fuzzy inputs and outputs and the partial rules in the mapping build a "fuzzy inference system" (FIS) or more popularly, "a rule-firing system." The FIS computations are shown in Figure 11.

It is very easy to understand how such an FIS works. The four rules are displayed in Figure 11 from top to bottom: On the left are the four input MFs representing the oil price (oil_P) as in Figure 10; on the right are the four output MFs representing the market share (MS) as in Figure 9.

Assume the expected price for next year to be 67$/barrel, as indicated by the cursor in the left-hand window. Consider the four MFs and the corresponding four rules. Rule 1 and Rule 4 listed in Equation (28) are not active ("fired"), as the oil price 67$/barrel is outside their respective range, i.e., the trapezoidal MF "SMALL P," or "LARGE P." Rule 2 and Rule 3 are "fired," i.e., partially active as the oil price lies within the range of their respective MF "REDUCED P" or "NOTICEABLE P."

The possibility grade of the output in each fired rule is limited above by the membership grade MG(Rule i; oil_P) ≤ 1, up to the level indicated by the gray surface.

The corresponding rule output for MS, indicated in the right window, thus appears as a truncated trapezoidal MF (dark surface) calculated by means of the minimum operator "min":

$$\text{MF(Output } i; \text{ MS}) = \min\left[\text{MG(Rule } i; \text{ oil\_P}), \text{MF(Rule } i; \text{ MS})\right]. \qquad (29)$$

The specific fuzzy-inference system (FIS) using the minimum operator "min" to fire rules as in Equation (29) is called the Mamdani FIS: It is most familiar for those who use control theory. Interesting textbooks in control theory are Kacprzyk (1997) and Passino and Yurkovich (1998). The Mamdani inference generally serves in engineering for the purpose of controlling technical devices, but it is very useful here, too, as well as in many economic or financial applications. An introduction to fuzzy control for decision theory is given in Chapter 7 of Bouyssou et al. (2000).

Note that some FIS utilize other operators, which I will not discuss here. Dubois and Prade (1996), and Fodor and Roubens (1994) are useful references.

Moreover, one of my previous papers in Kunsch and Fortemps (2002) describes an FIS using different fuzzy-inference operators. It is used in the economic calculus of projects in radioactive waste management.

The Mamdani inference thus provides two truncated trapezoidal, but not normalized, MFs, visible in the right-hand windows in Figure 11. Each shape corresponds to the partial conclusion of one of the two active rules. Those partial conclusions are no FN's because they are not normalized. This is not the end yet. First, a global fuzzy conclusion is obtained in the so-called aggregation step in the FIS: In the Mamdani FIS, the maximum operator "max" is used:

$$\text{MF(Global Output; MS)} = \max_i \left[\text{MG(Output } i; \text{ MS})\right]. \qquad (30)$$

The aggregation of the partial conclusions to a global conclusion is shown in the lowest window on the right-hand side of Figure 11. It is a composite MF, which is, of course, no fuzzy number (one reason is that this MF is not normalized: Can you give another reason?).

This MF is not usable as such: A crisp number is expected to be the global practical conclusion of the Mamdani FIS. The last step is thus defuzzification. In this case the center-of-gravity (COG) of this composite shape is computed to obtain one final value for the output. This gives in the example a market share MS (oil_P=67\$/barrel) = 7.79% appearing on top of the right-hand windows in Figure 11.

Looking at other values of the oil price in the range [20, 120]\$/barrel, the corresponding values of the market share MS(oil_P) are obtained. Figure 12 delivers the staircase-like MS values in function of the oil price in the universe of discourse.

Advanced Operations Research Techniques in Capital Budgeting 321

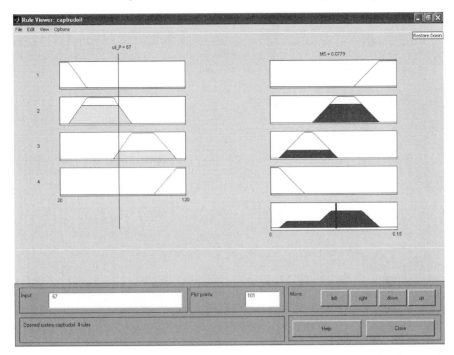

Fig. 11. The fuzzy-inference system (FIS) provides a market share (MS) of 7.8% when the expected oil price next year is 67$/barrel.

How do we decide which oil price to use as an input to the Mamdani FIS? It is the nice thing about fuzzy logic that this question can be answered just as well by building an additional FIS: The input data to this new FIS will now be given by some expert judgments on the oil price.

Assume that there are three experts, ex1, ex2, and ex3, whose oil price forecasts for next year are represented by the MFs of Figure 13, i.e., FN's of Gaussian or trapezoidal shapes. The first expert predicts a lower range pex1 between 35 and 45$/barrel; the two other experts forecast higher ranges pex2, respectively pex3, between 45 and 85$/barrel. Experts have been given credibility scores in the range [0, 1] resulting from an analysis of the accuracy of previous forecasts they have made on oil prices. A nice textbook on how to evaluate the value of expert judgments is Meyer and Booker (2001).

A high score indicates a high level of trust in the expertise. Assume the credibility scores to be 0.2 for ex1, 0.9 for ex2, and 0.75 for ex3. This indicates that the last two experts are quite credible, while the first one is not.

The Mamdani FIS is elaborated along three rules; the credibility scores serve as inputs to the partial rules:

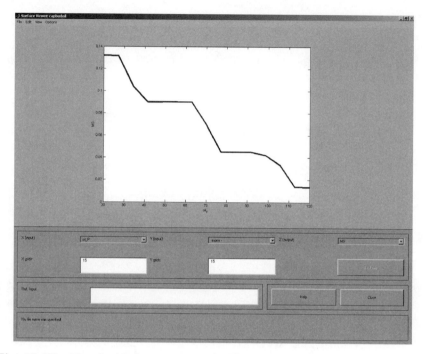

Fig. 12. The Mamdani inference system (FIS) delivers the evolution of the market share (MS) as a function of the oil price for the whole universe of discourse [20, 120]$/barrel.

$$\begin{aligned}&\text{Rule 1: If Expert is ex1, then oil\_P is pex1,}\\&\text{Rule 2: If Expert is ex2, then oil\_P is pex2,}\\&\text{Rule 3: If Expert is ex3, then oil\_P is pex3.}\end{aligned} \quad (31)$$

The activation of the Mamdani FIS is shown in Figure 14, resulting in a defuzzified oil price with the COG technique: oil_P = 60.4$/barrel.

4.4 Fuzzy Multicriteria Capital Budgeting

The fuzzy treatment in the case where only one fuzzy monetary criterion is being considered, i.e., the NPV value of each project cash flow streams, can be "in principle" extended to MCDA: A global score is obtained through aggregating several criteria representing multiple facets of the projects.

This is only true "in principle" because this statement would only be valid for MCDA procedures that aggregate partial preferences in an "arithmetic way" (addition/subtraction and multiplication/division), like in the weighed sum rule, or in utility-based approaches. Only elementary operations of fuzzy arithmetic can indeed produce FNs from FNs.

By contrast, outranking methodologies and in particular PROMETHEE (see Sections 3.2 and 3.3) are not reducible to fuzzy arithmetic, because they

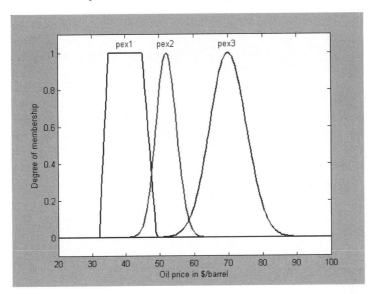

Fig. 13. The guesses of three experts for the dollar price of the oil barrel next year.

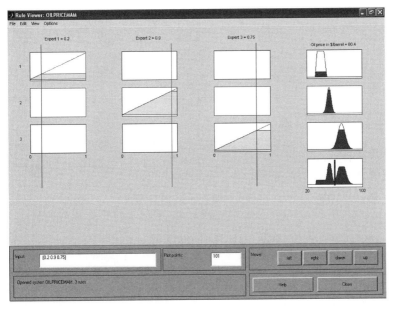

Fig. 14. The price of oil according to three expert analysts with credibility factors (0.2; 0.9; 0.75). The price resulting from the Mamdani inference and the COG defuzzifying is 60.4$/barrel, a value between the previsions of the two most credible experts.

require thresholds, nonlinear, or piecewise linear preference functions, etc. As a result, in general, the resulting global scores, e.g., the net flows Φs of PROMETHEE, will no longer be FNs, even if each partial score attached to each criterion is.

Nevertheless, the use of α-cuts and interval arithmetic as we explained in Section 4.2 is quite useful to cope with uncertainties in PROMETHEE, as has been demonstrated in Le Teno and Mareschal (1998). Fuzzy arithmetic provides a measure of the dispersion of results, beyond central values obtained with the crisp maximum MG values ($\alpha = 1$) of all partial scores. For example, in a maximizing problem, minimum scores (pessimistic view) and maximum scores (optimistic view) are obtained from α-cuts with $\alpha = 0$, 0.25, 0.5, 0.75, etc.

Example 4.4.1

Assume two projects with two maximizing criteria with the partial fuzzy scores represented by TFNs, shown in Table 5.

Table 5. Two Projects Evaluated in Two Fuzzy Criteria

Projects	PIN (%)	EIN (0–10)
Proj$_1$	[40, 50, 60]	[3.8, 4, 4.3]
Proj$_2$	[25, 28, 32]	[7, 8, 8.8]

A first calculation could be made using the central $\alpha = 1$ scores, i.e., Proj$_1$(50; 4) and Proj$_2$(28; 8).

If it is decided to consider only data for which MG ≥ 0.5, the α-cut with $\alpha = 0.5$ is then considered to provide a good measure of the score dispersion. This provides two sets of possible scores for each project, corresponding to two optimistic and two pessimistic assumptions, i.e.:

Proj$_{11}$(45; 3.9) (pessimistic) and Proj$_{12}$(55; 4.15) (optimistic),
Proj$_{21}$(26.5; 7.5) (pessimistic) and Proj$_{22}$(30; 8.4) (optimistic).

Two additional PROMETHEE calculations can be performed by combining two assumptions in the pairwise comparison, a pessimistic and an optimistic one, i.e., (Proj$_{11}$, Proj$_{22}$), (Proj$_{12}$, Proj$_{21}$): In this way the smallest and largest values of the net flows are obtained for both projects. In total for each project $i = 1, 2$, one will get three net flow values ($\Phi_i^p, \Phi_i^c, \Phi_i^o$), where the superscripts p, c, o refer to pessimistic, central, or optimistic net flows, respectively.

In case there are $n > 2$ projects, the same approach can be generalized for some α-cut by pairwise comparisons of projects: The optimistic (pessimistic) scores on each project $i = 1, n$ is compared to the pessimistic (optimistic) scores on any other project $j \neq i$.

After $2\frac{n(n-1)}{2} = n(n-1)$ pairwise comparisons, all optimistic and pessimistic flow values are calculated. At the end of the day, the triplets $(\Phi_i^p, \Phi_i^c, \Phi_i^o)$ are obtained for all projects $i = 1, 2, \ldots, n$.

To decide about the ranking of the projects in a portfolio, a definition of "greater than" must be found, as was done previously in Equation (24) for comparing FNs: Intervals are provided in this case. Simple comparison rules between a pair of intervals can be elaborated for project1, $I_1 = [\Phi_1^p, \Phi_1^o]$, and for project2, $I_2 = [\Phi_2^p, \Phi_2^o]$.

There are only three main possibilities: The two intervals are disjoint, or they are overlapping; in this second case there can be full or partial overlapping. Assume first the case that the interval I_1 is either located left or fully included in the central part of the interval I_2. The following simple comparison procedure seems to be sensible for ranking both projects:

$$\begin{aligned}
&\text{if Project}_1 \text{ does not overlap with Project}_2 \\
&\quad \text{Project}_2 \text{ is preferred to Project}_1 \\
&\text{else if } I_1 \subset I_2 \\
&\quad \text{Project}_1 \text{ is preferred to Project}_2 \\
&\text{else if } I_1 \not\subset I_2 \\
&\quad \text{Project}_2 \text{ is preferred to Project}_1.
\end{aligned} \qquad (32)$$

As a short explanation:

- If there is no overlapping: Because I_1 is in our assumption located left to I_2, Project$_1$ is ranked second;
- If I_1 is fully included in I_2, assume first that it is located in the central part of I_2: In a fuzzy sense both values are equal; nothing can be said about the ranking. But one can possibly argue that Project$_1$ has the smaller dispersion of net flows around its central value, and therefore it appears as being less risky: Project$_1$ is ranked first. This may not be a sufficient answer when the interval I_2 is rather large; several possibilities then exist with respect to the positioning of I_1 within I_2, when comparing central net flow values Φ_1^c, Φ_2^c: I_1 is located in the left corner (Project$_2$ is preferred); I_1 is rather central (as before Project$_1$ is preferred); I_1 is located in the right corner of interval I_2 (Project$_1$ is preferred), etc.
- If there is partial overlapping, Project$_2$ will again be ranked first, because I_2 is located by assumption right of I_1.

In case there are constraints, like budget constraints, things may become much more intricate. The knapsack approach explained in Section 3.3 can be used, but unfortunately, in general, three different portfolios of adopted projects will come out when using the pessimistic, central, and optimistic values for the net flows: In general, the nonvanishing binary variables will be different in each case: The three solutions are not part of a unique FN.

How in practice do we decide which projects should be either chosen or rejected? The problem can be handled by multiobjective integer linear programming as explained in Vincke (1992), considering the three pessimistic, central, and optimistic scores: The usually inaccessible "ideal solution," optimizing those three global scores at the same time is unattainable. It could also be handled as in the crisp case by goal programming in which the three goals are again the three global scores.

Other authors would like to use a different approach and consider fuzzy linear programming, as explained in Chapter 6 of Buckley et al. (2002). I stop the discussion here.

To be complete about uncertainties and the use of fuzzy logic, the uncertainties in the choice of technical parameters in the applied MCDA methodologies should also be mentioned. In PROMETHEE preference functions and thresholds, and, last but not least, the weights have to be chosen. Only a limited literature exists about how to deal with those issues using fuzzy logic. Fuzzy extensions of PROMETHEE have, however, been discussed in Goumas and Lygerou (2000) and Gelderman (2000). I will not go into details, because of the technicalities of some developments.

The problem of weights deserves some additional considerations, however. To discuss this point I will use the simple weighed sum rule.

Assume crisp normalized scores $0 \leq s_k^i \leq 1$, for criteria $k = 1, 2, \ldots, K$, for some projects $i = 1, 2, \ldots, k$, and the corresponding fuzzy weights $\overline{w}_k$, $k = 1, 2, \ldots, K$, assumed to be represented by TFNs. The weighed sum is used to obtain the fuzzy global score of the project $\overline{S}_i$:

$$\overline{S}_i = \sum_{k=1}^{K} \overline{w}_k s_k^i = (S_i^p, S_i^c, S_i^o), \tag{33}$$

$$\overline{w}_k = (a_k, b_k, c_k) \quad \text{with} \quad 0 \leq a_k < b_k < c_k < 1, \tag{34}$$

where the superscripts p, c, o again respectively refer to the pessimistic, central, and pessimistic evaluations of the global score.

This looks like trivial fuzzy arithmetic: S_i appears to be a TFN (or a triangular-shaped fuzzy number if the scores s_k^i are fuzzy as well, which we did not assume for simplicity). This rough result is, however, meaningless as the sum of weights is not equal to (crisp) "one"; therefore, there is no guarantee that the so-calculated global fuzzy score will stay in the interval [0, 1].

A not-so-trivial explanation of that difficulty is that there is a strange property about fuzzy arithmetic, namely that if A, B, C are fuzzy numbers:

$$\text{if } B = C - A, \quad \text{then} \quad A + B \neq C. \tag{35}$$

Example 4.4.2

Assume the following TFNs:

$A[5, 6, 7]$,
$C[10, 12, 14]$,
$B = (C - A) \, [3, 6, 9]$,

But $A + B = D$ [8, 12, 16] $\neq C$.

The consequence is that if we cannot impose that fuzzy weights will sum up to the crisp "one," i.e., the TFN (1, 1, 1) with zero spread, because the fuzzy addition will, on the contrary, amplify the spread of each term in the sum.

So, though Equation (33) is expected to provide an FN, the latter cannot be directly computed with fuzzy arithmetic. The more involved way to compute this FN is as follows:

1. Compute (33) with central values of the weights b_k. In this way the central value S_i^c is obtained.
2. Set up two linear programs for each project with constraint on the weights as follows:

For the pessimistic project score:

$$\max S_i^p = \min \sum_{k=1}^K w_k^p s_k^i,$$
$$\text{s.t.} \sum_{k=1}^K w_k^p = 1, \qquad (36)$$
$$a_k \leq w_k^p \leq c_k,$$

and for the optimistic project score:

$$\max S_i^o = \min \sum_{k=1}^K w_k^o s_k^i,$$
$$\text{s.t.} \sum_{k=1}^K w_k^o = 1, \qquad (37)$$
$$a_k \leq w_k^o \leq c_k.$$

Note that instead of choosing the optimistic and pessimistic values given by the α-cut $= 0$, other values of $\alpha > 0$ could be used.

Moreover, in cases where the scores s_k^i are themselves fuzzy, the usual interval arithmetic can be used. The optimistic and pessimistic global scores come out for the chosen α-cut. At the end of the day triplets $(S_i^p(\alpha), S_i^c, S_i^p(\alpha))$ are obtained for each project, where α refers to the α-cut.

As explained earlier in this section, comparisons can be made between projects, or other capital budgeting constraints can be introduced, using, for example, multi-objective integer linear programming to test the changes in solutions due to the weight uncertainties. Recent work of Mavrotas and Diakoulaki (2004) has been in developing similar ideas.

5 Treatment of Uncertainties Using Real Options

5.1 The Value of Flexibility

Until now we have come across static uncertainties, which are present at the very time a decision is made and are not changing over time. But time has a value, not only because there is a preference for now rather than for later, as manifested in the choice of a discount rate. There is also a value in reducing some uncertainties, by learning more about the project value, observing the changes in the economic environment, or the advancements in science, technology, etc.

As a simple example, consider the investment in electric cars mentioned in Example 4.3.1. The development of this business is crucially dependent on the knowledge about oil prices, and also current technological development of batteries, fuel cells, etc., for which it may be difficult to measure the exact impact at the present time. But let us assume that it becomes easier one year from now, because new information will be available. All those factors crucially affect the profitability or even the viability of the project. A prudent stance is therefore to cautiously develop the activity, keeping the flexibility to withdraw, to wait before investing for further expansion, or even to abandon the project if it becomes unprofitable.

The added value embedded in keeping open some options is in many cases far from being negligible. Value is added to the project because new degrees of flexibility are gained, e.g., from the possibility of changing the present stance in the face of fresh incoming information.

In the last 15 years, a promising approach has been developing around financial option (FO) theory to deal with the learning and flexibility aspects embedded in risky assets. Real option (RO) theory is used when the underlying risky asset is not a financial asset, like a stock, but it is a real asset, like an investment project.

In the main field of application of ROs are projects for which the DCF technique does provide a slightly negative or a slightly positive NPV, so that no clear-cut decision can be made: On a strict DCF basis, most projects of this type would not pass or would hardly pass the profitability test. Because of the value of their intrinsic flexibility, they are worth considering for further analysis: The RO value captures those flexibility aspects, and this may change the picture for the better.

Unfortunately, valuing ROs may become very intricate in the most general cases when partial differential equations must be solved. I do not have the space here to write a full treaty on this topic, and therefore readers are referred to excellent textbooks for details on option-valuation methodologies. But still, I hope to be able to meet the challenge of this chapter to present the basic theoretical background sufficient for understanding the principles of RO evaluation in simpler problems, and for providing herewith an incentive for further self-study.

For a technical introduction to options, see Hull (2000) and Beninga (2000). The practical use of ROs in management is described in Copeland and Antikarov (2001); ROs and financial flexibility is described in Dyson and Berry (1998) and Trigeorgis (1998).

Let me start with a definition taken from this literature, paraphrasing the usual definition of FOs, i.e., by replacing "financial asset" by "investment project":

> A real option (RO) is the right, but not the obligation, to take an action (e.g., deferring, expanding, contracting, or abandoning an investment project) at a predetermined cost called the exercise price, for a predetermined period of time – the life of the option.

An RO is a right and not an obligation. This right will only be exercised when it is useful to do so. For example, a growth option on buying a company will only be exercised when information arriving before the exercise time confirms the growth expectations. Because option theory helps modeling flexibility in decisions under uncertainty, an underlying risky asset with uncertain value is always present, either a stock (in the case of FOs) or a real asset, e.g., an irreversible investment in some risky project (in the case of ROs).

Therefore, ROs will only be useful under two conditions, which are related to both uncertainty and flexibility:

1. There are uncertainties on the real asset, and more can be learned about it over time. This would exclude irresolvable uncertainties, which are not reduced with learning.
2. The learning process adds flexibility value by revisiting the investment strategy.

With those conditions in mind, we can identify three large families of ROs for investing in real assets:

1. *Growth RO* in which growth opportunities are created when investing in some markets, e.g., in using the investment as an opportunity to penetrate those developing markets with more efficient products. Example of market-opportunity ROs are given in Chevalier et al. (2004) for mergers and acquisitions;
2. *Defer or waiting RO* in which key uncertainties can be resolved by learning or by doing, e.g., by performing preliminary investigations in a newly discovered oil field. Examples of waiting options can be found in Dixit and Pindyck (1994);
3. *Abandonment/shrinking RO* in which the risk of unfavorable development is considerably reduced by downsizing, divesting, or abandoning the real asset, e.g., by selling a piece of machinery for some residual value. Examples of divesting options are found in Chevalier et al. (2003).

In the following, simple examples of each type will be analyzed. Before going further, it is necessary to become familiar with the "jargon" that is usual for FOs and has also been adopted for discussing ROs. Note beforehand, however, that no market exists for ROs, like it does for FOs, so that an RO is indeed only a notional instrument.

- An RO is *exercised* if it is favorable at some time to make the subjacent decision, e.g., to invest. Otherwise, the RO is not exercised and it becomes worthless.
- When an *exercise time* is imposed at some expiry time T, the RO is a *European option*. In case there is a possibility of premature exercise before the expiry time T, the RO is an *American option*. Note that many ROs are American options, in the frequent case where early exercise of the option indeed increases the flexibility, and thus the project's intrinsic values. This is one of the reasons why the valuation of ROs can often be quite difficult: A closed-form formula like the famous Black–Scholes formula, described in Brealey and Meyers (1991) or in Chapter 16 of Beninga (2000), is only useful for European options, as it indeed assumes a fixed exercise time. A short introduction to the Black–Scholes formula is given in Section 5.2. Another reason is that many ROs are compound options, i.e., the exercise of one option may triggers a cascade of many ROs. Closed-form formulas are useless here. We will see that, when the time takes discrete values, the binomial-tree technique will render good services most of the time. Though the latter technique does not look at first sight like it is an OR technique, its origin is found in stochastic dynamic programming. Interested readers may check in Dixit and Pindyck (1994), Chapter 4, that the Bellman equation of dynamic programming in the discrete-time case converges to the partial differential equation, which constitutes the continuous-time foundation of option theory. It is shown with an example in Section 5.2 that the Black–Scholes formula is a particular case of this theory for European options.
- The *exercise price*, X, is the monetary value acting as a cutoff for exercising or not the RO, given some time-dependent project NPV, S.
- The RO is a *call option* if it is exercised when the investment value S is above the exercise price. Both Growth and Defer ROs are call options, because they capture growth opportunities associated with a project value expected to increase above the exercise price X. The *value of a call RO* (for an FO it would be the call price) is thus measured by the expected value of the positive difference $S - X$:

$$C = \max(S - X, 0). \qquad (38)$$

A positive call value C when $S > X$ will place the option "in the money," as it can be exercised with some benefit. Otherwise, C will be zero ("out of the money"), and the option will never be exercised, because it is worthless.
- The RO is a *put option* if it is exercised when the investment value is below the exercise price. Abandonment/divesting/shrinking ROs are thus put options: They allow withdrawing when the expectations are not fulfilled

and the project value falls below the exercise price. They thus provide a stop loss, or they reward the investor with some residual value. The *value of a put option* (for FOs it would be the put price) is thus measured by the expected positive difference $X - S$:

$$P = \max(X - S, 0). \tag{39}$$

The put option is "in the money" when $X > S$; otherwise, it is "out of the money" and worthless.

Call and put options are thus like the two sides of a coin, and they provide protection ("hedging") to the investors against missing growth opportunities or losing too much money in a risky project, respectively.

- Option theory distinguishes the risk-free rate of return r of the investment from the discount rate (DR), which is used to evaluate the NPV. It is indeed assumed that decisions are made in a risk-free world, i.e., that the decision makers are neither risk-prone nor risk-adverse, but that they are indifferent to risk. This is discussed in the Section 5.2.
- Note also that the option value (or option price) always increases with the risk associated with the investment, i.e., the *volatility* of its intrinsic value, measured by the standard deviation of the risky NPV. The RO value represents the actual value of the opportunity attached to this volatility.
- The RO value is expected to grow with (1) the time still to run till expiry, (2) an increase in the risk-free rate r because it represents the time-value of money, (3) the project NPV, and (4) the volatility of the NPV, creating new opportunity values.
- The RO value is expected to decrease with (1) a higher initial outlay, necessary to acquire the real asset, and (2) the exercise price X.

I am now in a position to discuss with practical and simple examples the binomial-tree technique, which is used to evaluate ROs.

5.2 Real Options and How to Evaluate Them

This section provides basic information on how options can be evaluated with the binomial-tree technique.

In order to keep the presentation as easy to follow as possible, the best way seems to avoid abstract theory by solving step by step a rather trivial numerical example of RO valuation: References are made to the cells in the spreadsheet shown in Figures 15 and 16.

Example 5.2.1 Abandonment and Growth Real Options

A manufacturing company considers buying a new machine to boost its present production relying on four rather old and less efficient machines. The expected NPV of the future free cash flows generated by the new machine is 100 AU (arbitrary accounting units) as shown in cell B8 of the spreadsheet in Figures 15 and 16. The discount rate is DR = 10% in cell D2.

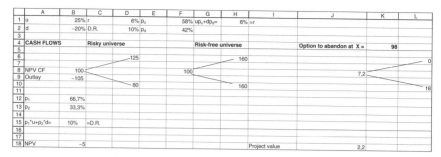

Fig. 15. A risky project taking into account an abandonment put real option at the end of year 1.

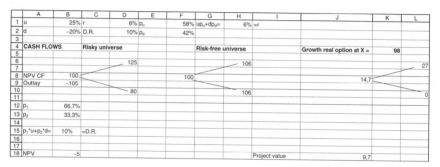

Fig. 16. A risky project taking into account a growth call real option at the end of year 1.

Unfortunately, the acquisition investment (initial outlay) is 105 AU (B9), so that the total NPV value is negative: $100 - 105 = -5$ AU (B18); the project should be rejected by the company according to the sole DCF test.

However, the NPV can in reality take two values next year: It can go up by $u = 25\%$ (B1) or it can go down by $d = -20\%$ (B2).

Therefore, the company can take the following stance:

- Should the project prove to be unprofitable in the down state d, it will be abandoned next year: The equipment will be sold at a residual value of 98 AU. A put abandonment RO is hidden behind this possibility of pulling out of the project in this instance, and thus the RO exercise price is $X = 98$ AU (K4).
- Should the project be in the up state u, it will be more profitable that the equipment residual value of 98 AU: The company will keep the machine running and even scrap the four old machines and replace them by four new machines. A call growth RO is given by this opportunity to increase in the productivity one go. The exercise price is again $X = 98$ AU (K4).

Consider first the put option in Figure 15. On the left-hand section of the spreadsheet called "Risky universe," it is noted that the NPV of future cash flows today can become next year either 125 AU (D6), corresponding to the up movement $u = 25\%$ (B1), with a probability $p_1 = 66.7\%$ (B12), i.e., two chances out of three, or 80 AU (D10), corresponding to a down movement $d = -20\%$ (B2), with a probability $p_2 = 33.3\%$ (B13), i.e., one chance out of three.

Assume now that, instead of investing in the risky project, the company would be indifferent to drop the project, and to invest instead in some risk-free asset, e.g., bonds with zero volatilities, bringing the risk-free rate of $r = 6\%$ (D1). After one year the value would become 106 AU for all states of the risky world (H6 and H10).

In such a universe characterized by indifference to risk, the expected values of the cash flows of both the risky assets and the risk-free asset would be the same, so that one would expect the following equation to hold:

$$p_u(1+u) + p_d(1+d) = 1 + r, \qquad (40)$$
$$p_u + p_d = 1,$$

where p_u and p_d are the probabilities in this risk-free world: They are different from the actual probabilities p_1 and p_2 to observe an up or down movement in the project value measured by the NPV of cash flows.

The general solution of this equation system is as follows:

$$p_u = \frac{r-d}{u-d}; \qquad p_d = 1 - p_u = \frac{u-r}{u-d}, \qquad (41)$$

which here gives $p_u = 58\%$ (F1), and $p_d = 42\%$ (F2).

The put RO is to abandon the project in year 1, in case the NPV of the cash flow in this year would be 80 AU(D10); the equipment would then be sold for the residual value of $X = 98$ AU (K4).

What is the present value of this option?

The calculation is shown on the right-hand section of the spreadsheet called "Option to abandon." It is made backward by starting at the end of the first year, at which time the option to abandon the project may be exercised: In this case the company will realize what was before only a possibility, and it will give up the project.

At end of year 1, if the NPV of the project is in upper state, i.e., $S = 125$ AU, the option will not be exercised, and its value is zero according to Equation (39), because $X < S$ (L6). In the option jargon, the RO is "out of the money," because it brings nothing more to the already rosy situation.

If the project is now in the down state, i.e., $S = 80$ AU, it is worth exercising the option, i.e., to receive the residual value of 98 AU instead of continuing the project. The RO value is exercised; it is "in the money," because according to Equation (39) it brings a positive value of $X - S = 98 - 80 = 18$ AU at the end of the first year (L10).

This put option provides an advantageous stop loss for the project. Without the option, and considering the initial outlay of 105 AU, the undiscounted loss in the down state would be $80 - 105 = -25$ AU, while with the option, the loss is reduced to $98 - 105 = -7$ AU.

The present value of the put option P is calculated as being the expected value of those two situations, i.e., 0 AU or 18 AU, discounted at the present time with r: It would be the price of the corresponding FO effectively traded on the derivative market. Because we are in the risk-free universe, the risk-free probabilities $p_u = 58\%$ (F1), $p_d = 42\%$ (F2) and the risk-free rate $r = 6\%$ (D1) are used:

$$P = \frac{p_u \cdot 0}{1+r} + \frac{p_d \cdot 18}{1+r} = q_u \cdot 0 + q_d \cdot 18 = 7.2 \text{ AU (\$J\$8)}, \qquad (42)$$

$$q_u = \frac{p_u}{1+r} \qquad q_d = \frac{p_d}{1+r}. \qquad (43)$$

q_u and q_d defined in Equation (43) are called the *state prices*: They are the prices today of one AU to be paid in the succeeding period either in the up state u, or in the down state d, as explained in Chapter 14 of Beninga (2000). Using them makes the backward calculation in Equation (42) more straightforward. It provides the option value at any node in the tree from the option values at the connected nodes in the following time period.

Adding the option value of the project to its negative NPV$=-5$ leads to a total positive value $7.2 - 5 = 2.2 > 0$ (J18): The project value with the abandonment RO, though small, is positive and is now passing the DCF test.

Now the growth call RO can also be easily evaluated, as shown in the spreadsheet of Figure 16.

The calculation procedure is as before, except that now it will be worth exercising the call option when the upper value $S = 125$ AU (D6) is observed in year 1. On the right of the diagram at year 1, the option values are given by $\max(S - X, 0)$ [Equation (38)], so that they will be 27 AU (L6) in the upper state and 0 AU (L10) in the down state. The present value of the call option is thus obtained by using the state prices q_u, q_d defined in Equation (43):

$$P = q_u 27 + q_d 0 = 14.7 \text{ AU (\$J\$8)}. \qquad (44)$$

The total value of the project with the call option is 9.7 AU (J18).

The binomial tree has only two branches in this example, spanning one year. Because in practice the asset value will take many possible values, it may be better to consider smaller time periods h. Therefore, it is worth investigating how the option values change for two periods of six months ($h = 0.5$ month), instead of one period of one year, four periods of one quarter ($h = 0.25$ month), etc.

The calculation for the call option and two six-month periods, i.e., $h = 0.5$ month, is shown in the spreadsheet in Figure 17.

The principle of the RO valuation is the same as before, except that all parameters are calculated at $h = 0.5$ year (F1), the elementary time step. This is easy for the DR and the risk-free rate:

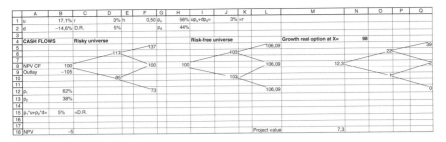

Fig. 17. The growth call RO with two periods of six months.

$$\mathrm{DR}(h) = \mathrm{DR}\frac{h}{1} = 5\%(\$D\$2); \qquad r(h) = r\frac{h}{1} = 3\%(\$D\$1). \qquad (45)$$

How do we represent the up (u) and down (d) movements over a six-month period given their annual values? There is a nice formula, which comes from the theory of random-walk processes explained in Chapter 15 of Beninga (2000): It allows a painless calculation. Assume that the NPV of such a random-walk process is characterized by an average value μ and a standard deviation σ: u and d can then be represented by the following equations:

$$u = e^{\mu+\sigma\sqrt{h}}, \qquad (46)$$
$$d = e^{\mu-\sigma\sqrt{h}}.$$

Looking at $u = +20\%$ and $d = -25\%$ for $h = 1$ year in the present example, it is found that they are compatible with $\mu = 0$ (no expected growth) and $\sigma = 22.31\%$. Thus, for $h = 0.5$

$$u = e^{\sigma\sqrt{h}} = 17.09\% \ (\$B\$1); \qquad d = \frac{1}{u} = e^{-\sigma\sqrt{h}} = -14.59\%(\$B\$2). \qquad (47)$$

Therefore, using Equation (41), the risk-free probabilities are $p_u = 56\%$ and $p_d = 44\%$, and the state prices are obtained by means of Equation (43).

Starting again at the end of the second six-month period in the spreadsheet of Figure 17, the call RO values are obtained in column Q, i.e., $(39 = 137 - 98, 2 = 100 - 98, 0)$ AU. Using (42), the RO values are obtained at the end of the first six-month period in column O, i.e., (22, 1). By backward application of the same formula, the present value of the RO is calculated to be 12.3 AU (M8), so that the project NPV including the call RO is $12.3 - 5 = 7.3$ AU (M18).

This process can be repeated when considering $h = 0.25$ year, i.e., four quarters. Figure 18 shows this computation for the call RO. It now results in $C = 12.7$, so that the project NPV including the call RO is $12.7 - 5 = 7.7$ AU.

If we now go on splitting the one-year period into shorter and shorter time intervals of diminishing length $h \to 0$, we come to the continuous case, which can be calculated with an analytical formula, the well-known Black–Scholes

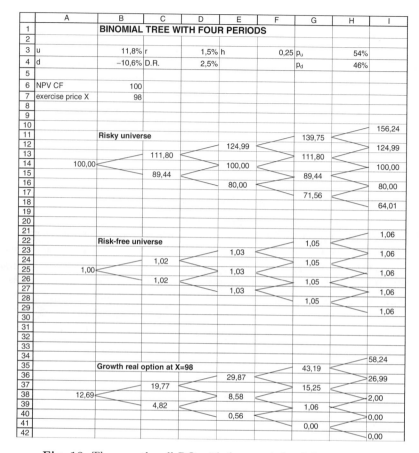

Fig. 18. The growth call RO with four periods of three months.

formula, explained in Chapter 16 of Beninga (2000). The intention is not to discuss here this closed-form formula for the continuous case in the valuation of European options, which admit no early exercise before the expiry time T.

In my opinion one of the most impressive achievements of modern finance and a fantastic, first unexpected, property is the following: The RO value, obtained with the binomial-tree technique applied to European options, converges to the Black–Scholes formula when the elementary time interval between branches in the tree $h \to 0$, i.e., the continuous case is obtained at the limit.

To be a little bit more precise, the Black–Scholes formula is given in Equation (48), considering T the remaining time from now to the expiry date, the call value C, the put value P, S the NPV of the project cash flows; the exercise price X, the risk-free rate r, and the volatility σ of the risky real asset:

$$C = SN(d_1) - Xe^{-rT}N(d_2),$$
$$d_1 = \frac{\ln(S/X) + (r + \frac{\sigma^2}{2})T}{\sigma\sqrt{T}}, \quad (48)$$
$$d_2 = d_1 - \sigma\sqrt{T},$$
$$P = C - S + Xe^{-rT}. \quad (49)$$

$N(x)$ is the value in x of the cumulative normal distribution with mean 0 and variance 1. $N(d_2)$ is shown to represent the probability to exercise the call option at exercise time T. It can immediately be seen that at time T, if the option is strongly "in the money," this probability will be growing to 1, so as $d_1 = d_2$ in $t = T$, the call price will be converging toward the call value $C = S - X > 0$, given in Equation (38). The same type of reasoning applies to the put option price $P = X - S > 0$ given in Equation (39). Relation (49) is called the put-call parity identity: It is always verified for the pair of call and put option values in the same problem.

Table 6 illustrates our previous statement for Example 5.2.1. Note that the RO values are not always growing monotonously with the number of time periods, but fluctuate before converging to the limit values given by the Black–Scholes formula.

Table 6. The Call and Put Options in Example 5.2.1. The Evaluations are Made with Binomial Trees Counting Different Numbers of Periods in the Discrete Case, and with the Black–Scholes Formula in the Continuous Case

Number of Periods (n)	Call RO Value	Put RO Value
1	14.7	7.2
2	12.3	4.7
4	12.7	5.0
Black–Scholes ($n = \infty$)	12.9	5.2

In all cases with a finite number of periods (n) and a European option, the put-call parity identity in Equation (50) generalizes the continuous case in Equation (49):

$$P + [\text{NPV}(\text{project}) = S] = C + \frac{X}{(1 + r/n)^n}. \quad (50)$$

Equations (49) and (50) reflect the complementary roles in the pair of call and put ROs, both with the exercise price X (see Brealey and Myers, 1991, Chapter 20, pages 488–490). This identity is useful to verify the price computation of European ROs, which are not exercised before their expiry dates.

5.3 A Real Option with Early Exercise

In many situations a RO will not be kept until its expiry date. In other words, American ROs offer the possibility of early exercise, contrary to the call and put options handled in the previous section. To evaluate such options, the Black–Scholes formula is not adapted, as it works exclusively with European options with a fixed exercise time.

An example of this kind is the waiting RO, which captures the value of waiting for new information. Details on early option exercise can be found in Beninga (2000), Chapter 19.

Example 5.3.1 A Waiting Real Option

A company makes plans for implementing a project, which is expected to provide cash flows for an indefinite time. The initial investment is 90 AU (arbitrary accounting units) and the present value of the project is 100 UA, assuming a 10% DR.

For the next years the annual cash flow is estimated to increase from its level in the year before by 20% with a probability $p_1 = 60\%$, or to decrease by 30% with a probability $p_2 = 40\%$. The indefinite expected cash flow is 10 UA/year. In the first year, the cash flow is thus expected to be either 7 AUs (down state), or 12 AU (up state). The risk-free rate is 5%.

Although the investment NPV = $100 - 90 = 10$ AU> 0, the company managers are finding an additional benefit in postponing the investment decision. In other words, a waiting call option increases the investment value and gives more flexibility in time.

Assume that the decision about the project must be taken at the latest two years from the present time. The evaluation of the call RO is made as before by means of a binomial tree over two time periods. Looking at the time evolution of the RO value, the managers will have to answer two questions regarding the project investment: (1) How much is the option worth? (2) When should the RO be exercised, possibly before the expiry time in two years?

The tree representing the project value can be established as in the upper part of figure 19. The cash flow in the first year are 7 AU and 12 AU, so that the project value after one year will be given by a perpetuity, calculated with DR = 10% to be either 70 or 120 AU. The cash flows in the next years are again increasing by $u = 20\%$, or decreasing by $d = 30\%$, so that the resulting annual cash flows in the second year are estimated to be 4.9, 8.4, or 14.4 AU, and the corresponding present values calculated with the DR = 10% at the end of the second year are thus 49, 84, or 144 AU.

In each year the project generates a profit, comparable to a dividend paid by a stock in the world of FOs. The return to the company owner is given in the upper state by $u = (12 + 120)/100 - 1 = 32\%$, and in the lower state by $d = (7 + 70)/100 - 1 = -23\%$.

By contrast with the previous example with European options, the presence of this yearly profit stream makes an early exercise of the call option

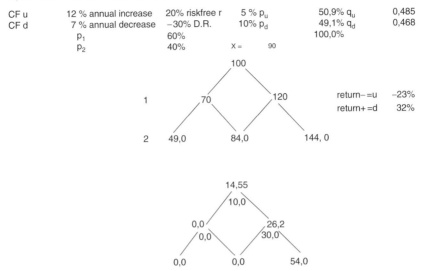

Fig. 19. An RO with early exercise.

possible. For those readers who are familiar with FO theory, note that an American call option without dividend-paying stock can never be prematurely exercised.

In the assumption of the risk-free world, the payoff of the risky project is exactly the same as a risk-free investment at the risk-free rate $r = 5\%$; one must have

$$p_u 32\% + p_d(-23\%) = 5\%,$$
$$p_d = 1 - p_u,$$
$$p_u = 50.9\%; \qquad p_d = 49.1\%, \qquad (51)$$
$$q_u = \frac{p_u}{1+r} = 0.485; \qquad q_d = \frac{p_d}{1+r} = 0.468,$$

where p_u and p_d are, as usual, respectively, the probability of moving up or down in the risk-free world of options, and q_u and q_d are the corresponding state prices in Equation (43).

We again work out the option value at the end of year 2 and then move backward to the present time, each time using the state prices to calculate the option value at the previous period according to Equation (42).

The lowest row in the tree in Figure 19 gives the NPV values of cash flows. The exercise price X of the call is the initial outlay for the project, i.e., 90 AU. Equation (39) gives $C = \max[\text{NPV}(\text{year 2}) - 90, 0]$, so that we obtain the respective option values (0, 0, 54) AU at end of year 2 shown in the lowest row in the lower tree in Figure 19. Only the branch in the upper state thus has a positive value equal to $144 - 90 = 54$ AU.

The RO value in year 1 for this branch is worked out as usual with Equation (42) as being

$$54q_u + 0q_d = 26.2 \text{ AU}. \tag{52}$$

The other branch gives zero. The positive value at end of year 1 is the value of the option at end of this branch, only if the RO is kept alive to year 2. What is, however, the price of the option at end of the same branch when it is exercised immediately, i.e., at end of year 1? In this case the option is worth the NPV at end of this branch minus the exercise price, i.e., $120 - 90 = 30$ AU > 26.2 AU. The option thus has more value dead than when alive, and it should be exercised at the end of year 1.

The value 30 AU is used to calculate in turn with the state prices the option value at top of the tree at present time, which results in 14.55 AU $> 100 - X = 10$, so that this time it is worth keeping the option open until at least year 1.

The procedure to verify possible early exercise of an RO is thus as follows: Work backward and check at each node of the binomial tree if the option is more worth if exercised at this time, or if kept alive to the next period. Keep whatever value is larger, record the early exercise possibility, and move again upward to the top of the tree in year 0.

The possibility of early exercise clearly shows why RO valuation is usually not feasible with Black–Scholes alone, because this formula imposes a fixed RO exercise time.

Another reason why the use of the binomial tree is necessary in more general cases is that any arbitrary evolution of the cash flows can be represented, and not only up and down states corresponding to the random-walk assumption in Equation (46).

5.4 Real Options and Multicriteria Analysis

In the existing literature the option value (OV) is simply added to the NPV: The resulting total value takes into account the flexibility given by the OV, but not available in plain DCF evaluations. Under those conditions the DCF is applied to the resulting total project value, including the OV.

The OV is added to the NPV for computing a complete profitability index, extending Equation (2):

$$\text{PIN}(\text{project } i) = \frac{[\text{NPV} + \text{OV}] \, (\text{project } i)}{|CF_0|} \tag{53}$$

This full PIN value may be used in a subsequent MCDA analysis. But, as the OV has been aggregated in the NPV, the additional information it represents gets lost. An easy remedy to this drawback is found by adding a new criterion to the analysis, representing the gained flexibility:

$$\text{FLEX}(\text{project } i) = \frac{\text{OV}(\text{project } i)}{|CF_0|} \tag{54}$$

Any of the MCDA procedures introduced in Section 3 can then be applied.

6 Conclusions

This chapter had the ambition of showing that capital budgeting is not just an old-fashioned cookbook of traditional recipes. Although I had to be very concise, I hope to have convinced my readers that, on the contrary capital budgeting benefits from a rich asset of modern techniques coming from operations research (OR).

I could only give a flavor of available OR techniques, without being, of course, exhaustive with respect to the many possibilities of combinations between them. I have limited my presentation to multicriteria decision analysis, fuzzy arithmetic, fuzzy-rule systems, and finally binomial trees used in real option evaluations, which are closely related to stochastic dynamic programming.

My main hope is to have awakened in my readers the desire to delve deeper into the existing literature about those different OR fields, for which the current research is still very active.

References

1. Bellman, R. E., and Zadeh, L. A. Decision-making in a fuzzy environment. *Management Science*, 17:141–164, 1970.
2. Beninga, S. *Financial Modeling*, 2nd edition. MIT Press, Cambridge, MA, 2000.
3. Besley, S., and Brigham, E. F. *Essentials of Managerial Finance*, 13th edition. Thomson Southwestern, Mason, OH, 2005.
4. Bouyssou, D., Marchant, T., Pirlot, M., Perny, P., Tsoukias, A., and Vincke, P., *Evaluation and Decision Models: A Critical Perspective*. Kluwer Academic, Dordrecht, 2000.
5. Brans, J. P., and Vincke, P. A preference ranking organisation method: The PROMETHEE Method for MCDM. *Management Science*, 31(6):647–656, 1985.
6. Brans, J. P., and Mareschal, B. The PROMCALC and GAIA decision support system. *Decision Support Systems*, 12(4):7–310, 1994.
7. Brans, J. P., and Mareschal, B. *Prométhée-Gaia: Une Méthodologie d'Aide à la Decision en Présence de Crières Multiples*. Éditions Ellipses, ULB, Brussels, 2002.
8. Brealey, R. A., and Myers, S. C. *Principles of Corporate Finance*, 4th edition. McGraw-Hill, Inc., New York, 1991.
9. Buckley, J. J., Esmali, E., Feuring, T. *Fuzzy Mathematics in Economics and Engineering*. Physica-Verlag, Heidelberg, 2002
10. Chevalier, A., Kunsch, P. L., and Brans, J. P. A contribution to the development of a strategic control and planning instrument in corporations. A divestiture case study. *Operational Research. An International Journal*, 3(1):25–40, 2003.
11. Chevalier, A., Kunsch, P. L., and Brans, J. P. A contribution to the development of a strategic control and planning instrument in corporations. An acquisition Case Study. *International Transactions on Operation Research*, 11:155–168, 2004.
12. Copeland, T., and Antikarov, V. *Real Options. A Practitioner's Guide*. Texere, New York, 2001.

13. Cox, D. E. *Fuzzy Logic for Business and Industry*. Charles River Media, Rockland, MA, 1995.
14. Dixit, A. K., and Pindyck, R. S. *Investment Under Uncertainty*. Princeton University Press, Princeton, NJ, 1994.
15. Dubois, D., and Prade, H. What are fuzzy rules and how to use them. *Fuzzy Sets and Systems*, 84:169–185, 1996.
16. Dyson, R. G., and Berry, R. H. The financial evaluation of strategic investments. In R. G. Dyson, and F. A. O'Brien, Editors, *Strategic Development: Methods and Models*. John Wiley & Sons, Chichester, 1998.
17. Fodor, J., and Roubens, M. *Fuzzy Preference Modelling and Multicriteria Decision Support*, Kluwer Academic Publishers, Dordrecht, 1994.
18. Fuzzy Logic Toolbox for Use with MATLAB: User's Guide, Version 2, The MATHWORKS Inc., Natick, MA, 2001.
19. Geldermann, J. Fuzzy outranking for environmental assessment. Case study: Iron and steel making industry. *Fuzzy Sets and Systems*, 115:45–65, 2000.
20. Goumas, M., and Lygerou, V. An extension of the PROMETHEE method for decision making in fuzzy environment: Ranking of alternative energy exploitation projects. *European Journal of Operational Research*, 123:606–613, 2000.
21. Hull, J. C. *Options, Futures, & Other Derivatives*, 4th edition. Prentice-Hall, Upper Saddle River, NJ, 2000.
22. Kacprzyk, J. *Multistage Fuzzy Control: A Model-Based Approach to Fuzzy Control and Decision-Making*. John Wiley and Sons, Chichester, 1997.
23. Keeney, R. L., and Raiffa, H. *Decisions with Multiple Objectives: Preferences and Value Tradeoffs*. John Wiley and Sons, New York, 1976.
24. Kunsch P. L., and Fortemps P. A fuzzy decision support system for the economic calculus in radioactive waste management. *Information Science*, 142:103–116, 2002.
25. Le Teno, J. F., and Mareschal, B. An interval version of PROMETHEE for the comparison of building products' design with ill-defined data on environmental quality. *European Journal of Operational Research*, 109:522–529, 1998.
26. Mavrotas, G., and Diakoulaki, D. A combined MCDA-MOMILP approach to assist in project selection under policy constraints and uncertainty in the criteria weights. *Proceedings of the MUDSM 2004 Conference*, Coimbra, September 22–24, 2004.
27. Meyer, M. A., and Booker, J. M. *Eliciting and Analysing Expert Judgment: A Practical Guide*. Society for Industrial and Applied Mathematics (SIAM), Philadelphia, 2001.
28. Mishan E. J. *Cost-Benefit Analysis*. Unwin Hyman, London, 1988.
29. Passino, K. M., and Yurkovich, S. *Fuzzy Control*. Addison-Wesley, Menlo Park, CA, 1998.
30. Roy B. *Méthodologie Multicritère d'Aide à la Décision*. Economica, Paris, 1985.
31. Roy, B., and Bouyssou, D. *Aide multicritère à la décision: Méthodes et Cas*. Economica, Paris, 1993.
32. Trigeorgis, L. Real options and interactions with financial flexibility. In R. G. Dyson, and F. A. O'Brien, Editors, *Strategic Development: Methods and Models*. John Wiley and Sons, Chichester, 1998, pages 299–331.
33. Vincke, P. *Multicriteria Decision Aid*, John Wiley and Sons, Chichester, 1992.

Financial Networks

Anna Nagurney

Dept. of Finance and Operations Management, Isenberg School of Management,
University of Massachusetts at Amherst, Amherst, MA 01003, USA
nagurney@gbfin.umass.edu

1 Introduction

Finance is a discipline concerned with the study of capital flows over space and time in the presence of risk. It has benefited from a plethora of mathematical and engineering tools that have been developed and utilized for the modeling, analysis, and computation of solutions in the present complex economic environment. Indeed, the financial landscape today is characterized by the existence of distinct sectors in economies, the proliferation of new financial instruments, with increasing diversification of portfolios internationally, various transaction costs, the increasing growth of electronic transactions, and a myriad of governmental policy interventions. Hence, rigorous methodological tools that can capture the complexity and richness of financial decision making today and that can take advantage of powerful computer resources have never been more important and needed for financial quantitative analyses.

This chapter focuses on financial networks as a powerful financial engineering tool and medium for the modeling, analysis, and solution of a spectrum of financial decision making problems ranging from portfolio optimization to multisector, multi-instrument general financial equilibrium problems, dynamic multiagent financial problems with intermediation, as well as the financial engineering of the integration of social networks with financial systems.

Note that throughout history, the emergence and evolution of various physical networks, ranging from transportation and logistical networks to telecommunication networks and the effects of human decision making on such networks, have given rise to the development of rich theories and scientific methodologies that are network-based (cf. Ford and Fulkerson, 1962; Ahuja et al., 1993; Nagurney, 1999; Guenes and Pardalos, 2003). The novelty of networks is that they are pervasive, providing the fabric of connectivity for our societies and economies, while, methodologically, network theory has developed into a powerful and dynamic medium for abstracting complex problems, which, at first glance, may not even appear to be networks, with associated nodes, links, and flows.

The topic of networks as a subject of scientific inquiry originated in the paper by Euler (1736), which is credited with being the earliest paper on *graph* theory. By a graph in this setting is meant, mathematically, a means of abstractly representing a system by its depiction in terms of vertices (or nodes) and edges (or arcs, equivalently, links) connecting various pairs of vertices. Euler was interested in determining whether it was possible to stroll around Königsberg (later called Kaliningrad) by crossing the seven bridges over the River Pregel exactly once. The problem was represented as a graph in which the vertices corresponded to land masses and the edges to bridges.

Quesnay (1758), in his *Tableau Economique*, conceptualized the circular flow of financial funds in an economy as a network; this work can be identified as the first paper on the topic of financial networks. Quesnay's basic idea has been utilized in the construction of financial flow of funds accounts, which are a statistical description of the flows of money and credit in an economy (see Cohen, 1987).

The concept of a network in economics, in turn, was implicit as early as the classical work of Cournot (1838), who not only seems to have first explicitly stated that a competitive price is determined by the intersection of supply and demand curves, but had done so in the context of two spatially separated markets in which the cost associated with transporting the goods was also included. Pigou (1920) studied a network system in the form of a transportation network consisting of two routes and noted that the decision-making behavior of the the users of such a system would lead to different flow patterns. Hence, the network of concern therein consists of the graph, which is directed, with the edges or links represented by arrows, as well as the resulting flows on the links.

Copeland (1952) recognized the conceptualization of the interrelationships among financial funds as a network and asked the question, "Does money flow like water or electricity?" Moreover, he provided a "wiring diagram for the main money circuit." Kirchhoff is credited with pioneering the field of electrical engineering by being the first to have systematically analyzed electrical circuits and with providing the foundations for the principal ideas of network flow theory. Interestingly, Enke in 1951 had proposed electronic circuits as a means of solving spatial price equilibrium problems, in which goods are produced, consumed, and traded, in the presence of transportation costs. Such analog computational devices were soon to be superseded by digital computers along with advances in computational methodologies, that is, algorithms, based on mathematical programming.

In this chapter, we further elaborate upon historical breakthroughs in the use of networks for the formulation, analysis, and solution of financial problems. Such a perspective allows one to trace the methodological developments as well as the applications of financial networks and provides a platform upon which further innovations can be made. One of the principal goals of this chapter is to highlight some of the major developments in financial engineering in the context of financial networks. Methodological tools that will be utilized to formulate and solve the financial network problems in this chapter are drawn from optimization, variational inequalities, as well as projected dynamical systems theory. We begin with a discussion of financial optimization problems within a network context and then turn to a range of financial network equilibrium problems.

2 Financial Optimization Problems

Network models have been proposed for a wide variety of financial problems characterized by a single objective function to be optimized as in portfolio optimization and asset allocation problems, currency translation, and risk management problems, among others. This literature is now briefly overviewed with the emphasis on the innovative work of Markowitz (1952, 1959) that established a new era in financial economics and became the basis for many financial optimization models that exist and are used to this day.

Note that although many financial optimization problems (including the work by Markowitz) had an underlying network structure, and the advantages of network programming were becoming increasingly evident (cf. Charnes and Cooper, 1958), not many financial network optimization models were developed until some time later. Some exceptions are several early models due to Charnes and Miller (1957) and Charnes and Cooper (1961). It was not until the last years of the 1960s and the first years of the 1970s that the network setting started to be extensively used for financial applications.

Among the first financial network optimization models that appear in the literature were a series of currency-translating models. Rutenberg (1970) suggested that the translation among different currencies could be performed through the use of arc multipliers. His network model was multiperiod with linear costs on the arcs (a characteristic common to the earlier financial networks models). The nodes of such generalized networks represented a particular currency in a specific period and the flow on the arcs the amount of cash moving from one period and/or currency to another. Christofides et al. (1979) and Shapiro and Rutenberg (1976), among others, introduced related financial network models. In most of these models, the currency prices were determined according to the amount of capital (network flow) that was moving from one currency (node) to the other.

Barr (1972) and Srinivasan (1974) used networks to formulate a series of cash management problems, with a major contribution being Crum's (1976) introduction of a generalized linear network model for the cash management of a multinational firm. The links in the network represented possible cash flow patterns and the multipliers incorporated costs, fees, liquidity changes, and exchange rates. A series of related cash management problems were modeled as network problems in subsequent years by Crum and Nye (1981) and Crum et al. (1983), and others. These papers further extended the applicability of network programming in financial applications. The focus was on linear network flow problems in which the cost on an arc was a linear function of the flow. Crum et al. (1979), in turn, demonstrated how contemporary financial capital allocation problems could be modeled as an integer-generalized network problem, in which the flows on particular arcs were forced to be integers.

It is important to note that in many financial network optimization problems the objective function must be nonlinear due to the modeling of the risk function and, hence, typically, such financial problems lie in the domain of nonlinear, rather than linear, network flow problems. Mulvey (1987) presented a collection of nonlinear financial network models that were based on previous cash flow and portfolio models in which the original authors (see, e.g., Rudd and Rosenberg, 1979, and Soenen, 1979) did not realize, and, thus, did not exploit the underlying network structure. Mulvey also recognized that the Markowitz (1952, 1959) mean-variance minimization problem was, in fact, a network optimization problem with a nonlinear objective function. The classical Markowitz models are now reviewed and cast into the framework of network optimization problems. See Figure 1 for the network structure of such problems. Additional financial network optimization models and associated references can be found in Nagurney and Siokos (1997) and in the volume edited by Nagurney (2003).

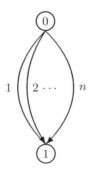

Fig. 1. Network structure of classical portfolio optimization.

Markowitz's model was based on mean-variance portfolio selection, where the average and the variability of portfolio returns were determined in terms of the mean and covariance of the corresponding investments. The mean is a measure of an average return and the variance is a measure of the distribution of the returns around the mean return. Markowitz formulated the portfolio optimization problem as associated with risk minimization with the objective function:

$$\text{minimize} \quad V = X^T Q X \qquad (1)$$

subject to constraints, representing, respectively, the attainment of a specific return, a budget constraint, and that no short sales were allowed, given by

$$R = \sum_{i=1}^{n} X_i r_i, \qquad (2)$$

$$\sum_{i=1}^{n} X_i = 1, \qquad (3)$$

$$X_i \geq 0, \quad i = 1, \ldots, n. \qquad (4)$$

Here n denotes the total number of securities available in the economy, X_i represents the relative amount of capital invested in security i, with the securities being grouped into the column vector X, Q denotes the $n \times n$ variance-covariance matrix on the return of the portfolio, r_i denotes the expected value of the return of security i, and R denotes the expected rate of return on the portfolio. Within a network context (cf. Figure 1), the links correspond to the securities, with their relative amounts $X_1, \ldots, X_n$ corresponding to the flows on the respective links: $1, \ldots, n$. The budget constraint and the nonnegativity assumption on the flows are the network conservation of flow equations. Since the objective function is that of risk minimization, it can be interpreted as the sum of the costs on the n links in the network. Observe that the network representation is abstract and does not correspond (as in the case of transportation and telecommunication) to physical locations and links.

Markowitz suggested that, for a fixed set of expected values r_i and covariances of the returns of all assets i and j, every investor can find an (R, V) combination that better fits his taste, solely limited by the constraints of the specific problem. Hence, according to the original work of Markowitz (1952), the efficient frontier had to be identified and then every investor had to select a portfolio through a mean-variance analysis that fitted his preferences.

A related mathematical optimization model (see Markowitz, 1959) to the one above, which can be interpreted as the investor seeking to maximize his returns while minimizing his risk can be expressed by the quadratic programming problem:

$$\text{maximize} \quad \alpha R - (1 - \alpha) V \qquad (5)$$

subject to

$$\sum_{i=1}^{n} X_i = 1, \qquad (6)$$

$$X_i \geq 0, \quad i = 1, \ldots, n, \qquad (7)$$

where α denotes an indicator of how risk-averse a specific investor is. This model is also a network optimization problem with the network as depicted in Figure 1, with Equations (6) and (7) again representing a conservation of flow equation.

A collection of versions and extensions of Markowitz's model can be found in Francis and Archer (1979), with $\alpha = 1/2$ being a frequently accepted value. A recent interpretation of the model as a multicriteria decision making model along with theoretical extensions to multiple sectors can be found in Dong and Nagurney (2001), where additional references are available. References to multicriteria decision making and financial applications can also be found in Doumpos et al. (2000).

A segment of the optimization literature on financial networks has focused on variables that are stochastic and have to be treated as random variables in the optimization procedure. Clearly, since most financial optimization problems are of large size, the incorporation of stochastic variables made the problems more complicated and difficult to model and compute. Mulvey (1987) and Mulvey and Vladimirou (1989, 1991), among others, studied stochastic financial networks, utilizing a series of different theories and techniques (e.g., purchase power priority, arbitrage theory, scenario aggregation) that were then utilized for the estimation of the stochastic elements in the network in order to be able to represent them as a series of deterministic equivalents. The large size and the computational complexity of stochastic networks, at times, limited their usage to specially structured problems where general computational techniques and algorithms could be applied. See Rudd and Rosenberg (1979), Wallace (1986), Rockafellar and Wets (1991), and Mulvey et al. (2003) for a more detailed discussion on aspects of realistic portfolio optimization and implementation issues related to stochastic financial networks.

3 General Financial Equilibrium Problems

We now turn to networks and their utilization for the modeling and analysis of financial systems in which there is more than a single decision maker, in contrast to the above financial optimization problems. It is worth noting that Quesnay (1758) actually considered a financial system as a network.

Thore (1969) introduced networks, along with the mathematics, for the study of systems of linked portfolios. His work benefited from that of Charnes and Cooper (1967), who demonstrated that systems of linked accounts could be represented as a network, where the nodes depict the balance sheets and the links depict the credit and debit entries. Thore considered credit networks, with the explicit goal of providing a tool for use in the study of the propagation of money and credit streams in an economy, based on a theory of the behavior of banks and other financial institutions. The credit network recognized that these sectors interact and its solution made use of linear programming. Thore (1970) extended the basic network model to handle holdings of financial reserves in the case of uncertainty. The approach utilized two-stage linear programs under uncertainty introduced by Ferguson and Dantzig (1956) and Dantzig and Madansky (1961). See Fei (1960) for a graph-theoretic approach to the credit system. More recently, Boginski et al.(2003) presented a detailed study of the stock market graph, yielding a new tool for the analysis of market structure through the classification of stocks into different groups, along with an application to the U.S. stock market.

Storoy et al. (1975), in turn, developed a network representation of the interconnection of capital markets and demonstrated how decomposition theory of mathematical programming could be exploited for the computation of equilibrium. The utility functions facing a sector were no longer restricted to being linear functions. Thore (1980) further investigated network models of linked portfolios, financial intermediation, and decentralization/decomposition theory. However, the computational techniques at that time were not sufficiently well-developed to handle such problems in practice.

Thore (1984) later proposed an international financial network for the Euro dollar market and viewed it as a logistical system, exploiting the ideas of Samuelson (1952) and Takayama and Judge (1971) for spatial price equilibrium problems. In this paper, as in Thore's preceding papers on financial networks, the micro-behavioral unit consisted of the individual bank, savings and loan, or other financial intermediary, and the portfolio choices were described in some optimizing framework, with the portfolios being linked together into a network with a separate portfolio visualized as a node and assets and liabilities as directed links.

Notably, the above-mentioned contributions focused on the use and application of networks for the study of financial systems consisting of multiple economic decision makers. In such systems, equilibrium was a central concept, along with the role of prices in the equilibrating mechanism. Rigorous approaches that characterized the formulation of equilibrium and the corresponding price determination were greatly influenced by the Arrow–Debreu economic model (cf. Arrow, 1951; Debreu, 1951). In addition, the importance of the inclusion of dynamics in the study of such systems was explicitly emphasized (see also Thore and Kydland, 1972).

The first use of finite-dimensional variational inequality theory for the computation of multisector, multi-instrument financial equilibria is due to Nagurney et al. (1992), who recognized the network structure underlying the subproblems encountered in their proposed decomposition scheme. Hughes and Nagurney (1992) and Nagurney and Hughes (1992) had, in turn, proposed the formulation and solution of estimation of financial flow of funds accounts as network optimization problems. Their proposed optimization scheme fully exploited the special network structure of these problems. Nagurney and Siokos (1997) then developed an international financial equilibrium model utilizing finite-dimensional variational inequality theory for the first time in that framework.

Finite-dimensional variational inequality theory is a powerful unifying methodology in that it contains, as special cases, such mathematical programming problems as nonlinear equations, optimization problems, and complementarity problems. To illustrate this methodology and its application in general financial equilibrium modeling and computation, we now present a multisector, multi-instrument model and an extension due to Nagurney et al. (1992) and Nagurney (1994), respectively. For additional references to variational inequalities in finance, along with additional theoretical foundations, see Nagurney and Siokos (1997) and Nagurney (2001, 2003).

3.1 A Multisector, Multi-Instrument Financial Equilibrium Model

Recall the classical mean-variance model presented in the preceding section, which is based on the pioneering work of Markowitz (1959). Now, however, assume that there are m sectors, each of which seeks to maximize his return and, at the same time, to minimize the risk of his portfolio, subject to the balance accounting and nonnegativity constraints. Examples of sectors include households, businesses, state and local governments, banks, etc. Denote a typical sector by j and assume that there are liabilities in addition to assets held by each sector. Denote the volume of instrument i that sector j holds as an asset by X_i^j, and group the (nonnegative) assets in the portfolio of sector j into the column vector $X^j \in R_+^n$. Further, group the assets of all sectors in the economy into the column vector $X \in R_+^{mn}$. Similarly, denote the volume of instrument i that sector j holds as a liability by Y_i^j, and group the (nonnegative) liabilities in the portfolio of sector j into the column vector $Y^j \in R_+^n$. Finally, group the liabilities of all sectors in the economy into the column vector $Y \in R_+^{mn}$. Let r_i denote the nonnegative price of instrument i, and group the prices of all the instruments into the column vector $r \in R_+^n$.

It is assumed that the total volume of each balance sheet side of each sector is exogenous. Recall that a *balance sheet* is a financial report that demonstrates the status of a company's assets, liabilities, and the owner's equity at a specific point of time. The left-hand side of a balance sheet contains the assets that a sector holds at a particular point of time, whereas the right-hand side accommodates the liabilities and owner's equity held by that sector at the same point of time. According to accounting principles, the sum of all assets is equal to the sum of all the liabilities and the owner's equity. Here, the term "liabilities" is used in its general form and, hence, also includes the owner's equity. Let S^j denote the financial volume held by sector j. Finally, assume that the sectors under consideration act in a perfectly competitive environment.

A Sector's Portfolio Optimization Problem

Recall that in the mean-variance approach for portfolio optimization, the minimization of a portfolio's risk is performed through the use of the variance-covariance matrix. Hence, the portfolio optimization problem for each sector j is the following:

$$\text{minimize} \quad \begin{pmatrix} X^j \\ Y^j \end{pmatrix}^T Q^j \begin{pmatrix} X^j \\ Y^j \end{pmatrix} - \sum_{i=1}^n r_i \left(X_i^j - Y_i^j \right) \qquad (8)$$

subject to

$$\sum_{i=1}^{n} X_i^j = S^j, \tag{9}$$

$$\sum_{i=1}^{n} Y_i^j = S^j, \tag{10}$$

$$X_i^j \geq 0, \ Y_i^j \geq 0, \quad i = 1, 2, \ldots, n, \tag{11}$$

where Q^j is a symmetric $2n \times 2n$ variance-covariance matrix associated with the assets and liabilities of sector j. Moreover, since Q^j is a variance-covariance matrix, one can assume that it is positive definite and, as a result, the objective function of each sector's portfolio optimization problem, given by the above, is strictly convex.

Partition the symmetric matrix Q^j as

$$Q^j = \begin{pmatrix} Q_{11}^j & Q_{12}^j \\ Q_{21}^j & Q_{22}^j \end{pmatrix},$$

where Q_{11}^j and Q_{22}^j are the variance-covariance matrices for only the assets and only the liabilities, respectively, of sector j. These submatrices are each of dimension $n \times n$. The submatrices Q_{12}^j and Q_{21}^j, in turn, are identical since Q^j is symmetric. They are also of dimension $n \times n$. These submatrices are, in fact, the symmetric variance-covariance matrices between the asset and the liabilities of sector j. Denote the ith column of matrix $Q_{(\alpha\beta)}^j$ by $Q_{(\alpha\beta)i}^j$, where α and β can take on the values of 1 and/or 2.

Optimality Conditions

The necessary and sufficient conditions for an optimal portfolio for sector j are that the vector of assets and liabilities, $(X^{j*}, Y^{j*}) \in K_j$, where K_j denotes the feasible set for sector j, given by (9) – (11), satisfies the following system of equalities and inequalities: For each instrument $i; i = 1, \ldots, n$, we must have

$$2(Q_{(11)i}^j)^T \cdot X^{j*} + 2(Q_{(21)i}^j)^T \cdot Y^{j*} - r_i^* - \mu_j^1 \geq 0,$$

$$2(Q_{(22)i}^j)^T \cdot Y^{j*} + 2(Q_{(12)i}^j)^T \cdot X^{j*} + r_i^* - \mu_j^2 \geq 0,$$

$$X_i^{j*} \left[2(Q_{(11)i}^j)^T \cdot X^{j*} + 2(Q_{(21)i}^j)^T \cdot Y^{j*} - r_i^* - \mu_j^1 \right] = 0,$$

$$Y_i^{j*} \left[2(Q_{(22)i}^j)^T \cdot Y^{j*} + 2(Q_{(12)i}^j)^T \cdot X^{j*} + r_i^* - \mu_2^j \right] = 0,$$

where μ_j^1 and μ_j^2 are the Lagrange multipliers associated with the accounting constraints, (9) and (10), respectively.

Let $\mathcal{K}$ denote the feasible set for all the asset and liability holdings of all the sectors and all the prices of the instruments, where $\mathcal{K} \equiv \{K \times R_+^n\}$ and $K \equiv \prod_{i=1}^{m} K_j$. The network structure of the sectors' optimization problems is depicted in Figure 2.

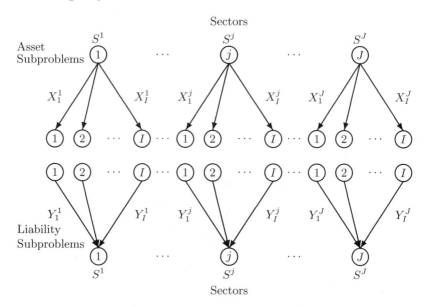

Fig. 2. Network structure of the sectors' optimization problems.

Economic System Conditions

The economic system conditions, which relate the supply and demand of each financial instrument and the instrument prices, are given by the following: For each instrument $i; i = 1, \ldots, n$, an equilibrium asset, liability, and price pattern, $(X^*, Y^*, r^*) \in \mathcal{K}$, must satisfy

$$\sum_{j=1}^{J}(X_i^{j*} - Y_i^{j*}) \begin{cases} = 0 & \text{if } r_i^* > 0, \\ \geq 0 & \text{if } r_i^* = 0. \end{cases} \qquad (12)$$

The definition of financial equilibrium is now presented along with the variational inequality formulation. For the derivation, see Nagurney et al. (1992) and Nagurney and Siokos (1997). Combining the above optimality conditions for each sector with the economic system conditions for each instrument, we have the following definition of equilibrium.

Definition 1. Multisector, Multi-Instrument Financial Equilibrium
A vector $(X^*, Y^*, r^*) \in \mathcal{K}$ is an equilibrium of the multisector, multi-instrument financial model if and only if it satisfies the optimality conditions and the economic system conditions (12), for all sectors $j; j = 1, \ldots, m$, and for all instruments $i; i = 1, \ldots, n$, simultaneously.

The variational inequality formulation of the equilibrium conditions, due to Nagurney et al. (1992) is given by the following.

Theorem 1. Variational Inequality Formulation for the Quadratic Model

A vector of assets and liabilities of the sectors, and instrument prices, $(X^*, Y^*, r^*) \in \mathcal{K}$, is a financial equilibrium if and only if it satisfies the variational inequality problem:

$$\sum_{j=1}^{m}\sum_{i=1}^{n} \left[2(Q_{(11)i}^{j})^{T} \cdot X^{j*} + 2(Q_{(21)i}^{j})^{T} \cdot Y^{j*} - r_{i}^{*}\right] \times \left[X_{i}^{j} - X_{i}^{j*}\right]$$

$$+ \sum_{j=1}^{m}\sum_{i=1}^{n} \left[2(Q_{(22)i}^{j})^{T} \cdot Y^{j*} + 2(Q_{(12)i}^{j})^{T} \cdot X^{j*} + r_{i}^{*}\right] \times \left[Y_{i}^{j} - Y_{i}^{j*}\right]$$

$$+ \sum_{i=1}^{n}\sum_{j=1}^{m} \left[X_{i}^{j*} - Y_{i}^{j*}\right] \times [r_{i} - r_{i}^{*}] \geq 0, \quad \forall (X, Y, r) \in \mathcal{K}. \tag{13}$$

For completeness, the *standard form* of the variational inequality is now presented. For additional background, see Nagurney (1999). Define the N-dimensional column vector $Z \equiv (X, Y, r) \in \mathcal{K}$, and the N-dimensional column vector $F(Z)$ such that

$$F(Z) \equiv D \begin{pmatrix} X \\ Y \\ r \end{pmatrix}, \quad \text{where} \quad D = \begin{pmatrix} 2Q & B \\ -B^{T} & 0 \end{pmatrix},$$

$$Q = \begin{pmatrix} Q_{11}^{1} & & & Q_{21}^{1} & & \\ & \ddots & & & \ddots & \\ & & Q_{11}^{J} & & & Q_{21}^{J} \\ Q_{12}^{1} & & & Q_{22}^{1} & & \\ & \ddots & & & \ddots & \\ & & Q_{12}^{J} & & & Q_{22}^{J} \end{pmatrix}_{2mn \times 2mn},$$

and

$$B^{T} = \begin{pmatrix} -\mathcal{I} \ldots -\mathcal{I} \ \mathcal{I} \ldots \mathcal{I} \end{pmatrix}_{n \times mn},$$

and $\mathcal{I}$ is the $n \times n$-dimensional identity matrix.

It is clear that variational inequality problem (13) can be put into standard variational inequality form: Determine $Z^* \in \mathcal{K}$, satisfying

$$\langle F(Z^*)^{T}, Z - Z^* \rangle \geq 0, \quad \forall Z \in \mathcal{K}. \tag{14}$$

3.2 Model with Utility Functions

The above model is a special case of the financial equilibrium model due to Nagurney (1994) in which each sector j seeks to maximize his utility function,

$$U^{j}(X^{j,j}, r) = u^{j}(X^{j}, Y^{j}) + r^{T} \cdot (X^{j} - Y^{j}),$$

which, in turn, is a special case of the model with a sector j's utility function given by the general form $U^j(X^j, Y^j, r)$. Interestingly, it has been shown by Nagurney and Siokos (1997) that, in the case of utility functions of the form $U^j(X^j, Y^j, r) = u^j(X^j, Y^j) + r^T \cdot (X^j - Y^j)$, of which the above-described quadratic model is an example, one can obtain the solution to the above variational inequality problem by solving the optimization problem:

$$\text{maximize} \sum_{j=1}^{J} u^j(X^j, Y^j) \tag{15}$$

subject to

$$\sum_{j=1}^{J}(X_i^j - Y_i^j) = 0, \quad i = 1, \ldots, n, \tag{16}$$

$$(X^j, Y^j) \in K_j, \quad j = 1, \ldots, m, \tag{17}$$

with Lagrange multiplier r_i^* associated with the ith "market clearing" constraint (16). Moreover, this optimization problem is actually a network optimization problem as revealed in Nagurney and Siokos (1997). The structure of the financial system in equilibrium is as depicted in Figure 3.

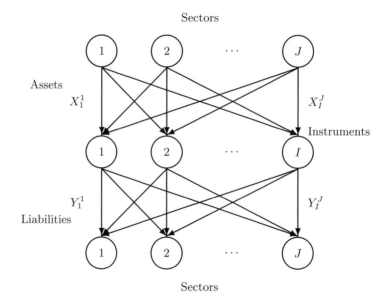

Fig. 3. The network structure at equilibrium.

3.3 Computation of Financial Equilibria

In this section, an algorithm for the computation of solutions to the above financial equilibrium problems is recalled. The algorithm is the modified projection method of Korpelevich (1977). The advantage of this computational method in the context of the general financial equilibrium problems is that the original problem can be decomposed into a series of smaller and simpler subproblems of network structure, each of which can then be solved explicitly and in closed form. The realization of the modified projection method for the solution of the financial equilibrium problems with general utility functions is then presented.

The modified projection method can be expressed as follows:

Step 0: Initialization

Select $Z^0 \in \mathcal{K}$. Let $\tau := 0$ and let γ be a scalar such that $0 < \gamma \leq 1/L$, where L is the Lipschitz constant (see Nagurney and Siokos, 1997).

Step 1: Computation

Compute $\bar{Z}^\tau$ by solving the variational inequality subproblem:

$$\langle (\bar{Z}^\tau + \gamma F(Z^\tau)^T - Z^\tau)^T, Z - \bar{Z}^\tau \rangle \geq 0, \quad \forall Z \in \mathcal{K}. \tag{18}$$

Step 2: Adaptation

Compute $Z^{\tau+1}$ by solving the variational inequality subproblem:

$$\langle (Z^{\tau+1} + \gamma F(\bar{Z}^\tau)^T - Z^\tau)^T, Z - Z^{\tau+1} \rangle \geq 0, \quad \forall Z \in \mathcal{K}. \tag{19}$$

Step 3: Convergence Verification

If $\max |Z_b^{\tau+1} - Z_b^\tau| \leq \epsilon$, for all b, with $\epsilon > 0$, a prespecified tolerance, then stop; else, set $\tau := \tau + 1$, and go to step 1.

For completeness, we now present the modified projection algorithm in which the function $F(Z)$ is in expanded form for the specific model.

The Modified Projection Method

Step 0: Initialization

Set $(X^0, Y^0, r^0) \in \mathcal{K}$. Let $\tau := 0$ and set γ so that $0 < \gamma \leq 1/L$.

Step 1: Computation

Compute $(\bar{X}^\tau, \bar{Y}^\tau, \bar{r}^\tau) \in \mathcal{K}$ by solving the variational inequality subproblem:

$$\sum_{j=1}^{J}\sum_{i=1}^{I}\left[\bar{X}_i^{j^\tau}+\gamma\left(-\frac{\partial U^j(X^{j^\tau},Y^{j^\tau},r^\tau)}{\partial X_i^j}\right)-X_i^{j^\tau}\right]\times\left[X_i^j-\bar{X}_i^{j^\tau}\right]$$

$$+\sum_{j=1}^{J}\sum_{i=1}^{I}\left[\bar{Y}_i^{j^\tau}+\gamma\left(-\frac{\partial U^j(X^{j^\tau},Y^{j^\tau},r^\tau)}{\partial Y_i^j}\right)-Y_i^{j^\tau}\right]\times\left[Y_i^j-\bar{Y}_i^{j^\tau}\right]$$

$$+\sum_{i=1}^{I}\left[\bar{r}_i^\tau+\gamma\left[\sum_{j=1}^{J}\left(X_i^{j^\tau}-Y_i^{j^\tau}\right)\right]-r_i^\tau\right]\times[r_i-\bar{r}_i^\tau],\quad\forall(X,Y,r)\in\mathcal{K}.$$

Step 2: Adaptation

Compute $(X^{\tau+1},Y^{\tau+1},r^{\tau+1})\in\mathcal{K}$ by solving the variational inequality subproblem:

$$\sum_{j=1}^{J}\sum_{i=1}^{I}\left[X_i^{j^{\tau+1}}+\gamma\left(-\frac{\partial U^j(\bar{X}^{j^\tau},\bar{Y}^{j^\tau},\bar{r}^\tau)}{\partial X_i^j}\right)-X_i^{j^\tau}\right]\times\left[X_i^j-X_i^{j^{\tau+1}}\right]$$

$$+\sum_{j=1}^{J}\sum_{i=1}^{I}\left[Y_i^{j^{\tau+1}}+\gamma\left(-\frac{\partial U^j(\bar{X}^{j^\tau},\bar{Y}^{j^\tau},\bar{r}^\tau)}{\partial Y_i^j}\right)-Y_i^{j^\tau}\right]\times\left[Y_i^j-Y_i^{j^{\tau+1}}\right]$$

$$+\sum_{i=1}^{I}\left[r_i^{\tau+1}+\gamma\left[\sum_{j=1}^{J}\left(\bar{X}_i^{j^\tau}-\bar{Y}_i^{j^\tau}\right)\right]-r_i^\tau\right]\times[r_i-r_i^{\tau+1}],\quad\forall(X,Y,r)\in\mathcal{K}.$$

Step 3: Convergence Verification:

If $\max_{i,j}|X_i^{j^{\tau+1}}-X_i^{j^\tau}|\leq\epsilon$; $\max_{i,j}|Y_i^{j^{\tau+1}}-Y_i^{j^\tau}|\leq\epsilon$; $\max_i|r_i^{\tau+1}-r_i^\tau|\leq\epsilon$, for all $i; i=1,\ldots,I$, and $j; j=1,\ldots,J$, with $\epsilon>0$, a prespecified tolerance, then stop; else, set $\tau:=\tau+1$, and go to step 1.

Convergence results are given in Nagurney et al. (1992); see also Nagurney and Siokos (1997).

An interpretation of the modified projection method as an adjustment process is now provided. The interpretation of the algorithm as an adjustment process was given by Nagurney (1999). In particular, at an iteration, the sectors in the economy receive all the price information on every instrument from the previous iteration. They then allocate their capital according to their preferences. The market reacts on the decisions of the sectors and derives new instrument prices. The sectors then improve upon their positions through the adaptation step, whereas the market also adjusts during the adaptation step. This process continues until no one can improve upon his position, and the equilibrium is reached, that is, the above variational inequality is satisfied with the computed asset, liability, and price pattern.

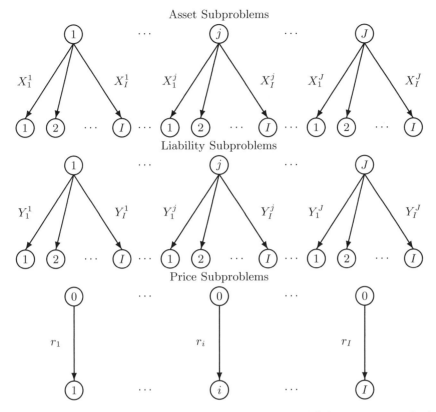

Fig. 4. Financial network subproblems induced by the modified projection method.

The financial optimization problems in the computation step and in the adaptation step are equivalent to separable quadratic programming problems, of special network structure, as depicted in Figure 4. Each of these network subproblems structure can then be solved, at an iteration, simultaneously, and exactly in closed form. The exact equilibration algorithm (see, e.g., Nagurney and Siokos, 1997) can be applied for the solution of the asset and liability subproblems, whereas the prices can be obtained using explicit formulas.

A numerical example is now presented for illustrative purposes and solved using the modified projection method, embedded with the exact equilibration algorithm.

Example 1: A Numerical Example

Assume that there are two sectors in the economy and three financial instruments. Assume that the "size" of each sector is given by $S^1 = 1$ and $S^2 = 2$. The variance-covariance matrices of the two sectors are

$$Q^1 = \begin{pmatrix} 1 & .25 & .3 & 0 & 0 & 0 \\ .25 & 1 & .1 & 0 & 0 & 0 \\ .3 & 1 & 1 & 0 & 0 & 0 \\ 0 & 0 & 0 & 1 & .2 & .3 \\ 0 & 0 & 0 & .2 & 1 & .5 \\ 0 & 0 & 0 & .3 & .5 & 1 \end{pmatrix}$$

and

$$Q^2 = \begin{pmatrix} 1 & 0 & .3 & 0 & 0 & 0 \\ 0 & 1 & .2 & 0 & 0 & 0 \\ .3 & .2 & 1 & 0 & 0 & 0 \\ 0 & 0 & 0 & 1 & .5 & 0 \\ 0 & 0 & 0 & .5 & 1 & .2 \\ 0 & 0 & 0 & 0 & .2 & 1 \end{pmatrix}.$$

The modified projection method was coded in FORTRAN. The variables were initialized as follows: $r_i^0 = 1$, for all i, with the financial volume S^j equally distributed among all the assets and among all the liabilities for each sector j. The γ parameter was set to 0.35. The convergence tolerance ε was set to 10^{-3}.

The modified projection method converged in 16 iterations and yielded the following equilibrium pattern:

Equilibrium Prices:

$$r_1^* = 0.34039, \quad r_2^* = 0.23805, \quad r_3^* = 0.42156,$$

Equilibrium Asset Holdings:

$$X_1^{1*} = 0.27899, \quad X_2^{1*} = 0.31803, \quad X_3^{1*} = 0.40298,$$

$$X_1^{2*} = 0.79662, \quad X_2^{2*} = 0.60904, \quad X_3^{2*} = 0.59434,$$

Equilibrium Liability Holdings:

$$Y_1^{1*} = 0.37081, \quad Y_2^{1*} = 0.43993, \quad Y_3^{1*} = 0.18927,$$

$$Y_1^{2*} = 0.70579, \quad Y_2^{2*} = 0.48693, \quad Y_3^{2*} = 0.80729.$$

The above results show that the algorithm yielded optimal portfolios that were feasible. Moreover, the market cleared for each instrument, since the price of each instrument was positive.

Other financial equilibrium models, including models with transaction costs, with hedging instruments such as futures and options, as well as international financial equilibrium models, can be found in Nagurney and Siokos (1997) and the references therein.

Moreover, with projected dynamical systems theory (see the book by Nagurney and Zhang, 1996), one can trace the dynamic behavior prior to an equilibrium state (formulated as a variational inequality). In contrast to classical dynamical systems, projected dynamical systems are characterized by a discontinuous right-hand side, with the discontinuity arising due to the constraint set underlying the application in question. Hence, this methodology allows one to model systems dynamically that are subject to limited resources, with a principal constraint in finance being budgetary restrictions.

Dong et al. (1996) were the first to apply the methodology of projected dynamical systems to develop a dynamic multisector, multi-instrument financial model, whose set of stationary points coincided with the set of solutions to the variational inequality model developed in Nagurney (1994), and then to study it qualitatively, providing stability analysis results. In the next section, the methodology of projected dynamical systems is illustrated in the context of a dynamic financial network model with intermediation (cf. Nagurney and Dong, 2002).

4 Dynamic Financial Networks with Intermediation

In this section, dynamic financial networks with intermediation are explored. As noted earlier, the conceptualization of financial systems as networks dates back to Quesnay (1758), who depicted the circular flow of funds in an economy as a network. His basic idea was subsequently applied to the construction of flow of funds accounts, which are a statistical description of the flows of money and credit in an economy (cf. Board of Governors, 1980; Cohen, 1987; Nagurney and Hughes, 1992). However, since the flow of funds accounts are in matrix form and, hence, two-dimensional, they fail to capture the dynamic behavior on a micro level of the various financial agents/sectors in an economy such as banks, households, insurance companies, etc. Moreover, as noted by the Board of Governors (1980) on page 6 of that publication, "the generality of the matrix tends to obscure certain structural aspects of the financial system that are of continuing interest in analysis," with the structural concepts of concern including financial intermediation.

Thore (1980) recognized some of the shortcomings of financial flow of funds accounts and instead developed network models of linked portfolios with financial intermediation, using decentralization/decomposition theory. Note that intermediation is typically associated with financial businesses, including banks, savings institutions, investment and insurance companies, etc., and the term implies borrowing for the purpose of lending, rather than for nonfinancial purposes. Thore also constructed some basic intertemporal models. However, the intertemporal models were not fully developed and the computational techniques at that time were not sufficiently advanced for computational purposes.

In this section, we address the dynamics of the financial economy, which explicitly includes financial intermediaries along with the "sources" and "uses" of financial funds. Tools are provided for studying the disequilibrium dynamics as well as the equilibrium state. Also, transaction costs are considered, since they bring a greater degree of realism to the study of financial intermediation. Transaction costs had been studied earlier in multisector, multi-instrument financial equilibrium models by Nagurney and Dong (1996a, 1996b) but without considering the more general dynamic intermediation setting.

The dynamic financial network model is now described. The model consists of agents with sources of funds, agents who are intermediaries, as well as agents who are consumers located at the demand markets. Specifically, consider m agents with sources of financial funds, such as households and businesses, involved in the allocation of their financial resources among a portfolio of financial instruments that can be obtained by transacting with distinct n financial intermediaries, such as banks, insurance and investment companies, etc. The financial intermediaries, in turn, in addition to transacting with the source agents, also determine how to allocate the incoming financial resources among distinct uses, as represented by o demand markets with a demand market corresponding to, for example, the market for real estate loans, household loans, or business loans, etc. The financial network with intermediation is now described and depicted graphically in Figure 5.

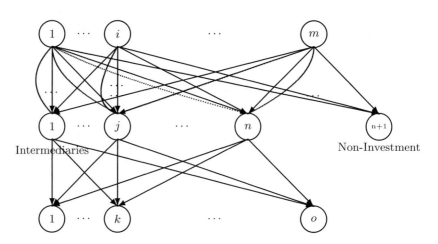

Fig. 5. The network structure of the financial economy with intermediation and with non-investment allowed.

The top tier of nodes in Figure 5 consists of the agents with sources of funds, with a typical source agent denoted by i and associated with node i. The middle tier of nodes in Figure 5 consists of the intermediaries, with a typical intermediary denoted by j and associated with node j in the network.

The bottom tier of nodes consists of the demand markets, with a typical demand market denoted by k and corresponding to the node k.

For simplicity of notation, assume that there are L financial instruments associated with each intermediary. Hence, from each source of funds node, there are L links connecting such a node with an intermediary node with the lth such link corresponding to the lth financial instrument available from the intermediary. In addition, the option of non-investment in the available financial instruments is allowed and to denote this option, construct an additional link from each source node to the middle tier node $n + 1$, which represents non-investment. Note that there are as many links connecting each top tier node with each intermediary node as needed to reflect the number of financial instruments available. Also, note that there is an additional abstract node $n+1$ with a link connecting each source node to it, which, as shall shortly be shown, will be used to "collect" the financial funds that are not invested. In the model, it is assumed that each source agent has a fixed amount of financial funds.

From each intermediary node, construct o links, one to each "use" node or demand market in the bottom tier of nodes in the network to denote the transaction between the intermediary and the consumers at the demand market.

Let x_{ijl} denote the nonnegative amount of the funds that source i "invests" in financial instrument l obtained from intermediary j. Group the financial flows associated with source agent i, which are associated with the links emanating from the top tier node i to the intermediary nodes in the logistical network, into the column vector $x_i \in R_+^{nL}$. Assume that each source has, at his disposal, an amount of funds S_i and denote the unallocated portion of this amount (and flowing on the link joining node i with node $n+1$) by s_i. Group then the x_i of all the source agents into the column vector $x \in R_+^{mnL}$.

Associate a distinct financial product k with each demand market, bottom-tiered node k, and let y_{jk} denote the amount of the financial product obtained by consumers at demand market k from intermediary j. Group these "consumption" quantities into the column vector $y \in R_+^{no}$. The intermediaries convert the incoming financial flows x into the outgoing financial flows y.

The notation for the prices is now given. Note that prices will be associated with each of the tiers of nodes in the network. Let ρ_{1ijl} denote the price associated with instrument l as quoted by intermediary j to source agent i, and group the first tier prices into the column vector $\rho_1 \in R_+^{mnL}$. Also, let ρ_{2j} denote the price charged by intermediary j and group all such prices into the column vector $\rho_2 \in R_+^n$. Finally, let ρ_{3k} denote the price of the financial product at the third or bottom-tiered node k in the network, and group all such prices into the column vector $\rho_3 \in R_+^o$.

We now turn to describing the dynamics by which the source agents adjust the amounts they allocate to the various financial instruments over time, the dynamics by which the intermediaries adjust their transactions, and those by which the consumers obtain the financial products at the demand markets.

In addition, the dynamics by which the prices adjust over time are described. The dynamics are derived from the bottom tier of nodes of the network on up since it is assumed that it is the demand for the financial products (and the corresponding prices) that actually drives the economic dynamics. The price dynamics are presented first and then the dynamics underlying the financial flows.

The Demand Market Price Dynamics

We begin by describing the dynamics underlying the prices of the financial products associated with the demand markets (see the bottom-tiered nodes). Assume, as given, a demand function d_k, which can depend, in general, upon the entire vector of prices ρ_3, that is,

$$d_k = d_k(\rho_3), \quad \forall k. \tag{20}$$

Moreover, assume that the rate of change of the price ρ_{3k}, denoted by $\dot{\rho}_{3k}$, is equal to the difference between the demand at the demand market k, as a function of the demand market prices, and the amount available from the intermediaries at the demand market. Hence, if the demand for the product at the demand market (at an instant in time) exceeds the amount available, the price of the financial product at that demand market will increase; if the amount available exceeds the demand at the price, then the price at the demand market will decrease. Furthermore, it is guaranteed that the prices do not become negative. Thus, the dynamics of the price ρ_{3k} associated with the product at demand market k can be expressed as

$$\dot{\rho}_{3k} = \begin{cases} d_k(\rho_3) - \sum_{j=1}^{n} y_{jk} & \text{if } \rho_{3k} > 0, \\ \max\{0, d_k(\rho_3) - \sum_{j=1}^{n} y_{jk}\} & \text{if } \rho_{3k} = 0. \end{cases} \tag{21}$$

The Dynamics of the Prices at the Intermediaries

The prices charged for the financial funds at the intermediaries, in turn, must reflect supply and demand conditions as well (and, as shall be shown shortly, also reflect profit-maximizing behavior on the part of the intermediaries who seek to determine how much of the financial flows they obtain from the different sources of funds). In particular, assume that the price associated with intermediary j, ρ_{2j}, and computed at node j lying in the second tier of nodes, evolves over time according to

$$\dot{\rho}_{2j} = \begin{cases} \sum_{k=1}^{o} y_{jk} - \sum_{i=1}^{m} \sum_{l=1}^{L} x_{ijl} & \text{if } \rho_{2j} > 0, \\ \max\{0, \sum_{k=1}^{o} y_{jk} - \sum_{i=1}^{m} \sum_{l=1}^{L} x_{ijl}\} & \text{if } \rho_{2j} = 0, \end{cases} \tag{22}$$

where $\dot{\rho}_{2j}$ denotes the rate of change of the jth intermediary's price. Hence, if the amount of the financial funds desired to be transacted by the consumers (at an instant in time) exceeds that available at the intermediary, then the price charged at the intermediary will increase; if the amount available is greater than that desired by the consumers, then the price charged at the intermediary will decrease. As in the case of the demand market prices, it is guaranteed that the prices charged by the intermediaries remain nonnegative.

Precursors to the Dynamics of the Financial Flows

First some preliminaries are needed that will allow the development of the dynamics of the financial flows. In particular, the utility-maximizing behavior of the source agents and that of the intermediaries are now discussed.

Assume that each such source agent's and each intermediary agent's utility can be defined as a function of the expected future portfolio value, where the expected value of the future portfolio is described by two characteristics: the expected mean value and the uncertainty surrounding the expected mean. Here, the expected mean portfolio value is assumed to be equal to the market value of the current portfolio. Each agent's uncertainty, or assessment of risk, in turn, is based on a variance-covariance matrix denoting the agent's assessment of the standard deviation of the prices for each instrument/product. The variance-covariance matrix associated with source agent i's assets is denoted by Q^i, is of dimension $nL \times nL$, and is associated with vector x_i, whereas intermediary agent j's variance-covariance matrix is denoted by Q^j, is of dimension $o \times o$, and is associated with the vector y_j.

Optimizing Behavior of the Source Agents

Denote the total transaction cost associated with source agent i transacting with intermediary j to obtain financial instrument l by c_{ijl} and assume that

$$c_{ijl} = c_{ijl}(x_{ijl}), \quad \forall i,j,l. \tag{23}$$

The total transaction costs incurred by source agent i, are thus equal to the sum of all the agent's transaction costs. His revenue, in turn, is equal to the sum of the price (rate of return) that the agent can obtain for the financial instrument times the total quantity obtained/purchased of that instrument. Recall that ρ_{1ijl} denotes the price associated with agent i/intermediary j/instrument l.

Assume that each such source agent seeks to maximize the net return while, simultaneously, minimizing the risk, with source agent i's utility function denoted by U^i. Moreover, assume that the variance-covariance matrix Q^i is positive semidefinite and that the transaction cost functions are continuously differentiable and convex. Hence, one can express the optimization problem facing source agent i as

$$\text{maximize} \quad U_i(x_i) = \sum_{j=1}^{n}\sum_{l=1}^{L} \rho_{1ijl} x_{ijl} - \sum_{j=1}^{n}\sum_{l=1}^{L} c_{ijl}(x_{ijl}) - x_i^T Q^i x_i, \tag{24}$$

subject to $x_{ijl} \geq 0$, for all j, l, and to the constraint

$$\sum_{j=1}^{n}\sum_{l=1}^{L} x_{ijl} \leq S^i, \qquad (25)$$

that is, the allocations of source agent i's funds among the financial instruments made available by the different intermediaries cannot exceed his holdings. Note that the utility function above is concave for each source agent i. A source agent may choose not to invest in any of the instruments. Indeed, as shall be illustrated through subsequent numerical examples, this constraint has important financial implications.

Clearly, in the case of *unconstrained* utility maximization, the gradient of source agent i's utility function with respect to the vector of variables x_i and denoted by $\nabla_{x_i} U_i$, where $\nabla_{x_i} U_i = (\partial U_i/\partial x_{i11}, \ldots, \partial U_i/\partial x_{inL})$, represents agent i's idealized direction, with the jl-component of $\nabla_{x_i} U_i$ given by

$$\rho_{1ijl} - 2Q^i_{z_{jl}} \cdot x_i - \frac{\partial c_{ijl}(x_{ijl})}{\partial x_{ijl}}, \qquad (26)$$

where $Q^i_{z_{jl}}$ denotes the z_{jl}th row of Q^i, and z_{jl} is the indicator defined as $z_{jl} = (l-1)n + j$. We return later to describe how the constraints are explicitly incorporated into the dynamics.

Optimizing Behavior of the Intermediaries

The intermediaries, in turn, are involved in transactions both with the source agents as well as with the users of the funds, that is, with the ultimate consumers associated with the markets for the distinct types of loans/products at the bottom tier of the financial network. Thus, an intermediary conducts transactions both with the "source" agents as well as with the consumers at the demand markets.

An intermediary j is faced with what is termed a *handling/conversion* cost, which may include, for example, the cost of converting the incoming financial flows into the financial loans/products associated with the demand markets. Denote this cost by c_j and, in the simplest case, one would have that c_j is a function of $\sum_{i=1}^{m}\sum_{l=1}^{L} x_{ijl}$, that is, the holding/conversion cost of an intermediary is a function of how much he has obtained from the various source agents. For the sake of generality, however, allow the function to, in general, depend also on the amounts held by other intermediaries and, therefore, one may write

$$c_j = c_j(x), \quad \forall j. \qquad (27)$$

The intermediaries also have associated transaction costs in regard to transacting with the source agents, which are assumed to be dependent on the type of instrument. Denote the transaction cost associated with intermediary j transacting with source agent i associated with instrument l by $\hat{c}_{ijl}$ and assume that it is of the form

$$\hat{c}_{ijl} = \hat{c}_{ijl}(x_{ijl}), \quad \forall i, j, l. \tag{28}$$

Recall that the intermediaries convert the incoming financial flows x into the outgoing financial flows y. Assume that an intermediary j incurs a transaction cost c_{jk} associated with transacting with demand market k, where

$$c_{jk} = c_{jk}(y_{jk}), \quad \forall j, k. \tag{29}$$

The intermediaries associate a price with the financial funds, which is denoted by ρ_{2j}, for intermediary j. Assuming that the intermediaries are also utility maximizers with the utility functions for each being comprised of net revenue maximization as well as risk minimization, then the utility maximization problem for intermediary agent j with his utility function denoted by U^j, can be expressed as

$$\begin{aligned}\text{maximize} \quad U_j(x_j, y_j) = & \sum_{i=1}^{m}\sum_{l=1}^{L} \rho_{2j} x_{ijl} - c_j(x) - \sum_{i=1}^{m}\sum_{l=1}^{L} \hat{c}_{ijl}(x_{ijl}) \\ & - \sum_{k=1}^{o} c_{jk}(y_{jk}) - \sum_{i=1}^{m}\sum_{l=1}^{L} \rho_{1ijl} x_{ijl} - y_j^T Q^j y_j,\end{aligned} \tag{30}$$

subject to the nonnegativity constraints: $x_{ijl} \geq 0$, and $y_{jk} \geq 0$, for all i, l, and k. Here, for convenience, we have $x_j = (x_{1j1}, \ldots, x_{mjL})$. The above bijective function expresses that the difference between the revenues minus the handling cost and the transaction costs and the payout to the source agents should be maximized, whereas the risk should be minimized. Assume now that the variance-covariance matrix Q^j is positive semidefinite and that the transaction cost functions are continuously differentiable and convex. Hence, the utility function above is concave for each intermediary j.

The gradient $\nabla_{x_j} U_j = (\partial U_j / \partial x_{1j1}, \ldots, \partial U_j / \partial x_{mjL})$ represents agent j's idealized direction in terms of x_j, ignoring the constraints, for the time being, whereas the gradient $\nabla_{y_j} U_j = (\partial U_j / \partial y_{j1}, \ldots, \partial U_j / \partial y_{jo})$ represents his idealized direction in terms of y_j. Note that the ilth component of $\nabla_{x_j} U_j$ is given by

$$\rho_{2j} - \rho_{1ijl} - \frac{\partial c_j(x)}{\partial x_{ijl}} - \frac{\partial \hat{c}_{ijl}(x_{ijl})}{\partial x_{ijl}}, \tag{31}$$

whereas the jkth component of $\nabla_{y_j} U_j$ is given by

$$-\frac{\partial c_{jk}(y_{jk})}{\partial y_{jk}} - 2Q_k^j \cdot y_j. \tag{32}$$

However, since both source agent i and intermediary j must agree in terms of the x_{ijl}, the direction (26) must coincide with that in (31), so adding both gives us a "combined force," which, after algebraic simplification, yields

$$\rho_{2j} - 2Q^i_{z_{jl}} \cdot x_i - \frac{\partial c_{ijl}(x_{ijl})}{\partial x_{ijl}} - \frac{\partial c_j(x)}{\partial x_{ijl}} - \frac{\partial \hat{c}_{ijl}(x_{ijl})}{\partial x_{ijl}}. \qquad (33)$$

The Dynamics of the Financial Flows Between the Source Agents and the Intermediaries

We are now ready to express the dynamics of the financial flows between the source agents and the intermediaries. In particular, define the feasible set $K_i \equiv \{x_i | x_{ijl} \geq 0, \forall i, j, l, \text{ and } (25) \text{ holds}\}$. Also let K be the Cartesian product given by $K \equiv \Pi_{i=1}^m K_i$ and define F^1_{ijl} as minus the term in (33) with $F^1_i = (F^1_{i11}, \ldots, F^1_{inL})$. Then the *best realizable* direction for the vector of financial instruments x_i can be expressed mathematically as

$$\dot{x}_i = \Pi_{K_i}(x_i, -F^1_i), \qquad (34)$$

where $\Pi_K(Z, v)$ is defined as

$$\Pi_K(Z, v) = \lim_{\delta \to 0} \frac{P_K(Z + \delta v) - Z}{\delta}, \qquad (35)$$

and P_K is the norm projection defined by

$$P_K(Z) = \mathrm{argmin}_{Z' \in K} \|Z' - Z\|. \qquad (36)$$

The Dynamics of the Financial Flows Between the Intermediaries and the Demand Markets

In terms of the financial flows between the intermediaries and the demand markets, both the intermediaries and the consumers must be in agreement as to the financial flows y. The consumers take into account in making their consumption decisions not only the price charged for the financial product by the intermediaries but also their transaction costs associated with obtaining the product.

Let $\hat{c}_{jk}$ denote the transaction cost associated with obtaining the product at demand market k from intermediary j. Assume that this unit transaction cost is continuous and of the general form

$$\hat{c}_{jk} = \hat{c}_{jk}(y), \quad \forall j, k. \qquad (37)$$

The consumers take the price charged by the intermediaries, which was denoted by ρ_{2j} for intermediary j, plus the unit transaction cost, in making their consumption decisions. From the perspective of the consumers at the demand markets, one can expect that an idealized direction in terms of the evolution of the financial flow of a product between an intermediary/demand market pair would be

$$(\rho_{3k} - \hat{c}_{jk}(y) - \rho_{2j}). \qquad (38)$$

On the other hand, as already derived above, one can expect that the intermediaries would adjust the volume of the product to a demand market according to (32). Now combining (32) and (38), and guaranteeing that the financial products do not assume negative quantities, yields the following dynamics:

$$\dot{y}_{jk} = \begin{cases} \rho_{3k} - \hat{c}_{jk}(y) - \rho_{2j} - \frac{\partial c_{jk}(y_{jk})}{\partial y_{jk}} - 2Q_k^j \cdot y_j & \text{if } y_{jk} > 0, \\ \max\{0, \rho_{3k} - \hat{c}_{jk}(y) - \rho_{2j} - \frac{\partial c_{jk}(y_{jk})}{\partial y_{jk}} - 2Q_k^j \cdot y_j\} & \text{if } y_{jk} = 0. \end{cases} \quad (39)$$

The Projected Dynamical System

Consider now the dynamic model in which the demand prices evolve according to (21) for all demand markets k, the prices at the intermediaries evolve according to (22) for all intermediaries j, the financial flows between the source agents and the intermediaries evolve according to (34) for all source agents i, and the financial products between the intermediaries and the demand markets evolve according to (39) for all intermediary/demand market pairs j, k.

Now let Z denote the aggregate column vector (x, y, ρ_2, ρ_3) in the feasible set $\mathcal{K} \equiv K \times R_+^{no+n+o}$. Define the column vector $F(Z) \equiv (F^1, F^2, F^3, F^4)$, where F^1 is as has been defined previously; $F^2 = (F_{11}^2, \ldots, F_{no}^2)$, with component $F_{jk}^2 \equiv (2Q_k^j \cdot y_j + \frac{\partial c_{jk}(y_{jk})}{\partial y_{jk}} + \hat{c}_{jk}(y) + \rho_{2j} - \rho_{3k})$, $\forall j, k$; $F^3 = (F_1^3, \ldots, F_n^3)$, where $F_j^3 \equiv (\sum_{i=1}^m \sum_{l=1}^L x_{ijl} - \sum_{k=1}^o y_{jk})$, and $F^4 = (F_1^4, \ldots, F_o^4)$, with $F_k^4 \equiv (\sum_{j=1}^n y_{jk} - d_k(\rho_3))$.

Then the dynamic model described by (21), (22), (34), and (39) for all k, j, i, l can be rewritten as the *projected dynamical system* defined by the following initial-value problem:

$$\dot{Z} = \Pi_\mathcal{K}(Z, -F(Z)), \quad Z(0) = Z_0, \quad (40)$$

where $\Pi_\mathcal{K}$ is the projection operator of $-F(Z)$ onto $\mathcal{K}$ at Z and $Z_0 = (x^0, y^0, \rho_2^0, \rho_3^0)$ is the initial point corresponding to the initial financial flows and the initial prices. The trajectory of (40) describes the dynamic evolution of and the dynamic interactions among the prices and the financial flows.

The dynamical system (40) is nonclassical in that the right-hand side is discontinuous in order to guarantee that the constraints in the context of the above model are not only nonnegativity constraints on the variables, but also a form of budget constraints. Here this methodology is applied to study financial systems in the presence of intermediation. A variety of dynamic financial models, but without intermediation, formulated as projected dynamical systems can be found in the book by Nagurney and Siokos (1997).

A Stationary/Equilibrium Point

The stationary point of the projected dynamical system (40) is now discussed. Recall that a stationary point Z^* is that point that satisfies

$$\dot{Z} = 0 = \Pi_{\mathcal{K}}(Z^*, -F(Z^*))$$

and, hence, in the context of the dynamic financial model with intermediation, when there is no change in the financial flows and no change in the prices. Moreover, as established in Dupuis and Nagurney (1993), since the feasible set $\mathcal{K}$ is a polyhedron and convex, the set of stationary points of the projected dynamical system of the form given in (40) coincides with the set of solutions to the variational inequality problem given by the following: Determine $Z^* \in \mathcal{K}$ such that

$$\langle F(Z^*)^T, Z - Z^* \rangle \geq 0, \quad \forall Z \in \mathcal{K}, \tag{41}$$

where in the model $F(Z)$ and Z are as defined above and recall that $\langle \cdot, \cdot \rangle$ denotes the inner product in N-dimensional Euclidean space, where here $N = mnL + no + n + o$.

Variational Inequality Formulation of Financial Equilibrium with Intermediation

In particular, variational inequality (41) here takes the following form: Determine $(x^*, y^*, \rho_2^*, \rho_3^*) \in \mathcal{K}$, satisfying

$$\sum_{i=1}^{m}\sum_{j=1}^{n}\sum_{l=1}^{L}\left[2Q_{z_{jl}}^i \cdot x_i^* + \frac{\partial c_{ijl}(x_{ijl}^*)}{\partial x_{ijl}} + \frac{\partial c_j(x^*)}{\partial x_{ijl}} + \frac{\partial \hat{c}_{ijl}(x_{ijl}^*)}{\partial x_{ijl}} - \rho_{2j}^*\right]$$
$$\times \left[x_{ijl} - x_{ijl}^*\right]$$
$$+ \sum_{j=1}^{n}\sum_{k=1}^{o}\left[2Q_k^j \cdot y_j^* + \frac{\partial c_{jk}(y_{jk}^*)}{\partial y_{jk}} + \hat{c}_{jk}(y^*) + \rho_{2j}^* - \rho_{3k}^*\right] \times \left[y_{jk} - y_{jk}^*\right]$$
$$+ \sum_{j=1}^{n}\left[\sum_{i=1}^{m}\sum_{l=1}^{L}x_{ijl}^* - \sum_{k=1}^{o}y_{jk}^*\right] \times \left[\rho_{2j} - \rho_{2j}^*\right]$$
$$+ \sum_{k=1}^{o}\left[\sum_{j=1}^{n}y_{jk}^* - d_k(\rho_3^*)\right] \times \left[\rho_{3k} - \rho_{3k}^*\right] \geq 0, \quad \forall (x, y, \rho_2, \rho_3) \in \mathcal{K}, \tag{42}$$

where $\mathcal{K} \equiv \{K \times R_+^{no+n+o}\}$ and $Q_{z_{jl}}^i$ is as defined following (26).

We now discuss the equilibrium conditions. First, note that if the rate of change of the demand price $\dot{\rho}_{3k} = 0$, then from (21) one can conclude that

$$d_k(\rho_3^*) \begin{cases} = \sum_{j=1}^{n} y_{jk}^* & \text{if } \rho_{3k}^* > 0, \\ \leq \sum_{j=1}^{n} y_{jk}^* & \text{if } \rho_{3k}^* = 0. \end{cases} \quad (43)$$

Condition (43) states that, if the price the consumers are willing to pay for the financial product at a demand market is positive, then the quantity consumed by the consumers at the demand market is precisely equal to the demand. If the demand is less than the amount of the product available, then the price for that product is zero. This condition holds for all demand market prices in equilibrium.

Note that condition (43) also follows directly from variational inequality (42) if one sets $x = x^*$, $y = y^*$; $\rho_2 = \rho_2^*$, and make the substitution into (42) and note that the demand prices must be nonnegative.

Observe now that if the rate of change of a price charged by an intermediary is zero, that is, $\dot{\rho}_{2j} = 0$, then (22) implies that

$$\sum_{i=1}^{m}\sum_{l=1}^{L} x_{ijl}^* - \sum_{k=1}^{o} y_{jk}^* \begin{cases} = 0 & \text{if } \rho_{2j}^* > 0, \\ \geq 0 & \text{if } \rho_{2j}^* = 0. \end{cases} \quad (44)$$

Hence, if the price for the financial funds at an intermediary is positive, then the market for the funds "clears" at the intermediary, that is, the supply of funds, as given by $\sum_{i=1}^{m}\sum_{l=1}^{L} x_{ijl}^*$, is equal to the demand of funds, $\sum_{k=1}^{o} y_{jk}^*$ at the intermediary. If the supply exceeds the demand, then the price at the intermediary will be zero. These are well-known economic equilibrium conditions, as are those given in (43). Of course, condition (44) could also be recovered from variational inequality (42) by setting $x = x^*$, $y = y^*$, and $\rho_3 = \rho_3^*$, making the substitution into (42), and noting that these prices must be nonnegative. In equilibrium, condition (44) holds for all intermediary prices.

On the other hand, if one sets $\dot{x}_i = 0$ [cf. (34) and (40)], for all i and $\dot{y}_{jk} = 0$ for all j,k [cf. (39) and (40)], one obtains the equilibrium conditions, which correspond, equivalently, to the first two summands in inequality (42) being greater than equal to zero. Expressed in another manner, we must have that the sum of the inequalities (45), (46), and (48) must be satisfied.

Optimality Conditions for All Source Agents

Indeed, note that the optimality conditions for all source agents i, since each K_i is closed and convex, and the objective function (24) is concave, can be expressed as (assuming a given ρ_{1jl}^*, for all i,j,l)

$$\sum_{i=1}^{m}\sum_{j=1}^{n}\sum_{l=1}^{L}\left[2Q_{z_{jl}}^i \cdot x_i^* + \frac{\partial c_{ijl}(x_{ijl}^*)}{\partial x_{ijl}} - \rho_{1ijl}^*\right] \times [x_{ijl} - x_{ijl}^*] \geq 0, \quad \forall x \in K. \quad (45)$$

Optimality Conditions for All Intermediary Agents

The optimality conditions for all the intermediaries j, with objective functions of the form (30), which are concave, and, given ρ_1^* and ρ_2^*, can, in turn, be expressed as

$$\sum_{i=1}^{m}\sum_{j=1}^{n}\sum_{l=1}^{L}\left[\frac{\partial c_j(x^*)}{\partial x_{ijl}} + \rho_{1ijl}^* + \frac{\partial \hat{c}_{ijl}(x_{ijl}^*)}{\partial x_{ijl}} - \rho_{2j}^*\right] \times \left[x_{ijl} - x_{ijl}^*\right]$$

$$+ \sum_{j=1}^{n}\sum_{k=1}^{o}\left[2Q_k^j \cdot y_j + \frac{\partial c_{jk}(y_{jk}^*)}{\partial y_{jk}}\right] \times \left[y_{jk} - y_{jk}^*\right] \geq 0, \forall x \in R_+^{mnL}, \forall y \in R_+^{no}.$$

(46)

Note that (46) provides a means for recovering the top-tiered prices, ρ_1^*.

Equilibrium Conditions for Consumers at the Demand Markets

Also, the equilibrium conditions for consumers at demand market k thus take the following form: For all intermediaries: $j; j = 1, \ldots, n$,

$$\rho_{2j}^* + \hat{c}_{jk}(y^*) \begin{cases} = \rho_{3k}^* & \text{if } y_{jk}^* > 0, \\ \geq \rho_{3k}^* & \text{if } y_{jk}^* = 0, \end{cases}$$

(47)

with (47) holding for all demand markets k, which is equivalent to $y^* \in R_+^{no}$ satisfying

$$\sum_{j=1}^{n}\sum_{k=1}^{o}(\rho_{2j}^* + \hat{c}_{jk}(y^*) - \rho_{3k}^*) \times (y_{jk} - y_{jk}^*) \geq 0, \quad \forall y \in R_+^{no}. \quad (48)$$

Conditions (47) state that consumers at demand market k will purchase the product from intermediary j if the price charged by the intermediary for the product plus the transaction cost (from the perspective of the consumers) does not exceed the price that the consumers are willing to pay for the product, that is, ρ_{3k}^*.

In Nagurney and Ke (2001), a variational inequality of the form (42) was derived in a manner distinct from that given above for a static financial network model with intermediation, but with a slightly different feasible set where it was assumed that the constraints (25) had to be tight, that is, to hold as an equality. Nagurney and Ke (2003), in turn, demonstrated how electronic transactions could be introduced into financial networks with intermediation by adding additional links to the network in Figure 5 and by including additional transaction costs and prices and expanding the objective functions of the decision makers accordingly. We discuss electronic financial transactions subsequently, when we describe the financial engineering of integrated social and financial networks with intermediation.

4.1 The Discrete-Time Algorithm (Adjustment Process)

Note that the projected dynamical system (40) is a continuous-time adjustment process. However, in order to further fix ideas and to provide a means of "tracking" the trajectory of (40), we present a discrete-time adjustment process, in the form of the Euler method, which is induced by the general iterative scheme of Dupuis and Nagurney (1993).

The statement of the Euler method is as follows:

Step 0: Initialization

Start with a $Z^0 \in \mathcal{K}$. Set $\tau := 1$.

Step 1: Computation

Compute Z^τ by solving the variational inequality problem:

$$Z^\tau = P_\mathcal{K}(Z^{\tau-1} - \alpha_\tau F(Z^{\tau-1})), \tag{49}$$

where $\{\alpha_\tau; \tau = 1, 2, \ldots\}$ is a sequence of positive scalars such that $\sum_{\tau=1}^{\infty} \alpha_\tau = \infty$, $\alpha_\tau \to 0$, as $\tau \to \infty$ (which is required for convergence).

Step 2: Convergence Verification

If $|Z_b^\tau - Z_b^{\tau-1}| \leq \epsilon$, for some $\epsilon > 0$, a prespecific tolerance, then stop: otherwise, set $\tau := \tau + 1$, and go to step 1.

The statement of this method in the context of the dynamic financial model takes the following form.

The Euler Method

Step 0: Initialization Step

Set $(x^0, y^0, \rho_2^0, \rho_3^0) \in \mathcal{K}$. Let $\tau = 1$, where τ is the iteration counter, and set the sequence $\{\alpha_\tau\}$ so that $\sum_{\tau=1}^{\infty} \alpha_\tau = \infty$, $\alpha_\tau > 0$, $\alpha_\tau \to 0$, as $\tau \to \infty$.

Step 1: Computation Step

Compute $(x^\tau, y^\tau, \rho_2^\tau, \rho_3^\tau) \in \mathcal{K}$ by solving the variational inequality subproblem:

$$\sum_{i=1}^{m}\sum_{j=1}^{n}\sum_{l=1}^{L}\left[x_{ijl}^\tau + \alpha_\tau(2Q_{z_{jl}}^i \cdot x_i^{\tau-1} + \frac{\partial c_{ijl}(x_{ijl}^{\tau-1})}{\partial x_{ijl}} + \frac{\partial c_j(x^{\tau-1})}{\partial x_{ijl}}\right.$$

$$\left. + \frac{\partial \hat{c}_{ijl}(x_{ijl}^{\tau-1})}{\partial x_{ijl}} - \rho_{2j}^{\tau-1}) - x_{ijl}^{\tau-1}\right] \times [x_{ijl} - x_{ijl}^\tau]$$

$$+ \sum_{j=1}^{n}\sum_{k=1}^{o}\left[y_{jk}^\tau + \alpha_\tau(2Q_k^i \cdot y_j^{\tau-1} + \hat{c}_{jk}(y^{\tau-1}) + \frac{\partial c_{jk}(y_{jk}^{\tau-1})}{\partial y_{jk}}\right.$$

$$+ \rho_{2j}^{\tau-1} - \rho_{3k}^{\tau-1}) - y_{jk}^{\tau-1}\Big] \times [y_{jk} - y_{jk}^{\tau}]$$

$$+ \sum_{j=1}^{n} \left[\rho_{2j}^{\tau} + \alpha_{\tau} (\sum_{i=1}^{m} \sum_{l=1}^{L} x_{ijl}^{\tau-1} - \sum_{k=1}^{o} y_{jk}^{\tau-1}) - \rho_{2j}^{\tau-1} \right] \times [\rho_{2j} - \rho_{2j}^{\tau}]$$

$$+ \sum_{k=1}^{o} \left[\bar{\rho}_{3k}^{\tau} + \alpha_{\tau} (\sum_{j=1}^{n} y_{jk}^{\tau-1} - d_k(\rho_3^{\tau-1})) - \rho_{3k}^{\tau-1} \right] \times [\rho_{3k} - \rho_{3k}^{\tau}] \geq 0,$$

$$\forall (x, y, \rho_2, \rho_3) \in \mathcal{K}.$$

Step 2: Convergence Verification

If $|x_{ijl}^{\tau} - x_{ijl}^{\tau-1}| \leq \epsilon$, $|y_{jk}^{\tau} - y_{jk}^{\tau-1}| \leq \epsilon$, $|\rho_{2j}^{\tau} - \rho_{2j}^{\tau-1}| \leq \epsilon$, $|\rho_{3k}^{\tau} - \rho_{3k}^{\tau-1}| \leq \epsilon$, for all $i = 1, \ldots, m; j = 1, \ldots, n; l = 1, \ldots, L; k = 1, \ldots, o$, with $\epsilon > 0$, a prespecified tolerance, then stop; otherwise, set $\tau := \tau + 1$, and go to step 1.

Note that the variational inequality subproblem encountered in the computation step at each iteration of the Euler method can be solved explicitly and in closed form since it is actually a quadratic programming problem and the feasible set is a Cartesian product consisting of the product of $\mathcal{K}$, which has a simple network structure, and the nonnegative orthants, R_+^{no}, R_+^{n}, and R_+^{o}, corresponding to the variables x, y, ρ_2, and ρ_3, respectively.

Computation of Financial Flows and Products

In fact, the subproblem in the x-variables can be solved using exact equilibration (see also Dafermos and Sparrow, 1969) noted in the discussion of the modified projection method, whereas the remainder of the variables can be obtained by explicit formulas, which are provided below for convenience.

In particular, compute, at iteration τ, the y_{jk}^{τ}, according to:

$$y_{jk}^{\tau} = \max\left\{0, y_{jk}^{\tau-1} - \alpha_{\tau}\left(2Q_k^i \cdot y_j^{\tau-1} + \hat{c}_{jk}(y^{\tau-1}) + \frac{\partial c_{jk}(y_{jk}^{\tau-1})}{\partial y_{jk}} + \rho_{2j}^{\tau-1} - \rho_{3k}^{\tau-1}\right)\right\},$$

$$\forall j, k. \qquad (50)$$

Computation of the Prices

At iteration τ, compute the ρ_{2j}^{τ} according to

$$\rho_{2j}^{\tau} = \max\left\{0, \rho_{2j}^{\tau-1} - \alpha_{\tau}\left(\sum_{i=1}^{m}\sum_{l=1}^{L} x_{ijl}^{\tau-1} - \sum_{k=1}^{o} y_{jk}^{\tau-1}\right)\right\}, \quad \forall j, \qquad (51)$$

whereas the ρ_{3k}^{τ} are computed explicitly and in closed form according to

$$\rho_{3k}^{\tau} = \max\left\{0, \rho_{3k}^{\tau-1} - \alpha_{\tau}\left(\sum_{j=1}^{n} y_{jk}^{\tau-1} - d_k\left(\rho_3^{\tau-1}\right)\right)\right\}, \quad \forall k. \qquad (52)$$

5 Numerical Examples

In this section, the Euler method is applied to several numerical examples. The algorithm was implemented in FORTRAN. For the solution of the induced network subproblems in x, we utilized the exact equilibration algoritm, which fully exploits the simplicity of the special network structure of the subproblems.

The convergence criterion used was that the absolute value of the flows and prices between two successive iterations differed by no more than 10^{-4}. For the examples, the sequence $\{a_\tau\} = 0.1\{1, \frac{1}{2}, \frac{1}{2}, \frac{1}{3}, \frac{1}{3}, \frac{1}{3}, \ldots\}$, which is of the form given in the intialization step of the algorithm in the preceding section. The numerical examples had the network structure depicted in Figure 6 and consisted of two source agents, two intermediaries, and two demand markets, with a single financial instrument handled by each intermediary.

The algorithm was initialized as follows: Since a single financial instrument was associated with each of the intermediaries, we set $x_{ij1} = S^i/n$ for each source agent i. All the other variables, that is, the initial vectors y, ρ_2, and ρ_3, were set to zero. Additional details are given in Nagurney and Dong (2002).

Example 2

The data for this example were constructed for easy interpretation purposes. The supplies of the two source agents were $S^1 = 10$ and $S^2 = 10$. The variance-covariance matrices Q^i and Q^j were equal to the identity matrices for all source agents i and all intermediaries j.

The transaction cost functions faced by the source agents associated with transacting with the intermediaries were given by

$$c_{111}(x_{111}) = 0.5x_{111}^2 + 3.5x_{111}, \quad c_{121}(x_{121}) = 0.5x_{121}^2 + 3.5x_{121},$$

$$c_{211}(x_{211}) = 0.5x_{211}^2 + 3.5x_{211}, \quad c_{221}(x_{221}) = 0.5x_{221}^2 + 3.5x_{221}.$$

The handling costs of the intermediaries, in turn, were given by

$$c_1(x) = 0.5 \left(\sum_{i=1}^{2} .5x_{i11} \right)^2, \quad c_2(x) = 0.5 \left(\sum_{i=1}^{2} x_{i21} \right)^2.$$

The transaction costs of the intermediaries associated with transacting with the source agents were respectively given by

$$\hat{c}_{111}(x_{111}) = 1.5x_{111}^2 + 3x_{111}, \quad \hat{c}_{121}(x_{121}) = 1.5x_{121}^2 + 3x_{121},$$

$$\hat{c}_{211}(x_{211}) = 1.5x_{211}^2 + 3x_{211}, \quad \hat{c}_{221}(x_{221}) = 1.5x_{221}^2 + 3x_{221}.$$

The demand functions at the demand markets were

$$d_1(\rho_3) = -2\rho_{31} - 1.5\rho_{32} + 1000, \quad d_2(\rho_3) = -2\rho_{32} - 1.5\rho_{31} + 1000,$$

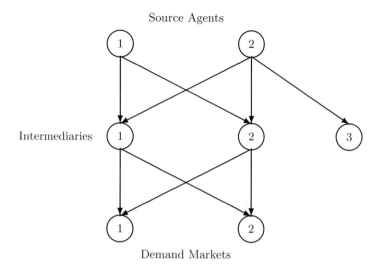

Fig. 6. The financial network structure of the numerical examples.

and the transaction costs between the intermediaries and the consumers at the demand markets were given by

$$\hat{c}_{11}(y) = y_{11} + 5, \quad \hat{c}_{12}(y) = y_{12} + 5, \quad \hat{c}_{21}(y) = y_{21} + 5, \quad \hat{c}_{22}(y) = y_{22} + 5.$$

It was assumed for this and the subsequent examples that the transaction costs as perceived by the intermediaries and associated with transacting with the demand markets were all zero, that is, $c_{jk}(y_{jk}) = 0$, for all j, k.

The Euler method converged and yielded the following equilibrium pattern:

$$x^*_{111} = x^*_{121} = x^*_{211} = x^*_{221} = 5.000,$$

$$y^*_{11} = y^*_{12} = y^*_{21} = y^*_{22} = 5.000.$$

The vector ρ_2^* had components $\rho_{21}^* = \rho_{22}^* = 262.6664$, and the computed demand prices at the demand markets were: $\rho_{31}^* = \rho_{32}^* = 282.8106$.

The optimality/equilibrium conditions were satisfied with good accuracy. Note that in this example, the budget constraint was tight for both source agents, that is, $s_1^* = s_2^* = 0$, where $s_i^* = S^i - \sum_{j=1}^{n} \sum_{l=1}^{L} x^*_{ijl}$, and, hence, there was zero flow on the links connecting node 3 with top-tier nodes 1 and 2. Thus, it was optimal for both source agents to invest their entire financial holdings in each instrument made available by each of the two intermediaries.

Example 3

The following variant of Example 2 was then constructed to create Example 3. The data were identical to that in Example 2 except that the supply for each source sector was increased so that $S^1 = S^2 = 50$.

The Euler method converged and yielded the following new equilibrium pattern:

$$x_{111}^* = x_{121}^* = x_{211}^* = x_{221}^* = 23.6832,$$
$$y_{11}^* = y_{12}^* = y_{21}^* = y_{22}^* = 23.7247.$$

The vector ρ_2^* had components $\rho_{21}^* = \rho_{22}^* = 196.0174$, and the demand prices at the demand markets were $\rho_{31}^* = \rho_{32}^* = 272.1509$.

It is easy to verify that the optimality/equilibrium conditions again were satisfied with good accuracy. Note, however, that unlike the solution for Example 2, both source agent 1 and source agent 2 did not invest his entire financial holdings. Indeed, each opted not to invest the amount 23.7209, and this was the volume of flow on each of the two links ending in node 3 in Figure 6.

Since the supply of financial funds increased, the price for the instruments charged by the intermediaries decreased from 262.6664 to 196.1074. The demand prices at the demand markets also decreased, from 282.8106 to 272.1509.

Example 4

Example 3 was then modified as follows: The data were identical to that in Example 3 except that the first diagonal term in the variance-covariance matrix Q^1 was changed from 1 to 2.

The Euler method again converged, yielding the following new equilibrium pattern:

$$x_{111}^* = 18.8676, \quad x_{121}^* = 23.7285, \quad x_{211}^* = 25.1543, \quad x_{221}^* = 23.7267,$$
$$y_{11}^* = y_{12}^* = 22.0501, \quad y_{21}^* = y_{22}^* = 23.7592.$$

The vector ρ_2^* had components $\rho_{21}^* = 201.4985$, $\rho_{22}^* = 196.3633$, and the demand prices at the demand markets were $\rho_{31}^* = \rho_{32}^* = 272.6178$.

5.1 The Integration of Social Networks with Financial Networks

As noted by Nagurney et al. (2007), globalization and technological advances have made major impacts on financial services in recent years and have allowed for the emergence of electronic finance. The financial landscape has been transformed through increased financial integration, increased cross-border mergers, and lower barriers between markets. Moreover, as noted by several authors, boundaries between different financial intermediaries have become less clear (cf. Claessens and Jansen, 2000; Claessens et al., 2003; G-10, 2001).

For example, during the period 1980–1990, global capital transactions tripled, with telecommunication networks and financial instrument innovation being two of the empirically identified major causes of globalization with regards to international financial markets (Kim, 1999). The growing importance of networks in financial services and their effects on competition have been also addressed by Claessens et al. (2003). Kim (1999) argued for the necessity of integrating various theories, including portfolio theory with risk management, and flow theory in order to capture the underlying complexity of the financial flows over space and time.

At the same time that globalization and technological advances have transformed financial services, researchers have identified the importance of social networks in a plethora of financial transactions (cf. Nagurney et al., 2007 and the references therein), notably in the context of personal relationships. The relevance of social networks within an international financial context needs to be examined both theoretically and empirically. It is clear that the existence of appropriate social networks can affect not only the risk associated with financial transactions but also the transaction costs.

Given the prevalence of networks in the discussions of globalization and international financial flows, it seems natural that any theory for the illumination of the behavior of the decision makers involved in this context as well as the impacts of their decisions on the financial product flows, prices, appreciation rates, etc., should be network-based. Recently, Nagurney et al. (2007) took on a network perspective for the theoretical modeling, analysis, and computation of solutions to international financial networks with intermediation in which they explicitly integrated the social network component. They also captured electronic transactions within the framework since that aspect is critical in the modeling of international financial flows today.

Here, that model is highlighted. This model generalizes the model of Nagurney and Cruz (2003) to explicitly include social networks.

As in the model of Nagurney and Cruz (2003), the model consists of L countries, with a typical country denoted by l or $\hat{l}$, I "source" agents in each country with sources of funds, with a typical source agent denoted by i, and J financial intermediaries with a typical financial intermediary denoted by j. As noted earlier, examples of source agents are households and businesses, whereas examples of financial intermediaries include banks, insurance companies, investment companies, and brokers, where we now include electronic brokers, etc. Intermediaries in the framework need not be country-specific but, rather, may be virtual.

Assume that each source agent can transact directly electronically with the consumers through the Internet and can also conduct his financial transactions with the intermediaries either physically or electronically in different currencies. There are H currencies in the international economy, with a typical currency being denoted by h. Also, assume that there are K financial products, which can be in distinct currencies and in different countries with a typical financial product (and associated with a demand market) being denoted by k. Hence, the financial intermediaries in the model, in addition to transacting with the source agents, also determine how to allocate the incoming financial resources among distinct uses, which are represented by the demand markets with a demand market corresponding to, for example, the market for real estate loans, household loans, or business loans, etc., which, as mentioned, can be associated with a distinct country and a distinct currency combination. Let m refer to a mode of transaction with $m = 1$ denoting a physical transaction and $m = 2$ denoting an electronic transaction via the Internet.

The depiction of the *supernetwork* (see also, e.g., Nagurney and Dong, 2002) is given in Figure 7. As this figure illustrates, the supernetwork is comprised of the social network, which is the bottom-level network, and the international financial network, which is the top-level network. Internet links to denote the possibility of electronic financial transactions are denoted in the figure by dotted arcs. In addition, dotted arcs/links are used to depict the integration of the two networks into a supernetwork.

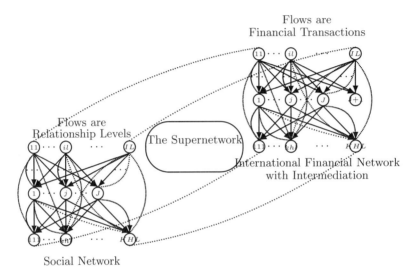

Fig. 7. The multilevel supernetwork structure of the integrated international financial network / social network system.

The supernetwork in Figure 7 consists of a social and an international financial network with intermediation. Both networks consist of three tiers of decision makers. The top tier of nodes consists of the agents in the different countries with sources of funds, with agent i in country l being referred to as agent il and associated with node il. There are, hence, IL top-tiered nodes in the network. The middle tier of nodes in each of the two networks consists of the intermediaries (which need not be country-specific), with a typical intermediary j associated with node j in this (second) tier of nodes in the networks. The bottom tier of nodes in both the social network and in the financial network consists of the demand markets, with a typical demand market for product k in currency h and country $\hat{l}$ associated with node $kh\hat{l}$. There are, as depicted in Figure 7, J middle- (or second) tiered nodes corresponding to the intermediaries and KHL bottom- (or third) tiered nodes in the international financial network. In addition, we add a node $J+1$ to the middle tier of nodes in the financial network only in order to represent the possible non-investment (of a portion or all of the funds) by one or more of the source agents, as was also done in the model in the previous section.

Note that the network in Figure 7 includes classical physical links as well as Internet links to allow for electronic financial transactions. Electronic transactions are possible between the source agents and the intermediaries, the source agents and the demand markets, as well as the intermediaries and the demand markets. Physical transactions can occur between the source agents and the intermediaries and between the intermediaries and the demand markets.

Nagurney et al. (2007) described the behavior of the decision makers in the model, and allow for multicriteria decision making, which consists of profit maximization, risk minimization (with general risk functions), as well as the maximization of the value of relationships. Each decision maker is allowed to weigh the criteria individually. The dynamics of the interactions are discussed and the projected dynamical system derived. The Euler method is then used to track the dynamic trajectories of the financial flows (transacted either physically or electronically), the prices, as well as the relationship levels until the equilibrium state is reached.

Acknowledgments

The writing of this chapter was supported, in part, by NSF Grant no. IIS 0002647 under the MKIDS Program. This support is gratefully acknowledged.

References

1. Ahuja, R. K., Magnanti, T. L., and Orlin, J. B. *Network Flows*. Prentice Hall, Upper Saddle River, NJ, 1993
2. Arrow, K. J. An extension of the basic theorems of classical welfare economics. *Econometrica*, 51:1305–1323, 1951.
3. Barr, R. S. The multinational cash management problem: A generalized network approach. University of Texas, Austin, Working paper, 1972.
4. Board of Governors *Introduction to Flow of Funds*. Flow of Funds Section, Division of Research and Statistics, Federal Reserve System, Washington, DC, June, 1980.
5. Boginski, V., Butenko, S., and Pardalos, P. M. On structural properties of the market graph. In A. Nagurney, Editor, *Innovations in Financial and Economic Networks*. Edward Elgar Publishing, Cheltenham, 2003.
6. Charnes, A., and Cooper, W. W. Nonlinear network flows and convex programming over incidence matrices. *Naval Research Logistics Quarterly*, 5:231–240, 1958.
7. Charnes, A., and Cooper, W. W. *Management Models and Industrial Applications of Linear Programming*. John Wiley & Sons, New York, 1961.

8. Charnes, A., and Cooper, W. W. Some network characterizations for mathematical programming and cccounting approaches to planning and control. *The Accounting Review*, 42:24–52, 1967.
9. Charnes, A., and Miller, M. Programming and financial budgeting. Symposium on Techniques of Industrial Operations Research, Chicago, June, 1957.
10. Christofides, N., Hewins, R. D., and, Salkin, G. R. Graph theoretic approaches to foreign exchange operations. *Journal of Financial and Quantitative Analysis*, 14:481–500, 1979.
11. Claessens, S. and Jansen, M., Editors *Internationalization of Financial Services*. Kluwer Academic Publishers, Boston, 2000.
12. Claessens, S., Dobos, G., Klingebiel, D., and Laeven, L. The growing importance of networks in finance and their effects on competition. In A. Nagurney, Editor, *Innovations in Financial and Economic Networks*. Edward Elgar Publishing, Cheltenham, 2003.
13. Cohen, J. *The Flow of Funds in Theory and Practice*. Kluwer Academic Publishers, Dordrecht, 1987.
14. Copeland, M. A. *A Study of Moneyflows in the United States*. National Bureau of Economic Research, New York, 1952.
15. Cournot, A. A. *Researches into the Mathematical Principles of the Theory of Wealth*, 1838; English translation, Macmillan, London, 1897.
16. Crum, R. L. Cash management in the multinational firm: A constrained generalized network approach. University of Florida, Working paper, 1976.
17. Crum, R. L., and Nye, D. J. A network model of insurance company cash flow management. *Mathematical Programming Study*, 15:86–101, 1981.
18. Crum, R. L., Klingman, D. D., and Tavis, L. A. Implementation of large-scale financial planning models: Solution efficient transformations. *Journal of Financial and Quantitative Analysis*, 14:137–152, 1979.
19. Crum, R. L., Klingman, D. D., and Tavis, L. A. An operational approach to integrated working capital planning. *Journal of Economics and Business*, 35:343–378, 1983.
20. Dafermos, S. C., and Sparrow, F. T. The traffic assignment problem for a general network. *Journal of Research of the National Bureau of Standards*, 73B:91–118, 1969.
21. Dantzig, G. B., and Madansky, A. On the solution of two-stage linear programs under uncertainty. In *Proceedings of the Fourth Berkeley Symposium on Mathematical Statistics and Probability*. University of California Press, Berkeley, 1961, pages 165–176.
22. Debreu G. The coefficient of resource utilization. *Econometrica*, 19:273–292, 1951.
23. Dong, J., and Nagurney, A. bicriteria decision-making and financial equilibrium: A variational inequality perspective. *Computational Economics*, 17:29–42, 2001.
24. Dong, J., Zhang, D., and Nagurney, A. A projected dynamical systems model of general financial equilibrium with stability analysis. *Mathematical and Computer Modelling*, 24:35–44, 1996.
25. Doumpos, M., Zopounidis, C., and Pardalos, P. M. Multicriteria sorting methodology: Application to financial decision problems. *Parallel Algorithms and Applications*, 15:113–129, 2000.
26. Dupuis, P., and Nagurney, A. Dynamical systems and variational inequalities. *Annals of Operations Research*, 44:9–42, 1993.

27. Enke, S. Equilibrium among spatially separated markets. *Econometrica*, 10:40–47, 1951.
28. Euler, L. Solutio problematis ad geometriam situs pertinentis. *Commetarii Academiae Scientiarum Imperialis Petropolitanae*, 8:128–140, 1736.
29. Fei, J. C. H. The study of the credit system by the method of linear graph. *The Review of Economics and Statistics*, 42:417–428, 1960.
30. Ferguson, A. R., and Dantzig, G. B. The allocation of aircraft to routes. *Management Science*, 2:45–73, 1956.
31. Ford, L. R., and Fulkerson, D. R. *Flows in Networks*. Princeton University Press, Princeton, NJ, 1962.
32. Francis, J. C., and Archer, S. H. *Portfolio Analysis*. Prentice Hall, Englewood Cliffs, NJ, 1979.
33. G-10 Report on Consolidation in Financial Services. Bank for International Settlements, Switzerland, 2001.
34. Guenes, J., and Pardalos, P. M. Network optimization in supply chain management and financial engineering: An annotated bibliography. *Networks*, 42:66–84, 2003.
35. Hughes, M., and Nagurney, A. A network model and algorithm for the estimation and analysis of financial flow of funds. *Computer Science in Economics and Management*, 5:23–39, 1992.
36. Korpelevich, G. M. The extragradient method for finding saddle points and other problems. *Matekon*, 13:35–49, 1977.
37. Kim, H. M. *Globalization of International Financial Markets*. Ashgate, Hants, 1999.
38. Markowitz, H. M. Portfolio Selection. *The Journal of Finance*, 7:77–91, 1952.
39. Markowitz, H. M. *Portfolio Selection: Efficient Diversification of Investments*. John Wiley & Sons, New York, 1959.
40. Mulvey, J. M. Nonlinear networks in finance. *Advances in Mathematical Programming and Financial Planning*, 1:253–271, 1987.
41. Mulvey, J. M., and Vladimirou, H. Stochastic network optimization models for investment planning. *Annals of Operations Research*, 20:187–217, 1989.
42. Mulvey, J. M., and Vladimirou, H. Solving multistage stochastic networks: An application of scenario aggregation. *Networks*, 21:619–643, 1991.
43. Mulvey, J. M., Simsek, K. D., and Pauling, B. A stochastic network optimization approach for integrated pension and corporate financial planning. In A. Nagurney, Editor, *Innovations in Financial and Economic Networks*. Edward Elgar Publishing, Cheltenham, 2003.
44. Nagurney, A. Variational inequalities in the analysis and computation of multisector, multi-instrument financial equilibria. *Journal of Economic Dynamics and Control*, 18:161–184, 1994.
45. Nagurney, A. *Network Economics: A Variational Inequality Approach*, 2nd edition. Kluwer Academic Publishers, Dordrecht, 1999.
46. Nagurney, A. Finance and variational inequalities. *Quantitative Finance*, 1:309–317, 2001.
47. Nagurney, A. *Innovations in Financial and Economic Networks*. Edward Elgar Publishing, Cheltenham, 2003.
48. Nagurney, A., and Cruz, J. International financial networks with electronic transactions. In A. Nagurney, Editor, *Innovations in Financial and Economic Networks*. Edward Elgar Publishing, Cheltenham, 2003, pages 135–167.

49. Nagurney, A., and Dong, J. Network decomposition of general financial equilibria with transaction costs. *Networks*, 28:107–116, 1996a.
50. Nagurney, A., and Dong, J. General financial equilibrium modelling with policy interventions and transaction costs. *Computational Economics*, 9:363–384, 1996b.
51. Nagurney, A., and Dong, J. *Supernetworks: Decision-Making for the Information Age*. Edward Elgar Publishers, Cheltenham, 2002.
52. Nagurney, A., and Hughes, M. Financial flow of funds networks. *Networks*, 22:145–161, 1992.
53. Nagurney, A., and Ke, K. Financial networks with intermediation. *Quantitative Finance*, 1:441–451, 2001.
54. Nagurney, A., and Ke, K. Financial networks with electronic transactions: Modeling, analysis, and computations. *Quantitative Finance*, 3:71–87, 2003.
55. Nagurney, A., and Siokos, S. Variational inequalities for international general financial equilibrium modeling and computation. *Mathematical and Computer Modelling*, 25:31–49, 1997.
56. Nagurney, A., and Zhang, D. *Projected Dynamical Systems and Variational Inequalities with Applications*. Kluwer Academic Publishers, Boston, MA, 1996.
57. Nagurney, A., Cruz, J. M., and Wakolbinger, T. The co-evolution and emergence of integrated international financial networks and social networks: Theory, analysis, and computations. In R. J. Cooper, K. P. Donaghy, and G. J. D. Hewings, Editors, *Globalization and Regional Economic Modeling*. Springer, Berlin, 2007, pages 183–226.
58. Nagurney, A., Dong, J., and Hughes, M. Formulation and computation of general financial equilibrium. *Optimization*, 26:339–354, 1992.
59. Pigou, A. C. *The Economics of Welfare*. Macmillan, London, 1920.
60. Quesnay, F. *Tableau Economique*, 1758; reproduced in facsimile with an introduction by H. Higgs by the British Economic Society, 1895.
61. Rockafellar, R. T., and Wets, R. J.-B. Scenarios and policy in optimization under uncertainty. *Mathematics of Operations Research*, 16:1–29, 1991.
62. Rudd, A., and Rosenberg, B. Realistic portfolio optimization. *TIMS Studies in the Management Sciences*, 11:21–46, 1979.
63. Rutenberg, D. P. Maneuvering liquid assets in a multi-national company: Formulation and deterministic solution procedures. *Management Science*, 16:671–684, 1970.
64. Samuelson, P. A. Spatial price equilibrium and linear programming. *American Economic Review* 42:283–303, 1952.
65. Shapiro, A. C., and Rutenberg, D. P. Managing exchange risks in a floating world. *Financial Management*, 16:48–58, 1976.
66. Soenen, L. A. *Foreign Exchange Exposure Management: A Portfolio Approach*. Sijthoff and Noordhoff, Germantown, MD, 1979.
67. Srinivasan, V. A transshipment model for cash management decisions. *Management Science*, 20:1350–1363, 1974.
68. Storoy, S., Thore, S., and Boyer, M. Equilibrium in linear capital market networks. *The Journal of Finance*, 30:1197–1211, 1975.
69. Takayama, T., and Judge, G. G. *Spatial and Temporal Price and Allocation Models*. North-Holland, Amsterdam, 1971.
70. Thore, S. Credit networks. *Economica*, 36:42–57, 1969.
71. Thore, S. Programming a credit network under uncertainty. *Journal of Money, Banking, and Finance*, 2:219–246, 1970.

72. Thore, S. *Programming the Network of Financial Intermediation*. Universitetsforlaget, Oslo, 1980.
73. Thore, S. Spatial models of the Eurodollar market. *Journal of Banking and Finance*, 8:51–65, 1984.
74. Thore, S., and Kydland, F. Dynamic for flow-of-funds networks. In S. Eilon and T. R. Fowkes, Editors, *Applications of Management Science in Banking and Finance*. Gower Press, London, 1972, pages 259–276.
75. Wallace, S. Solving stochastic programs with network recourse. *Networks*, 16:295–317, 1986.

Part IV

Mergers, Acquisitions, and Credit Risk Ratings

The Choice of the Payment Method in Mergers and Acquisitions

Alain Chevalier[1] and Etienne Redor[2]

[1] ESCP EAP, 79 Avenue de la République, 75543 Paris Cedex 11, France
chevalier@escp-eap.net
[2] ESCP EAP, 79 Avenue de la République, 75543 Paris Cedex 11, France
etienne.redor@escp-eap.net

1 Introduction

Mergers and acquisitions are major events in a firm's life. It is not surprising, that numerous studies aim at explaining this phenomenon. Since the late 1970s, the number of studies following the evolution of the number of deals. Among the various points that have been studied by researchers in finance, we can identify the study of the motivations of mergers and acquisitions (the market power, the hubris hypothesis, the economy of scale and the economy of scope, the managerial hypothesis, etc.), the short- and long-term performances for target's and bidder's shareholders, the vast merger waves, and the choice of the payment methods.

The latter has been the subject of numerous research as well as empirical and theoretical studies. A part of the literature on this issue focuses on the impact of the choice of the payment methods on the shareholders' wealth. Concerning the returns earned by the bidder, Travlos (1987) reported negative abnormal returns when the operation is financed with stocks but positive abnormal returns when it is financed with cash. In addition, Antoniou and Zhao (2004) showed that the bidders' returns are lower when the operation is financed in stocks than in case of alternative combined and cash offers. Similarly, many empirical studies reveal that the target's returns are higher in cash offers than in stock offers (Huang and Walkling, 1987; Franks et al., 1988; Eckbo and Langohr, 1989), confirming that the choice of the payment method has an impact on the profitability of a takeover.

A second part of the literature has tried to explain the choice of the payment method by managers. Many theories and many models have thus been developed; primary among these are informational asymmetry models. As their name indicates, they are based on the principle that there is asymmetry between the information owned by the managers and the other agents of the market. In other words, managers have access to private information concerning the firm's stocks value and its investment opportunities, whereas external investors don't. Therefore, if he strikes a deal with an another agent having more information than him, a non-informed agent may have to face heavy adverse selection problems because he does not know whether the other agent has positive or negative information (for a very famous case of adverse selection and a study of the signals allowing to solve this kind of problem, see Akerlof, 1970).

In some countries, taxation is also a factor that managers have to take into account when they determine the payment method used to finance the acquisition. Taxation of capital gains is in this case immediate for cash acquisitions, whereas it is postponed for stock acquisitions. If the option allowing one to postpone this taxation is important for the target's shareholders, the bidding firm can be incited to offer stocks to finance the deal. Similarly, a bidding firm can profit from the tax losses and tax credits of a target if the deal is financed with stocks.

In the same way as the models based on informational asymmetry and on taxation, a family of theories has developed, dealing on the one hand with the managerial ownership, and on the other hand with external control. The first branch comes from the fact that the choice of the payment method influences the ownership structure of the firms concerned. One can therefore suppose that a manager will not finance an acquisition with stocks if he owns a large proportion of the bidder's stocks, to avoid having his stake diluted in the combined firm. Conversely, one can suppose that the manager of the target will prefer an acquisition to be financed with stocks if he owns a large stake in the target, so as to retain power in the merged entity, and therefore to increase the probability of keeping a managerial job in the combined firm.

The second branch of theories deals with the control of managers by shareholders. Admittedly, a shareholder owning a small part of the firm's share cannot afford to control every action of the manager, because it would be too costly in money and in time. However, investors who own a large part of the firm's shares are incited to control the managers' actions, that is, to check the projects' quality in which the manager invests as well as the way these investments are financed. Since mergers and acquisitions financed with stocks are, on average, not well thought of by the market (they destroy value for the bidder's shareholders), large shareholders can put pressure on the bidder's managers so that they use cash rather than stocks to finance their acquisitions. In the same family of theories, we can also find the free cash flow theory. According to Jensen (1986), bidding firms owning large free cash flow or having a sufficient debt capacity will be predisposed to undertake cash acquisitions rather than stock acquisitions. In this theory, debt plays a positive role, because it allows one to reduce the agency phenomena and increases the managers' control.

Various studies also allow one to show that past performance, investment opportunities, and business cycles are elements that could influence the choice of the payment method used to finance the merger or the acquisition.

According to Weston et al. (1990),[1] a firm having an efficient excess management team can use its excess managerial resources by acquiring a firm managed inefficiently because of a lack of resources of this kind. In this case, it can be incited to use cash to convince the target's shareholders to participate in the deal and to replace the inefficient management by managers with planning and monitoring capacities which will be beneficial to the target (Trautwein, 1990). Similarly, a bidder having good investment opportunities is supposed to have a higher probability to use stocks to finance its deal, because this payment method is less compelling than debt. Indeed, the use of debt requires the payment of cash flows at regular intervals. Thus, the higher the investment opportunities of the bidding firm, the more likely it will be willing to use stocks to be able to benefit from these opportunities. Conversely, the lower its investment opportunities, the more likely it will have interest in financing the operation with debt, because the repayments of the debt will not be invested in negative net present value projects. Some studies have also indicated that business cycles influence the choice of the payment method, the probability of a stock offer increasing with the general economic activity.

The optimal capital structure hypothesis suggests that managers can choose the payment method used to finance a merger or an acquisition according to the capital structure of the merged firm after the acquisition. In this case, cash-rich firms will use cash and cash-poor firms will use stocks. Moreover, this theory suggests that the bidders of leveraged firms will use stocks and that the bidders of firms with a debt capacity will use debt.

A last family of theories deals with the period of time necessary to the completion of the deal. The delays of control theory stipulate that a hostile offer is more likely to be financed with cash, because a stock offer needs to obtain authorizations. Obtaining these authorizations can take a long time, which offers time for rivals to get organized, to obtain information, and to potentially intercede. As for the target, it has more time to organize its defense and to eventually find a white knight.

The minimization of the renegotiation costs is close to the delays of control theory. The idea in this case is that since the delay between the operation announcement and its completion is long, there is an incentive for managers to ask for an ex-post renegotiation of the contract terms, if the value of the bidder's offer has sensibly changed in comparison with the target's value during this period. The risk is particularly high when it is a stock offer. As this renegotiation has a cost, the payment method can be chosen so as to minimize the renegotiation costs of both firms at the completion of the deal.

The aim of this survey is to successively present these theories and the different empirical studies that allow one to confirm or invalidate them. The first two sections are devoted to asymmetric information models. The first

---

[1] Quoted by Yanne Gourvil (2002): "La croissance externe et l'emploi dans les entreprises absorbantes," Cahier de recherche no 2002-07, GREFIGE, Université Nancy 2. This article is available at the following Internet address: http://www.univ-nancy2.fr/GREFIGE/cahier2002/Gourvil02-07.pdf.

one will more particularly describe the origin of these models and the precursory work of Myers and Majluf (1984), and the second one will present the complementary work carried out by Hansen (1987), Fishman (1989), Eckbo et al. (1990), and Berkovitch and Narayanan (1990). Then, in a third part, we will look into the impact of taxation on the choice of the payment method. The fourth part of this study will present managerial ownership and outside control theories, and then the fifth one will be devoted to the fact that past performance, investment opportunities, and business cycles can influence the choice of the payment method. A sixth part will study the models binding the capital structure to the payment method. Finally, before presenting theories explaining the choice of the payment method in acquisitions of nonpublic firms, we will study the delay of completion theory.

2 The Origin of the Asymmetric Information Models and Myers and Maljuf's Model (1984)

The asymmetric information models that allow one to explain the choice of the payment method in mergers and acquisitions have initially been developed by Hansen (1987) and by Fishman (1989) and are based on Myers and Maljuf's work (1984). They suppose asymmetric information on both sides, which means that both the target and the bidder have private information on their own value. According to these informational models, the abnormal returns are significantly negative when the operation is financed with stocks but do not show that the performance of the investment is low, since the choice of the payment method reveals private information of the bidder concerning the value or the synergies of both firms.

2.1 The Origin of the Informational Asymmetry Models

As Leland and Pyle (1977) showed, a market in which there would be no information transfers between agents could not work very well. Actually, if we consider the financing of projects of variable quality, the market value should correspond to projects of average quality since lenders cannot distinguish between good and bad projects. To be financed, the good-quality projects need information transfers to take place. This information can be obtained by shareholders simply by observing the managers because their actions reveal information about the quality of the project.

According to Ross (1977), managers who can have an informational advantage are incited to signal their private information through the choice of their debt level. For firms with low cash flows, it is costly to have high debt levels, because the probability of bankruptcy is higher than for firms whose cash flows are high. Therefore, managers of firms with high cash flows can signal this information to the market by issuing enough debt. In Leland and Pyle (1977), under certain conditions, managers of valuable firms signal their quality by retaining a large stake in the firm and so will use more debt than managers of low-value firms. Debt financing allows the manager to retain a large part of the ownership of the firm, but holding a large stake in the firm is costly for risk-averse managers. This large stake is less costly to a manager of a high-quality firm than to a manager of a low-quality firm, and therefore the proportion of stocks held by insiders is a signal concerning the firm's quality. The question of the capital structure is also bound to the phenomena of signals. As Blazenko (1987) showed, high-quality projects are signaled by the use of debt or cash. Indeed, risk-averse managers will not use cash if they are compelled to issue debt in order to obtain cash, unless they are sure that the quality of the stocks to be acquired justifies the personal risk of losing their job. In other words, since managers are not in favor of the use of cash as a payment method, the fact that they use it nevertheless can be a signal concerning the potential of the stocks that will be acquired. Numerous other studies of asymmetric information have been developed since the works of Ross (1977) and of Leland and Pyle (1977) (see, for example, Heinkel, 1982; Vermaelen, 1984; John, 1987; Ravid and Sarig, 1991; Brick et al., 1998; Persons, 1994, 1997; and McNally, 1999).

Similarly, as a result, for some authors, the choice of the payment method used in mergers and acquisitions can signal different kinds of information for investors. In a perfect market, without asymmetric information and taxes, the payment method of an acquisition is not important (Modigliani and Miller, 1958): The division and the level of gains are equivalent whether the operation is financed with stocks only or with a mix of cash and stocks. In 1977, Miller extended this result to a world in which tax exists. However, due to asymmetry problems and to nonzero transaction costs, the choice of the payment method has an impact on the success or failure of the deal and on the returns associated.

2.2 The Work of Myers and Majluf (1984)

In a world of asymmetric information between managers and investors, firms that raise external capital to finance their new projects have to face problems of adverse selection. Firms with low investment opportunities can issue stocks looking like those issued by firms with high investment opportunities. As a result, stocks from low investment opportunities firms will be overvalued, whereas stocks from high investment opportunities firms will be undervalued.

Myers and Majluf (1984) showed that, in a world of asymmetric information, the choice of the payment method by the bidding firm in acquisitions can reveal information concerning the bidder. The managers who own information and want to act in the interest of their current shareholders will use stocks if they are overvalued. They will put some positive net present value investments aside if the stocks necessary to finance the operation are undervalued by the market. Thus, the decision to finance an investment by stocks will be interpreted by the market as bad news, so that the stock price of the firm will decrease at the announcement of the acquisition. Moreover, investors are incited to decrease their valuation of a stock offer, for fear of acquiring overvalued stocks. Conversely, when a cash offer is announced, the assets of the bidder will be considered as being undervalued, which constitutes a positive signal for investors.

This hypothesis is confirmed by Travlos (1987) who, through annual data between 1972 and 1981, showed that the deals financed with stock offers result in significantly negative abnormal returns for the bidders' shareholders. It is also confirmed by Tessema (1989) who, in a similar study, concluded that the market considers a stock offer as being less attractive than a cash offer. The negative abnormal returns observed in stock offers are not due to the realization of negative net present value deals but to the fact that the positive effects of the deal are offset by the negative informational effects caused by the use of stocks.

Asquith and Mullins (1986), Masulis and Korwar (1986), and Mikkelson and Partch (1986) highlighted a decrease of the stock price when new stocks are issued. This fall of the stock price is consistent with the Myers and Maljuf hypothesis (1984) of negative information transfers when stocks are offered.

However, as Chang (1998) pointed out, when firms offer stocks to acquire privately held firms, that is, held by a small number of shareholders, the asymmetric information problems described by Myers and Majluf (1984) can be reduced thanks to the revelation of private information by the bidder to the target's shareholders.

Myers and Majluf (1984) argued that the underinvestment problem can be avoided by the issue of a less risky asset, that is, less sensitive to errors in its value estimation. Given this underinvestment problem, there is a hierarchy of preferences or a "pecking order" in the issue of new financing: Informational asymmetries between new investors and the managers who maximize the wealth of their current shareholders make the stock issue more costly than the debt issue. So the riskier the financing, the more negative the effect on the stocks, which is why managers use internal financing first, then debt, and they finally issue stocks (Myers, 1984). In accordance with this hypothesis, Nayar and Switzer (1998) showed that the riskier the debt used in the acquisition, the less favorable the effects for the bidder's shareholders.

Sung (1993) showed that, in the 1980s, everything else being equal, cash offers were mainly chosen by relatively cash-rich firms, whereas stock offers were rather chosen by cash-generating firms in comparison with their sector. This result is consistent with the "pecking order" theory, but not with the theory of the signal.

The Myers and Maljuf (1984) model has been the subject of numerous empirical studies, whose results seem to be mixed. Whereas the results of Amihud et al. (1990) and Chaplinsky and Niehaus (1993) also go in the way of the "pecking order" theory, the results of Korajczyk et al. (1991) seem to contradict it. Moreover, the tests of Rajan and Zingales (1995), of Jung et al.

(1996), of Helwege and Liang (1996), of Shyam-Sunder and Myers (1999), and of Fama and French (2002) are consistent with this theory on certain points, but opposite to it on others.

According to Nayar and Switzer (1998), whereas some managers will effectively try to use overvalued stocks to acquire stocks from the target, others use stocks to undertake an acquisition that is really profitable, and not to exploit an unjustified overvaluation of their stocks. Unfortunately, the market being unable to distinguish between these two scenarios, every firm that offers stocks in an acquisition will suffer a decrease of their stock price at the announcement of the acquisition.

As a consequence, potential bidders may abandon potentially profitable deals to avoid this adverse selection problem and bidders using cash avoid this problem. In addition, Nayar and Switzer (1998) showed, in accordance with Myers and Maljuf's predictions (1984), according to which the use of debt reduces the effects of adverse selection associated with stock-only offers, that the average abnormal returns of offers that include debt as a payment method are higher than those observed with other payment methods. Finally, once the impact of the taxation and the risk of the debt have been controlled, the higher the proportion of stocks used in the acquisition, the more favorable the reaction of the market. This can indicate that the issuing of stocks to counterbalance the increase of the debt level linked to the deal improves the shareholders' wealth. This increase of the number of stocks, if it is accompanied by a debt issue, is not seen as an attempt by managers to exploit the private information to use a possible stock overvaluation. On the contrary, the market can see acquisitions as a phenomenon in which payment methods allow the realization of risk-sharing adjustments.

According to Jensen and Ruback (1983), since most tender offers are financed with cash while most mergers are financed with stocks, the informational theory suggests that the returns of tender offers are higher than those of mergers. Thus, the difference between mergers and tender offers underlined in empirical studies may only reflect the various informational effects, according to the payment method used in the acquisition.

Moeller et al. (2003) showed that acquisitions undertaken by small firms are profitable, but that they realize small acquisitions, the returns of which are low (altogether, the returns of small firms reach 8 billion dollars in their sample). On the contrary, large firms suffer high losses (around 226 billion dollars in their sample). Acquisitions are therefore value-destroying on average for the bidder's shareholders, since the losses suffered by large firms in acquisitions are significantly higher than the returns earned by small firms. These results seem to contradict the idea according to which the bidder's decision to finance the deal with stocks is a signal pointing out that the bidder's managers think that their firm's stocks are overvalued. Actually, if this explanation were valid, there would be no reason to find a difference between the abnormal returns of small and large firms.

The asymmetric information models are also called into question by Cornett and De (1991) who, through a study of interstate banks mergers between 1982 and 1986, showed that the bidder's shareholders have positive abnormal returns that are significant at the 1% level, in cash offers, stock offers, and combined offers. This result contradicts the previous studies on this issue and seems to be inconsistent with the theory of informational asymmetry. The authors suggested two explanations. The first one might be that the role of informational asymmetry is not as important in the banking sector as in the nonbanking sector. The second possible explanation is that a stock offer is good news for a bank, because it is a signal indicating that the bidder is efficient in asset management. Indeed, before an interstate merger can take place in the banking sector, the bidder's asset management has to be examined and approved by organisms of control.

Conversely, Harris et al. (1987) showed, with empirical evidence, that for bidders, cash offers result in higher post-acquisition performance than stock offers. This result is consistent with the theory of informational asymmetry.

According to Yook (2003), the difference between the information held by the managers and the one held by the external investors in an acquisition is "more complex than the one existing in the new security offering market." In his view, the asset value of the merged firm is therefore the main source of informational asymmetry on the market of corporate acquisition. Managers often claim that the deal creates value for the shareholders. However, as the quantity of synergy is nonmeasurable by shareholders, an acquisition generates a lot of uncertainty, as far as the operating performance of the merged entity is concerned. Thus, according to the author, informational asymmetry in the takeover market mainly concerns the synergies of the deal and the merged entity valuation rather than the bidder's assets. Even though the payment method probably affects the value of the merged firm, it is probably not the only element taken into account to make the acquisition.

Managers must first of all decide whether they want to acquire a particular firm or not, and then the payment method must be chosen. However, since managers are in a good position to estimate the impact of an acquisition on their firm's products, markets, strategies, investment opportunities, etc., they give information on expected synergies by their choice of the payment method. In other words, they will offer cash if they think that their valuation of the synergy is higher than the market's one. Therefore, the payment method in acquisitions gives information about the bidder's real valuation of the merged firm's assets as well as on the valuation of the assets in place.

Finally, a study by Davidson and Cheng (1997) based on 219 deals completed in the United States between 1981 and 1987 showed results that are inconsistent with the informational asymmetry. The authors thought that after the impact of the premium has been controlled, the payment method necessary remains a significant determinant of the returns, if a cash offer effectively reduces the informational asymmetry costs borne by the target's shareholders. They pointed out that the targets acquired through cash offers earn on average 39.93% of the premium, while those acquired through stock offers earn on average 29.25%. However, the authors showed that once the impact of the non-distributed cash flows of the multiple offers, of the premium, and of the relative size of the target compared to the bidder have been controlled, the payment method is not linked to the abnormal returns. Thus, for the authors, cash offers do not contain any additional information, they do not create additional value either, and, finally, they do not seem to reduce the informational asymmetry problems more than stock offers.

3 The Informational Asymmetry Models Subsequent to Myers and Maljuf (1984)

Many other models of informational asymmetry have been developed following Myers and Maljuf's work (1984). Among them, we find Hansen's model of bargaining under asymmetric information model (1987) and its extension developed by Eckbo et al. (1990), Fishman's preemptive bidding model (1989), and Berkovitch and Narayanan's model of informational asymmetry with competition (1990). The aim of this section is to present these models as well as the main empirical studies run to test their validity (Table 1).

3.1 Hansen's Model (1987)

Hansen (1987) imagined the case of a bidding firm with a monopolistic access to information concerning the true value of the merger. The optimal strategy for the bidder in this case is to make one offer only. In cash offers and when the target holds private information concerning the state of its assets, a problem can occur: The target will only accept to sell its stocks if their value is inferior to the bidder's offer. So as to protect itself from adverse selection phenomena, the bidder must base its optimal offer on "expected value conditional on the offer being accepted." Thus, the target using the information at its disposal will not always accept the offer, and as a consequence, the deal will not always take place.

Then, the bidding firm can use its own stocks instead of cash, because stocks have a contingent pricing effect, which, at the same cost for the bidder, encourages the target to accept all the offers that it would accept in cash. Actually, the main difference between a cash offer and a stock offer is that the value of a stock offer depends on the acquisition returns, in opposition to a cash offer. However, if we admit that the bidding firm can have private information about its own value, then a double problem is raised: The bidding firm will not offer stocks if the target underestimates the value of the offer (that is, if the bidding firm has information allowing it to think that its stocks are more worthy than the target thinks). Once again, it is an adverse selection problem that encourages the target to lower its valuation of the bidder's stocks.

Table 1. Empirical Studies Testing the Validity of the Models of Hansen (1987) and of Fishman (1989)

Tested Models	Authors	Conclusions of the Empirical Studies	Samples (Years)	Consistent/ Inconsistent
Hansen (1987)	Noronha and Sen (1995)	The probability of a stock offer has a negative correlation with the debt to asset ratio and a positive one with the leverage of the bidding firm.	74 US acquisitions (1980–1988)	Consistent
	Martin (1996)	No obvious relationship between the relative size of the target comparatively to the bidder and the payment method used in mergers and acquisitions.	846 US acquisitions (1978–1988)	Inconsistent
	Grullon et al. (1997)	In the banking sector, the larger the relative size of the target comparatively to the bidders size, the higher the probability of a stock or stock and cash financing.	146 US banks (1981–1990)	Consistent
	Houston and Ryngaert (1997)	High elasticity (that is, a stock financing) is more likely when the target is large and when the correlation between the returns of the target and those of the bidding firm are high.	209 US mergers (1985–1992)	Consistent
	Gosh and Ruland (1998)	The relative size of the target does not differ in a significant way according to the payment method.	212 US acquisitions (1981–1988)	Inconsistent
	Zhang (2001)	The larger the target in comparison with the size of the bidding firm, the higher the probability of a stock financing.	103 UK acquisitions (1990–1999)	Consistent
	Chemmanur and Paeglis (2003)	The relative size of the targets taken over with cash was not significantly different from that of the targets taken over with stocks.	437 US takeovers (1975–1995)	Inconsistent
Fishman (1989)	Franks et al. (1988)	The competition is higher in cash offers than in stock offers [see also Cornu and Isakov (2000)].	2500 UK/US acquisitions (1955–1985)	Inconsistent
	Cornu and Isakov (2000)	The probability to have competition in a hostile deal is weaker after a cash offer than after a stock offer. As a result, the returns of the initial bidder are higher when the deal is financed with cash.	86 acquisitions UK (1995–1996)	Consistent
	Chemmanur and Paeglis (2003)	The use of cash dissuades the potential bidders from entering into the competition.	437 acquisitions US (1975–1995)	Consistent

Nevertheless, equilibrium can occur (in the case of a double asymmetry) if the acquirer offers stocks when they are overvalued and if it offers cash when the stocks are undervalued. The target will interpret the payment method as well as the size of the bid as a signal indicating the bidder's value, and the bidder, conscious of the target's interpretation, will make the optimum choice as far as the payment method and the size of the bid are concerned, to confirm the target's valuation. In this model, no mixed offer can be observed, because the bidders use the foregone synergy cost to signal their value to the target. The first implication of this model is that there is a link between the probability for the bidder to offer cash only or stocks only, and the bidder's and target's sizes, as well as their level of debt. However, Hansen's study (1987), based on a comparison of means in a sample made up of 46 deals financed with stocks only and 60 ones financed with cash only, poorly confirmed his model.

Within the framework of Hansen's model (1987), the probability of a stock offer is inversely correlated to the relative size of the bidder in comparison with the size of the target, because contingent pricing characteristics of stocks depend on the relative size of the target's asset compared to the target. The larger the equity of the target compared to the bidder, the higher the contingent pricing effect of stocks.

Thus, this model predicts that the larger the target, the more important the informational asymmetry problems. Consequently, if the target represents an important weight comparatively to the bidding firm, the latter will more probably use stocks as a payment method. Martin (1996) tested this hypothesis, according to which the larger the bidding firm, the lower the probability of a stock financing, and the larger the target, the higher the probability of cash financing. He showed that the relative size of the target, measured by the ratio of the sum paid for the acquisition to the market value of the bidding firm, in the 20 days preceding the announcement, is not significant at the 5% level. These results therefore suggest that there is no obvious relationship between the relative size of the target comparatively to the bidder and the payment method used in mergers and acquisitions.

However, Grullon et al.'s study (1997) contradicted the empirical results of Martin (1996) since they found that, in the banking sector, the larger the relative size of the target comparatively to the bidder's size, the higher the probability of a stock or stock and cash financing. Another study, carried out by Ghosh and Ruland (1998), confirmed Martin's results (1996). Indeed, through a logit model, they found that the relative size of the target does not differ significantly according to the payment method. They accounted for this result by the fact that managers of large targets will prefer to obtain stocks in order to retain power in the merged firm. Conversely, managers of bidding firms have a strong incentive to offer cash to finance their acquisition if they do not want to dilute their current ownership. According to Ghosh and Ruland (1998), those two opposite incentives show that there is no clear link between the relative size of the two firms and the payment method used to finance the acquisition. On the other hand, Zhang (2001), through a sample made up of deals taking place in the United Kingdom, showed that the larger the target in comparison with the size of the bidding firm, the higher the probability of a stock financing.

Hansen's predictions (1987) were also confirmed by Noronha and Sen (1995) who showed that the probability of a stock offer has a negative correlation with the debt-to-asset ratio and a positive one with the leverage of the bidding firm, and by Houston and Ryngaert (1997), who showed that high elasticity (that is, a stock financing) is more likely when the target is large and when the correlation between the returns of the target and those of the bidding firm are high.

3.2 Fishman's Model (1989)

In his model, Fishman (1989) emphasized the role of the payment method in preemptive bidding for the control of the same firm by several rivals. Indeed, if a potential bidder makes an offer, other potential bidders will then study the offer, obtain information concerning the potential profitability of the offer, and perhaps enter into the competition. Therefore, a preemptive bid has to avoid this competition, because in the case of a competition between several potential bidders, the target's returns increase whereas the bidder's returns decrease as the competition goes on (Berkovitch and Narayanan, 1990; Bradley et al., 1988; De et al., 1996). If a firm competes with a bidder who proposed an important initial valuation for the target, this firm may have to face a small probability to win and small expected returns if it wins this competition. Thus, if the initial bidder sends a signal of high valuation, it can discourage competition. The fact that a high offer shows a high valuation and that this can serve to get ahead of the competition was demonstrated by Fishman (1988), on samples made up of offers in cash only.

Contrary to Hansen's model (1987), the target, like the bidder, is supposed to have access to private information on the profitability of the acquisition. In this case, stock offers become an interesting alternative to cash offers. Suppose the bidder offers an important sum if the target's information indicates that the acquisition is profitable and a low payment in the opposite case. This leads the target to make an efficient decision given its information. On the other hand, if the information is not verifiable, the offer is not feasible. The alternative is then to resort to a stock offer. Rather than making the offer dependent on the target's information concerning its future cash flows, a stock offer makes the value of the offer dependent on the cash flows themselves. If it is correctly built, a stock offer will encourage the target to make an efficient decision. The value of a cash offer, contrary to a stock offer, is not dependent on the target's future cash flows.

Thus, a target may make a decision regardless of the information concerning its cash flows and a cash offer may not imply an efficient decision.

Fishman (1989) developed a model of preemptive bid. At the equilibrium, stocks are offered by the bidding firms that have a low valuation of the target, and cash is offered by the firms that have a high valuation of it. The advantage of a cash offer is that at the equilibrium, it discourages a potential competition by sending a signal of high valuation. This model has various implications:

- The results expected by a bidder are lower if it uses stocks rather than cash in its first offer.
- The probability of a competition between different bidders is higher if the initial offer is in stocks rather than in cash.
- The probability for the target's management to refuse an offer is higher if the payment method used is stocks rather than cash.
- The more important the cost for studying the target, the more likely the initial offer will be in cash, and the less likely a competition between bidders will be.

In both models, the interest of using stocks is linked to their contingent effect. The difference between the two models lies in the interest of using cash. For Hansen (1987), bidders will pay in cash if they think that stocks are undervalued. For Fishman (1989), they pay in cash in order to signal an important valuation of the target and to discourage competition from other potential bidders, but Fishman (1989) does not propose empirical tests of his model in his paper. However, Franks et al. (1988) reported that in the 1955–1985 period, contrary to Fishman's predictions (1989), the competition is higher in cash offers than in stock offers. Chowdhry and Nanda (1993) also supposed that the bidding firm and the target own private information, but they only work on 100% cash and 100% stocks deals. They show that the existence of bondholders in the bidding firm and the use of debt as a payment method allow the bidder to make more aggressive offers, since a part of the acquisition's costs is borne by the existing bondholders. This can deter potential bidders from entering in competition. However, a cost can exist when the competition occurs in spite of the dissuasive effect. In this case, indeed, as a result, a bidding war between bidders happens, which can lead the various potential bidders to offer an excessive premium, that is, higher than the target's valuation.

Dodd's results (1980) seem to verify Fishman's predictions (1989) rather than Hansen's ones (1987). Cornu and Isakov (2000) showed that cash offers are more frequently associated with a competition between bidders than stock offers, because they are, by definition, more aggressive, since they are often used in hostile deals. On the other hand, they underlined the fact that cash offers allow one to signal a high valuation by the bidder even if a competition between bidders happens. Thus, cash offers have a more important dissuasive power than stock offers. In addition, they found, on both the theoretical and empirical levels, that the probability to have a competition in a hostile deal is weaker after a cash offer than after a stock offer. As a result, the returns of the initial bidder are higher when the deal is financed with cash. These results confirm Fishman's model (1989).

Chemmanur and Paeglis (2002), through a model that directly tests the informational asymmetry on both sides concerning the choice of the payment method by the bidder in acquisitions, showed that the bidders using stocks are indeed overvalued, and they found elements allowing one to think that the bidders using cash are undervalued. They proved that these estimation errors influence the choice of the payment method. Concerning the targets, they showed that preemptive bidding considerations are contingent upon payment ones. Their study also showed that cash dissuades the potential bidders from entering into the competition. Finally, they found results in contradiction with Hansen's theory of the relative size (1987).

On the basis of Fishman's work (1989), Nayar and Switzer (1998) studied the use of debt as a payment method used by bidding firms in their acquisitions. Fishman (1989) indeed asserted that debt also has a contingent price effect because, as in case of stocks, stockholders are compelled to receive a security from the firm after the deal. Although debt is less sensitive than stocks, the final value of the security attained by the target's shareholders depends on the future profitability of the deal. Therefore, there is an important distinction between acquisitions financed with debt and those financed with cash, since, whatever the way the money has been raised (capital increase, debt, or self-financing), the shareholders no longer dispose of stocks after the deal, and therefore a cash offer has no contingent impact on the price.

3.3 Eckbo, Giammarino, and Heinkel (1990)

Eckbo et al.'s model (1990) is an extension of the model developed by Hansen (1987). It claims that an informational asymmetry on both sides between the bidder and the target can lead to an optimal mix of cash and stocks as a payment method. The authors showed a separating equilibrium for which the true post-acquisition value of the bidding firm is revealed to the target by the composition of the mixed offer and where this value is increasing and convex in the amount of cash used in the offer. They argue that the abnormal returns of the bidding firm are made up of two parts: a component related to synergy revaluation and a component for the signal. According to this model, a cash offer results in no signal. The abnormal returns of the bidding firm are linked to the revaluation of synergy. In the case of stock offers, it is the opposite. Abnormal returns result from the signal of the stock offer. Only in a mixed offer can both a signal effect and a synergy revaluation effect take place. In addition, they argued that the target's stock price increases at the acquisition announcement, by an amount that is independent from the chosen payment method. This result is due to the fact that the bidder is compelled to make an offer that is acceptable for every kind of targets and, as a consequence, there is no separation between the various targets.

On the contrary, according to Fishman (1989), the reaction of the target's stock price at the announcement of the acquisition is ambivalent. The stock price of the target will react positively to the cash offer, only if the preemptive cash offer, is superior to the expected value of the target. Such ambivalence also exists in Hansen's model (1987).

In the previous models developed by Fishman (1989), Hansen (1987), and Eckbo et al. (1989), the bidders, whose private information is positive, use cash as a payment method, which can explain why the bidders' stock prices react more favorably to the announcement of a cash offer than to the announcement of a stock offer.

In the last part of their study, Eckbo et al. (1989) have empirically tested their model on a sample of 182 Canadian deals, among which 56 are mixed offers. The abnormal returns observed are positive and significantly higher for mixed offers than for cash-only and stock-only offers. Therefore, the empirical results do not allow one to confirm the model's predictions.

3.4 Berkovitch and Narayanan's Model of Informational Asymmetry with Competition (1990)

Bradley et al. (1988) showed that, when there is competition for the target in a tender offer, the average abnormal returns are lower for the bidding firm and that they are higher for the target in case of competition.

Berkovitch and Narayanan's model (1990) studied the role of the payment method in the competition between bidders and its effects on the returns of the target's and the bidder's shareholders. Their theory is consistent with the previous works. In this model, there are two types of bidders: high-type bidders and low-type bidders. The merged firm value is higher for high-type bidders than for low-type bidders. A potential bidder makes an offer with a given payment method, and this offer can be rejected or accepted by the target. If the offer is rejected, there is a time period during which no new offer can be realized by the existing bidders. During this period, other potential bidders can enter into the competition. If it is actually the case, there is a competition between the two potential bidders and, the highest offer can be rejected or accepted by the target. If the offer is rejected, the process is repeated after a new time period.

Thus, this model comes within a framework of informational asymmetry, where the target earns a higher sum if it is acquired by a high-type bidder but earns a higher proportion of synergies if it is acquired by a low-type bidder. This result is due to the fact that the low-type bidder will have to face a higher competition than a high-type bidder and that it will be ready to offer the target a higher proportion of the created synergies.

If the bidder is conscious of the kind of bidders he belongs to, then there is a unique separating sequential equilibrium in which the high-type bidder uses a higher amount of cash, and the low-type bidder uses a higher proportion of stocks. The value of the offer is the same as in the case of symmetric information. Since the fraction of synergy offered by low-type bidders is higher than the one offered by high-type bidders, the latter have no incentive to imitate the former by offering stocks. Similarly, since the value of the offer made by low-type bidders is lower than the one realized by high-type bidders, the former have no incentive to imitate the latter by offering cash. As in the models of informational symmetry, the offers are accepted without delay.

Berkovitch and Narayanan's model (1990) allows one to study the interaction between the informed bidders and the uninformed target. The dynamic structure of the model allows one to show that

- In mixed offers, the more important the amount of cash, the higher the abnormal returns of the target and of the bidder.
- The proportion of synergy captured by the target decreases with the level of total synergies. However, the higher the proportion of cash, the lower the proportion of synergy captured by the target.
- The dollar amount of synergy captured by the target increases with the level of total synergy and the proportion of cash. When the competition (whether it is real or not) increases, the proportion of synergy captured by the target increases. The target's payoff is higher when the competition is real than when it is potential.
- When the potential competition increases, the amount of cash used to finance the deal also increases. Moreover, the amount of cash as a proportion of the total offer increases if the target's payoff is a concave function of the synergy.
- In the case of real competition, all firms but the lowest type make cash-only offers. Because of this, cash-only offers are more profitable than mixed and stock-only offers.

De et al. (1996) showed that cash-only offers and stock-only offers are more competitive than mixed offers. Moreover, the authors did not prove that cash offers generate less competition than stock offers. Thus, these results are inconsistent with Fishman (1989) and with Berkovitch and Narayanan (1990).

4 The Impact of Taxation on the Choice of the Payment Method

Numerous authors emphasize the fact that the choice of the payment method is influenced by taxation (Table 2). The advantage linked to the taxation of a given payment method corresponds to a disadvantage for another one. Indeed, cash offers are considered immediately taxable for the target's shareholders. Thus, a cash offer requires the payment of a higher premium in order to compensate for the tax burden. This additional premium may be offset by the tax advantage linked to the possibility of writing up the value of the purchased assets to market value for depreciation purposes. On the contrary, stock offers are generally nontaxable until the stocks are sold. In order to benefit from this advantage, the offer needs to be composed of at least 50% stock. Thus, although the deals financed with cash have an advantage over

Table 2. Empirical Studies Testing the Validity of the Models of Eckbo et al. (1990), of Berkovitch and Narayanan (1990), and the Taxation's Hypothesis

Tested Models	Authors	Conclusions of the Empirical Studies	Samples (Years)	Consistent/ Inconsistent
Eckbo et al. (1990)	Eckbo et al. (1990)	The empirical results do not allow one to confirm the model's predictions.	182 Canadian acquisitions (1964–1982)	Inconsistent
Berkovitch and Narayanan (1990)	De et al. (1996)	Cash-only offers and stock-only offers are more competed than mixed offers. Moreover, the authors do not achieve to prove that cash offers generate less competition than stock offers.	958 U.S. tender offers (1962–1988)	Inconsistent
Taxation	Wansley et al. (1983)	The higher returns obtained by the target's shareholders in cash deals offset the impact of complementary taxes (see also Huang and Walking, 1987).	203 U.S. acquisitions (1970–1978)	Consistent
	Franks et al. (1988)	There is no clear evidence that capital gains taxation is the main reason for the choice of the payment method. The target's abnormal returns in tender offers are higher than the ones in mergers, even after the impact of the payment method has been controlled (see also Suk and Sung, 1997).	2500 UK/U.S. acquisitions (1955–1985)	Inconsistent
	Noronha and Sen (1995)	The propensity to realize stock offers is positively linked to the accumulated tax credits of the target.	74 U.S. acquisitions (1980–1988)	Consistent
	Carleton et al. (1983)	Lower dividend payout ratios or lower market-to-book ratios increase the probability of being acquired in cash.	61 U.S. targets (1976–1977)	Consistent
	Gilson et al. (1988)	No direct link between tax benefits and the payment method (quoted by Yook 2003).		Inconsistent
	Niden (1986)	No relation between the tax situation of the target's shareholders and the payment method (quoted by Amihud et al., 1990).		Inconsistent
	Auerbach and Reishus (1988)	The tax savings due to the use of the target's losses and credits are not significant enough to explain the payment method.	318 M&A (1968–1983)	Inconsistent
	Eckbo and Langhor (1989)	The post-expiration premium is practically the same, whether the deal has been financed with stocks or with cash.	French deals (1972–1982)	Inconsistent
	Brown and Ryngaert (1991)	Stocks are used for the tax advantage and mixed offers often use almost the maximum possible amount of cash, while enjoying the tax free status.	342 U.S. mergers (1981–1986)	Consistent

the deals financed with stocks from the bidder's point of view, they require the payment of a higher premium. The amortization of this goodwill will artificially decrease the bidder's earnings. From the taxation point of view, and without any knowledge of the signal effect, the bidder's shareholders will prefer a cash offer if the premium offered to the target's shareholders is not higher than the taxation advantages of the deal. The managers will be favorable to a stock offer in order to avoid the artificial decrease of the returns linked to the amortization of the goodwill (Blackburn et al., 1997).

The attractiveness of a target increases if it has accumulated tax losses and tax credits. As Noronha and Sen (1995) as well as Brown and Ryngaert (1991) concluded, the continuity of interests is required for a firm to inherit from the favorable aspects of taxation. From a legal point of view, two conditions are needed. First, the majority of the target's stocks has to be acquired in exchange for the bidder's stocks. Thus, the target's shareholders will partially be owners of the merged entity. The second condition is the continuity of the target's operations. The acquisition has to have a legitimate goal, which will, thus be proved, in case of a continuity of the target's activity. If these conditions are verified, the merger becomes exempt of tax: The capital losses or gains of the target's shareholders can be postponed and the tax attributes of the target can be inherited. The notion of continuity of interests also applies to the taxation of firms. In a nontaxable deal, the tax credits not used by the target and the loss carryovers can be deduced from the taxable earnings of the future merged firm, since the target shareholders have kept sufficient ownership. In a taxable offer, ownership rights are considered sold, and the bidder has the right to set up the depreciation basis of the assets purchased.

American tax laws allow one to carry back the net operating losses 3 years and to carry them forward 15 years. The present value of this carryover is weak unless the firm has been profitable enough before and after the losses. However, the value of these tax characteristics increases when the losses are transferred to a bidding firm that has important earnings before tax.

The payment method also affects the accounting treatment and has tax implications for the bidder. The main accounting treatments for acquisitions are the pooling of interest and the purchase accounting. The first technique allows the acquirer to register the acquisition in the group's accounts on the basis of the book value. The stocks issued by the bidder are then registered in the balance sheet for the book value of the target's equity. For example, in France, for this technique to be used, the deal needs to have been made in one time, to include at least 90% of the target's equity, to have been paid through the issue of new stocks, and not to be called into question during the two years following its realization. The IAS norms also demand that bidders cannot be identified. This technique has been abolished in the United States. On the contrary, when the operation is financed with more than 10% cash or assimilated (contingent value right), the purchase accounting technique applies: The target's assets and liabilities are reevaluated and the goodwill is amortized on a more or less long period among the intangible assets. This technique is used in the Unites States to register takeovers.

From the conceptual point of view, these two methods should be equivalent since they do not affect the future cash flows of the bidding firm: The amortization of the goodwill reduces the declared returns, but it is not deductible from taxation. However, as Huang and Walkling (1987) emphasized, the choice of the payment method can be linked to the rewarding method used to compensate the managers. Thus, the managers whose compensation is linked to book performance measures will prefer to avoid the amortization of the goodwill.

According to Erickson and Wang (1999), some managers who use stocks as a payment method can be incited to use discretionary accruals aggressively, in order to temporarily and artificially increase the stock value of their firm, and therefore to reduce the effective cost of their acquisitions. Their empirical study confirmed this hypothesis and showed that the returns linked to book manipulations increase with the relative size of the bidder. If managers do have recourse to discretionary accruals, a decrease in the operating performance should occur, because accounting procedures compel accruals to inverse as time goes by. By examining the returns of pre-acquisition and post-acquisition over a sample of 947 acquisitions, Loughran and Vijh (1997) showed that the target's shareholders make gains when they sell their stocks not a long time after the acquisition's effective date, but those who retain the stocks received as a payment method see their gains decrease over time. Even worse, for one of their subsamples they found that the shareholders suffer losses. On the contrary, Heron and Lie (2002) showed in their sample that there is no proof of these phenomena concerning the three years preceding the acquisition. Moreover, contrary to Erickson and Wang (1999), they found no differences in the use of discretionary accruals between the different payment method used by the bidder (cash, stocks, and mixed offers).

Wansley et al. (1983) jointly studied the taxation and the payment method. They showed that differences in returns exist between the different payment methods used: The average abnormal return at the announcement day in cash offers is significantly superior to the one observed in stock offers. The residuals in cash offers are superior to the one in stock offers from the 23rd day preceding the announcement, and this difference increases until the day of announcement.

During the 41 working days following the announcement of the acquisition, they found average cumulated abnormal returns of 33.54% for the target when the deal is financed with cash, of 17.47% when the deal is financed with stocks, and of 11.77% for mixed offers. They accounted for this important difference between cash offers and stock offers by the impact of taxation. Indeed, they explained that the higher returns obtained by the target's shareholders in cash deals offset the impact of complementary taxes.

Harris et al. (1987) also showed that cash offers produce higher abnormal returns for targets. Through a sample made up of 2,500 acquisitions that have taken place in the United Kingdom and in the United States between

1955 and 1985, they showed that, in both countries, cash-only offers and stock-only offers have been the most frequent payment methods in mergers and acquisitions. They explained this result by the fact that the shareholders who care about capital gains taxation will be ready to accept stock offers, and that the others will accept cash: That is what they called the tax and transaction cost efficiency. On the other hand, they emphasized that there is no clear evidence that capital gains taxation is the main reason for the choice of the payment method.

Huang and Walkling (1987) confirmed the previous works thanks to a sample of 204 operations during the 1977–1982 period by showing that average cumulated abnormal returns for cash offers are of 29.3%, whereas they are of 14.4% in stock offers and are of 23.3% in mixed offers. They also explained this result by the impact of taxation.

According to Brown and Ryngaert's model (1991), the taxation aspect plays an important role in the determination of the payment method. The bidding firm takes into consideration the target's valuation of the stocks offered by the bidder as well as the tax consequences of the offer given the payment method. The equilibrium is consistent with the returns observed for the bidder, but the model also allows one to make different predictions from those that only consider the informational role of the payment method. For example, since the use of stocks is explained by tax advantages only, stocks should not be used in taxable deals. Moreover, this model supposes that every non taxable deal, that is, stock and mixed offers, reveals negative information concerning the bidder. The bidders who have a low valuation for their firm thus use at least 50% stocks to avoid the taxation of the deal, and the bidders who have a high valuation of their firm use cash to avoid the stocks issued to be undervalued. The empirical results they presented are consistent with the idea according to which stocks are used for the tax advantage: Only 7 taxable deals out of the 342 constituting the sample used stocks and only 12 taxable deals used securities that could be transformed into stocks. Out of the 131 nontaxable deals, 86 were stock offers and 45 were mixed offers (34 were deals using more than 50% stock). Thus, mixed offers often use almost the maximum possible amount of cash, while enjoying the tax-free status. This result shows that taxation plays an important role in the choice of the payment method in the United States. Moreover, in accordance with the model's predictions, the results show that the abnormal returns of mixed offers and stock offers are negative. The abnormal returns associated with cash offers are zero and are significantly higher than those associated with stock offers or mixed offers. On the other hand, these results are not consistent with the idea according to which bidders signal a higher asset value through a more important use of cash in nontaxable deals: The results associated with mixed offers are not noticeably different from those associated with stock-only offers.

The hypothesis of the role of taxation in the choice of the payment method was also verified by Noronha and Sen (1995), since they showed that the propensity to realize stock offers is positively linked to the cumulated tax credits of the target.

Carleton et al. (1983) modeled the decision of the bidder concerning the payment method according to the target's characteristics. They showed that lower dividend payout ratios or lower market to book ratios increase the probability of being acquired in cash. Low market-to-book ratios linked to the use of cash are inconsistent with tax effect, since high market-to-book ratios represent important potential tax savings due to higher depreciation. On the contrary, this result is consistent with a second tax aspect, since the book value is close to the basis on which capital gains liabilities of the target's shareholders are calculated.

On the other hand, Gilson et al. (1988) showed that there is no direct link between tax benefits and the payment method. Niden (1986) found no relation between the tax situation of the target's shareholders and the payment method; for Auerbach and Reihus (1988), the tax savings due to the use of the target's losses and credits are not significant to explain the payment method.

If we consider the tax advantage linked to the use of debt, then a stock offer can have a negative impact on the stock price of the firm (Modigliani and Miller, 1963; DeAngelo and Masulis, 1980; Masulis, 1980a, 1980b). In this view, according to Nayar and Switzer (1998), the use of debt securities can entail tax advantages, since the interests of the debt offered to shareholders are deductible for the bidder. The bidder can offer either cash or debt to avoid a stock price decrease that would occur in case of a stock offer, but an offer with debt will, however, be preferred if the bidder needs an important tax reduction. Thus, according to Nayar and Switzer (1998), a debt issue is a signal to the market that the firm anticipates its ability to exploit the tax reduction linked to the payment of the new debt's interests. They confirmed their hypothesis through an empirical study that showed that for the firms using debt in their offer, the higher the tax rate, the more positive the market's reaction.

Contrary to the suggestions of the informational and tax hypotheses, Franks et al. (1988) and Suk and Sung (1997) showed that the target's abnormal returns in tender offers are higher than the ones in mergers, even after the impact of the payment method has been controlled. The latter also showed that there is no relation between the offer premium and the institutional ownership of the target in cash offers and that there is no difference in premiums between cash offers and stock offers, even after the institutional ownership and other variables linked to taxation have been controlled. These results are also inconsistent with the informational and tax hypotheses.

On the contrary, Eckbo and Langohr (1989) showed, through a French sample of deals realized between 1972 and 1982, that the informational hypothesis seems to dominate the tax hypothesis. Indeed, they showed that the average premium is 17.2% in stock offers, and 73.3% in cash offers whereas the post-expiration premium is practically the same, whether the deal has been financed with stocks (23.7%) or with cash (22.5%). This result is inconsistent with the tax theory, because the post-expiration premium should be higher for cash offers, if the firms using this payment method pay a higher premium in order to compensate for the negative impact of taxation.

5 The Theories Linked to Managerial Ownership and to Outside Monitoring

Most empirical studies run on the impact of the announcement of a merger or acquisition on the shareholder's wealth show that the bidder's shareholders earn less than the target's shareholders. Some studies even show that acquisitions are beneficial only to the target's shareholders (Langetieg, 1978; Dodd, 1980; Morck et al., 1990). One of the explanations generally proposed is the agency theory. Jensen and Meckling (1976) defined the agency relationship "as a contract under which one or more persons (the principal(s)) engage another person (the agent) to perform some service on their behalf which involves delegating some decision making authority to the agent." Then, conflicts of interest can occur between managers and shareholders, due to the separation of ownership and control. Such is the case when the managers do not own all the stocks of the firm they run: They can be incited no to serve the shareholders' interests anymore, but to give more importance to personal interests because of a lack of internal monitoring.

This theory relies on the empire-building theory, which suggests that managers deliberately overpay the acquisition. Their compensation and their power are functions of the increase of their responsibility inside the firm, that is, of the asset quantity they control; they launch the takeover process only to maximize their own welfare, to the detriment of their shareholders (Rhoades, 1983; Williamson, 1964; Jensen, 1986). Thus, according to this theory, managers are inclined to try to increase the firm's size (even beyond the optimal size) and not to optimize its profitability. Believing that all the mergers and acquisitions deals are motivated by shareholder wealth creation therefore seems to be illusory. Nevertheless, different elements can influence the managerial motivations in mergers and acquisitions. Among these elements, we find in particular the managerial ownership and the outside monitoring. The managerial ownership theory stipulates that managers' interests converge toward those of shareholders as the managerial ownership increases. In other words, the more the manager owns stocks of the bidding firm, the more he will be incited to make wealth-creating deals for shareholders.

Amihud and Lev (1981) argued that imperfect monitoring allows managers to realize actions that are in their own interest and not necessary in the shareholders' interest. Managers can thus have the will to diversify the firm's activities in order to reduce the risk of losing their job (preventing the firm where they work to be acquired, diversifying risk, and minimizing financial bankruptcy cost, etc.). This incentive is contrary to the shareholders' wealth maximization, since the shareholders can reduce their own risk by diversifying their portfolio for a lower cost in an integrated market. From this perspective, they showed in addition that the constitution of conglomerates is more likely when shareholding is widespread, since, in those cases, managers are able to pursue strategies that serve their own interest. Thus, a better control of managers' actions could reduce agency costs. Gompers and Metrick (2001) showed on that matter that a firm's profits are higher when the proportion of stocks held by institutional investors is high. This result is consistent with Amihud and Lev's theory on monitoring (1981), since an important quantity of stocks held by institutional investors means a more effective control of firms by shareholders. Ang et al. (2000) showed that agency costs are more important when a nonshareholder manager runs the firm and that they vary inversely with the proportion of stocks he holds. Moreover, they found evidence that these costs increase with the number of nonmanager shareholders. Finally, they proved that outside monitoring by banks allows one to reduce these costs.

5.1 The Managerial Ownership Theory

Managerial ownership refers to the number of stocks held by management and the insiders. Many authors think that the choice of the payment method used in a merger or an acquisition depends on the managerial ownership of the two firms implied in the deal (cf. Table 3). Thus, it is often supposed that the more stocks the bidder's management owns before the deal, the more likely cash will be used to avoid dilution phenomena. On the contrary, the more stocks the target's managers own, the more likely stocks will be used to allow managers to retain power in the future merged firm.

Moreover, by increasing the debt and by using the means that allow one to reduce the number of stocks held by the public, the owner-manager increases the probability to retain the control of the firm as well as the profits linked, since exchanging debt for external stocks reduces the proportion of investors' control rights that can be acquired by a raider.

A stock offer increases the number of stocks held by outsiders. The proportion of stocks held by the managers therefore decreases (if we suppose that the managers do not increase their ownership in the firm). Thus, this reduction implies a decrease of the firm's value according to the adverse selection model of Leland and Pyle (1977) and to the agency theory of Jensen and Meckling (1976). Indeed, according to them, the more important the decrease of stocks held by managers, the more the stock value will decrease. Travlos (1987) found evidence that is consistent with these hypotheses.

Stulz (1988) studied the relationship between the choice of the payment method and the managerial ownership of the bidding firm. He showed that the use of new debts reduces the firm's attractiveness for corporate raiders. The owner-managers who want to retain the control of their firm will therefore prefer to use cash in order to avoid the threat of a hostile takeover. However, Harris and Raviv (1988) noticed that increasing the debt also reduces the probability to retain the control of the firm, since the bankruptcy risk increases

Table 3. Empirical Studies Testing the Validity of the Theories Linked to the Managerial Ownership, to Outside Monitoring, to the Positive Impact of Debt and to the Free Cash Flows

Tested Models	Authors	Conclusions of the Empirical Studies	Samples (Years)	Consistent/ Inconsistent
Managerial ownership	Martin (1996)	A lower managerial ownership is negatively correlated with the probability of a stock financing, although the relationship is not linear.	846 U.S. acquisitions (1978–1988)	Consistent
	Amihud et al. (1990)	In cash offers, the five most important managers hold close to 11% of the firms stocks, whereas in stock offers, they hold less than 7%.	209 U.S. acquisitions (1981–1983)	Consistent
	Ghosh and Ruland (1998)	In stock offers, the average managerial ownership of the target, is significantly higher, whereas in cash offers, it is the bidders managerial ownership that is higher.	212 U.S. acquisitions (1981–1988)	Consistent
	Yook et al. (1999)	The higher the managerial ownership of the bidding firm, the more inclined to use cash the bidder will be. Moreover, manager-owners more often sell their stocks before the announcement of a stock offer than before a cash offer.	309 U.S. acquisitions (1979–1988)	Consistent
Outside monitoring	Martin (1996)	The higher the proportion of institutional investors, the lower the probability of stock financing. However, the existence of external stockholders does not seem to have an impact on the choice of the payment method.	846 U.S. acquisitions (1978–1988)	Consistent/ Inconsistent
Positive impact of debt	Maloney et al. (1993)	The bidders returns vary positively and significantly with the preexisting level of leverage and with leverage changes.	428 mergers (1962-1982) 389 acquisitions (1982-1986) 173 acquisitions (1978-1990)	Consistent
Free cash flow	Martin (1996)	The more the bidding firm has cash, the less likely stock financing.	846 U.S. acquisitions (1978–1988)	Consistent
	Zhang (2001)	The higher is the dividend payment (which is a measure of the available free cash flow), the more likely the deal will be financed by cash.	103 UK acquisitions (1990–1999)	Consistent
	Noronha and Sen (1995)	The amount of free cash flow of the bidder is directly linked to the probability of cash financing.	74 U.S. acquisitions (1980–1988)	Consistent

and because of the increased restrictiveness of loan covenants, and the greater commitment for future cash payments. Stulz (1988) also admitted that his argument has to be somewhat nuanced, since an increase in debt reduces the total value of stocks in circulation, so that it becomes cheaper for a bidder that has an increasing marginal cost, to acquire stocks. Moreover, Stulz (1988) showed that the more important the proportion of stocks held by the bidding firm, the less likely the acquisition is to be financed with stocks. Indeed, in this case, the bidder's management is not inclined to offer stocks as a payment method, so as not to dilute its control after the deal. Moreover, he showed that the probability of a hostile deal is low if the proportion of control rights held by the target's management is high, since the target, which owns an important fraction of his stocks, will want more rights after the deal.

Martin (1996), through a sample made up with 846 deals, showed that a lower managerial ownership is negatively correlated to the probability of a stock financing, although the relationship is not linear. He showed indeed that the managers who own less than 5% of the stocks are not concerned by the effects of control dilution. On the other hand, when they own between 5 and 25% of the stocks, they consider the dilution an important element in the choice of the payment method. Beyond 25%, the link between managerial ownership and the probability of cash financing is no longer verified.

Amihud et al. (1990) established a link between the preferences of the bidding firm's managers and the payment method. Through a sample of 209 American acquisitions in the 1981–1983 period, they showed that in cash offers, the five more important managers held close to 11% of the firm's stocks, whereas in stock offers, they held less than 7%. Thus, managers who own an important fraction of stocks will prefer to finance the acquisition by cash rather than by stocks, because they value having control of the firm and do not want to increase the risk of losing it after the merger. Moreover, they showed that the negative abnormal return of a stock offer mainly concerns firms whose managers did not own a lot of the bidder's stocks. According to these results, the extent of managerial ownership in firms announcing new stocks issues therefore affects abnormal returns.

Song and Walkling (1993) showed that the managerial ownership of target firms is significantly lower than the managerial ownership of nontarget firms from the same sector. In other words, a firm with low managerial ownership is more likely to be the target of a deal. They also showed that a target whose managerial ownership level is low has a higher probability to be competed by several bidders: The average managerial ownership in competed offers is around 6.4%, which seems to be very different from the 18.7% observed for nontarget firms in the same sector. Moreover, they emphasized through the use of a logistic regression that the probability to be a target is inversely correlated to the managerial ownership, which means that firms whose managerial ownership level is high have a lower probability to be the subject of an attack. Finally, they showed that for successful deals in which the managerial ownership of the target is low, the average abnormal return is 29.5%, whereas in the case of unsuccessful offers, it only reaches 5.2%. They explained this result through a regression that indicates that there is a positive and strong relationship between the target's managerial ownership and the abnormal return in competed but finally successful deals. In other cases (unsuccessful deals, noncompeted deals), this relationship is not significant.

According to Blackburn et al. (1997), if managers undertake deals that are not in the interest of their shareholders, then the ownership structure of the firm plays an important part in case of mixed offers. The managers who own few stocks will not suffer much from the negative re-evaluation of stocks described by Eckbo et al. (1990). They can therefore accept a mixed offer even if their firm is overvalued, as long as their private profit is higher than the loss of value of the stocks they hold. On the contrary, managers who own an important stake in the firm will be less likely to realize a mixed offer if their firm is overvalued. Thus, lower returns are expected in mixed offers realized by firms that are not run by their owner. In addition, consistently with Amihud et al. (1990), Blackburn et al. (1997) showed that the manager-controlled firms that make stock offers suffer significant losses, but contrary to Amihud et al. (1990), they also showed that the firms run and controlled by their managers suffer significant losses. The negative signal associated with stocks, use for firms run and controlled by their managers is not eased by the fact that managers do not want to dilute, or even, in some cases, to lose control of their firm by investing in a deal that is not wealth-creating.

Ghosh and Ruland (1998) studied the way target and bidder managers' preferences for control rights influence the choice of the payment method in a takeover. In a study including 212 deals in the United States between 1981 and 1988, they showed that in stock offers, the average managerial ownership of the target is significantly higher, whereas in cash offers, the bidder's managerial ownership is higher. Thus, the bidding firms will prefer to use cash when their managerial ownership is high, and, conversely, targets will prefer to use stocks if the management of the target wants to obtain influence in the merged firm. The results show that the managerial ownership of the target is more important than the managerial ownership of the bidding firm to account for the choice of the payment method.

Yook et al. (1999) confirmed the result according to which the higher the managerial ownership of the bidding firm, the more inclined to use cash the bidder will be. In addition, they showed that manager-owners more often sell their stocks before the announcement of a stock offer than before a cash offer. This result is consistent with both the management control and the informational asymmetry theories. Moreover, they showed a negative relationship between the sale of stocks by owner-managers before the announcement and the abnormal return at the announcement. In other words, the more stocks that are sold by insiders before the announcement of the deal, the higher the abnormal losses realized by the bidder. So, this result implies that insiders have access to information on which they base the sale of their stocks. The authors concluded the existence of informational asymmetry phenomena on the market of corporate control and that they can influence the choice of the payment method.

5.2 The Outside Monitoring Hypothesis

According to Berle and Means (1932), the passivity of shareholders cannot be prevented. Firms grow so much that they call up new shareholders to raise capital. As a result, every shareholder owns a small part of the firm's stocks. If a shareholder decided to play an active role in the monitoring of managers, he would only get a small part of the profit (the other shareholders would benefit from this monitoring, too), but he would bear all the costs alone. Thus, the passivity serves the personal interest of every shareholder, even if managers' control assures collective gains. In the same way, a shareholder's vote is not likely to have any influence on the realization or not of a project, which is why numerous shareholders become apathetic and accept the management's propositions without evaluating them with care.

However, the shareholders will keep on controlling the managers' actions if the returns linked to the managers' control are superior to the costs they have to bear. A shareholder who owns an important stake in the firm will be more likely to accept to control the manager's actions than a shareholder who owns only a small stake in the firm. Thus, for Jensen (1991), active external investors such as blockholders and institutional investors benefit to the firm because they are incited to undertake a costly control of the firm's functioning. Black (1992) thought that institutional investors can control the managers' actions in order to bring the managers' interests into line with the shareholders' interests. For example, some institutional investors have contacts with top managers and can thus influence the characteristics of the deal. Since numerous empirical studies show that a stock offer has a negative impact on the wealth of the bidder's shareholders, the probability of a stock offer should be lower when the proportion of institutional investors or of stockholders increases. Martin (1996) partially confirmed this hypothesis since he showed that the higher the proportion of institutional investors, the lower the probability of a stock financing. However, the existence of external blockholders does not seem to have an impact on the choice of the payment method.

5.3 The Free Cash Flows Hypothesis

Jensen (1986) defined the free cash flows as cash flows in excess of those required to finance the positive net present value investment projects of the firm. The shareholders will prefer managers to distribute this excess of cash rather than to invest in projects with low returns. However, it is possible that managers have incentives to retain this supplementary cash in order to make the firm grow beyond the optimal level, their compensation often being correlated to the sales rather than to the profit. From this perspective, tender offers are both evidence of the conflict between managers and shareholders and a solution to the problem. Therefore, this theory suggests that bidders with high free cash flows or with a sufficient debt capacity will be inclined to realize cash offers rather than stock offers.

The increase of the leverage allows one to explain the higher returns in cash operations. Indeed, this increase does not only make managers work harder for fear of the bankruptcy of the firm, but it also allows one to limit the agency costs of free cash flows by reducing the cash flows freely usable by managers. According to Jensen (1986, 1988), Harris and Raviv (1990), and Stulz (1990), the debt has a positive impact, since it allows one to reduce the agency phenomena and increases the managers' control. The profit linked to this theory is all the more important as the bidding firm does not have many growing opportunities, has an unused supplementary debt capacity, and has important free cash flows, since in this case, the probability that this deal results in low wealth creation, or even in value destruction, is higher. Thus, the use of debt can allow managers to signal that the future returns of the acquisition will be high enough for them to be able to repay the interests linked to the debt, and by this way, they will not have to pay an excessive price for the target. Henceforth, resorting to debt allows managers to show that they are not motivated by agency phenomena.

Maloney et al. (1993) studied the relationship between the bidding firm's returns in the announcement period and the leverage level before the deal, as well as the changes in the leverage after the deal. They showed that the bidder's returns vary positively and significantly with the pre-existing level of leverage and with leverage changes. Then they concluded that debt improves the managers' decisions.

Yook (2003) tested this hypothesis through the changes in Standard and Poor's debt rating. When the bidding firms offering cash are divided into two sub-groups, one in which the rating has been reduced, and the other in which it has not changed, those that have suffered a decrease obtain an abnormal return in two days ($t = -1$ and $t = 0$) significantly higher than the one obtained by the other group, which corroborates the theory of the benefit of the debt. This result contradicts the signal theory because the rating of the bidding firm should remain unchanged or improve if the synergies of the deal offset the increase of the financial risk. On the contrary, when bidding firms offering stocks are grouped together in three groups according to whether the rating is reduced, remains unchanged, or is improved, the results strongly contrast with the ones of cash-financed bidders. The abnormal return is positive and significant for the group with an upgraded rating, whereas it is negative for the group whose rating is reduced. This result is explained by synergies phenomena according to the author and implies that cash acquisitions and stock acquisitions are two distinct kinds of deals, with different sources of value creation. Cash offers are used in hostile financial deals and stock offers in friendly strategic deals.

According to Mayer and Walker (1996), the choice of the payment method depends on the financial conditions of the target and of the bidder. The financing cost of a cash acquisition includes an increase of the interests and therefore of the bankruptcy risk. An additional debt or a decrease of cash or of liquid assets can be costly for a firm that already owns debt. The propensity to use cash is lower when the free cash flows and liquidity decrease and when the leverage increases. In addition, a change in the monetary policy can have a positive impact on interest rates and on the credit availability, which can affect the choice of the payment method. For example, a higher interest rate and a narrower credit market are problematic for the liquidity and will make a cash offer less attractive.

Firms with large liquidity but few internal growth opportunities are supposed to own important cash flows. Martin (1996) confirmed Jensen's free cash flow theory, since he underlined that the more the bidding firm has cash, the less likely a stock financing will be. Zhang (2001) showed that the higher the dividend payment is (which is a measure of the available free cash flow according to the author), the more likely the deal will be financed by cash.

Noronha and Sen's results (1995) are also consistent with Jensen's free cash flow theory, since they showed that the amount of free cash flow of the bidder is directly linked to the probability of cash financing. Actually, Shrieves and Pashley (1984) showed that the firms that use cash increase their leverage, whereas those using stocks do not. However, before the acquisition, the authors showed no differences in the leverage of bidding firms using stocks and those using cash.

6 The Past Performances, the Investment Opportunities, and the Business Cycles

Indicators of the past performances of the bidder and of the target also have an impact on the choice of the payment method used to finance the deal. A firm run inefficiently is a potential source of returns for the bidding firm. In this case, the latter will be more likely to finance its deal with cash in order to ensure the control of the target and to replace the inefficient management team. Conversely, if the bidding firm has realized particularly important stock performances recently, the use of stocks can turn out to be attractive for the target and the bidder.

Investment opportunities of the bidding firm also seem to influence the choice of the payment method. Indeed, the more investment opportunities the bidding firm will have, the more it will be incited to use stocks to finance its investments, because this payment method gives the manager more freedom than debt, which will allow one to fully profit from the investment opportunities. Conversely, debt is beneficial for firms with low investment opportunities, because debt requires the payment of cash flows that cannot be used to invest in low-profit projects. Finally, business cycles are also supposed to influence the choice of the payment method in such a way that the probability of stock financing increases when the general business activity increases.

6.1 The Theories of the Inefficient Managers' Replacement and of the Financial Characteristics

According to Zhang (2001), the fact that the target realizes low pre-acquisition performances means that the target is run inefficiently. In such a case, the bidding firm is more inclined to use cash, so as to replace the target's management. The author was not able to confirm this hypothesis in an empirical study in which the target's performances before the deal are measured through its return on equity. He showed a positive correlation between the return on equity of the target and the probability of cash financing. This result is inconsistent with Grullon et al. (1997), who showed that in the banking sector, a cash payment is more likely when the target realized low performances in the past.

On the other hand, Zhang (2001) thought that the good stock performance of the bidding firm is an incentive for it to use stocks, because this payment method becomes cheaper. Moreover, this good stock performance can be attractive for the target, which can be incited to accept the offer. In an empirical study, he showed a positive link between the stock performances of the bidding firm (measured through the market-to-book ratio) and the probability of a stock financing. This result is consistent with previous works of Carleton et al. (1983) showing the same result.

Carleton et al. (1983) also showed that firms with higher dividend payout ratios have a higher probability to be acquired with stock than with cash. They studied the target's characteristics thanks to data distinguishing three kinds of firms: non-acquired firms, firms acquired with stocks, and firms acquired with cash. They showed, among other things, that acquired firms are generally smaller than non-acquired firms, which are smaller than bidding firms. They also showed that the price earnings ratios of acquired firms are lower than those of other firms, that cash-acquired firms have a lower dividend payout ratio, that acquired firms use less debt than the average in their sector, and that acquired firms are more profitable. On the other hand, there are no significant differences between financial elements according to the payment method chosen. The probability for a given firm to be acquired increases as its liquidity, its leverage, its price earning ratio, or its size decreases. On the other hand, the dividend payout ratio and the book-to-market value do not seem to have an impact on the probability for a firm to be acquired. A higher size and a higher level of liquidity reduce the probability of an acquisition, even though these effects are significant only for stock-financed acquisitions. According to the authors, these results are consistent with the tax hypothesis and the theory of managerial preferences.

The study of Carleton et al. (1983) was completed by Chaney et al. (1991), who studied the bidding firm's characteristics. They showed that the bidding firms using stocks are different from the bidding firms using cash. The firms that pay in stocks are larger, have higher price earning ratios, and have lower debt equity and return on assets ratios. Conversely, the ones paying in cash are smaller, with higher leverage and return on assets ratios.

Finally, Jung et al. (1996) and Baker and Wurgler (2002), among others, showed that the firms whose market value on book value ratio is high have a higher probability to issue stocks.

6.2 The Investments Opportunities Theory

The link between the investment opportunities of a firm and its corporate finance activities has been the subject of studies for many years. As early as 1977, Myers emphasized the link between the existence of growing opportunities and the corporate borrowing activity. According to Jung et al. (1996), managers with growing objectives prefer to raise capital thanks to stocks, because it gives them more freedom in their use than debt. Actually, contrary to stock issuing, debt compels the manager to repay cash flows. Therefore, he cannot use them to invest in low-profit projects. Consequently, debt financing maximizes the value of firms with low investment opportunities. Conversely, the freedom in the use of the funds obtained with stock is high for firms with good investment opportunities, because it makes them more likely to take full advantage of their investment opportunities. Thus, they supposed that firms with good investment opportunities are more likely to issue stocks in order to finance their acquisitions and that, on the contrary, firms with low investment opportunities finance their acquisitions by issuing debt. Moreover, they showed that the firms financing their acquisitions with stocks have interesting investment opportunities and experience an important growth in their assets during the period between the year preceding the stock issue and the end of the year following it.

Using this perspective, Martin (1996) studied the relationship between the payment method used in mergers and acquisitions and the growing opportunities of the firm. He showed that both the target's and the bidder's growing opportunities are important determinants of the payment method. Actually, he showed that more important investment opportunities, measured by either using Tobin's Q or the average annual growth rate of sales in the five-year period preceding the acquisition, lead to a higher probability for the deal to be financed with stocks.

6.3 The Business Cycles Hypothesis

According to this hypothesis, the business cycles are supposed to have an impact on the payment method used in acquisitions. Thus, in Brealey et al. (1976), Taggart (1977), Marsh (1982), and Choe et al. (1993), an increase in the general economic activity resulted in an increase in the probability of stock financing. According to the latter, one can explain this phenomenon by lower adverse selection costs, by even more promising investment opportunities and by a decrease of uncertainty concerning the assets in place. This hypothesis is weakly empirically verified by Martin (1996).

Lucas and McDonald (1990) proposed a dynamic model of stock issuing that took into account adverse selection phenomena. However, in their work, firms do not have the possibility to finance their projects with debt. Their model predicted that stock issuing tends to follow a general increase of the market. However, contrary to Choe et al.'s model (1993), it does not predict a relationship between returns, business cycles, or the stocks issues on the one hand, and the reaction of the firm's stock price at the announcement of the stock issues on the other hand, once the stock price's variations before the deal have been taken into account.

7 The Optimal Structure of Capital

According to Murphy and Nathan (1989), the payment method can be chosen so as to optimize the capital structure of the merged firm after the acquisition. In this case, the choice of the payment method can be crucial for the success of a deal, since it influences the financial structure of the merged firm: A cash acquisition reduces the liquidity of the merged firm, whereas a stock offer reduces its leverage. The managers who organize a merger or an acquisition had better choose the payment method according to the optimal capital structure of the firm after the acquisition. Thus, according to this theory, cash-rich firms will use cash and cash-poor firms will use stocks. Moreover, the bidding firms in debt will use stocks, whereas the bidders with a debt capacity will use debt. The authors reported that returns in the announcement period are positive (whatever the payment method), as long as the payment method is consistent with the concept of the optimal capital structure. Conversely, Travlos and Papaioannou (1991) showed that the abnormal returns of cash and stock offers are not affected by the market's perception of changes in the optimal capital structure. This result is consistent with the hypothesis according to which the payment method in acquisitions conveys precious information to the market. The returns or losses come from the informational effects associated with the payment method, whatever the financial changes implied by the deal.

In most cases, an acquisition reduces the variance of the firm's returns and also its underlying debt capacity. According to the theory of the increase of the underlying debt capacity, sometimes also called the coinsurance effects theory, the firms that merge and do not have a perfect correlation with their returns decrease their default probability (Lewellen, 1971).

This increase in the debt capacity associated with the deductibility of interest can be an incentive to realize a tender offer for managers who want to maximize the wealth of their shareholders. Moreover, Higgins and Schall (1975) and Galai and Masulis (1976) showed that this coinsurance effect leads to an increase in the market value of the firm's debts, and therefore to a decrease of the market value of its stocks, which benefits bondholders at the detriment of shareholders. Thus, the debt level of the merged firm can be increased beyond the individual levels of the firms, without increasing the risk of default. In such a case, a stock offer increases this effect and leads to a more important wealth transfer between shareholders and bondholders, which results in an important decrease in the stock price. Conversely, a cash offer could offset the decrease of the bidder's stock price caused by coinsurance phenomena. The impact on the stock price would then be zero.

On the contrary, Jensen and Meckling (1976) thought that there is a real incentive for shareholders of highly leveraged firms to appropriate the bondholders' wealth by investing in projects increasing the firm's risk. Therefore, there is an incentive for shareholders to acquire firms that increase the variability of the bidding firm's cash flows. Although the deal does not generate synergies, the bidding firm's shareholders could thus obtain a positive abnormal return to the detriment of bondholders. The bondholders' loss, which can be explained by the increase in the risk of default of the existing bonds, would cause a wealth transfer from bondholders to stockholders in this case. According to Chowdhry and Nanda (1993), in the case of a fight for control of a firm, the existence of bondholders allows the bidding firm to offer more than the target's valuation, provided they can be expropriated by the use of new debts whose seniority is equal to or greater than that of the existing debt.

Unfortunately, empirical studies do not really allow one to conclude on the existence of such a wealth transfer between shareholders and bondholders in acquisitions. On the one hand, Kim and McConnell (1977), Asquith and Kim (1982), and Dennis and McConnell (1986) showed that there is no significant wealth transfer from shareholders to bondholders, and Travlos (1987) even showed that, on the contrary, bondholders suffer a small loss. Conversely, Eger (1983) and Settle et al. (1984) showed that the bondholders' wealth is positively affected by the acquisitions, these latter generating synergies and/or a diversification effect for bondholders.

Nonowner-managers can undertake deals reducing the variability of the cash flows and of the firm's returns. If such is the case without a capital restructuring, the wealth of the target's shareholders will decrease. From the capital structure perspective, it appears that shareholders will prefer a cash offer, whereas managers will tend to offer stocks, even more so if the manager's situation is inversely linked to the fluctuations of the firm's returns and cash flows.

In the banking sector, as Grullon et al. (1997) showed that the higher the capital ratio of the bidding firm, the higher the probability of stock financing. According to the authors, this result is linked to the norms of capital sufficiency in force in the banking sector.

8 The Theories Linked to the Delays of Achievement of the Deal

The waiting time for an authorization to undertake a stock offer is longer than in the case of a cash offer. A bidding firm planning to realize a hostile takeover is therefore incited to use cash to finance its acquisition so as not to give the target time to organize its defense.

This waiting time also influences the probability of renegotiations of the contract's terms. In the case of a stock offer in a fixed ratio, the stock prices of both firms may evolve between the announcement and the completion of the deal, so that one of the two parts may have a strong incentive to ask for a renegotiation of the offer's terms. The payment method can thus also be chosen so as to minimize the renegotiation costs.

8.1 The Control Delays Hypothesis

According to Martin (1996), the acquisition method can play an important role in the choice of the payment method. A tender offer financed with cash is subject to the William Act and can begin a few days after the deal's announcement. Conversely, a stock offer, whether it is a tender offer or a merger, has to respect the Securities Act of 1933 and compels the bidding firm to obtain authorization from the Securities and Exchange Commission (see Gilson, 1986). This process can take several months, and yet the speed of the transaction has an impact on the success of a hostile deal: Since the process is longer in the case of a stock offer, the management of the target has more time to organize a response. This delay also allows rival bidders to enter into the competition for the target.

Finally, this delay allows the management of the target to give information about its own value to his favorite bidder (if it is a different bidder from the one that realized the initial offer). This information can come from an upward revision of the estimated cash flows and from a reduction of the uncertainty borne by the privileged bidder. In this case, the privileged bidder can offer a higher premium. Therefore, a hostile stock offer is less likely to succeed than a hostile cash offer.

The results Martin (1996) presented are consistent with this hypothesis. According to him, the desire to realize the deal as quickly as possible, because of the actual or potential competition from other bidding firms and the different regulations applicable according to the payment method finally chosen, encourages managers to use cash in tender offers.

On the other hand, Noronha and Sen (1995) showed that the hostile or friendly character of the offer is not a function of the payment method.

8.2 The Hypothesis of the Negotiations Costs Minimization

According to Officer (2004), the payment method used to finance an acquisition can be chosen so as to minimize the renegotiation costs of both firms at the time of the deal. As the period between the announcement of the deal and its completion is quite long, managers have an ex-post incentive to ask for a renegotiation of the contract's terms if the value of the bidder's offer has changed in comparison with the target's value during this period. The risk of having to renegotiate an offer is particularly high when the bidding firm offers a fixed number of its own stocks to acquire the target's stocks. Indeed, in this case, a decrease of the bidder's stock value reduces the value of the compensation paid to the targets' shareholders if the elasticity of the offer is high, that is, if it is sensitive to the changes of value of the merged firm. To avoid such costly renegotiations of the contract's terms, the bidding firm can protect the target from a decrease of the compensation, in case of a fall in the bidding firm's stock price, by including, for example, a protection such as a collar in the offer. Although the renegotiation ex-ante of such a financial tool is costly for both parties, Officer (2004) thought that the collar is used as a contractual system that allows one to reduce the ex-ante expected costs of negotiations during the bid period.

However, the ex-ante expected costs of negotiation between the deal announcement and its completion will be lower if the elasticity of the offer matches the relative sensitivity of the bidder and the target to economic shocks during the period of the bid. In other words, if both firms have to face very different economic shocks, or if the sensitivity of their market value is very different for the same shock, then a high elasticity offer (a stock offer) will need an ex-post renegotiation. Because of this, the will to structure an offer that minimizes the renegotiation costs will lead the bidder and the target to choose a payment method so that changes in the value of the offer will correspond to the changes in the value of the target's assets. The firm will use cash, if it is not financially compelled, or a protection (of the collar type, for example) if the target's and bidder's stocks prices are not sufficiently correlated to allow an offer in stocks only without a renegotiation.

Officer (2004) tested this theory in the same article. He showed that the strongest determinants of the offer's structure (cash or stocks, use of a collar or not) are the bidder's and the target's market-related stock return volatilities. Thus, this result tends to show that merging firms take into account the historical differences in the sensitivity of their market values to general economic shocks. Not to take these differences into account could have the consequence of overpaying or underpaying the targets' assets, which could require costly negotiations once again during the period when the result of the deal is uncertain. This result is thus consistent with the hypothesis of the minimization of the renegotiations costs.

9 The Acquisition of Nonpublic Firms

Most studies carried out until now (and therefore summed up in this survey, cf. Table 4) have dealt with the acquisition of public firms by other public

Table 4. Empirical Studies Testing the Validity of the Theories Linked to the Past Performances, the Investment Opportunities, the Business Cycles, the Optimal Structure of Capital, the Delays of Achievement of the Deal and the Acquisition of Nonpublic Firms

Tested Models	Authors	Conclusions of the Empirical Studies	Samples (Years)	Consistent/ Inconsistent
Inefficient managers replacement	Zhang (2001)	Positive correlation between the return on equity of the target and the probability of cash financing.	103 UK acquisitions (1990–1999)	Inconsistent
	Grullon et al. (1997)	In the banking sector, a cash payment is more likely when the target realized low performances in the past.	146 U.S. banks (1981–1990)	Consistent
Performance of the bidder	Carleton et al. (1983)	There is a positive link between the stock performances of the bidding firm (measured through the market-to-book ratio) and the probability of a stock financing (see also Zhang, 2001).	61 U.S. targets (1976–1977)	Consistent
Investment opportunities	Martin (1996)	More important investment opportunities, measured by using either Tobins Q or the average annual growth rate of sales in the five-year period preceding the acquisition, lead to a higher probability for the deal to be financed with stocks.	846 U.S. acquisitions (1978–1988)	Consistent
Optimal structure of capital	Murphy and Nathan (1989)	Returns in the announcement period are positive (whatever the payment method), as long as the payment method is consistent with the concept of the optimal capital structure (quoted by Nayar and Switzer, 1998).		Consistent
	Travlos and Papaioannou (1991)	The abnormal returns of cash and stock offers are not affected by the market's perception of changes in the optimal capital structure.	57 U.S. acquisitions (1972–1981)	Consistent
Control delays	Martin (1996)	The results presented by Martin (1996) are consistent with this hypothesis.	846 U.S. acquisitions (1978–1988)	Consistent
Negotiation cost minimization	Officer (2004)	The strongest determinants of the offer structure (cash or stocks, use of a collar or not) are the bidders and the target's market-related stock return volatilities. Thus, this result tends to show that merging firms take into account the historical differences in the sensitivity of their market values to general economic shocks.	1,366 U.S. acquisitions (1991–1999)	Consistent
Acquisition of nonpublic firms	Faccio et al. (2003)	The different theories designed to explain why the abnormal returns are higher in nonpublic acquisitions than in public acquisitions are not empirically validated.	4,429 European acquisitions (1996–2001)	Inconsistent

firms. Numerous authors have shown a negative reaction to the announcement of a stock offer. For example, Travlos (1987) and Asquith et al. (1987) showed higher returns for the bidder in a cash offer than in case of a stock offer. As Agrawal et al. (1992) showed, this negative impact of a stock offer does not only concern the announcement period, but can also be observed during the period following the deal. Conversely, the recent research has shown that, when the target is a nonpublic firm, a stock offer has a positive impact and a cash offer has no impact on the bidder's wealth. Indeed, for Chang (1998), the acquisition of a nonpublic firm thanks to an offer mainly in stocks tends to create outside blockholders. According to Shleifer and Vishny (1986), this can be positive for the bidder's shareholders, because these blockholders can generate an effective monitoring of the performance or can ease mergers, which can increase the bidding firm's value. Moreover, when firms offer stocks to acquire firms held by a small number of shareholders, the informational asymmetry problems described, among others, by Myers and Majluf (1984) can be lowered by the disclosure of private information to the target's shareholders. The author also explains that the shareholder of a privately held firm owning an important quantity of stocks had better examine the bidder's prospects with caution, because, at the end of the deal, he will hold a large ownership in the merged firm. Therefore, the shareholders of privately held firms send a positive signal to the market if they accept a stock offer.

Similarly, more recently, Fuller et al. (2002) showed, through a sample of 3,135 deals, that when target firms are subsidiaries or privately held firms, the bidders' returns are significantly higher than when the target is a nonpublic firm, whatever the payment method. They also confirmed that the bidder's returns are higher when the deal is financed with stocks than when it is financed with cash. The authors also showed that, for public firms, when the relative size of the targets increases, the returns increase in cash offers and decrease in stock offers and that they are not changed in mixed offers. On the contrary, concerning subsidiaries and privately held acquisitions, there is a positive relation between the relative size of the target and the positive abnormal returns of the bidding firm. According to these authors, this difference between the market's reaction in nonpublic acquisitions and in public acquisitions is due to the creation of blockholders, to a liquidity effect, and to taxation. The liquidity effect comes from the fact that the privately held firms and the subsidiaries cannot be sold as easily as public firms. This poor liquidity makes these investments less attractive and less valuable than similar investments (Koeplin et al., 2000).

According to Hansen and Lott (1996), public firms' shareholders own diversified portfolios holding stocks from other firms, but these portfolio cannot include privately held firms' stocks. This is why they thought that given the rule of value conservation, shareholders do not care if managers overpay public firms' acquisitions, because the losses suffered by the bidders are wealth transfers to the target's shareholders. The acquisition of a nonpublic firm is different because the stocks of such firms are not part of the bidder's shareholders portfolio, so that the shareholder of the bidding firm will demand that only creating value acquisitions of privately held firms should be realized.

Faccio et al. (2006) tested these different hypotheses. First, they tested if the public firm effect could be due to the creation of blockholders. Using a proxy for the creation of blockholders, they showed that contrary to Chang's hypothesis, in a slightly higher proportion, the acquirers of nonpublic firm structure their deal so as to create blockholders in the bidding firm more often than the acquirers of nonpublic firms do. Whether a blockholder is created or not, the abnormal return of the nonpublic firms is positive and is significantly different from zero. It is also significantly higher than the abnormal return of public target firms. Moreover, the average abnormal return of the bidding firm is not statistically different from zero whether there is a blockholder or not. The public firm effect is therefore not a blockholder creation effect.

Faccio et al. (2006) then tested the bidder size effect, according to which large bidders try to acquire public firms, whereas smaller bidders try to acquire nonpublic firms. Not surprisingly, they showed that the acquirers of public firms tend to be larger than the acquirers of nonpublic firms. Although small acquirers as well as large acquirers suffer losses when they acquire public firms, small acquirers have lower performance in their sample. On the contrary, small acquirers have better performance than large acquirers when they buy nonpublic firms. However, both small and large acquirers have positive and significant abnormal returns when they acquire nonpublic firms. Finally, both small and large nonpublic bidding firms earn significantly higher returns than those earned by small and large acquirers of public firms. The public firm effect is not a disguised size effect.

Faccio et al. (2006) also offered a test of the Hansen and Lott theory (1996). According to this hypothesis, shareholders should not worry about the fact that managers overpay their acquisition of public firms' subsidiaries. In their sample, 95% of the parent companies of subsidiaries are public. However, they showed a significantly positive average return for the subsidiaries. The result is inconsistent with the Hansen and Lott theory (1996).

Since the results are not consistent either with Fuller et al.'s hypothesis of illiquidity premium (2002), the study carried out by Faccio et al. (2006) contradicted one whole theories allowing to explain the abnormal returns observed in the acquisitions of nonpublic firms. Thus, until now, research has not succeeded in understanding why the abnormal returns are higher in nonpublic acquisitions than in public acquisitions.

10 Conclusions

Since the late 1970s, finance researchers have been studying the elements influencing the choice of the payment methods in mergers and acquisitions. This study aimed at presenting the main theories developed until now concerning this issue and their main empirical tests.

A first group of theories is linked to the debt of the two merging firms. Thus, some authors have suggested that the payment method was chosen to optimize the capital structure of the merged firm after the completion of the deal. Others such as Jensen have shown the positive role that debt could play on the actions of the managers. Thus, in addition to the outside monitoring and the managerial ownership, debt may encourage managers to act in the interest of the shareholders. The legislature also has a large impact on the choice of the payment method. Indeed, in some countries, taxation on the capital gains realized is immediate in cash offers and postponed in stock offers, which makes cash less attractive for the target's shareholders, who are unwilling to be taxed immediately. On the contrary, the legislature makes stock offers less interesting for the bidding firm, which is impatient to realize its acquisition, because in this case, the delays are higher. The past and future performances of both firms, as well as those of the market, are also elements that can influence this choice. Finally, the informational asymmetry problems between the managers of both firms and the competition between the different bidders are also supposed to influence the choice of the payment method, because in these two cases, the use of stocks is a disadvantage in comparison with the use of cash.

The main result of this survey is that empirical studies often do not allow one to validate these different theories because they show contradictory results. The need for knowledge concerning this issue is therefore very important, all the more so as the last research has raised more questions than it has solved problems. Thus, for example, the different theories designed to explain why the abnormal returns are higher in nonpublic acquisitions than in public acquisitions are not empirically validated. Future research will therefore have to answer this question.

The choice of the payment method in mergers and acquisitions is a very difficult issue because it is influenced by a lot of variables. Before he makes a decision, the manager has to take into consideration his own interests, the legislation, and the past and future performance of both firms. This could probably explain why empirical studies are often contradictory.

Rejecting all existing theories on account of the fact that some empirical studies do not allow one to validate them would without doubt be a mistake. On the contrary, studying the reasons why in some cases those theories are validated and in other cases rejected (by examing the composition of the sample in terms of friendly/hostile mergers and acquisitions, target status, value of the deal, etc.) would certainly be fruitful.

References

1. Agrawal, A., Jaffe, J., and Mandelker, G. The post-merger performance of acquiring firms: A re-examination of an anomaly. *Journal of Finance* 47:1605–1621, 1992.

2. Akerlof, G. The market for "lemons": Quality uncertainty and the market mechanism. *Quarterly Journal of Economics* 84:488–500, 1970.
3. Amihud, Y., and Lev, B. Risk reduction as a managerial motive for conglomerate mergers. *Bell Journal of Economics* 12:605–617, 1981.
4. Amihud, Y., Lev, B., and Travlos, N. Corporate control and the choice of investment financing: The case of corporate acquisitions. *Journal of Finance*, 45:603–616, 1990.
5. Ang, J., Cole, R., and Lin, J. 2000 Agency costs and ownership structure. *Journal of Finance*, 55:81–106.
6. Antoniou, A., and Zhao, H. Long-run post takeover stock return: The impact of overlapping return, takeover premium, and method of payment. Centre for Empirical Research in Finance (CERF), Durham Business School, Working paper, 2004.
7. Asquith, P., Bruner, R., and Mullins, D. Merger returns and the form of financing. *Proceedings of the Seminar on the Analysis of Security Prices*, 1987.
8. Asquith, P., and Kim, E. The impact of merger bids on the participating firm's security holders. *Journal of Finance*, 37:1209–1228, 1982.
9. Asquith, P., and Mullins, D. Equity issues and offering dilution. *Journal of Financial Economics*, 15:61–89, 1986.
10. Auerbach, A., and Reihus, D. The impact of taxation on mergers and acquisitions. In A. Auerbach, Editor. *Mergers and Acquisitions*. The University of Chicago Press, Chicago, IL, 1988.
11. Baker, M., and Wurgler, J. Market timing and capital structure. *Journal of Finance*, 57:1–32, 2002.
12. Berkovitch, E., and Narayanan, M.P. Competition and the medium of exchange in takeovers. *Review of Financial Studies*, 3:153–174, 1990.
13. Berle, A., and Means, G. *The Modern Corporation and Private Property*. Macmillan, New York, 1932.
14. Black, B. Agents watching agents: The promise of institutional investor voice. *UCLA Law Review*, 39:811–893, 1992.
15. Blackburn, V., Dark, F., and Hanson, R. Mergers, method of payment and returns to manager- and owner-controlled firms. *Financial Review*, 32:569–589, 1997.
16. Blazenko, G. Managerial preference, asymmetric information, and financial structure. *Journal of Finance*, 42:839–862, 1987.
17. Bradley, M., Desai, A., and Kim, H. Synergic gains from corporate acquisitions and their division between the stockholders of target and acquiring firms. *Journal of Financial Economics*, 21:3–40, 1988.
18. Brealey, R., Hodges, S., and Capron, D. The return on alternative sources of finance. *Review of Economics and Statistics*, 58:469–477, 1976.
19. Brick, I., Frierman, M., and Kim, Y. Asymmetric information concerning the variance of cash flows: The capital structure choice. *International Economic Review*, 39:745–761, 1998.
20. Brown, D., and Ryngaert, M. The mode of acquisition in takeovers: Taxes and asymmetric information. *Journal of Finance*, 46:653–669, 1991.
21. Carleton, W., Guilkey, D., Harris, R., and Stewart, J. An empirical analysis of the role of the medium of exchange in mergers. *Journal of Finance*, 38:813–826, 1983.

22. Chaney, P., Lovata, L., and Philipich, K. Acquiring firm characteristics and the medium of exchange. *Quarterly Journal of Business and Economics*, 30:55–69, 1991.
23. Chang, S. Takeovers of privately held targets, methods of payment, and bidder returns. *Journal of Finance*, 53:773–784, 1998.
24. Chaplinsky, S., and Niehaus, G. Do inside ownership and leverage share common determinants? *Quarterly Journal of Business and Economics*, 32:51–65, 1993.
25. Chemmanur, T., and Paeglis, I. The choice of the medium of exchange in acquisitions: A direct test of the double sided asymmetric information hypothesis. Carroll Scholl of Management, Boston College and John Molson Scholl of Business, Concordia University, Working paper, 2002.
26. Choe, H., Masulis, R., and Nanda, V. Common stock offerings across the business cycle: Theory and evidence. *Journal of Empirical Finance*, 1:3–31, 1993.
27. Chowdhry, B., and Nanda, V. The strategic role of debt in takeover contests. *Journal of Finance*, 48:731–745, 1993.
28. Cornett, M., and De, S. Medium of payment in corporate acquisitions: evidence from interstate bank mergers. *Journal of Money, Credit and Banking*, 23:767–776, 1991.
29. Cornu, P., and Isakov, D. The deterring role of the medium of payment in takeover contests: Theory and evidence from the UK. *European Financial Management*, 6:423–440, 2000.
30. Davidson, W., and Cheng, L. Target firm returns: Does the form of payment affect abnormal returns? *Journal of Business Finance and Accounting*, 24:465–479, 1997.
31. De, S., Fedenia, M., and Triantis, A. Effects of competition on bidder returns. *Journal of Corporate Finance*, 2:261–282, 1996.
32. DeAngelo, H., and Masulis, R. Optimal capital structure under corporate and personal taxation. *Journal of Financial Economics*, 8:3–29, 1980.
33. Dennis, D., and McConnell, J. Corporate mergers and security returns. *Journal of Financial Economics*, 16:143–187, 1986.
34. Dodd, P. Merger proposals, management discretion and stockholder wealth. *Journal of Financial Economics*, 8:105–138, 1980.
35. Eckbo, E., and Langohr, H. Information disclosure, method of payment, and takeover premiums: Public and private tender offers in France. *Journal of Financial Economics*, 24:363–403, 1989.
36. Eckbo, E., Giammarino, R., and Heinkel, R. Asymmetric information and the medium of exchange in takeovers: Theory and tests. *Review of Financial Studies*, 3:651–675, 1990.
37. Eger, C. An empirical test of the redistribution effect in pure exchange mergers. *Journal of Financial and Quantitative Analysis*, 18:547–572, 1983.
38. Erickson, M., and Wang, S. Earnings management by acquiring firms in stock for stock mergers. *Journal of Accounting and Economics*, 27:149–176, 1999.
39. Faccio, M., McConnell, J., and Stolin, D. Returns to acquirers of listed and unlisted targets. *Journal of Financial and Quantitative Analysis*, 41:197–220, 2006.
40. Fama, E., and French, K. Testing trade-off and pecking order predictions about dividends and debt. *Review of Financial Studies*, 15:1–33, 2002.
41. Fishman, M. A theory of preemptive takeover bidding. *RAND Journal of Economics*, 19:88–101, 1988.

42. Fishman, M. Preemptive bidding and the role of the medium of exchange in acquisitions. *Journal of Finance*, 44:41–57, 1989.
43. Franks, J., Harris, R., and Mayer, C. Means of payment in takeovers: Results for the United Kingdom and the United States. In A. Auerbach, Editor. *Corporate Takeovers: Causes and Consequences*. The University of Chicago Press, Chicago, 1988.
44. Fuller, K., Netter, J., and Stegemoller, M. What do returns to acquiring firms tell us? Evidence from firms that make many acquisitions. *Journal of Finance*, 57:1763–1793, 2002.
45. Galai, D., and Masulis, R. The option pricing model and the risk factor of stock. *Journal of Financial Economics*, 3:53–81, 1976.
46. Ghosh, A., and Ruland, W. Managerial ownership, the method of payment for acquisitions, executive job retention. *Journal of Finance*, 53:785–798, 1998.
47. Gilson, R. *The Law and Finance of Corporate Acquisition*. The Foundation Press, Mineola, New York, 1986.
48. Gilson, R., Scholes, M., and Wolfson, M. Taxation and the dynamics of corporate control: The uncertain case for tax motivated acquisitions. In L. Lowenstein, S. Rose-Ackerman, and J. Coffee, Editors. *Knights, Raiders, and Targets: The Impact of the Hostile Takeover*. Oxford University Press, New York, 1988.
49. Gompers, P., and Metrick, A. Insitutional investors and equity prices. *Quarterly Journal of Economics*, 116:229–259, 2001.
50. Grullon, G., Michaely, R., and Swary, I. Capital adequacy, bank mergers, and the medium of payment. *Journal of Business Finance and Accounting*, 24:97–124, 1997.
51. Hansen, R. A theory for the choice of exchange medium in mergers and acquisitions. *Journal of Business*, 60:75–95, 1987.
52. Hansen, R., and Lott, J. Externalities and corporate objectives in a world with diversified shareholder/consumers. *Journal of Financial and Quantitative Analysis*, 31:43–68, 1996.
53. Harris, M., and Raviv, A. Corporate control contests and capital structure. *Journal of Financial Economics*, 20:55–86, 1988.
54. Harris, M., and Raviv, A. Capital structure and the informational role of debt. *Journal of Finance*, 45:321–349, 1990.
55. Harris, R., Franks, J., and Mayer, C. Means of payment in takeovers: Results for the UK and US. In A. Auerbach, Editor, *Corporate Takeovers: Causes and Consequences*. The University of Chicago Press, Chicago, 1987.
56. Heinkel, R. A theory of capital structure relevance under imperfect information. *Journal of Finance*, 37:1141–1150, 1982.
57. Helwege, J., and Liang, N. Is there a pecking order? Evidence from a panel of IPO firms. *Journal of Financial Economics*, 40:429–458, 1996.
58. Heron, R., and Lie, E. Operating performance and the method of payment in takeovers. *Journal of Financial and Quantitative Analysis*, 37:137–155, 2002.
59. Higgins, R., and Schall, L. Corporate bankruptcy and conglomerate merger. *Journal of Finance*, 30:93–114, 1975.
60. Houston, J., and Ryngaert, M. Equity issuance and adverse selection: A direct test using conditional stock offers. *Journal of Finance*, 52:197–219, 1997.
61. Huang, Y.-S., and Walkling, R. Target abnormal returns associated with acquisition announcements: Payment, acquisition form, and managerial resistance. *Journal of Financial Economics*, 19:329–349, 1987.

62. Jensen, M. Agency costs of free cash flow, corporate finance, and takeovers. *American Economic Review*, 76:323–339, 1986.
63. Jensen, M. Takeovers: Their causes and consequences. *Journal of Economic Perspectives*, 2:21–48, 1988.
64. Jensen, M. Corporate control and the politics of finance. *Journal of Applied Corporate Finance*, 4:13–33, 1991.
65. Jensen, M., and Meckling, W. Theory of the firm: Managerial behavior, agency costs and ownership structure. *Journal of Financial Economics*, 3:305–360, 1976.
66. Jensen, M., and Ruback, R. The market for corporate control: The scientific evidence. *Journal of Financial Economics*, 11:5–50, 1983.
67. John, K. Risk-shifting incentives and signalling through corporate capital structure. *Journal of Finance*, 42:623–641, 1987.
68. Jung, K., Kim, Y., and Stulz, R. Timing, investment opportunities, managerial discretion, and the security issue decision. *Journal of Financial Economics*, 42:159–185, 1996.
69. Kim, E., and McConnell, J. Corporate mergers and the co-insurance of corporate debt. *Journal of Finance*, 32:349–365, 1977.
70. Koeplin, J., Sarin, A., and Shapiro, A. The private company discount. *Journal of Applied Corporate Finance*, 12:94–101, 2000.
71. Korajczyk, R., Lucas, D., and McDonald, R. The effect of information releases on the pricing and timing of equity issues. *Review of Financial Studies*, 4:685–708, 1991.
72. Langetieg, T. An application of a three-factor performance index to measure stockholder gains from merger. *Journal of Financial Economics*, 6:365–383, 1978.
73. Leland, H., and Pyle, D. Informational asymmetries, financial structure, and financial intermediation. *Journal of Finance*, 32:371–387, 1977.
74. Lewellen, W. A pure financial rationale for the conglomerate merger. *Journal of Finance*, 26:521–537, 1971.
75. Loughran, T., and Vijh, A. Do long-term shareholders benefit from corporate acquisitions? *Journal of Finance*, 52:1765–1790, 1997.
76. Lucas, D., and McDonald, R. Equity issues and stock price dynamics. *Journal of Finance*, 45:1019–1043, 1990.
77. Maloney, M., McCormick, R., and Mitchell, M. Managerial decision making and capital structure. *Journal of Business*, 66:189–217, 1993.
78. Marsh, P. The choice between equity and debt: An empirical study. *Journal of Finance*, 37:121–144, 1982.
79. Martin, K. The method of payment in corporate acquisitions, investment opportunities, and management ownership. *Journal of Finance*, 51:1227–1246, 1996.
80. Masulis, R. The effects of capital structure change on security prices: A study of exchange offers. *Journal of Financial Economics*, 8:139–178, 1980a.
81. Masulis, R. Stock repurchase by tender offer: An analysis of the causes of common stock price changes. *Journal of Finance*, 35:305–319, 1980b.
82. Masulis, R., and Korwar, A. Seasoned equity offerings: An empirical investigation. *Journal of Financial Economics*, 15:91–118, 1986.
83. Mayer, W., and Walker, M. An empirical analysis of the choice of payment method in corporate acquisitions during 1980 to 1990. *Quarterly Journal of Business and Economics*, 35:48–65, 1996.

84. McNally, W. Open market stock repurchase signaling. *Financial Management*, 28:55–67, 1999.
85. Mikkelson, W., and Partch, M. Valuation effects of security offerings and the issuance process. *Journal of Financial Economics*, 15:31–60, 1986.
86. Miller, M. Debt and taxes. *Journal of Finance*, 32:261–275, 1977.
87. Modigliani, F., and Miller, M. The cost of capital, corporation finance and the theory of investment. *American Economic Review*, 48:261–297, 1958.
88. Modigliani, F., and Miller, M. Corporate income taxes and the cost of capital: A correction. *American Economic Review*, 53:433–443, 1963.
89. Moeller, S., Schlingemann, F., and Stulz, R. Do shareholders of acquiring firms gain from acquisitions? Dice Centre Working paper 2003-4, 2003.
90. Morck, R., Shleifer, A., and Vishny, R. Do managerial objectives drive bad acquisitions? *Journal of Finance*, 45:31–48, 1990.
91. Murphy, A., and Nathan, K. An analysis of merger financing. *Financial Review*, 24:551–566, 1989.
92. Myers, S. Determinants of corporate borrowing. *Journal of Financial Economics*, 5:147–175, 1977.
93. Myers, S. The capital structure puzzle. *Journal of Finance*, 39:575–592, 1984.
94. Myers, S., and Majluf, N. Corporate financing and investment decisions when firms have information that investors do not have. *Journal of Financial Economics*, 13:187–221, 1984.
95. Nayar, N., and Switzer, J. Firm characteristics, stock price reactions, and debt as a method of payment for corporate acquisitions. *Quarterly Journal of Business and Economics*, 37:51–64, 1998.
96. Niden, C. The role of taxes in corporate acquisitions: effects on premium and type of consideration. University of Chicago, Working paper, 1986.
97. Noronha, G., and Sen, N. Determinants of the medium of payment in corporate acquisitions. *Journal of Applied Business Research*, 11:15–23, 1995.
98. Officer, M. Collars and renegotiation in mergers and acquisitions. *Journal of Finance*, 59:2719–2743, 2004.
99. Persons, J. Signaling and takeover deterrence with stock repurchases: Dutch auctions versus fixed price tender offers. *Journal of Finance*, 49:1373–1402, 1994.
100. Persons, J. Heterogeneous shareholders and signaling with share repurchases. *Journal of Corporate Finance*, 3:221–249, 1997.
101. Rajan, R., and Zingales, L. What do we know about capital structure? Some evidence from international data. *Journal of Finance*, 50:1421–1460, 1995.
102. Ravid, A., and Sarig, O. Financial signalling by committing to cash outflows. *Journal of Financial and Quantitative Analysis*, 26:165–180, 1991.
103. Rhoades, S. *Power, Empire Building, and Mergers*. Lexington Book, Lexington MA, 1983.
104. Ross, S. The determination of financial structure: the incentive-signalling approach. *Bell Journal of Economics*, 8:23–40, 1977.
105. Settle, J., Petry, G., and Hsia, C. Synergy, diversification, and incentive effects of corporate merger on bondholder wealth: Some evidence. *Journal of Financial Research*, 7:329–339, 1984.
106. Shleifer, A., and Vishny, R. Large shareholders and corporate control. *Journal of Political Economy*, 94:461–488, 1986.
107. Shrieves, R., and Pashley, M. Evidence on the association between mergers and capital structure. *Financial Management*, 13:39–48, 1984.

108. Shyam-Sunder, L., and Myers, S. Testing static tradeoff against pecking order models of capital structure. *Journal of Financial Economics*, 51:219–244, 1999.
109. Song, M., and Walkling, R. The impact of managerial ownership on acquisition attempts and target shareholder wealth. *Journal of Financial and Quantitative Analysis*, 28:439–457, 1993.
110. Stulz, R. Managerial control of voting rights: Financing policies and the market for corporate control. *Journal of Financial Economics*, 20:25–54, 1988.
111. Stulz, R. Managerial discretion and optimal financing policies. *Journal of Financial Economics*, 26:3–27, 1990.
112. Suk, D., and Sung, H. The effects of the method of payment and the type of offer on target returns in mergers and tender offers. *Financial Review*, 32:591–607, 1997.
113. Sung, H. The effects of overpayment and form of financing on bidder returns in mergers and tender offers. *Journal of Financial Research*, 16:351–365, 1993.
114. Taggart, R. A model of corporate financing decisions. *Journal of Finance*, 32:1467–1484, 1977.
115. Tessema, A. The role of medium of exchange in acquisitions. *Mid American Journal of Business*, 4:39–45, 1989.
116. Trautwein, F. Merger motives and prescriptions. *Strategic Management Journal*, 11:283–295, 1990.
117. Travlos, N. Corporate takeover bids, methods of payment, and bidding firm's stock returns. *Journal of Finance*, 42:943–963, 1987.
118. Travlos, N., and Papaioannou, G. Corporate acquisitions: Method of payment effects, capital structure effects, and bidding firms' stock returns. *Quarterly Journal of Business and Economics*, 30:3–22, 1991.
119. Vermaelen, T. Repurchase tender offers, signaling, and managerial incentives. *Journal of Financial and Quantitative Analysis*, 19:163–184, 1984.
120. Wansley, J., Lane, W., and Yang, H. Abnormal returns to acquired firms by type of acquisition and method of payment. *Financial Management*, 12:16–22, 1983.
121. Weston, J., Chung, K., and Hoag, S. *Mergers, Restructuring, and Corporate Control*. Prentice Hall, Englewood Cliffs, NJ, 1990.
122. Williamson, O. *The Economics of Discretionary Behaviour: Managerial Objectives in a Theory of the Firm*. Prentice Hall, Englewood Cliffs, NJ, 1964.
123. Yook, K. Larger return to cash acquisitions: Signaling effect or leverage effect? *Journal of Business*, 76:477–498, 2003.
124. Yook, K., Gangopadhyay, P., and McCabe, G. Information asymmetry, management control, and method of payment in acquisitions. *Journal of Financial Research*, 22:413–427, 1999.
125. Zhang, P. What really determines the payment methods in M&A deals. Manchester School of Management, Working paper, 2001.

An Application of Support Vector Machines in the Prediction of Acquisition Targets: Evidence from the EU Banking Sector

Fotios Pasiouras[1], Chrysovalantis Gaganis[2], Sailesh Tanna[3], and Constantin Zopounidis[2]

[1] School of Management, University of Bath, Claverton Down, Bath, BA2 7AY, UK f.pasiouras@bath.ac.uk
[2] Financial Engineering Laboratory, Department of Production Engineering and Management, Technical University of Crete, University Campus, Chania, 73100, Greece bgaganis@yahoo.com kostas@dpem.tuc.gr
[3] Department of Economics, Finance and Accounting, Faculty of Business, Environment and Society, Coventry University, Priory Street, Coventry, CV1 5FB, UK s.tanna@coventry.ac.uk

1 Introduction

In recent years, support vector machines (SVMs) have been applied in several problems in finance and accounting, such as credit rating (Huang et al., 2004; Hardle et al., 2004; Lee, 2007), bankruptcy prediction (Hardle et al., 2005; Min and Lee, 2005; Salcedo-Sanz et al., 2005; Shin et al., 2005; Gaganis et al., 2005; Min et al., 2006; Wu et al., 2007), financial time-series forecasting (Tay and Cao, 2001, 2002; Cao, 2003; Huang et al., 2005; Pai and Lin, 2005), and auditing (Doumpos et al., 2005). In general, the results from these studies are quite promising. The purpose of the present study is to illustrate the application of SVMs in the development of classification models for the prediction of acquisition targets.

Over the last 30 years, a number of empirical studies have developed classification models using publicly available information to identify potential acquisition targets. From a methodological perspective, discriminant analysis (Simkowitz and Monroe, 1971; Stevens, 1973; Barnes, 1990) and logit analysis (Dietrich and Sorensen, 1984; Barnes, 1998, 1999; Powell, 2001) have dominated the field. Other techniques that have been applied are artificial neural networks (Cheh et al., 1999), rough sets (Slowinski et al., 1997), the recursive partitioning algorithm (Espahbodi and Espahbodi, 2003), and multicriteria decision aid (MCDA) (Zopounidis and Doumpos, 2002; Doumpos et al., 2004; Pasiouras et al., 2006).

SVMs were only recently introduced in the field by Pasiouras et al. (2005), who developed a nonlinear model using the RBF kernel and compared it with models developed with various other techniques.[1] Hence, the use of the methodology in the prediction of acquisition targets is still in its infancy. In the present study, we use a sample of EU banks to investigate the relative performance of both linear and nonlinear SVMs models with a polynomial and an RBF kernel.

We focus on the EU banking industry for two reasons. First, only a few recent studies (Pasiouras et al., 2005, 2006) have developed prediction models specifically designed for banks, whereas previous studies that focused on non-financial sectors excluded banks from the analysis due to differences in their financial statements and the environment in which they operate. Second, over the last decade the European financial sector has witnessed a large number of mergers and acquisitions (M&As) that significantly transformed the market. As Altunbas and Ibanez (2004) pointed out, "According to most bankers and academics, [...], the process of banking integration seems far from completed and is expected to continue reshaping the European financial landscape in the years to come" (p. 7). Hence, several parties could be interested in the development of classification models capable of predicting acquisition targets in the banking industry. For instance, Tartari et al. (2003) pointed out that the prediction of acquisitions is of major interest to stockholders, investors, and creditors, and generally to anyone who has established a relationship with the target firm. Obviously, the managers of the banks are also among those who have an increased interest in the development of prediction models. Furthermore, the results of this study would be of particular interest to academics and researchers who work on the prediction of acquisitions and bankruptcy, and other classification problems in finance.

The rest of the chapter is organized as follows. Section 2 discusses the recent trends in M&As in the EU banking industry. Section 3 provides a review of the literature. Section 4 outlines the main concepts of support vector machines, while Section 5 describes the data and variables. Section 6 presents the empirical results, and Section 7 discusses the concluding remarks.

2 M&As Trends in the EU Banking Industry

The level of M&As in the European banking sector has been relatively high in recent years, resulting in the number of banks operating in the EU being reduced by 25% between 1997 and 2003 (European Central Bank-ECB, 2004). The M&A activity was very intense during the late 1990s, although it became considerably weaker since 2001. More detailed, ECB (2005a) indicates that over the entire period between 1992 and 2004, the highest activity, with respect to the number of M&As, was recorded during 1999 followed by 2000. Not surprisingly, the value of M&As also reached a peak in 1999, with 1998 and

---

[1] Discriminant analysis, logit analysis, utilities additives discriminants, multi-group hierarchical discrimination, nearest neighbors, classification and regression trees.

2000 recording relatively high figures as well. However, both the number and especially the value of M&As declined after 2000. Campa and Hernando (2006) indicated that the average monthly volume fell from 21.1 billion Euros in 2000 to 5.5 billion Euros in 2003, although it slightly increased to 6.6. billion Euros in 2004.

Panel A of Table 1 provides a breakdown of 2,153 M&As of credit institutions in the EU-15 that were recorded by the European Central Bank (ECB) between 1995 and the first half of 2000 (ECB, 2000). Panel B provides similar data from ECB (2006a) for 694 M&As in the EU-25 banking sector that occurred between 2000 and the first half of 2006. Direct comparisons between the data in Panels A and B should be treated with some caution; first, due to the different countries that they cover; second, because data for the 2000 report (Panel A) were collected from EU central banks and supervisory authorities, while data for the 2006 report were obtained from the Bureau Van Dijk data provider.

Table 1. Number of M&As in the EU Banking Sector

Panel A: M&As Among EU-15 Credit Institutions, 1995–2000							
	1995	1996	1997	1998	1999	2000A	Total
Domestic M&As	275	293	270	383	414	172	1,807
M&As within EEA	20	7	12	18	27	23	107
M&As with third country	31	43	37	33	56	39	239
Total	326	343	319	434	497	234	2,153
Panel B: M&As in the EU Banking Sector, 2000–2006							
	2000	2001	2002	2003	2004	2005	2006H1	Total
Domestic M&As								
MU-12	58	45	69	68	45	58	16	359
EU-25	70	65	74	73	61	65	21	429
Cross-border EU M&As								
MU-12	27	17	19	18	18	21	9	129
EU-25	54	32	36	27	28	31	13	221
M&As with third country								
MU-12	1	5	2	3	1	8	3	23
EU-25	4	7	5	8	2	12	6	44
Total								
MU-12	86	67	90	89	64	87	28	511
EU-25	128	104	115	108	91	108	40	694

Sources: Panel A: ECB (2000), p. 10; Panel B: ECB (2006a), p. 66.

However, irrespective of the panel and the initial source of information, the data indicate that during both periods domestic M&As were more common than cross-border ones. In more detail, ECB (2006a) indicated that between 1993 and 2003, the number of domestic M&As accounted for 80% of total consolidation activity within the EU. As Walner and Raes (2005) pointed out, the dominance of domestic consolidation in total was even more remarkable in terms of value, accounting for a share of about 90% or more in 13 out of the 17 years between 1987 and 2003 and falling below 70% only in 1989. However, a substantial number of EU cross-border M&As were recorded in 2005 and early 2006 (ECB, 2006a).

As it concerns the disaggregation of cross-border deals, the ECB (2000) report indicated that between 1995 and 2000, 5% of the total M&As occurred within the European Economic Area (EEA) and 11% with banks from a third country. During this period, most European banks have chosen to expand into Latin America (e.g., banks from Netherlands, Spain, Portugal, and Italy), Southeast Asia (e.g., banks from Netherlands), and Central and Eastern Europe (e.g., banks from the Netherlands, Ireland) probably in the search for markets offering higher margins or because of historical connections (ECB, 2000). Nevertheless, in some cases they have also expanded into developed markets such as the United States (e.g., banks from Germany). However, Panel B from the 2006 report shows that over the most recent years the number of EU cross-border deals was higher than the one of deals with third countries. In more detail, as the ECB (2006a) report indicated a substantial number of EU cross-border M&As were completed in 2005 such as the ones of Unicredit-HypoVereinsbank, ABN-AMRO – Banca Antonveneta, and Foreningssparbanken-Hansabank. Nevertheless, M&As involving institutions outside the EU also played a significant role in 2005 and early 2006. For example, British banks were involved in deals in South Africa, Malaysia, and Korea, while Austrian banks were involved in deals in Romania.

3 Literature Review

Prior literature related to the present chapter can be classified in two broad categories. The first consists of studies that focus on the prediction of acquisition targets. The second consists of studies that examine various aspects of M&As among banking institutions in the EU. In the sections that follow, we discuss in turn each category.

3.1 Predicting Acquisition Targets

Over the last years around 30 studies have proposed the development of quantitative models to predict acquisition targets. These studies were published between 1971 and 2006, and more than half of them appeared over the last 10 years. With the exception of Pasiouras et al. (2005, 2006), which pooled a sample of banks over the 15 EU member states, all the remaining studies have focused on individual countries. Obviously the most widely studied country is the United States (13), followed by the UK (10). The only other individual counties that have been examined are Greece (3) and Canada (2).

Some of these studies search for the best predictive variables of acquisitions (Bartley and Boardman, 1990; Walter, 1994; Cudd and Duggal, 2000), develop industry-specific models (Kim and Arbel, 1998), or reexamine methodological issues, and usually employ a single classification technique (Palepu, 1986; Barnes, 1998, 1999). They have considered, inter alia, the impact of alternative forms of variables, such as raw versus industry-adjusted (Cudd and Duggal, 2000; Barnes, 2000; Pasiouras et al., 2005) or historical versus current cost data (Bartley and Boardman, 1990; Walter, 1994); the impact of alternative ways of calculating the cutoff point (Palepu, 1986; Barnes, 1998, 1999; Powell, 2001; Pasiouras et al., 2005); and the impact of the proportion of acquired and non-acquired firms in training and validation samples (Palepu, 1986; Pasiouras et al., 2005).

Other studies have searched for the most effective empirical method for prediction. Hence, alternative methods have been compared on the basis of their prediction accuracy (Barnes, 2000; Doumpos et al., 2004), or new methods have been introduced in the prediction of acquisition targets and compared with existing methods (Cheh et al., 1999; Slowinski et al., 1997; Zopounidis and Doumpos, 2002; Espahbodi and Espahbodi, 2003; Pasiouras et al., 2005).

Table 2 presents the distribution of the methods these studies have used to develop the acquisition prediction models. As mentioned above, some studies have employed more than one method of prediction, and so the total frequency of methods exceeds the number of studies. Following the pioneering studies in bankruptcy predictions, researchers in the prediction of acquisition targets initially employed discriminant analysis (DA). The statistical assumptions of DA motivated other researchers to employ logit analysis (LA), while more recently, nonparametric techniques such as rough sets (RS), artificial neural networks (ANN), probabilistic neural networks (PNN), multicriteria decision aid (MCDA), recursive partitioning algorithm (RPA), support vector machines (SVMs), and nearest neighbors (NN) were used. Finally, more recently, Cheh et al. (1999), Tartari et al. (2003), and Pasiouras et al. (2005) proposed the combination of individual models into integrated ones. While DA and LA have been most frequently employed for comparing prediction models, it is worthwhile pointing out that very few studies have attempted to simultaneously compare several classification techniques in predicting acquisitions (Espahbodi and Espahbodi, 2003; Pasiouras et al., 2005). The common consensus emerging from these studies, despite the advanced nature of classification techniques that have been recently used, is that the prediction of acquisition targets remains a difficult task, since no technique has been found to clearly outperform the others.

3.2 Acquisitions in the EU Banking Industry

Recent studies on the EU banking sector have focused on examining the scale and operating efficiency of the merging institutions (Vander Vennet, 1996; Altunbas et al., 1997; Huizinga et al., 2001; Diaz et al., 2004; Altunbas and Ibanez, 2004; Campa and Hernando, 2006), the effect of M&A announcements on the share prices of the financial institutions (Tourani-Rad and Van Beek,

Table 2. Frequency of Appearance of Classification Techniques

Method	Freq.	Method	Freq.
Discriminant analysis	20	Support vector machines	1
Logistic regression	14	PAIRCLAS	1
UTADIS	5	Probit analysis	1
MHDIS	3	Probabilistic neural networks	1
Artificial neural networks	3	Nearest neighbors	1
Recursive partitioning algorithm	2	Mahalanobis distance	1
Rough sets	2	Majority voting	1
Stacked generalization	2		

1999; Cybo-Ottone and Murgia, 2000; Beitel and Schiereck, 2001; Beitel et al., 2004; Lepetit et al., 2004; Campa and Hernando, 2006), and the impact on the takeover premium paid (Dunis and Klein, 2005).

In one of the earliest studies, Vander Vennet (1996) reported that domestic mergers among equal-sized partners significantly increased the performance of the merged banks, while improvement in cost efficiency was also found in cross-border acquisitions. Furthermore, domestic takeovers were found to be influenced predominantly by defensive and managerial motives such as size maximization. Altunbas et al. (1997) examined the cost implications from hypothetical cross-border bank mergers in the EU. They indicated that the greatest opportunities for cost savings would appear to be generated by mergers between German and Italian banks, while mergers between French and German banks would likely result in substantial cost increases. Huizinga et al. (2001) found evidence of substantial unexploited scale economies and large X-inefficiencies in European banking. Comparing merging banks with their nonmerging peers, they found that large merging banks exhibit a lower degree of profit efficiency than average, while small merging banks exhibit a higher level of profit efficiency than their peer group. Their dynamic merger analysis indicated that the cost efficiency of merging banks is positively affected by the merger, while the relative degree of profit efficiency is only marginally improved. Finally, they found that deposit rates tend to increase following a merger, suggesting that the merging banks were unable to exercise greater market power. Diaz et al. (2004) examined the bank performance derived from both the acquisition of another bank and the acquisition of nonbanking financial entities in the EU. The results show an increase in the acquirer's long-term profitability, which is more significant for bank acquisitions than for nonbank acquisitions. Altunbas and Ibanez (2004) found that, on average, bank mergers in the EU result in improved return on capital. The results also indicate that, for domestic deals, it could be quite costly to integrate dissimilar institutions in terms of their loan, earnings, cost, deposits, and size strategies. For cross-border deals, differences in merging banks in their loan and credit risk strategies are conducive to higher performance, while diversity in their capital, cost structure, as well as technology and innovation investment strategies are counterproductive from a performance standpoint.

Using event study methodology, Tourani Rad and Van Beek (1999) found that targets' shareholders experience significant positive abnormal returns, while abnormal returns to bidder's shareholders are not significant. Furthermore, the results suggest that returns to bidders are more positive when the bidder is larger and more efficient. Cybo-Ottone and Murgia (2000) found a positive and significant increase in value for the average merger at the time of the deal's announcement. However, the results are mainly driven by the significant positive abnormal returns associated with the announcement of domestic deals between two banks and by product diversification of banks into insurance. Deals that occur between banks and securities firms and between domestic and foreign institutions do not gain a positive market's expectation. In a recent study, Beitel and Schiereck (2001) found that the shareholders of the targets obtain a positive and significant revaluation of their shares, while effects for bidders are mostly insignificant. Taken as a whole, M&As create value on an aggregate basis. In a latter study, Beitel et al. (2004) examined the same data set but with a different objective. The authors analyze the impact of 13 factors such as relative size, profitability, stock efficiency, market-to-book ratio, prior target stock performance, stock correlation, M&A experience of bidders, and method of payment on M&As success of European bank mergers and acquisitions, in an attempt to identify those factors that lead to abnormal returns to target shareholders, bidders, shareholders, and the combined entity of the bidder and the target around the announcement date of the M&A. Their results show that many of these factors have significant explanatory power, leading the authors to conclude that the stock market reaction to M&A announcements can be at least partly forecasted. Lepetit et al. (2004) examined stock market reactions to bank M&As, by distinguishing between different types of M&As. The results showed that there is, on average, a positive and significant increase in the value of target banks and that the market distinguishes among the different types of M&As. In another relatively recent study, Scholtens and Wit (2004) investigated the announcement effect of large bank mergers on the European and U.S. stock markets. They found that mergers result in small positive abnormal returns, as well as that target banks realize significantly higher returns than bidders. They also documented the existence of differences between the announcement effects of European and U.S. bank mergers.

Campa and Hernando (2006) examined both the shareholder returns and changes in the operating performance of financial institutions. They found that merger announcements imply positive excess returns to the shareholders of the target firm around the announcement date, and a slight positive excess return during the three months prior to the announcement. Returns to shareholders of the acquiring firms were essentially zero around the announcement, while one year after the announcement excess returns were not significantly zero for either targets or acquirers. With regard to the change in the operating performance of the acquired banks, Campa and Hernando (2006) found evidence of substantial improvements in their return on equity and efficiency; however, they pointed out that these improvements are not correlated with the excess returns earned by shareholders upon announcement of the deal.

Dunis and Klein (2005) followed a somewhat different underlying approach. They consider an acquisition as an option of potential benefits and applied a real option pricing theory model to examine whether mergers in their sample were possibly overpaid. The results show that the option premium exceeds the actual takeover premium, suggesting that the acquisitions in the sample are not, on average, overpaid. Their further analysis shows, assuming that the option premium equals the takeover premium, that at least one of the following is true: (1) The implicitly assumed volatility is too low; (2) the assumed time to maturity is very short; and (3) the assumption of subsequent market performance is too optimistic.

4 Support Vector Machines

While the preceding literature review serves to highlight a complex set of factors influencing mergers and acquisitions, the approach in developing prediction models rests on determining whether information from the outcomes, as reflected in the data prior to an acquisition, can provide signals of that impending event. In a dichotomous classification setting, that is, to predict one or the other class from a combined set of two classes (e.g., acquisitions and non-acquisitions), the development of a support vector machines model, as with other models of prediction, begins with the design of a training sample $T = \{\mathbf{x}_i, d_i\}$, $i = 1, 2, \ldots, n$, where $\mathbf{x}_i \in \mathbb{R}^m$ is the input information for the training object i on a set of m independent variables and $d_i \in \{-1, +1\}$ is the corresponding outcome (dependent variable). Formally, the aim of the analysis is the development of a function $f(\mathbf{x}) \to d$ that distinguishes between the two classes. In the simplest case, $f(mathbx)$ is defined by the hyperplane $\mathbf{xw} = \gamma$ as follows:

$$f(\mathbf{x}) = \text{sgn}(\mathbf{xw} - \gamma),$$

where $\mathbf{w}$ is the normal vector to the hyperplane and γ is a constant. Since f is invariant to any positive rescaling of the argument inside the sign function, the canonical hyperplane is defined by separating the classes by a "distance" of at least 1. The analysis of the generalization performance of the decision function $f(\mathbf{x})$ has shown that the optimal decision function f is the one that maximizes the margin induced in the separation of the classes, which is $2/\|\mathbf{w}\|$ (Vapnik, 1998).

Hence, given a training sample of n observations, the maximization of the margin can be achieved through the solution of the following quadratic programming problem:

$$\min \tfrac{1}{2}\mathbf{w}^\top \mathbf{w} + C\mathbf{e}^\top \mathbf{y},$$
$$\text{s.t.} \quad \mathbf{D}(\mathbf{Xw} - \mathbf{e}\gamma) + \mathbf{y} \geq \mathbf{e}, \qquad (1)$$
$$\mathbf{y} \geq \mathbf{0}, \ \mathbf{w}, \ \gamma \in \mathbb{R},$$

where $\mathbf{D}$ is an $n \times n$ matrix such that $D_{ii} = d_i$ and $D_{ij} = 0$, $\forall i \neq j$, $\mathbf{X}$ is an $n \times m$ matrix with the training data, $\mathbf{e}$ is a vector of ones, $\mathbf{y}$ is an $n \times 1$ vector of positive slack variables associated with the possible misclassification of the training objects when the classes are not linearly separable, and $C > 0$ is a parameter used to penalize the classification errors.

From the computational point of view, instead of solving the primal problem (1), it is more convenient to consider its dual Lagrangian formulation:

$$\max \mathbf{e}^\top \mathbf{u} - \tfrac{1}{2}\mathbf{u}^\top \mathbf{D}\mathbf{X}\mathbf{X}^\top \mathbf{D}\mathbf{u},$$
$$\text{s.t.} \quad \mathbf{e}^\top \mathbf{D}\mathbf{u} = 0,$$
$$\mathbf{0} \leq \mathbf{u} \leq C\mathbf{e}.$$

The decision function is then expressed in terms of the dual variables $\mathbf{u}$ as follows:

$$f(\mathbf{x}) = \operatorname{sgn}(\mathbf{x}\mathbf{X}^\top \mathbf{D}\mathbf{u} - \gamma).$$

Burges (1998) highlighted two reasons for using the Lagrangian formulation of the problem. The first is that the inequality constraints will be replaced by constraints on the Lagrange multipliers themselves, which will be easier to handle. The second is that in this reformulation of the problem, the training data will only appear (in the actual training and test algorithms) in the form of dot products between vectors. The latter is a crucial issue allowing generalizing of the procedure to the nonlinear case. Therefore, to generalize a linear SVMs model to a nonlinear one, the problem data are mapped to a higher-dimensional space H (feature space) through a transformation of the form $\mathbf{x}_i \mathbf{x}_j^\top \rightarrow \phi(\mathbf{x}_i)\phi^\top(\mathbf{x}_j)$. The mapping function ϕ is implicitly defined through a symmetric positive definite kernel function $K(\mathbf{x}_i, \mathbf{x}_j) = \phi(\mathbf{x}_i)\phi^\top(\mathbf{x}_j)$. Various kernel functions exist, such as the polynomial kernel, the radial basis function (RBF) kernel, the sigmoid kernel, etc. (Schölkopf and Smola, 2002). The representation of the data using the kernel function enables the development of a linear model in the feature space H. Since H is a nonlinear mapping of the original data, the developed model is nonlinear in the original input space. The model is developed by applying the above linear analysis to the feature space H.

Several computational procedures have been proposed to enable the fast training of SVMs models. In this study we use the proximal SVMs methodology proposed by Fung and Mangasarian (2001), where the primal problem (1) is transformed to the following optimization problem:

$$\min \tfrac{1}{2}\left(\mathbf{w}^\top \mathbf{w} + \gamma^2\right) + \tfrac{1}{2}C\mathbf{y}^\top \mathbf{y},$$
$$\text{s.t.} \quad \mathbf{D}(\mathbf{Xw} - \mathbf{e}\gamma) + \mathbf{y} = \mathbf{e}. \qquad (2)$$

This new formulation only involves equality constraints, thus enabling the construction of a closed-form optimal solution directly from the training data, without requiring the use of any quadratic programming algorithm.

As previously mentioned, we explore the development of both linear and nonlinear SVMs models with a polynomial and an RBF kernel. The width of the RBF kernel was selected through a cross-validation analysis to ensure the proper specification of this parameter. A similar analysis was also used to specify the trade-off constant C. All the data used during model development were normalized to zero mean and unit variance.

5 Data and Variables

5.1 Data

The data set we employ consists of 168 commercial banks, operating in the EU-15, that were acquired between 1998 and 2002, as well as of 566 non-acquired ones.[2] Table 3 presents the number of observations by country and year. This data set was constructed as follows.

Table 3. Observations in Sample by Country and Year

	1998		1999		2000		2001		2002		Total	
	A	NA	A	NA	A	NA	A	NA	A	NA	A	NA
Austria	2	1	0	1	0	3	1	2	1	13	4	20
Belgium	3	3	0	1	3	1	0	1	3	13	9	19
Denmark	0	0	2	3	2	8	3	2	3	28	10	41
Finland	0	0	0	0	1	0	1	0	0	3	2	3
France	10	4	9	9	7	15	3	8	6	105	35	141
Germany	3	3	3	2	4	12	5	3	1	68	16	88
Greece	0	0	3	0	4	0	0	0	1	6	8	6
Ireland	0	1	1	1	0	0	0	1	0	8	1	11
Italy	1	0	5	4	14	4	3	2	9	37	32	47
Luxembourg	1	3	1	2	7	6	7	4	2	42	18	57
Netherlands	0	4	1	0	1	1	0	0	0	18	2	23
Portugal	0	1	0	0	4	0	2	0	0	8	6	9
Spain	3	2	3	2	6	4	1	1	4	32	17	41
Sweden	0	0	0	0	0	0	1	0	0	4	1	4
UK	1	2	1	4	3	2	1	4	1	44	7	56
Total	24	24	29	29	56	56	28	28	31	429	168	566

A = acquired, NA = non-acquired.

The acquired banks were first identified in three databases: Bankscope, Zephyr, and BANKERSalmanac.com. In order to be included in the sample,

---

[2] This sample or subsamples thereof has been used in the past in the studies of Pasiouras et al. (2005, 2006).

acquired banks had to meet the following criteria: (1) They were acquired between January 1, 1998, and December 31, 2002. This time period was chosen because it offers a large sample without sacrificing the stability of a short time period, thereby minimizing the effect of economic variations that could bias the study;[3] (2) the acquisition represented the purchase of 50% or more of ownership of the acquired bank;[4] (3) they operated in one of the 15 EU countries (Austria, Belgium, Denmark, Finland, France, Germany, Greece, Ireland, Italy, Luxembourg, Netherlands, Portugal, Spain, Sweden, UK); (4) all were classified as commercial banks in the Bankscope database;[5] (5) all had financial data available for two years[6] before the year of acquisition (i.e., for acquisitions that occurred in 1998, the earliest year considered would be 1996) in Bankscope.

In order to be included in the sample, non-acquired banks had to (1) operate in one of the 15 EU countries, (2) be classified as commercial banks in the Bankscope database, (3) financial data available (in Bankscope) for the entire period 1996–2002. This requirement was placed for two reasons. First, it ensures than an acquired bank, if not identified in the sources of acquisitions, could not be wrongly considered as non-acquired. (It is obvious that if a bank was 100% acquired in a given year, e.g., 1998, it could not have had data from the following years, i.e., 1999, 2000, etc.). Second, with available data for all non-acquired banks for all years, the possibility of randomly matching with an acquired bank in any fiscal year is ensured.

An important issue of concern in evaluating the classification ability of a model is to ensure that it has not overfit the training (estimation) data set. As Stein (2002) mentioned, "A model without sufficient validation may only be a hypothesis." Prior research shows that when classification models are

---

[3] In general, collecting a large sample requires a long time span. However, the literature suggests that acquisition likelihood models are not robust over time (Barnes, 1990; Powell, 1997), as the economic environment, the characteristics of firms, and the motives for acquisitions change over time. Espahbodi and Espahbodi (2003) argued that in order to minimize the time-series distortion in the models, it would be essential to limit the analysis to the shortest period of time possible. Nevertheless, an adequate number of acquired banks for development and validation of the models is required. As Beitel and Schiereck (2001) pointed out, during the period 1998–2000, more M&A deals occurred in the EU banking industry than during the previous 14 years. Therefore, because of the large number of mergers during the selected period, it is possible to obtain an adequate set of recent observations that does not span an extremely large period of time.

[4] The Bankscope and BANKERSalamanac.com provide information only for full acquisitions and, therefore, we had to rely only on Zephyr for the selection of data relative to majority acquisitions. Consequently, whether our list is complete or not depends highly on the availability of information in Zephyr.

[5] The reason only commercial banks are included is to avoid comparison problems between different types of banks (e.g., cooperative, investment, etc.).

[6] Data for two years prior to the acquisition were necessary to allow us to calculate the variable GROWTH discussed in Section 5.2.

used to reclassify the observations of the training sample, the classification accuracies are "normally" biased upward. Thus, it is necessary to classify a set of observations that were not used during the development of the model, using some kind of testing sample.

As Barnes (1990) pointed out, given inflationary effects, technological, and numerous other reasons, including changing accounting policies, it is unreasonable to expect the distributional cross-sectional parameters of financial ratios to be stable over time. Thus, a superior approach would require that the model be tested against a future period, as this approach more closely reflects a "real-world" setting. As Espahbodi and Espahbodi (2003) mentioned "After all, the real test of a classification model and its practical usefulness is its ability to classify objects correctly in the future. While cross-validation and bootstrapping techniques reduce the over-fitting bias, they do not indicate the usefulness of a model in the future." Therefore, in the present study, in order to consider the case of a drifting population (i.e., change of population over time) and determine if the variables in the prediction model and their coefficients remain stable over other time periods, the data set was split into two distinct samples.

The first set includes 137 banks that were acquired between 1998 and 2001 and an equal number of randomly selected, non-acquired banks matched by fiscal year.[7] This set of companies was used to estimate the model (i.e., training sample). The second set includes 31 banks acquired during 2002 and the remaining 429 non-acquired banks not used for model development (i.e., 566 non-acquired banks in the initial observation set minus 137 banks used in the training sample). This set was used to examine the out-of-time and out-of-sample performance of the models (i.e., validation). Financial statements from the most recent year prior to the acquisition were used in the analysis (i.e., the first year before acquisition for the acquired banks and the same fiscal year for the non-acquired ones).

5.2 Selection of Variables

Table 4 presents a list of the variables included in the models, while in the discussion that follows we outline the relevance between these variables and banks' M&As.

EQAS is a bank's equity to assets ratio, used as a measure of capital strength. Prior literature suggests that capital strength may be of particular importance in the acquisition decision, with most empirical studies indicating

---

[7] In addition to time, researchers usually match firms on the basis of their size. However, if a characteristic is used as a matching criterion, its effects are obviously excluded from the analysis (Hasbrouck, 1985). For example, matching by size prevents analysis of the effects of size on the likelihood of acquisition. Since the literature suggests that size is an important explanatory variable in acquisitions, it was preferred in this study to use it as an independent variable rather than as a matching characteristic.

Table 4. Financial Variables Used in the Models

Category	Acronym	Calculation
Capital strength	EQAS	Equity/total assets
Profit efficiency	ROAA	Return on average assets
Cost efficiency	COST	Cost to income ratio
Loan activity	LOANS	Net loans/total assets
Liquidity	LIQCUST	Liquid assets/customer & short-term funding
Growth	GROWTH	Total assets annual change
Size	TASSET	Total assets
Market power	MSHARE	Deposits market share

a negative relationship between capital ratios and the probability of being acquired (Hannan and Rhoades, 1987; Moore, 1996; Wheelock and Wilson, 2000). Harper (2000) argued that "the key factor driving mergers and acquisitions in financial systems is the industry's need to rationalize its use of capital" (p. 68). This argument is based on the belief that today risks are traded on markets rather than absorbed through capital held on a balance sheet. Hence, in order to remain competitive, banks face the need either to release surplus capital or to raise the rate of return to the capital they retain. This can be achieved through M&As. Banks may also undertake M&As to meet capital regulatory requirements. In a recent study, Valkanov and Kleimeier (2007) examined a sample of 105 U.S. and European bank mergers from 1997 to 2003 and found that U.S. target banks are better capitalized than their acquirers and non-acquired peers and that U.S. banks maintain higher capital levels than European banks. They suggest that U.S. banks strategically raise their capital levels to avoid regulatory scrutiny.

ROAA is a bank's return on average total assets and is a measure of profitability. The inefficient management hypothesis states that acquisitions serve to drive out bad management that is not working in shareholder interests (Manne, 1965). Hence, poorly managed banks are likely targets for acquirers who think that they can manage more efficiently the assets of the acquired bank and increase profits and value (Hannan and Rhoades, 1987). The results from empirical studies are mixed. Hannan and Rhoades (1987) for the United States, and Pasiouras and Zopounidis (2008) for Greece, found no evidence to support the argument that poorly managed banks are more likely to be acquired. Pasiouras and Gaganis (2007a) found a positive relationship between profitability and banks' acquisition likelihood for Germany, and a negative one for France, UK, Italy, and Spain. However, only the latter two were statistically significant. Pasiouras and Gaganis (2007b), who examined Asian banking, estimated three binary logistic models to determine the probability, respectively, of being acquired (i.e., target) versus not being involved in an acquisition, of being an acquirer versus not being involved in an acquisition, and of being an acquirer rather than a target. While ROAA was not significant in the first two models, it was positive and significant in the third, indicating that banks with low profitability would be likely acquisition targets of a bank that could operate them differently from the current managers and produce higher profits. Other studies, such as the ones of Moore (1996) and Wheelock and Wilson (2000) for the United States, found profitability to be negatively related to the acquisition probability.

COST is a bank's cost to income ratio and serves as a measure of efficiency in expense management.[8] A bank characterized by efficient expense management will have a low cost to income ratio, while a bank with poor cost efficiency will be characterized by a high cost to income ratio. This variable is also related to the inefficient management hypothesis, as profits are not only affected by the ability of managers to generate revenue but also by their ability to manage expenses. Focarelli et al. (1999) for Italy and Wheelock and Wilson (2000) for the United States found that less cost-efficient banks are more likely to be acquired. Similar results were obtained by Pasiouras and Gaganis (2007a) for the French and German banking sectors.

LOANS is a bank's net loans to total assets ratio and is a measure of loan activity. Under the intermediation approach (Sealey and Lindley, 1977), loans are considered the first of a bank's basic outputs (securities being the second) and usually make up a great percentage of total assets.[9] Hannan and Rhoades (1987) argued that a high level of loans might be an indicator of aggressive behavior by the target bank and strong market penetration, hence making the bank an attractive target. On the other hand, a low level of loan activity may indicate a bank with conservative management that an aggressive acquiring bank could turn around to increase returns. The results of previous studies are mixed. While most of the studies suggest a negative relationship (Hannan and Rhoades, 1987; Moore, 1996; Pasiouras and Zopounidis, 2008), this is not significant in all cases. The results in Wheelock and Wilson (2000, 2004) are also mixed, with total loans to total asset being negatively correlated, but not statistically significant in some instances, and positively correlated but not always statistically significant in other instances, depending on the specification of the estimated model.

---

[8] An alternative would be to use the cost to average assets ratio. We reestimated the models with the use of the cost to average assets ratio, but this has not significantly affected the classification accuracies. We report the results with cost to income ratio, which is the one used in most previous studies. The importance of the cost to income ratio is also highlighted by its use as a key indicator of efficiency in expense management in the reports of the European Central Bank (2004, 2005a, b, 2006a, b, c, d). The results with the cost to average assets ratio are available from the authors upon request.

[9] Data from the European Central Bank report (2004) on the stability of the EU banking sector indicate that the share of customers' loans in total assets was approximately 50% in 2003, highlighting the importance of loans for EU banks.

LIQCUST is a bank's liquid assets[10] to customer and short-term funding ratio and is a measure of liquidity. Liquidity can be an additional factor influencing the attractiveness of a bank as acquisition target; however, it is difficult to determine a priori what the direction of the influence will be. Excess liquidity may signal a lack of investment opportunities or a poor allocation of assets, making these banks targets because of their good liquidity position (Walter, 1994), while it is also possible that banks are acquired because they have moved into liquidity difficulties. Consequently, the empirical results are mixed. Wheelock and Wilson (2000) indicated that low liquidity makes U.S. banks less attractive takeover targets. Pasiouras and Gaganis (2007b) also found that acquired banks in Asia were in a position to meet a higher percentage of customer and short-term funds (i.e., more liquid). By contrast, Pasiouras and Gaganis (2007a) found a negative relationship between banks' liquidity and their acquisition likelihood, which is statistically significant in four of the five banking sectors in their study. Finally, Pasiouras and Zopounidis (2008) in their study on Greece also reported a negative relationship between liquidity and acquisition likelihood, although not a statistically significant one.

GROWTH is the annual change in a bank's total assets and is a measure of growth. Again, there is no conclusive evidence as to whether growth has a positive or negative impact on the acquisition likelihood. For instance, Kocagil et al. (2002) referred to previous empirical evidence that suggests that some banks whose growth rates were relatively high have experienced problems because their management and/or structure was not able to deal with and sustain exceptional growth. Hence, it is possible that a firm constrained in this way could be an attractive acquisition target for a firm with surplus resources or management available to help (Barnes, 1999). On the other hand, Moore (1996) argued that a slow-growing bank may attract a buyer seeking to increase the value of the franchise by accelerating the bank's growth rate. Hannan and Rhoades (1987) found growth to be positively correlated to inside market acquisitions and negatively correlated to outside market characteristics but insignificant in all cases. Pasiouras and Gaganis (2007b) did not find a statistically significant relationship between growth and the probability of being involved in an acquisition, either as a target or as an acquirer. However, Moore (1996) and Pasiouras and Zopounidis (2008) found asset growth to be negatively correlated to the acquisition likelihood. Pasiouras and Gaganis (2007a) also found growth to be negatively correlated to the acquisition likelihood but statistically significant only in the case of Germany and Spain.

TASSET corresponds to a bank's total assets and is a measure of size. Size is related to both synergy (i.e., economies of scale and scope) and agency M&A motives (i.e., managers' self-interest, such as empire building, salary, prestige, etc.) and can influence acquisitions through several channels. The

---

[10] According to BankScope definitions, liquid assets refer to assets that can be easily converted to cash such as cash itself, balances with central banks, as well as government bonds.

bank's size may also have a negative influence on the acquisition likelihood because large banks are more expensive to acquire, with greater resources to fight an unwanted acquisition, while once acquired, it is more difficult to absorb in the existing organization of the acquirer. The empirical results of previous studies are mixed. Hannan and Rhoades (1987) and Moore (1996) found size to be insignificant; however, Wheelock and Wilson (2000) found that smaller banks were more likely to be acquired than larger banks, while Wheelock and Wilson (2004) found that that the probability of engaging in mergers increases with bank size. Pasiouras and Zopounidis (2008) found total assets to be negatively related to the acquisition likelihood but not statistically significant in all cases.

MSHARE is the market share of the bank and is calculated by dividing the deposits of the bank by the total deposits of the banking sector over the same year. Moore (1996) argued that regulatory concerns about anticompetitive effects could reduce the probability of acquiring of banks with a high market share. Furthermore, there might not be large enough acquirers to take over banks with a considerable market share. Finally, a small share could reflect a lack of success in the market and thereby increase the acquisition likelihood, consistent with the inefficient management hypothesis. The Group of Ten (2001) also pointed out market share as one of the most important motivating factors for within-country, within-segment mergers in the financial sector. The empirical results are mixed. Hannan and Rhoades (1987) found that market share has a positive and highly significant impact on the probability of acquisition from outside the market, but plays no statistically significant role in explaining the likelihood of a within-market acquisition. However, Moore (1996) found market share to be statistically significant and negatively related with the probability of acquisition in both in-market and out-of-market acquisitions. Finally, Pasiouras and Zopounidis (2008) also found market share to be statistically significant and negatively correlated to the acquisition likelihood in the Greek banking sector.

5.3 Country-Relative Financial Variables

Following Barnes (1990, 1999, 2000), Cudd and Duggal (2000), and Asterbo and Winter (2001), we use industry-relative variables to account for industry differences. However, since all the firms in our sample are drawn from one industry (i.e., commercial banking) but from various countries, we actually use country-relative variables, which, as in Pasiouras et al. (2005, 2006), are calculated by dividing the ratios of a specific bank with the corresponding averages of the commercial banking sector for the country where the bank operates.

Standardizing by country average enhances comparability, particularly as the levels of profitability, liquidity, cost efficiency, and other aspects of banks'

6 Empirical Results

Table 5 presents descriptive statistics (mean and standard deviation) and the results of of the Kruskal–Wallis test for mean differences in the variables between the two groups of banks (i.e., acquired and non-acquired). These results correspond to the training sample and indicate that non-acquired banks were better capitalized, on average, over the period 1998–2001. This might suggest that acquired banks were characterized by a lack of financial strength that attracted buyers capable of infusing capital (Moore, 1996; Wheelock and Wilson, 2000) or that they had skilful managers capable of operating successfully with high leverage, thus making them attractive targets (Wheelock and Wilson, 2000). However, the lower profitability (ROAA) and lower efficiency in expense management (COST) of the acquired banks seem to support the inefficient management hypothesis.

Table 5. Descriptive Statistics and Kruskal–Wallis Test (Training Sample)

	Non-acquired		Acquired		Kruskal–Wallis
	Mean	Std. Dev.	Mean	Std. Dev.	Chi-square
EQAS	2.064	1.479	1.819	1.465	7.588* (0.006)
ROAA	1.800	1.629	1.054	1.632	20.287* (0.000)
COST	0.956	0.253	1.122	0.297	19.997* (0.000)
LOANS	1.040	0.550	0.987	0.537	0.114 (0.736)
LIQCUST	1.317	0.973	1.223	0.988	1.528 (0.216)
GROWTH	−0.124	3.184	−0.763	3.821	0.099 (0.753)
TASSET	0.456	0.870	0.563	1.077	0.005 (0.944)
MSHARE	0.580	1.188	0.828	1.512	0.070 (0.791)

Notes: p-values in parentheses.
* Statistically significant at the 1% level.

While the remaining variables do not appear to be significantly different between the two groups of banks, we include them all in the development of the SVM models. We follow this approach because univariate statistical significance does not necessarily predict how a variable will contribute in a multivariate model. Table 6 presents the classification results of the three SVM models, showing that they all achieve average classification accuracies slightly above 65% in the training sample and only slightly below 65% in the holdout sample, thus confirming the overall robustness of the results. However, all the models correctly predict a higher percentage of the acquired banks in both samples, the only exception being the SVM model with the RBF kernel, which correctly classifies a higher percentage of non-acquired banks rather than the acquired ones. In general, the models exhibit more balanced Type I and Type II errors in the holdout sample.

Table 6. Classification Results (in %)

	Kernel Type	Non-acquired	Acquired	Average
Training	Linear	58.28	74.19	66.23
	Polynomial	58.04	74.19	66.12
	RBF	62.70	67.74	65.22
Validation	Linear	60.58	68.61	64.60
	Polynomial	60.58	68.61	64.60
	RBF	67.15	61.31	64.23

However, the results do indicate a fair amount of misclassification, around 35% in all cases, although it should be noted that this is not inconsistent with previous studies that have in general found the prediction of acquisitions to be a difficult task (Palepu, 1986; Barnes, 1998, 1999, 2000; Powell, 2001; Espahbodi and Espahbodi, 2003; Pasiouras et al., 2005, 2006). A direct comparison with the results of previous studies is not appropriate because of differences in the data sets (Kocagil et al., 2002; Gupton and Stein, 2002), the industry under investigation, the methods used to validate the models, and so on. Nevertheless, a tentative comparison indicates that the range of accuracy in our study is comparable to other studies that validate the models in a period later than the one used for development. Pasiouras et al. (2005) summarized the results of seven such studies[12] and indicated that the average classification

---

[12] These studies are the ones of Palepu (1986), Walter (1994), Barnes (1998, 2000), Cudd and Dougal (2000), Powell (2001), and Espahbodi and Espahbodi (2003). For some studies, the average classification accuracy was calculated by Pasiouras et al. (2005) on the basis of the reported data, for comparison reasons because only the overall accuracy was reported in the original articles. Some of these studies developed only one model (Palepu, 1986), while others developed several models using various techniques (Espahbodi and Espahbodi, 2003; Barnes, 2000) or data specifications (Walter, 1994; Cudd and Dougal, 2000).

accuracy is between 42.6% and 67%. In their study, Pasiouras et al. (2005) reported average classification accuracies between 47.9% and 61.7%, while in a latter study (Pasiouras et al., 2006) the corresponding figures are 61.6% and 65.7%.[13]

It should be mentioned at this point that perfect prediction models are difficult to be developed even in the bankruptcy prediction literature where, as Barnes (1999) noted failing firms have definitely inferior or abnormal performance compared to healthy firms. The identification of acquisition targets is potentially much more difficult because there might not be consistency across firms and across time as to the characteristics of targets (Barnes, 1999).

Figure 1 presents the receiver operating characteristic (ROC) curves for the three models.[14] The ROC curve plots the percentage of "hits" (i.e., true positives) of the model on the vertical axis, and the 1-specificity, or percentage of "false alarms" (i.e., false positives), on the horizontal axis. Hence, it offers a comprehensive analysis of all possible errors and all cutoff points. The result is a bowed curve rising from the 45-degree line to the upper left corner. The sharper the bend and the closer to the upper left corner, the higher the accuracy of the model. The area under the curve (AUC) measure (Table 7) can be considered as an averaging of the misclassification rates over all possible choices of the various cutoff points and can therefore be used to compare different classification models when no information regarding the costs or severity of classification costs is available. The three curves are almost identical, while the AUC equals 0.683 in the case of the linear and polynomial models and is slightly lower in the case of the RBF model (i.e., 0.679). Hence, no matter which evaluation method is being used, the differences among the models are only marginal.

Table 7. Area Under the Curve (AUC) Statistics

	AUC	Standard Error	Asymptotic Sig.	Asymptotic 95% Confidence Interval
Linear	0.683	0.046	0.001	[0.592, 0.774]
Polynomial	0.683	0.046	0.001	[0.592, 0.774]
RBF	0.679	0.048	0.001	[0.584, 0.773]

---

[13] However, Pasiouras et al. (2006) used a cross-validation technique for testing the models and consequently did not consider the population drifting over time.

[14] The ROC curves were initially used in signal detection theory by Peterson et al. (1954) and psychology by Tanner and Swets (1954). They were latter applied in numerous studies in medicine, while more recently Sobehart and Keenan (2001) suggested their use in rating models. Other studies in accounting and finance that have used ROC curves to evaluate their models are those of Nargundkar and Priestley (2003) and Hayden (2003) in credit risk modeling, Gaganis et al. (2005) in bankruptcy prediction, Rodriguez and Rodriguez (2006) in the prediction of sovereign debt rescheduling, and Gaganis et al. (2007) in auditing.

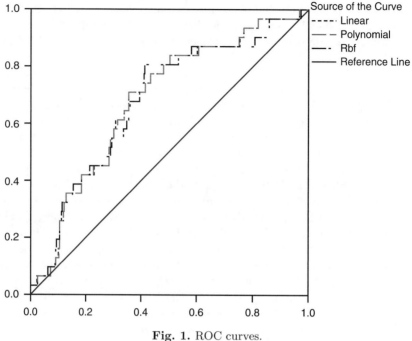

Fig. 1. ROC curves.

7 Conclusions

In this study we examined the relative efficiency of both linear and nonlinear support vector machines (SVM) models with polynomial and RBF kernels in the development of classification models for the prediction of bank acquisition targets.

The data set consisted of 168 commercial banks, operating in the EU-15, that were acquired between 1998 and 2002, as well as of 566 non-acquired ones. This data set was split in two subsets. The first one, consisting of acquisitions from the 1998–2001 period, was used for the development of the models (i.e., training). The second one, consisting of acquisitions from 2002, was used to test the out-of-time and out-of-sample performance of the models (i.e., validation). We used eight financial variables reflecting the following bank characteristics: capital strength, profitability, efficiency in expense management, loan activity, liquidity, size, growth, and market power. Since the sample was drawn from 15 EU countries, the ratios were transformed to country-relative ratios (by dividing the values of the variables of the individual banks with the corresponding average values of the commercial banking industry in the country where the banks operated). The models were evaluated in terms of their classification accuracy as well as with ROC analysis. In both cases the differences between the models were only marginal.

Future research could be directed in at least three directions. The first involves the inclusion of additional nonfinancial variables, such as ownership type, manager's experience, or technological capacity that were not available in the present study. The second could be the employment of alternative techniques that have been applied recently in other classification problems in finance, such as multidimensional scaling and probabilistic neural networks, but whose efficiency in the prediction of acquisition targets has not yet been examined. The third avenue for further research could be to combine SVMs with alternative classification techniques into an integrated model that has received limited attention, despite the promising results of Tartari et al. (2003) in acquisitions prediction, McKee and Lensberg (2002) in bankruptcy prediction, and Doumpos (2002) in credit risk assessment.

References

1. Altunbas, Y., and Ibanez, D. M. Mergers and acquisitions and bank performance in Europe: The role of strategic similarities, European Central Bank, Working paper no. 398, October, 2004.
2. Altunbas, Y., Molyneux, P., and Thorton, J. Big-bank mergers in Europe: An analysis of the cost implications. *Economica*, 64:317–329, 1997.
3. Asterbo, T., and Winter J. K. More than a dummy: The probability of failure, survival and acquisition of firms in financial distress, University of Waterloo, Working paper, 2001.
4. Barnes, P. The prediction of takeover targets in the U.K. by means of multiple discriminant analysis. *Journal of Business Finance & Accounting*, 17(1):73–84, 1990.
5. Barnes, P. Can takeover targets be identified by statistical techniques?: Some UK evidence. *The Statistician*, 47(4):573–591, 1998.
6. Barnes, P. Predicting UK takeover targets: Some methodological issues and an empirical study. *Review of Quantitative Finance and Accounting*, 12(3):283–301, 1999.
7. Barnes, P. The identification of U.K. takeover targets using published historical cost accounting data. Some empirical evidence comparing logit with linear discriminant analysis and raw financial ratios with industry-relative ratios. *International Review of Financial Analysis*, 9(2):147–162, 2000.
8. Bartley, J. W., and Boardman, C. M. The relevance of inflation adjusted accounting data to the prediction of corporate takeovers. *Journal of Business Finance and Accounting*, 17(1):53–72, 1990.
9. Beitel, P., and Schiereck D. Value creation at the ongoing consolidation of the European banking market. Presented at the Xth International Conference on Banking and Finance, Tor Vergata University Rome, December 5–7, 2001.
10. Beitel, P., Schiereck, D., and Wahrenburg, M. Explaining M&A success in European banks. *European Financial Management*, 10:109–140, 2004.
11. Burges, C. J. C. A tutorial on support vector machines for pattern recognition. *Data Mining and Knowledge Discovery*, 2:121–167, 1998.

12. Campa, J. M., and Hernando, I. M&As performance in the European financial industry. *Journal of Banking and Finance*, 30(12):3367–3392, 2006.
13. Cao, L. Support vector machines experts for time series forecasting. *Neurocomputing*, 51(April):321–339, 2003.
14. Cheh, J. J., Weinber, R. S., Yook, K. C. An application of an artificial neural network investment system to predict takeover targets. *The Journal of Applied Business Research*, 15 (4):33–45, 1999.
15. Cudd, M., and Duggal, R. Industry distributional characteristics of financial ratios: An acquisition theory application. *The Financial Review*, 41, 105–120, 2000.
16. Cybo-Ottone, A., and Murgia, M. Mergers and shareholder wealth in European banking. *Journal of Banking and Finance*, 24(6):831–859, 2000.
17. Diaz, B. D., Olalla, M. G., and Azorfa, S. S. Bank acquisitions and performance: evidence from a panel of European credit entities. *Journal of Economics and Business*, 56(5):377–404, 2004.
18. Dietrich, J. M., and Sorensen, E. An application of logit analysis to prediction of merger targets. *Journal of Business Research*, 12, 393–402, 1984.
19. Doumpos, M. A stacked generalization framework for credit risk assessment. *Operational Research: An International Journal*, 2(2):261–278, 2002.
20. Doumpos, M., Gaganis, C., and Pasiouras, F. Explaining qualifications in audit reports using a support vector machine methodology. *Intelligent Systems in Accounting, Finance and Management*, 13(4):197–215, 2005.
21. Doumpos, M., Kosmidou, K., and Pasiouras, F. Prediction of acquisition targets in the UK: A multicriteria approach. *Operational Research: An International Journal*, 4(2):191–211, 2004.
22. Dunis, C.L., and Klein, T. Analysing mergers and acquisitions in European financial services: An application of real options. *The European Journal of Finance*, 11(4):339–355, 2005.
23. Espahbodi, H., and Espahbodi, P. Binary choice models for corporate takeover. *Journal of Banking and Finance*, 27(4):549–574, 2003.
24. European Central Bank. Mergers and acquisitions involving the EU banking industry: Facts and implications, December, 2000.
25. European Central Bank. Financial stability review, December, 2004.
26. European Central Bank. Financial stability review, June, 2005a.
27. European Central Bank. Financial stability review, December, 2005b.
28. European Central Bank. EU banking structures, October, 2006a.
29. European Central Bank. EU banking sector stability, November, 2006b.
30. European Central Bank. Financial stability review, June, 2006c.
31. European Central Bank. Financial stability review, December, 2006d.
32. Focarelli, D., Panetta, F., and Salleo, C. Why do banks merge?. Paper no. 361, Banca d'Italia Research Department, 1999.
33. Fung, G., and Mangasarian, O. L. Proximal support vector machine classifiers, In F. Provost, and R. Srikant, Editors, *Proceedings of the 7th ACM SIGKDD International Conference on Knowledge Discovery and Data Mining*. ACM Press, New York, 2001, 77–86.
34. Gaganis, C., Pasiouras, F., and Doumpos, M. Probabilistic neural networks for the identification of qualified audit opinions. *Expert Systems with Applications*, 32:114–124, 2007.

35. Gaganis, C., Pasiouras, F., and Tzanetoulakos, A. A comparison and integration of classification techniques for the prediction of small UK firms failure. *The Journal of Financial Decision Making*, 1(1):55–69, 2005.
36. Goddard, J., Molyneux, P., and Wilson, J. O. S. *European Banking: Efficiency, Technology and Growth*. John Wiley & Sons, New York, 2001.
37. Group of Ten. Report on consolidation in the financial sector, January 25, 2001; available at http://www.imf.org/external/np/g10/2001/01/Eng/.
38. Gupton, G. M., and Stein, R. M. LossCalc^{TM}: Moody's model for predicting loss given default (LGD). Moody's Investors Service Global Credit Research, Special comment, February, 2002.
39. Hannan, T., and Rhoades, S. Acquisition targets and motives: The case of the banking industry. *The Review of Economics and Statistics*, 69(1):67–74, 1987.
40. Hardle, W., Moro, R. A., and Schafer, D. Rating companies with support vector machines. German Institute for Economic Research, Discussion paper 416, 2004.
41. Hardle, W., Moro, R. A., and Schafer, D. Predicting bankruptcy with support vector machines. School of Business and Economics, Humboldt-Universitaet zu Berlin, Discussion paper 2005-009, 2005.
42. Harper, I. R. Mergers in financial services: Why the rush?. *The Australian Economic Review*, 33(1):67–72, 2000.
43. Hasbrouck, J. The characteristics of takeover targets: Q and other measures. *Journal of Banking and Finance*, 9(3):351–362, 1985.
44. Hayden, E. Are credit scoring models sensitive with respect to default definitions? Evidence from the Austrian market. University of Vienna, Working paper, 2003.
45. Huang, W., Nakamori, Y., and Wang, S.-H. Forecasting stock market movement direction with support vector machine. *Computers & Operations Research*, 32(10):2513–2522, 2005.
46. Huang, Z., Chen, H., Hsu, C.-J., Chen, W.-H., and Wu S. Credit rating analysis with support vector machines and neural networks: A market comparative study. *Decision Support Systems*, 37(4):543–558, 2004.
47. Huizinga, H. P., Nelissen, J. H. M., and Vander Vennet, R. Efficiency effects of bank mergers and acquisitions in Europe. Tinbergen Institute, Discussion paper, 2001-088/3, 2001.
48. Kim, W. G., and Arbel, A. Predicting merger targets of hospitality firms (a Logit model). *International Journal of Hospitality Management*, 17(3):303–318, 1998.
49. Kocagil, A. E., Reyngold, A., Stein, R. M., and Ibarra, E. Moody's RiskCalc^{TM} model for privately-held U.S. banks. Moody's Investors Service, Global Credit Research, July, 2002.
50. Lee, Y.-C. Application of support vector machines to corporate credit rating prediction. *Expert Systems with Applications*, 33(1):67–74, 2007.
51. Lepetit, L., Patry, S., and Rous, P. Diversification versus specialization: An event study of M&As in the European banking industry. *Applied Financial Economics*, 14(9):663–669, 2004.
52. Manne, H. G. Mergers and the market for corporate control. *Journal of Political Economy*, 73(April): 110–120, 1965.
53. McKee, T. E., and Lensberg, T. Genetic programming and rough sets: A hybrid approach to bankruptcy classification. *European Journal of Operational Research*, 138(2):436–451, 2002.

54. Min, J. H., Lee, Y.-C. Bankruptcy prediction using support vector machine with optimal choice of kernel function parameters. *Expert Systems with Applications*, 28(4):603–614, 2005.
55. Min, S.-H., Lee, J., and Han, I. Hybrid genetic algorithms and support vector machines for bankruptcy prediction. *Expert Systems with Applications*, 31(3):652–660, 2006.
56. Moore, R. R. Banking's merger fervor: Survival of the fittest? Federal Reserved Bank of Dallas Financial Industry Studies, December, 9–15, 1996.
57. Nargundkar, S., and Priestley, J. Model development techniques and evaluation methods for prediction and classification of consumer risk in the credit industry. In P. Zhang, Editor, *Neural Networks for Business Forecasting*. IRM Press, Hershey, PA, 2003.
58. Pai, P.-F., and Lin, C.-S. A hybrid ARIMA and support vector machine model in stock price forecasting. *Omega*, 33(6):497–505, 2005.
59. Palepu, K. G. Predicting takeover targets: A methodological and empirical analysis. *Journal of Accounting and Economics*, 8:3–35, 1986.
60. Pasiouras, F., and Gaganis, C. Are the determinants of banks' acquisitions similar across EU countries? Empirical evidence from the principal banking sectors. *Applied Economics Letters* (forthcoming, 2007a).
61. Pasiouras, F., and Gaganis, C. Financial characteristics of banks involved in acquisitions: Evidence from Asia. *Applied Financial Economics*, 17(4):329–341, 2007b.
62. Pasiouras, F., and Zopounidis, C. Consolidation in the Greek banking sector: Which banks are acquired? *Managerial Finance* (forthcoming, 2008).
63. Pasiouras, F., Tanna, S., and Zopounidis C. *Application of Quantitative Techniques for the Prediction of Bank Acquisition Targets*, World Scientific, Singapore, 2005.
64. Pasiouras, F., Tanna, S., and Zopounidis, C. The identification of acquisition targets in the EU banking industry: an application of multicriteria approaches. *International Review of Financial Analysis* (forthcoming, 2006).
65. Peterson, W., Birdsall, T., and Fox, W. The theory of signal detection. IRE Professional Group on Information Theory, PGIT-4, 171–212, 1954.
66. Platt, H. D., and Platt, M. B. Development of a class of stable predictive variables: The case of bankruptcy prediction. *Journal of Business Finance and Accounting*, 17:31–51, 1990.
67. Platt, H. D., and Platt, M. B. A note on the use of industry-relative ratios in bankruptcy prediction. *Journal of Banking and Finance*, 15:1183–1194, 1991.
68. Powell, R. G. Modelling takeover likelihood. *Journal of Business Finance and Accounting*, 24(7–8):1009–1030, 1997.
69. Powell, R. G. Takeover prediction and portfolio performance: A note. *Journal of Business Finance and Accounting*, 28(7–8):993–1011, 2001.
70. Rodriguez, A., and Rodriguez, P. N. Understanding and predicting sovereign debt rescheduling: A comparison of the areas under receiver operating characteristic curves. *Journal of Forecasting* (forthcoming, 2006).
71. Salcedo-Sanz, S., Fernández-Villacañas, J.-L., Segovia-Vargas, M. J., and Bousoño-Calzón, C. Genetic programming for the prediction of insolvency in non-life insurance companies. *Computers & Operations Research*, 32(4):749–765, 2005.
72. Schölkopf, B., and Smola, A. *Learning with Kernels: Support Vector Machines, Regularization, Optimization and Beyond.* MIT Press, Cambridge, MA, 2002.

73. Scholtens, B., and Wit, R. Announcement effects of bank mergers in Europe and the US. *Research in International Business and Finance*, 18(2):217–228, 2004.
74. Sealey, C., and Lindley, J. T. Inputs, outputs and a theory of production and cost at depositary financial institutions. *Journal of Finance*, 32, 1251–1266, 1977.
75. Shin, K.-S., Lee, T. S., and Kim, H.-J. An application of support vector machines in bankruptcy prediction model. *Expert Systems with Applications*, 28(1):127–135, 2005.
76. Simkowitz, M., and Monroe, R. J. A discriminant analysis function for conglomerate mergers. *Southern Journal of Business*, 38:1–16, 1971.
77. Slowinski, R., Zopounidis, C., and Dimitras, A. I. Prediction of company acquisition in Greece by means of the rough set approach. *European Journal of Operational Research*, 100(1):1–15, 1997.
78. Sobehart, J., and Keenan, S. Measuring default accurately. *Risk*, 11(41):31–33, 2001.
79. Stein, R. M. Benchmarking default prediction models: Pitfalls and remedies in model validation. Moody's KMV, Technical report 020305, 2002.
80. Stevens, D. L. Financial characteristics of merged firms: A multivariate analysis. *Journal of Financial and Quantitative Analysis*, 8(March):149–158, 1973.
81. Tanner, W., and Swets, J. A decision-making theory of visual detection. *Psychological Review*, 61:401–409, 1954.
82. Tartari, E., Doumpos, M., Baourakis, G., and Zopounidis, C. A stacked generalization framework for the prediction of corporate acquisitions. *Foundations of Computing and Decision Sciences*, 28(1):41–61, 2003.
83. Tay, F. E. H., and Cao, L. J. Application of support vector machines in financial time series forecasting. *Omega*, 29(4):309–317, 2001.
84. Tay, F. E. H., and Cao, L. J. Modified support vector machines in financial time series forecasting. *Neurocomputing*, 48(1–4):847–861, 2002.
85. Tourani-Rad, A., and Van Beek, L. Market valuation of European bank mergers. *European Management Journal*, 17(5):532–540, 1999.
86. Valkanov, E., and Kleimeier, S. The role of regulatory capital in international bank mergers and acquisitions. *Research in International Business and Finance*, 21(1):50–68, 2007.
87. Vander Vennet, R. The effect of mergers and acquisitions on the efficiency and profitability of EC credit institutions. *Journal of Banking and Finance*, 20(9):1531–1558, 1996.
88. Vapnick, V. N. *Statistical Learning Theory*. Wiley, New York, 1998.
89. Walkner, C., and Raes, J.-P. Integration and consolidation in EU banking: An unfinished business. European Commission, Directorate-General for Economic and Financial Affairs, Economic papers no. 226, April, 2005
90. Walter, R. M. The usefulness of current cost information for identifying takeover targets and earning above-average stock returns. *Journal of Accounting, Auditing and Finance*, 9:349–377, 1994.
91. Wheelock, D. C., and Wilson, P. W. Why do banks disappear? The determinants of U.S. bank failures and acquisitions. *The Review of Economics and Statistics*, 82(1):127–138, 2000.
92. Wheelock, D. C., and Wilson, P. W. Consolidation in U.S. banking: Which banks engage in mergers? *Review of Financial Economics*, 13(1–2):7–39, 2004.

93. Wu, C.-H., Tzeng, G-H., Goo, Y-J., and Fang, W.-C. A real-valued genetic algorithm to optimize the parameters of support vector machines for predicting bankruptcy. *Expert Systems with Applications*, 32(2):397–408, 2007.
94. Zopounidis, C., and Doumpos, M. Multi-group discrimination using multi-criteria analysis: Illustrations from the field of finance. *European Journal of Operational Research*, 139(2):371–389, 2002.

Credit Rating Systems: Regulatory Framework and Comparative Evaluation of Existing Methods

Dimitris Papageorgiou[1], Michael Doumpos[1], Constantin Zopounidis[1], and Panos M. Pardalos[2]

[1] Technical University of Crete, Department of Production Engineering and Management, Financial Engineering Laboratory, University Campus, 73100 Chania, Greece papageorgiou.d@gmail.com, {mdoumpos; kostas}@dpem.tuc.gr
[2] Department of Industrial and Systems Engineering, Center for Applied Optimization, University of Florida, 303 Weil Hall, P.O. Box 116595, Gainesville, FL 32611-6595, USA pardalos@cao.ise.ufl.edu

1 Introduction

The credit industry has experienced a rapid growth during the last two decades. The fulfillment of modern economic needs has been a lucrative opportunity for credit institutions. Having redefined their credit policy, creditors offer a series of credit products, from credit cards to a wide range of loans. However, the potential losses from the excessive credit play a crucial role in a bank's viability and profitability perspectives. The recent bank failures in the United States, East Asia, and Latin America have raised the level of concern. As defined by the Basel Committee on Banking Supervision (BCBS, 2003), one category of credit exposures is the corporate ones, which today constitute one of the major issues of risk management for the banking industry.

Debt financing is perhaps the most popular organizational policy for raising capital. This capital is either invested, aiming at corporate growth, or is used to fulfill corporate obligations. Thus, credit institutions are called upon on a daily basis to decide whether or not to finance a finite number of firms. However, the problem of credit risk assessment is clearly more complex. Credit risk does not automatically cease to exist by rejecting the applications of firms that do not fulfill capital granting requirements. Most firms are classified as medium-grade debtors. Therefore, the assessment of an obligor's probability of default is a crucial input in order to estimate the potential future credit losses.

The aforementioned issue is addressed by employing quantitative techniques usually referred to as *credit scoring techniques*. These methodological approaches are used to develop models that classify the obligors into homogeneous groups representing different levels of credit risk while they are providing a score[1] for every firm being evaluated. Numerous quantitative techniques are currently employed by the credit scoring industry today.

In the present study, the relative performance of several classification techniques from different scientific fields is being evaluated. The conducted comparative analysis also investigates some important aspects of the process used to develop credit rating systems, including the predictive performance of the models, the variable selection process, the robustness of the models, and the sample selection effects. The objective is to analyze the development procedure of credit scoring models and consequently draw useful conclusions regarding the parameters that affect the efficiency of such models.

The rest of the chapter is organized as follows. Section 2 gives an introduction to credit rating systems. The theoretical background and the definitions related to credit risk and credit rating systems are discussed, the development process of a credit scoring model is analyzed, and the regulatory framework concerning credit risk assessment and management, as defined by the Basel's Committee on Banking Supervision provisions, is presented. Section 3 describes the experimental setup. The data and the methodologies used are presented together with a literature review of prior studies in the credit scoring. The obtained results are also discussed. Finally, Section 4 concludes the chapter, summarizes the main findings of this research, and proposes some future research directions.

2 Credit Rating Systems

2.1 The Basel Committee Scope

The Basel Committee on Banking Supervision consists of senior representatives of bank supervisory authorities and central banks from Belgium, Canada, France, Germany, Italy, Japan, Luxembourg, Spain, Sweden, Switzerland, the United Kingdom, and the United States. Issuing a series of provisions, the Committee has set a regulatory framework, aiming to secure the international convergence of the supervisory regulations governing the capital adequacy of banks. The framework consists of three mutually related pillars designed and defined to contribute to safer and more stable banking systems at national and international levels.

The first pillar (Minimum Capital Requirements) is related to the procedure of calculating the required level of capital, which must be reserved by financial institutions, as a security against the different undertaken risks. The second pillar (Supervisory Review Process) defines the procedures that must be adopted (1) from the supervisors in order to evaluate how well banks are

---

[1] This score can be accordingly processed in order to represent the firm's probability of default.

assessing their capital needs and (2) from the bank's risk managers for ensuring that the bank has reserved a sufficient capital to support its risks. Finally, the third pillar (Market Discipline) requires banks to provide disclosures with how senior management and the board of directors assess and manage the various types of risk.

Concerning corporate credit risk, under the New Basel Capital Accord (BCBS, 2001a), the financial institutions may choose between two approaches for calculating their capital needs: the standardized approach and the internal ratings-based (IRB) approach. In the IRB approach, institutions are allowed to use their own measures for the key drivers of credit risk as primary inputs to the capital calculations; thus, this approach aligns capital requirements to banks' internal risk measurement and management practices more closely (BCBS, 2001b).

2.2 Credit Rating Systems

A corporate credit rating system initially helps a financial institution in deciding whether or not to finance a specific firm, providing the firm's probability of default (PD). This probability is a measurement of the firm's creditworthiness. The system analyzes some predetermined characteristics of the firm (financial and nonfinancial) and classifies it into one of some predefined credit risk grades. Each grade is defined by a range of PD's values or by an average PD and represents a specific level of credit risk.[2] The Basel Committee states that a qualifying credit rating system must effectively distinguish the level of credit risk across the entire spectrum – from borrowers that are virtually risk-free to those in default. Thus, such a system must have a minimum of six to nine grades for performing borrowers and a minimum of two for nonperforming borrowers (BCBS, 2001b).

Once the firm's probability of default has been defined, the system provides an estimation of the losses that the bank would likely suffer if the firm proves to be inconsistent with its loan obligations. These losses are defined not only by the borrower's probability of default but also by the transaction characteristics. Thus, credit rating systems developed under the IRB approach are also called two-dimensional systems, since they provide an estimate of the borrower's probability of default at a first stage, and an estimate of the possible loan losses at a second one.

The New Basel Capital Accord provides quite a standardized procedure for calculating the possible loan losses in contrast with the rating assignment process that can be accomplished by (BCBS, 2001a) (1) expert judgment-based processes, (2) using external ratings (by rating agencies), and (3) quantitative techniques (scoring models). The distinctions among these three

---

[2] If average PDs are used, all borrowers within the same grade are assigned the same PD. The Basel Committee claims that average PD estimations are preferable to using PD ranges for each grade (BCBS, 2001b).

approaches may be less precise in practice. For example, personal expertise plays a crucial role in developing a credit scoring model, whereas external ratings can be the starting point or a benchmark in an expert's rating assignment.

Regarding internal credit analysts' rating assignments, they may sometimes prove to be extremely sound, but actually they cannot be adequately evaluated within a specific framework due to the subjectiveness of the process. Concerning the external ratings, even if considered to be quite reliable, they are available only for large corporations. However, this type of firm is rarely included in the loan portfolio of a commercial bank. The usual customers of such banks are small or middle-size firms (SMEs). In addition, the substantial difference between internal and external rating assignments is that agencies keep a neutral attitude toward the transactions between a borrower and a lender (Treacy and Carey, 2000).

Therefore, via an efficiently developed credit scoring model, the aforementioned problems cease, to some extent, to exist. The advantage of these models is that they provide a "score" for every firm being evaluated. In fact, this score can be accordingly calibrated to represent the firm's probability of default. In addition, credit scoring models are "embedding" all the past information in a mathematical model, which can numerically establish which factors are important in explaining credit risk (Saunders and Cornett, 2003). Finally, a credit scoring model is developed under a more subjective manner, reflecting the bank's policy in managing credit risk.

2.3 The Development Process

The development framework of a credit rating system that assigns ratings based on a credit scoring model consists of three stages, which are governed by a bidirectional relationship (Figure 1). The first stage is related with data collection and preparation. The database is originated by observed borrowers' characteristics with known creditworthiness. These characteristics or alternative criteria are defined by the bank's credit experts and must clearly demonstrate the firm's financial condition. The evaluation criteria should be both quantitative and qualitative. Financial ratios, measuring the company's profitability, liquidity, leverage etc., are usually considered to be the quantitative criteria. Qualitative criteria are related to the company's commercial activity, management quality, development perspectives, position in the firm's industry sector, credit history, etc. The Basel Committee states that a credit scoring model should be developed, irrespective of the data source employed, using a minimum observation period of at least five years (BCBS, 2001b), so as to incorporate the specific conditions that govern the general economic environment.

Credit scoring can essentially be considered as a classification problem. The objective is the development of a model that optimally classifies the credit customers into different risk groups. The criteria are the independent variables of the problem, while the customer's given creditworthiness is the

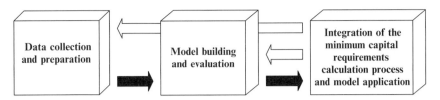

Fig. 1. The development process of a credit rating system.

dependent one. The final step before the implementation of the second stage of the development process is the segregation of the database into two samples: the training and the test sample.

In the second stage, the credit scoring model is built and evaluated for its predictability. First, it must be determined which quantitative technique will be used for the model's development. The technique can be a statistical one (i.e., linear discriminant analysis), a nonparametric one (i.e., neural networks), or an expert system which tries to simulate a credit analyst's decision making (Comptroller of the Currency Administrator of National Banks, 2001). The method's choice is driven by several factors (i.e., by the ability of the system's developers and users to comprehend the method's functionality and outputs, by the available data, etc.). A common approach for selecting the most suitable method is to develop various models based on different quantitative techniques and finally choose the one that satisfies some specific criterion.

In the training phase, the tuning of the method's parameters is accomplished, aiming primarily to achieve the highest classification accuracy. Actually, the training sample is used for the development of a model that tries to accurately address the relationship between credit risk and the observed borrowers' characteristics. In the evaluation phase, the test sample is used in order to derive useful conclusions for the model's performance and stability. A potential model must minimize the deviations between the given and the estimated classification of the test sample's observations.

The second stage of the development process is considered to be the most important. The model is built through a series of choices, and many expert interventions are involved in order to derive the "optimal solution" (Liu and Schumann, 2002). For example, it must be decided which variables will be used as the input ones, which quantitative technique will be selected for the model's training, how the database will be split, under which evaluation criteria the analysis of the model's performance will be conducted, etc.

Once the model that produces scores and ordinally ranks the companies in the database is constructed, the next step is to define the rating system's grades and associate an estimated average PD to its grade. This can be accomplished empirically by the credit analysts and the developers of the system, through a proper processing of the model's outputs or via mapping techniques. In mapping approaches, an agency's (i.e., Moody's or S&P's), grades and the corresponding default rates are used as a benchmark. Usually each internal grade is equated with an external one and the model's outputs are associated, in various ways, with the external default rates in order to derive average PD estimations for each internal grade (in Moody's Investor Services, 2000, various mapping approaches are presented). Work related to the aforementioned problem was conducted by Fernades (2005), while Carey and Hrycay (2001) examined the properties of the major methods currently used to estimate average default probabilities for each grade.

In the third stage, the calculation procedure of the minimum capital requirements, which is analyzed in the following paragraph, is intergraded in the model and the developed rating system is finally applied in practice. During the system's operation in real situations, the experts must monitor its functionality, interfering when a problem occurs.

2.4 Calculating the Minimum Capital Requirements

As already mentioned, credit risk assessment involves two major issues: the estimation of the potential borrower's PD and the calculation of the possible losses that may be incurred due to the credit granting. The implementation process of the latter issue is clearly defined by the Basel Committee's revised framework on international convergence of capital measurement and capital standards (BCBS, 2004).

A bank can usually forecast the average level of credit losses it expects to experience. These losses are referred to as expected losses (EL) and can be estimated by multiplying the obligor's PD with the exposure at default (EAD) and the the loss given default (LGD, expressed as a percentage of EAD):

$$EL = PD \times LGD \times EAD. \qquad (1)$$

The IRB approach can be implemented under the foundation or the evolutionary methodology. When the former one is adopted, LGD and EAD are determined through the application of standard supervisory rules, while in the evolutionary approach a bank is allowed to use its own internal assessments of these components.

Losses above expected levels are usually referred to as unexpected losses (UL). The Basel Committee requires banks to hold capital only against the unexpected losses. However, they have to demonstrate in the supervisory authority that they have built adequate provisions against expected ones (BCBS, 2005). In the IRB approach, regulatory capital requirements for unexpected losses are derived from risk weight functions, which are based on the so-called asymptotic risk factor (ASFR) model (Gordy, 2003; Aas, 2005).

The ASFR model framework assumes that a bank's credit portfolio is completely diversified. This means that idiosyncratic risks associated with individual exposures tend to cancel out one another and that only systematic risk has a material effect on portfolio losses. The great advantage of the ASFR model is that it allows banks to calculate capital charges on a decentralized loan-by-loan basis and then aggregate these in a portfolio level. This characteristic has been deemed vital for making the IRB framework applicable to a wide range of countries and institutions (BCBS, 2005).

Total credit risk-weighted assets, which are a crucial input for calculating minimum capital requirements,[3] are computed at an exposure level as a function of the three aforementioned parameters (PD, LGD and EAD) plus a maturity (M) and an asset correlation (R) parameter:

$$RWA = K(\text{PD}, \text{LGD}, M, R) \times 12.5 \times \text{EAD}.$$

The term K represents the capital requirement for a specific exposure and is calculated using the following formula:

$$K = \left[\text{LGD} \times N\left(\frac{G(\text{PD})}{\sqrt{1-R}} + N^{-1}(0.999)\sqrt{\frac{R}{1-R}} \right) - \text{PD} \times \text{LGD} \right]$$
$$\times \frac{1 + (M - 2.5)\beta}{1 - 1.5\beta},$$

where N is the cumulative standard normal distribution function and N^{-1} its inverse, R the asset correlation parameter, and β a smoothed maturity adjustment:

$$\beta = \left[0.11852 - 0.05478 \log(\text{PD}) \right]^2.$$

The asset correlation term is needed as an input in the ASFR model in order for the borrower's dependence on the general state of the economy to be specified, and it is computed by a specific formula defined by the Basel Committee (see BCBS, 2005, for details). The formula incorporates two empirical observations (BCBS, 2005). First, asset correlations decrease with increasing PD, which means that the higher the probability of default, the higher the idiosyncratic risk components of an obligor. Second, asset correlations increase with firm size, which means that larger firms are more closely related to the general conditions in the economy, while smaller firms are more likely to default due to idiosyncratic reasons.

Maturity effects are incorporated in the Basel Committee's model as a function of both maturity and probability of default. The function's form is based on the following considerations (BCBS, 2005). First, long-term borrowers are riskier than the short-term borrowers; downgrades are more likely to happen for long-term borrowers. As a consequence, the capital requirement should increase with maturity. Second, low-PD borrowers have more potential to downgrade than high-PD borrowers. Thus, maturity adjustments should be higher for low-PD borrowers.

---

[3] A bank must hold capital equivalent to at least 8% of its total risk-weighted assets, which are determined by multiplying the capital requirements for market risk and operational risk by 12.5 (i.e., the reciprocal of the minimum capital ratio 8%) and adding the resulting figures to the sum of risk-weighted assets for credit risk.

As mentioned above, the asymptotic capital formula has been derived under the assumption that the bank's credit portfolio is perfectly diversified. In real-world portfolios, though, there is always a residual of undiversified idiosyncratic risk components. If this residual risk is ignored, then the true capital requirements will be underestimated. Therefore, the Basel Committee proposes the calculation, on a portfolio level, of the so-called granularity adjustments (the calculation procedure is presented in BCBS, 2001b). Granularity is a measurement of the portfolio's concentration risk; the additional risk resulting from increased exposure to one-obligor or groups of correlated obligors. These adjustments can be either negative or positive, depending on the portfolio's diversification level.

2.5 The Specifications for a Qualifying System

This section presents the basic specifications that a system must meet. The analysis is mainly based on the requirements set in the Basel Committee's consultative document, "The Internal Ratings-Based Approach" (2001b).

1. Meaningful differentiation of risk. A credit rating system must be primarily designed to distinguish risk rather than to minimize regulatory capital requirements. The grades of the rating system must be properly defined so as to represent different levels of credit risk. In addition, borrowers within the same grade must be treated differently according to their transaction's characteristics. Finally, the exposures must be meaningfully distributed across grades in order for excessive concentration in any particular grade to be prevented.[4]
2. Continuous evaluation of borrowers. Companies that are included in the bank's credit portfolio should be, at a minimum, annually rerated. Banks must be capable of gathering, prioritizing and analyzing new information about their customers' economic progress.
3. Oversight of the rating system's operation. The involved parties in bank's credit risk management must constantly monitor the credit rating system for its proper functioning and correctness. A firm's estimated PD should not be significantly different from the actual/realized PD. In addition, the system must be subjected to sufficient and efficient controls (i.e., stress tests). Furthermore, an adequate feedback mechanism between the responsible parties should exist in order to accomplish system integrity.
4. Correct selection of the evaluation criteria. The bank must demonstrate to the regulatory authority that under the assessment criteria used, it can properly analyze and estimate a firm's creditworthiness. In the credit risk assessment process, a firm is evaluated for its future performance based on current information. Therefore, a bank must take into consideration

---

[4] Specifically, the Committee is proposing that no more than 30% of the gross exposures should fall in any single borrower grade.

other criteria that are related to the perspectives of the company's operational sector or to the business's ability to overcome a probable economic instability.
5. Collecting a substantial database. The database used for the development and evaluation of a credit rating system must be extremely representative. It must incorporate historical data on borrowers' characteristics (accounting as well as qualitative data), past ratings, past PDs estimations, credit migrations, actual creditworthiness, payment history, etc. A diverse database enables a bank to develop an efficient credit rating system, to enhance the predictive power of the used one, and to address the key factors that are strongly related to the credit risk estimation process.

2.6 Current Models for Estimating Private Firms' PDs

Rating agencies, apart from evaluating large or publicly traded corporations, have also designed models for estimating private firms' default risk. The most known credit scoring models are the RiskCalc developed by Moody's and the CreditModel developed by S&P. Both of them require as input only financial ratios and can be accordingly calibrated in order to operate in specific industries and regions. Actually, the aforementioned models are suites of localized credit scoring models, functioning under a common framework. RiskCalc uses the probit model as a quantitative tool (Moody's Investor Services, 2000), while CreditModel uses Proximal Support Vector Machines as a rating estimator (Standard and Poor's, 2006b).

Another known model, which assesses a private firm's credit risk under a market-based approach, is the KMV's Private Firm Model (KMV, 2001). Market or structural models are based on the pioneer work of Merton (1974). Generally, a Merton-based model incorporates the option nature of a firm's equity to derive the market value and the volatility of the firm's assets, using the Black–Scholes formula (1973). Once these two parameters have been estimated, they are combined with the firm's contractual liabilities (considered as the firm's debt), leading to a single measurement that represents the firm's probability of default.

The Private Firm Model is based on the structural framework used by KMV for producing public firms expected default frequencies, alternative default rates, or PDs. The market value and the volatility of the firm's assets are estimated under the Merton approach, in order for the firm's "distance to default" to be determined. The distance to default measure represents the number of standard deviations (or distance) between the market value of a firm's assets and its relevant liabilities (Moody's KMV, 2004). Then a default database is used to derive an empirical distribution relating/mapping the distance to default to an expected default frequency.

Since, by definition, private firms do not have publicly traded equity and debt, price series of these financial assets are not available; thus, the market value and the volatility of a public firm's assets cannot be estimated. For this reason, Private Firm Model uses public market information from comparable/similar public firms together with the firm's financial statement in order to estimate the aforementioned parameters (Moody's Investor Services, 2003).

After the merger of Moody's Investor Services and KMV in 2002, a new model was developed (Moody's KMV, 2004). The new model incorporates both the market-based approach, as used in the KMV's Private Firm Model, and the financial statement-based approach of the original RiskCalc model. Firm-specific (idiosyncratic) information, more predominant in RiskCalc, is blended with market-sector based information, more predominant in Private Firm Models, leading to more accurate credit risk assessments.

2.7 Benefits from the Use of Credit Rating Systems

The development and the implementation of efficient credit rating systems provide some important benefits to a financial institution. The most important ones are cited bellow.

1. They are a reliable supplemental tool for the loan approval process.
2. They confine the credit analyst's subjectiveness.
3. Through them the credit institutions can estimate the future loan losses and proceed to the appropriate hedging activities.
4. They are embedding a bank's credit culture, defining a common framework for the evaluation of all customers. This characteristic is quite essential because small business loan applications are usually evaluated on a regional level.
5. Through them a bank can define the transaction's characteristics (i.e., the type or the amount of the impeding financing), minimizing the level of risk undertaken.
6. They reduce the time and cost needed for the loan approval process. The credit analysts are intensively evaluating only those customers with specific characteristics.

To sum up, credit rating systems can contribute to the increment of a bank's profitability. Through them a bank can monitor the progress of its loan portfolio and redefine its credit policy, whatever that implies.

3 Comparison of Classification Methods for the Development of Credit Rating Models

3.1 Literature Review

Several methods have been proposed in the literature for the development of credit scoring models; thus, many comparative studies have been conducted investigating the relative performance of these techniques. It is generally accepted that the dominant statistical techniques include linear discriminant analysis (LDA) and logistic regression (LR), while artificial neural networks (NN) is the most frequently used nonparametric approach (Altman and Saunders, 1998).

Desai et al. (1996) compared NN with the two aforementioned statistical techniques for scoring loan applicants and concluded that NN significantly outperforms LDA but is only marginally better than logistic regression. West (2000) conducted a more thorough research, investigating the relative performance of NN. He compared five neural network models with a set of commonly used classification techniques (see Table 1), claiming that while the multilayer perceptron is the most commonly used neural network model, the mixture-of-experts and radial basis function neural networks should be considered for credit scoring applications.

Malhotra and Malhotra (2002) analyzed the beneficial aspects of using a neuro-fuzzy system in credit risk assessment. The authors performed a comparison with LDA, reporting that the neuro-fuzzy model achieves higher classification accuracy regarding the potential loan defaulters. They also referred to a critical issue of the credit scoring procedure, the interpretability of the decision-making process. They claimed that a user can easily understand the output of a neuro-fuzzy model, an advantage offered by the embedded fuzzy rules. Baesens et al. (2003b) also developed interpretable credit scoring models by applying three neural networks rule extraction approaches. They investigated the performance of the developed models using as a benchmark the popular C4.5 decision-tree algorithm and logistic regression. The results obtained show that two rule extraction techniques yield higher test set classification accuracy than the benchmark models.

Ong et al. (2005) used genetic programming to develop a decision tree and compared it with other commonly used classification techniques using two real world data sets. The authors concluded by claiming that genetic programming is more suitable for tackling credit scoring problems because (1) no assumptions need to be made regarding the available data and (2) it automatically selects the important variables. Huang et al. (2004) applied support vector machines and backpropagation neural networks into two bond rating data sets. They demonstrated that support vector machines can achieve accuracy comparable to that of neural networks. Callindo and Tamayo (2000) applied probit analysis, k-nearest neighbors, neural networks, and classification and regression trees to a mortgage loan data set, concluding that the latter approach provides the best results. Finally, Doumpos et al. (2002) applied the MHDIS method (Zopounidis and Doumpos, 2000) to a corporate loan database and compared the developed model with traditional statistical and econometric techniques. The results obtained illustrate that the MHDIS method can be considered an efficient approach for developing credit risk assessment models. Other comparative studies have been conducted by Piramuthu (1999) and Liu and Schumann (2002).

In most of the aforementioned research either a limited number of classification techniques are being evaluated or the data sets used are significantly small. Hence, the issue of identifying the appropriate technique for developing a credit scoring model has not been thoroughly examined. The most extensive research has been conducted by Baesens et al. (2003a). Various classification techniques have been compared using an adequate number of data sets of sufficient size (see Table 1). The authors concluded that a few techniques were clearly inferior to others, while least-squares support vector machines, neural networks, linear discriminant analysis, and logistic regression provide very good results.

Except for studies in which the performance of several methods is being compared, some studies have investigated other important aspects of a credit scoring model's development process. For example, Liu and Schumann (2005) applied four feature selection techniques in order to improve the model's simplicity, speed, and accuracy. The experiment was conducted using four classification methods, concluding that after feature selection only the accuracy of the k-nearest-neighbors model was improved. The authors claimed that this result may be caused by the nature of the classification techniques. The M5 decision-tree algorithm selects the most important features during its building process, while neural networks and logistic regression assign small weights to the irrelevant variables. In addition, the results obtained show that, in general, the reduction in the number of features decreases the training time and simplifies the final models. The aforementioned issue was also addressed by Fritz and Hosemann (2000). After having applied some variable reduction techniques, whether based on the classification algorithm used or not, they stated that better results should be achieved with variable sets selected depending on the specific training method.

3.2 Methods

Since the pioneer works of Altman (1968) and Ohlson (1980), traditional and econometric techniques have been widely used in credit risk assessment and financial distress prediction problems. Their deficiencies, however, with regard to the progress made in other fields have led to the development of a plethora of new classification approaches. This Section analyzes the methods used in the present study.

Linear Discriminant Analysis

Given a training data set, linear discriminant analysis (LDA) leads to the development of a linear combination of the input variables, maximizing the ratio of between-group to within-group variance. The model takes the following form:

$$f(\mathbf{x}) = w_0 + w_1 x_1 + w_2 x_2 + \cdots + w_n x_n, \qquad (2)$$

Table 1. Description of Prior Credit Scoring Studies

Study	Application	Methods	Data Sets	Data Size	Variables
Desai et al. (1996)	Loan applicants	NN, LDA, LR	3	962/918/853	18
Piramuthu (1999)	Credit cards, corporate loans, bank failure	NN, NFS	3	690/48/118	15/18/19
West (2000)	Credit applicants	NN, LDA, LR, CART, K-NN, KD	2	1,000/690	24/14
Fritz and Hosemann (2000)	Corporate loans	LDA, K-NN, GP, NN, DT	1	3,599	98
Galindo and Tamayo (2000)	Home mortgage loans	PA, CART, NN, K-NN	1	4,000	24
Liu and Schumann (2000)	Credit insurance	LDA, LR, K-NN, NN, DT	1	65,367	14
Malhotra and Malhotra (2002)	Loan applicants	NFS, LDA	1	790	2
Doumpos et al. (2002)	Corporate loans	MHDIS, LDA, LR, PA	1	1411	11
Baesens et al. (2003a)	Credit data	NN, C4.5	3	1,000/3,123/7,190	20/33/33
Baesens et al. (2003b)	Credit data	LDA, QDA, LR, LP, SVM, LS-SVM, NN, BNC, C4.5, K-NN	8	690–11,700	16–33
Huang et al. (2004)	Bond rating	SVM, NN, LR	2	74/265	21
Ong et al. (2005)	Credit applicants	GP, NN, CART, C4.5, RS, LR	2	690/1,000	14/20
Liu and Schumann (2005)	Credit insurance	DT NN, LR, K-NN	1	38,283	72

LDA: linear discriminant analysis; LR: logistic regression; K-NN: k-nearest neighbors; NN: neural networks; PNN: probabilistic neural networks; NFS: neuro-fuzzy systems; CART: classification and regression trees; LP: linear programming; SVM: support vector machines; LS-SVM: least-squares support vector machines; GP: genetic programming; BNC: Bayesian network classifiers; RS: rough sets; PA: probit analysis; DT: decision trees; KD: kernel density.

where $f(\mathbf{x})$ represents the discriminant score, w_0 is a constant factor, and w_j is the associated coefficient of the corresponding input variable x_j ($j = 1,\ldots,n$). Each firm is classified into one of the two groups by comparing the firm's discriminant score with a cutoff value.

In order to determine the optimal cutoff point, the prior probabilities of class membership and the misclassification costs must be specified. However, the specification of these parameters is a very subjective decision. Therefore, most researchers assume that the misclassification costs are equal and that the prior probabilities are equal with the proportion of the default and nondefault firms in the sample (Balcaen and Ooghe, 2004). An alternative approach is to determine the optimal cutoff point via trial-and-error procedures (Doumpos and Zopounidis, 2002), choosing as optimal the cutoff value at which a criterion regarding classification accuracy is maximized. In this study, the optimal cutoff point was determined by applying the Kolmogorov–Smirnoff distance.

The constant term w_0 and the coefficient vector $\mathbf{w} = (w_1, w_2, \ldots, w_n)^\top$ are obtained with the assumption that the input variables follow a multivariate normal distribution and that the covariance matrices of the groups are equal. Empirical evidence has shown that especially for default firms the normality condition is violated (Min and Lee, 2005). Furthermore, if the covariance matrices are not equal, quadratic terms are incorporated in the discriminant function; hence, quadratic discriminant analysis needs to be used. However, linear discriminant analysis has been reported to be a more robust approach than quadratic discriminant analysis even when the aforementioned assumption is violated (Lee et al., 2002).

Logistic Regression

The fundamental assumption of the logistic approach to discrimination is that the log of the group-conditional densities can be expressed as a linear combination of the input variables as follows:

$$\log \frac{p}{1-p} = f(\mathbf{x}) = w_0 + w_1 x_1 + w_2 x_2 + \cdots + w_n x_n.$$

After taking exponentials on both sides, it is assumed that the probability of a dichotomous outcome follows the cumulative logistic distribution function:

$$p = \frac{1}{1 + e^{-f(\mathbf{x})}}.$$

The model's parameters (i.e., the intercept term w_0 and the coefficient vector $\mathbf{w}$) are estimated using maximum-likelihood procedures.

The major advantage of logistic regression is that it overcomes the statistical assumptions of linear discriminant analysis. In addition, because a logit model's outputs are between 0 and 1, it is considered to have a nice probabilistic interpretation. However, logistic regression has a serious drawback. Logit models are extremely sensitive to the problem of multi-collinearity; therefore, the inclusion of highly correlated variables must be avoided.

Neural Networks

Despite the fact that neural networks originate from the field of artificial intelligence, there is an enormous amount of cited work applying neural networks to business-related problems (see Vellido et al., 1999, and Smith and Gupta, 2000, for an extensive review). The most commonly used neural networks for classification problems are the multilayer feed-forward ones. The aforementioned neural networks are typically composed of an input layer, one or more hidden layers, and an output layer, each consisting of several neurons or alternative units (Figure 2). Each neuron receives a total input of the weighted sum of the incoming signals s_k multiplied by interconnection weights w_k plus a bias term w_0. The neuron's output is computed by transforming the neuron's input using an activation function F, and it is transmitted to the neurons of the subsequent layer:

$$\text{output} = F\left(\sum_k w_k s_k + w_0\right).$$

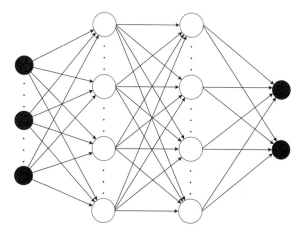

Fig. 2. The architecture of a multilayer perceptron network with two hidden layers.

Previous studies have shown that for classification problems, neural networks with one hidden layer are capable of achieving a desired level of accuracy (Patuwo et al., 1993; Subramanian et al., 1993; Bishop, 1995).

In the present study the known sigmoid function was used as the transfer one, while the popular backpropagation algorithm (Rumelhart et al., 1986) was used for the training of the developed models. The name "back-propagation" stems from the fact that network's error is propagated backwards, via an iterative process, in order for the "optimal" network's weights to be determined. Six multilayer feed-forward neural models were applied to the available data sets; with 1 hidden layer and 8, 12 and 16 neurons and with 2 hidden layers with 6, 8 and 12 neurons in each layer.

The main advantages of neural networks are considered to be (1) their ability to model complex, nonlinear data relationships and (2) the fact that they do not require any assumptions about the distribution of the data used. However, their inabilities (1) to address the relevance between the independent variables and the model's output and (2) to generate a set of rules to express the model's operation have been reported as their main drawbacks (Vellido et al., 1999). For these reasons, neural networks are often characterized as "black boxes."

Classification and Regression Trees

Classification and regression trees (CART; Breiman et al., 1984), is a nonparametric technique based on the philosophy of recursive partitioning algorithms. The studies of Frydman et al., (1985), Marais et al. (1985), and Srinivasan and Kim (1986) are the first ones in which the CART methodology had been applied in financial classification problems.

Given a training data set, the nonparametric approach leads to the development of a classification tree that consists of a finite number of nodes. In each node a decision rule is assigned (Figure 3). Before the development process, the prior probabilities and the misclassification costs must be determined. The tree-building process starts at the root node. The CART approach tries to find the best variable from the input ones and the corresponding split point, which will be assigned in the root node in order for two child nodes to be created. All the values that a variable takes in the training sample are considered as possible split points. The variable that maximizes a splitting function, a formula that measures the "purity" of the two child nodes, is selected as the best one. The most commonly used splitting functions are the Gini criterion and the Twoing rule. Then, each child node is assigned a class according to the rule used that minimizes the misclassification costs. The process of node splitting, followed by the class assignment procedure, is repeated for each new developed node (child node) until the final tree is developed.

The constructed final tree is generally very overfit; either there is only one observation or there are just a few observations in each terminal node. Although large trees are highly accurate for the training data, they provide poor results when applied in new/different data sets. In order to reduce the number of tree nodes, usually pruning techniques are employed (see Esposito et al., 1997, for a comprehensive review). The "cost-complexity" pruning method, in conjunction with a 10-fold cross-validation (Stone, 1974), was used in this study.

The main advantages of the CART approach are that (1) it can handle categorical/qualitative variables, (2) it leads to the development of interpretable models, (3) it makes no distributional assumptions about the dependent and independent variables, and (4) it can handle data sets with missing values.

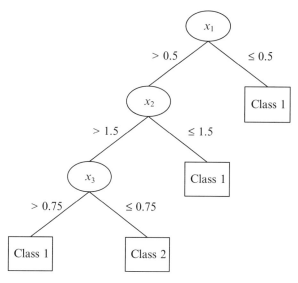

Fig. 3. The representation of a CART tree.

Linear Programming

Linear programming (LP) is probably one of the most commonly used techniques in the credit scoring industry today. Mangasarian (1968) was the first to propose the use of linear programming for solving classification problems, while Freed and Glover (1981a and 1981b) and Hand (1981) proposed that linear programming can be used even when the two groups are not linearly separable, using as objectives the minimization of the maximum error or the sum of absolute errors.

In the present study the following linear programming formulation was used:

$$\min \frac{1}{m_1} \sum_{i \in C_1} \sigma_i + \frac{1}{m_2} \sum_{i \in C_2} \sigma_i,$$
$$\text{s.t.} \quad \mathbf{w}^\top \mathbf{x}_i + w_0 + \sigma_i \geq 1, \quad \forall i \in C_1,$$
$$\mathbf{w}^\top \mathbf{x}_i + w_0 - \sigma_i \leq -1, \quad \forall i \in C_2,$$
$$\mathbf{w} \in \mathbb{R}, \sigma_i \geq 0,$$

where σ_i is the classification error for the firm i, and m_k is the number of training examples from class C_k.

By solving the above linear program, a discriminant hyperplane of the following form is constructed:

$$f(\mathbf{x}) = w_0 + \mathbf{w}^\top \mathbf{x}.$$

If the score for a firm is below (above) zero, then the firm is classified in the default (nondefault) group.

Proximal Support Vector Machines

Classical support vector machines (SVM) were proposed by Vapnik (1995) to solve classification and function estimation problems. The SVM approach to a two-class classification problem attempts to create a boundary (solid line) between the two different classes (Figure 4). The boundary is oriented by two linear planes, which are defined in such a way so that the distance between them (margin) is maximal. Given a training data set $\{\mathbf{x}_i, y_i\}_{i=1}^m$ with input data $\mathbf{x}_i \in \mathbb{R}^n$ and corresponding class labels $y_i \in \{-1, +1\}$, the following quadratic program is solved:

$$\min \ \frac{1}{2}\mathbf{w}^\top \mathbf{w} + C \sum_{i=1}^m \sigma_i,$$

$$\text{s.t.} \ y_i[\mathbf{w}^\top \phi(\mathbf{x}_i) + w_0] \geq 1 - \sigma_i, \quad i = 1, \ldots, m,$$

$$\mathbf{w}, w_0 \in \mathbb{R}, \quad \sigma_i \geq 0.$$

The variables σ_i are slack variables that are needed in order to allow misclassifications, while the positive real constant C is a trade-off parameter between the classification errors and the margin. In the nonlinear separable case, a transformation function $\phi(\mathbf{x})$ is used. This is referred as the *kernel function* and is used to map the input data to a high-dimensional feature space where linear separation can be accomplished.

By solving the above quadratic programming, a separating plane (classifier) is created in the primal weight space which takes the following form:

$$f(\mathbf{x}) = \text{sgn}[\mathbf{w}^\top \phi(\mathbf{x}) + b].$$

SVMs have been reported as a robust approach, especially when large data sets are used. This advantage stems from the fact that only a subset of the training data (called the support vectors) is used to define the decision function while the rest of the points can be discarded.

In this study, proximal support vector machines were used instead of the classical ones. Proximal SVMs were introduced by Fung and Mangasarian (2001). By modifying the quadratic programming problem of the standard SVMs, a simpler alternative approach is developed that requires smaller computational effort, since proximal SVMs require only the solution of a system of linear equations. In addition, in proximal SVMs the two linear planes cease to be bounded, are pushed as far as possible, and are placed where the points of each class are clustered:

$$\min_{\mathbf{w}, w_0, \sigma_i} \ \frac{1}{2}(\mathbf{w}^\top \mathbf{w} + w_0^2) + \frac{1}{2} C \sum_{i=1}^m \sigma_i^2,$$

$$\text{s.t.} \ y_i[\mathbf{w}^\top \phi(\mathbf{x}_i) + w_0] = 1 - \sigma_i, \quad i = 1, \ldots, m.$$

In the analysis we use proximal SVMs with linear, quadratic, and RBF kernels (these will be denoted as LPSVM, QPSVM, and RBF PSVM, respectively).

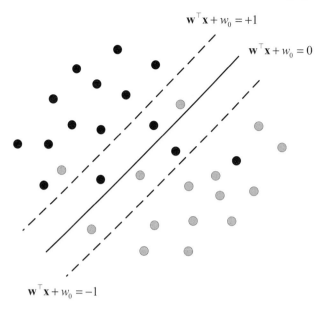

Fig. 4. The standard support vector machine classifier.

UTADIS

The UTADIS method (Zopounidis and Doumpos, 2002) originates from the field of multicriteria decision aid. The objective of the UTADIS methodology is to develop an *ordinal* classification model in the form of an additive value function:

$$V(\mathbf{x}) = \sum_{j=1}^{n} w_j v_j(x_j),$$

where $w_j \geq 0$ is weight of attribute j, $v_j(x_j)$ is the marginal value function of the attribute j, and $V(\mathbf{x}) \in [0,1]$ is the global value (score) of the evaluated observation. The marginal value functions can be either linear or nonlinear and provide two main advantages. First, they allow the modeling of nonlinear data relationships; second, the developed model can handle qualitative variables. For the simple two-class classification problem, an observation is assigned to one of the predetermined classes by comparing its global utility value with a cutoff point. The estimation of the marginal value functions, the weights of the attributes, and the cutoff value are accomplished through linear programming techniques. A detailed analysis of the linear programming formulation incorporated in the UTADIS method can be found in Doumpos and Zopounidis (2002).

3.3 Data

The data used in this study include 38,105 firms from two different business sectors (industrial and construction firms) corresponding to the period 1998–2003. The firms have been classified in two groups, default and nondefault ones, while 33 financial and nonfinancial variables were used in the analysis (see Table 2). The initial data were split in two databases according to the firms' operational sector. From each database, five data sets were created. The training sample for the first data set includes the observations from year the 1998 and the test sample the observations from the years 1999–2003. The observations from the years 1998–1999 are considered to be the training sample for the second data set, while the observations from the years 2000–2003 are considered to be the test sample. Following the above repetitive procedure, the five data sets for each business sector were constructed. Table 3 presents the number of default and nondefault firms for each year and for each business sector. It is clearly demonstrated that the samples used are considerably unbalanced regarding the proportion of default and nondefault observations.

The initial data were split in two databases according to the firm's type mainly for the following reasons. First, through the development of industry specific credit rating systems, the homogenous firms are being evaluated under a common framework. Firms within the same sector appear to have similar characteristics and business activities. Second, the sample size of the two sectors is quite different; for the industrial sector it is quite large, while for the construction sector it appears to be rather small. So, data segregation enables us to examine the methods' performance in conjunction with the size of the data used.

3.4 Experimental Setup

For data sets for each of the two business sectors, the experimental analysis is performed in five steps, each of which focuses on different aspects of the model development process. The experimental settings used in the analysis are summarized in Table 4.

Step 1: Initially, all methods were tested using the initial data sets, with the available attributes (henceforth, the complete set of attributes will be denoted as A1).

Step 2: At a second stage, the effectiveness of a simple attribute selection process is tested. In order to reduce the number of the initial variables, a simple univariate test was performed to select variables with high discriminating power. The test was based on the area under the receiver operating characteristic curve of each variable (AUC). Using the test of DeLong et al. (1988), for each data business sector, we selected variables with AUC significantly different than 0.5, at the 5% confidence level. Henceforth, this

Table 2. Attributes of the analysis

Attributes	Sign	Type
x_1 Imports	+	B
x_2 Exports	+	B
x_3 Representations	+	B
x_4 Premises	+	I
x_5 Banks	+	I
x_6 Age	+	I
x_7 Personnel in year $t-1$	+	I
x_8 Return on equity	+	C
x_9 Profit before income tax/total assets	+	C
x_{10} Gross profit margin	+	C
x_{11} Operating profitability	+	C
x_{12} Net profit margin (before income tax)	+	C
x_{13} Depreciation/net fixed assets	+	C
x_{14} Equity/total assets	+	C
x_{15} Interest expenses/net sales	−	C
x_{16} Collection period	−	C
x_{17} Payable period	−	C
x_{18} Inventory turnover	−	C
x_{19} Net sales/total assets	+	C
x_{20} Net sales/current liabilities	+	C
x_{21} Current ratio	+	C
x_{22} Quick ratio	+	C
x_{23} Net sales	+	C
x_{24} Value of default events over the last three years/most recent sales	−	C
x_{25} Number of default events over the last three years	−	I
x_{26} Protested bills over the last three years	−	B
x_{27} Uncovered checks over the last three years	−	B
x_{28} Payment orders over the last three years	−	B
x_{29} Seizures over the last three years	−	B
x_{30} Real estate auctions over the last three years	−	B
x_{31} Movable property auctions over the last three years	−	B
x_{32} Bankruptcy petitions over the last three years	−	B
x_{33} Most recent year with default events	−	B

Notes: The sign represents the attribute's relationship with credit risk. Attributes with a positive sign are positively correlated with the creditworthiness of an obligor (as the value of the attribute increases, the probability of default decreases). In contrast, attributes with a negative sign are negatively correlated with the level of creditworthiness. The type of the attributes is defined as follows: B = binary, I = integer, C = continuous.

reduced set of variables will be denoted as A2. The selected variables were used to build and test new models in order to compare the differences as opposed to the full models of step 1.

Table 3. Sample Observations by Sector, Year and Default Group

Sector		1998	1999	2000	2001	2002	2003	Total
Industrial	Nondefault	4,428	4,852	5,226	5,428	5,529	5,484	30,947
	Default	234	296	262	220	179	142	1,333
	Total	4,662	5,148	5,488	5,648	5,708	5,626	32,280
Construction	Nondefault	799	875	965	1,036	956	949	5,580
	Default	41	53	45	44	31	31	245
	Total	840	928	1,010	1,080	987	980	5,825

Table 4. Summary of the Experimental Settings

Setting	Scope	Input Attributes
1	Full models developed with the complete data	Full attribute set
2	Univariate attribute selection	Full attribute set
3	Multivariate attribute selection	Full attribute set From step 2
4	Effect of balanced samples	From step 2
5	Stability analysis	From step 2

Step 3: In this step the most important variables were selected by employing attribute selection procedures that take into account the method used to build the models. The goal of this procedure was to examine if the methods are capable of identifying the relevant variables and if the use of these variables improves the predictive power of the developed models. For LDA and LR, the forward stepwise selection process was performed. Regarding NN, a sensitivity analysis as described in Masters (1993) and Noble et al. (2000) was implemented. For LP and proximal SVMs, a bootstrapping was used to estimate the confidence intervals for the coefficients of the models' variables (500 bootstrap samples). Variables with coefficients not significantly different from zero at the 5% level were excluded as being not important. Finally, regarding UTADIS, as mentioned before, the method has the advantage of producing the weights of the variables used. By ranking the weights from the highest to the lowest, the most important variables with a total weight of at least 85% were selected.[5] For each of the above methodologies, the process is repeated twice: First, using the initial variables and, second using the set of variables A2. Henceforth, the corresponding sets of variables resulting from these two steps will be denoted by A3 and A4.

Step 4: As already noted, the samples used in the analysis are considerably imbalanced with regard to the number of cases in default and the nondefault observations. This imbalance may affect the accuracy of the resulting models. The objective of this step was to examine if the use

---

[5] Step 3 is not implemented for CART, because the use of pruning implicitly performs variable selection.

of balanced samples leads to the development of more accurate models. Using each data set's training samples (with the A2 sets of variables), 50 new balanced training data samples were created (at random) while their test samples have remained the same. Each of these new samples includes all the default observations, matched to an equal number of randomly selected nondefault observations. Henceforth, these tests on the use of balanced samples will be denoted as BS.

Step 5: In this step the stability of the developed models is examined to perturbations of the test data. In this analysis the test data are perturbed with normally distributed multiplicative noise for the nonbinary variables $x_4 - x_{25}$. Four levels of noise were selected, with zero mean and standard deviation 0.1, 0.2, 0.3, and 0.5. A noise level k (10, 20, 30, 50%) indicates that 99% of the time a variable will be randomly increased or decreased by between 0 and $2.58k\%$. For the binary variables $x_1 - x_3$, the corresponding noise was introduced with random perturbation of the data, whereas the variables $x_{26} - x_{33}$ have been left unchanged. For each noise level, 100 tests were performed. Similarly to step 4, this analysis is performed for the A2 set of variables, which was found to produce better models.

3.5 Results

The results of the experimental evaluation are analyzed in terms of both the average classification accuracy and the AUC of the resulting models. To facilitate the presentation, the results are averaged over the five tests described in Section 3.3. In all the following tables, the results of the best methodology are denoted in boldface. To check the statistical significance of the best methodology's results, a t-test was performed. The results that are significantly different at the 5% level are marked with an asterisk.

Industrial Sector

Concerning the average classification accuracy, it can be seen from Table 5 that NN with one hidden layer and eight neurons provides the best results when the complete set of variables is used (column A1), while quadratic PSVMs appears to be the least efficient methodology. The aforementioned neural model is significantly superior only to CART and the nonlinear PSVMs. The rest of the columns report the differences observed for the different settings used in the analysis. As illustrated in the second column of the same table, the use of the A2 set of variables improved the accuracy of most models. In contrast, the variable selection process that was employed in the third step of the experiment gave improved results only in the case of RBF PSVMs (columns A3-A1 and A4-A2). Finally, observe that the development of the models via the balanced samples provides less accurate results in most cases (an improvement is observed only in the case of nonlinear PSVMs).

Table 5. Average Classification Accuracy (Industrial Sector)

Method	A1	A2-A1	A3-A1	A4-A2	A4-A1	BS-A2
NN (1, 8)	**81.64**	0.08	−1.26	−1.13	−1.05	−0.70
NN (1, 12)	81.36	0.32	−1.04	−1.03	−0.71	−0.78
NN (1, 16)	81.38	0.05	−0.93	−0.68	−0.63	−0.72
NN (2, 6)	81.48	0.16	−0.99	−0.61	−0.46	−0.49
NN (2, 8)	81.43	0.42	−0.88	−1.00	−0.58	−0.82
NN (2, 12)	81.48	0.24	−1.10	−0.86	−0.62	−0.97
CART	80.44*	0.22	-	-	-	−0.50
LDA	80.95	0.46	0.67	−0.02	0.44	−0.32
LR	81.49	0.06	−0.06	−0.11	−0.05	−0.42
LP	81.46	−0.05	−1.32	−0.67	−0.72	−0.42
LPSVM	81.32	−0.18	−0.20	0.03	−0.15	−0.06
QPSVM	74.86*	0.32	4.49	4.39	4.71	0.57
RBF PSVM	79.85*	0.09	0.70	0.88	0.97	1.29
UTADIS	81.19	0.12	−0.60	−1.15	−1.03	−0.29
Average	80.74	0.16	−0.19	−0.15	0.01	−0.33

As far as the AUC is concerned, UTADIS exhibits the best performance (Table 6). In addition, similarly to the average classification accuracy, the univariate selection process based on the ROC curves improves the results of most models (column A2-A1). Furthermore, the use of variable sets A3 and A4 provides better results compared to the use of sets A1 and A2, respectively, in the cases of LR and nonlinear PSVMs, while the use of balanced samples generally leads to the development of inferior models.

Table 6. Area Under the ROC Curve (Industrial Sector)

Method	A1	A2-A1	A3-A1	A4-A2	A4-A1	BS-A2
NN (1, 8)	89.38*	−0.06	−1.20	−0.48	−0.54	−0.05
NN (1, 12)	89.48*	0.11	−1.34	−0.74	−0.63	−0.67
NN (1, 16)	89.50*	0.23	−1.26	−0.78	−0.54	−1.11
NN (2, 6)	89.20*	0.17	−0.86	−0.44	−0.27	0.21
NN (2, 8)	89.20*	0.01	−0.91	−0.33	−0.32	0.15
NN (2, 12)	89.41*	−0.01	−1.24	−0.53	−0.55	−0.60
CART	84.21*	0.44	-	-	-	0.21
LDA	88.24*	0.65	0.36	−0.23	0.42	0.44
LR	89.63*	0.05	0.13	0.12	0.17	−0.36
LP	89.69*	0.02	−1.51	−0.30	−0.28	−0.40
LPSVM	89.47*	0.06	−0.10	−0.06	0.00	−0.20
QPSVM	82.17*	0.48	4.70	4.17	4.65	−0.89
RBF PSVM	86.25*	0.42	1.04	0.56	0.99	3.02
UTADIS	**90.06**	−0.04	−2.02	−4.08	−4.13	−0.23
Average	88.27	0.18	−0.32	−0.24	−0.08	−0.03

The results on the stability of the classification results to perturbations of the test data are given in Tables 7 and 8, for the classification accuracy and the AUC, respectively. The results involve the changes compared to the models developed with the variable set A2 without noise. The average rate of change (column R) is also reported. The obtained results show that CART is the most stable methodology both in terms of its classification accuracy as well as in terms of the AUC. On the other hand, quadratic PSVMs appear to be the least stable classifier. These findings are valid for both evaluation criteria.

Table 7. Changes in Classification Accuracy with Different Noise Levels (Industrial Sector)

Method	10%	20%	30%	50%	R
NN (1, 8)	−0.42	−0.86	−1.46	−2.77	−0.69
NN (1, 12)	−0.18	−0.48	−1.07	−2.26	−0.57
NN (1, 16)	−0.52	−1.01	−1.55	−2.58	−0.64
NN (2, 6)	−0.31	−0.74	−1.30	−2.45	−0.61
NN (2, 8)	−0.37	−0.78	−1.36	−2.54	−0.64
NN (2, 12)	−0.28	−0.71	−1.32	−2.60	−0.65
CART	−0.32	−0.49	−0.85	−1.78	−0.44
LDA	−0.96	−1.36	−1.88	−3.24	−0.81
LR	−0.15	−0.64	−1.23	−2.33	−0.58
LP	−0.56	−1.01	−1.56	−2.65	−0.66
LPSVM	−0.45	−0.90	−1.45	−2.66	−0.67
QPSVM	−0.71	−2.09	−3.67	−7.75	−1.94
RBF PSVM	−0.42	−1.00	−1.72	−3.31	−0.83
UTADIS	−0.26	−0.63	−1.04	−2.02	−0.51
Average	−0.42	−0.91	−1.53	−2.92	−0.73

Construction Sector

Concerning the classification accuracy of the models, neural networks with 2 hidden layers and 12 neurons in each layer yields the best results, whereas quadratic PSVMs appears to be the least efficient approach. The aforementioned neural topology is significantly superior to LDA, LR, LP, nonlinear PSVMs, and UTADIS (see Table 9). Regarding the univariate variable selection process based on the ROC analysis, no general conclusion can be drawn. Furthermore, it is observed that in almost all cases the development of the models using the sets of variables A3 and A4 bears no improvement (with the exception of quadratic PSVMs). Finally, the process of model building through balanced samples appears to be insufficient.

The results of Table 10 demonstrate that the neural network with two layers of six neurons yields the best performance regarding the AUC. Its performance differs significantly compared to CART, LDA, LP, nonlinear PSVMs,

Table 8. Changes in AUC with Different Noise Levels (Industrial Sector)

Method	10%	20%	30%	50%	R
NN (1, 8)	−0.17	−0.70	−1.30	−2.65	−0.66
NN (1, 12)	−0.16	−0.70	−1.35	−2.81	−0.70
NN (1, 16)	−0.21	−0.74	−1.35	−2.77	−0.69
NN (2, 6)	−0.18	−0.65	−1.18	−2.38	−0.60
NN (2, 8)	−0.18	−0.67	−1.22	−2.46	−0.62
NN (2, 12)	−0.18	−0.70	−1.28	−2.61	−0.65
CART	−0.30	−0.60	−0.95	−1.88	−0.47
LDA	0.22	−0.34	−0.98	−2.33	−0.58
LR	−0.22	−0.69	−1.26	−2.66	−0.66
LP	−0.26	−0.82	−1.49	−3.17	−0.79
LPSVM	−0.25	−0.74	−1.33	−2.82	−0.71
QPSVM	−0.64	−1.95	−3.91	−9.28	−2.32
RBF PSVM	−0.12	−0.71	−1.58	−4.07	−1.02
UTADIS	−0.24	−0.67	−1.24	−2.74	−0.68
Average	−0.21	−0.76	−1.46	−3.19	−0.80

Table 9. Average Classification Accuracy (Construction Sector)

Method	A1	A2-A1	A3-A1	A4-A2	A4-A1	BS-A2
NN (1, 8)	78.93	0.33	−0.03	−1.06	−0.73	−3.77
NN (1, 12)	79.16	−0.40	−0.36	−0.53	−0.92	−4.25
NN (1, 16)	78.27	0.44	0.49	−0.42	0.02	−4.85
NN (2, 6)	78.60	−0.06	0.00	−0.44	−0.50	−2.47
NN (2, 8)	78.99	−0.59	−0.29	−0.23	−0.82	−3.17
NN (2, 12)	**79.63**	−0.50	−0.80	−0.88	−1.38	−5.23
CART	78.15	0.30	-	-	-	−1.96
LDA	78.73*	0.29	−1.31	−1.32	−1.03	−3.38
LR	77.64*	−0.23	−0.25	0.00	−0.23	−1.97
LP	77.31*	−0.14	−2.88	0.95	0.81	−1.48
LPSVM	76.68	−0.05	−4.23	−0.94	−0.98	−1.00
QPSVM	69.80*	1.63	1.87	3.36	4.99	−0.52
RBF PSVM	72.17*	4.28	−0.21	−1.94	2.34	1.05
UTADIS	77.20*	−0.02	−0.19	−1.09	−1.11	−0.12
Average	77.23	0.38	−0.63	−0.35	0.03	−2.37

and UTADIS. It is further observed that quadratic PSVMs is substantially inferior to other methodologies. However, in contrast to the accuracy rates, the use of the A2 set of variables generally enhances the predictive power of the initial models. In addition, it is noted that the process of selecting the most important variables via procedures that depend on the training method employed yields better results for the cases of NN, LR, and quadratic PSVMs. Finally, it is observed that only UTADIS and non-linear PSVMs provide better results when balanced samples are used.

Table 10. Area Under the ROC Curve (Construction Sector)

Method	A1	A2-A1	A3-A1	A4-A2	A4-A1	BS-A2
NN (1, 8)	84.74	0.26	0.14	0.49	0.74	−3.82
NN (1, 12)	84.11	0.61	0.71	0.77	1.38	−4.67
NN (1, 16)	83.35	1.45	1.25	0.65	2.10	−5.42
NN (2, 6)	**85.04**	0.06	0.07	0.40	0.46	−2.61
NN (2, 8)	84.92	0.08	0.07	0.50	0.58	−3.64
NN (2, 12)	84.95	−0.14	−0.06	0.66	0.52	−5.34
CART	81.81*	−1.12	−	−	−	−2.23
LDA	84.52*	0.23	0.00	−0.43	−0.20	−3.49
LR	83.07	0.73	0.78	0.64	1.36	−3.17
LP	82.04*	0.76	−4.52	−1.42	−0.66	−1.86
LPSVM	81.54	1.63	−4.83	−2.44	−0.81	−1.87
QPSVM	73.58*	1.45	1.29	4.46	5.91	0.50
RBF PSVM	79.81*	0.93	−3.72	−1.25	−0.32	3.64
UTADIS	83.06*	−0.20	−2.43	−1.54	−1.74	0.14
Average	82.61	0.48	−0.87	0.11	0.72	−2.42

Regarding the average classification accuracy criterion, RBF PSVMs models appears to be the most stable (Table 11), whereas for the area under curve criterion, UTADIS is the most stable methodology (Table 12). For both measures, quadratic PSVMs seems to lead to models with the least stable performance.

Table 11. Changes in Classification Accuracy with Different Noise Levels (Construction Sector)

Method	10%	20%	30%	50%	R
NN (1, 8)	−0.24	−0.40	−0.59	−0.92	−0.23
NN (1, 12)	−0.07	−0.20	−0.71	−0.96	−0.24
NN (1, 16)	0.05	0.10	0.12	−0.04	−0.01
NN (2, 6)	−0.57	−0.87	−0.99	−1.18	−0.30
NN (2, 8)	−0.14	−0.35	−0.48	−0.77	−0.19
NN (2, 12)	−0.18	−0.60	−0.96	−1.28	−0.32
CART	0.01	0.03	0.05	0.04	0.01
LDA	−0.49	−0.69	−0.88	−1.37	−0.34
LR	−0.17	0.24	0.24	−0.54	−0.14
LP	−0.16	−0.02	−0.06	−0.08	−0.02
LPSVM	0.73	0.72	0.59	−0.02	−0.01
QPSVM	−1.19	−1.87	−3.19	−5.55	−1.39
RBF PSVM	0.76	1.88	1.95	1.23	0.31
UTADIS	0.62	0.81	0.87	0.80	0.20
Average	−0.07	−0.09	−0.29	−0.76	−0.19

Table 12. Changes in AUC with Different Noise Levels (Construction Sector)

Method	10%	20%	30%	50%	R
NN (1, 8)	−0.14	−0.24	−0.72	−0.74	−0.19
NN (1, 12)	−0.06	0.09	−0.29	−0.24	−0.06
NN (1, 16)	0.13	0.24	−0.04	−0.63	−0.16
NN (2, 6)	−0.07	−0.13	−0.35	−0.51	−0.13
NN (2, 8)	−0.10	−0.16	−0.39	−0.52	−0.13
NN (2, 12)	−0.07	0.00	−0.17	−0.30	−0.07
CART	0.01	0.03	0.05	0.04	0.01
LDA	−0.06	−0.13	−0.47	−0.74	−0.19
LR	−0.08	−0.16	−0.33	−1.13	−0.28
LP	−0.10	−0.26	−0.57	−1.38	−0.34
LPSVM	−0.05	−0.06	−0.22	−0.78	−0.19
QPSVM	−0.80	−1.82	−3.15	−5.85	−1.46
RBF PSVM	−0.25	−0.15	−0.53	−1.85	−0.46
UTADIS	0.11	0.26	0.54	0.31	0.08
Average	−0.11	−0.18	−0.48	−1.02	−0.26

4 Conclusions and Future Perspectives

In developing a credit scoring model, the consistency of the evaluation criteria is of high importance. The above analysis has led to the conclusion that no methodology is clearly superior to the others. However, neural networks seems to provide better results in several settings, while multicriteria additive models developed with the UTADIS method also provide competitive results and stable performance. In addition, there are clear indications that the quadratic PSVMs is the least stable and efficient methodology. In both sectors the aforementioned approach exhibits the worst performance regarding the evaluation criteria used.

No certain conclusions can be drawn concerning the variable selection process through the training method employed. It must be noted that in several cases the aforementioned variable selection process provided conflicting results. In contrast, the implementation of a simple nonparametric univariate test (ROC analysis) seems to enhance the predictive power of the developed models. This finding is strengthened by the fact that in the majority of cases the combined selection of the most important variables (on the basis of both ROC analysis and the estimations of the methods themselves) yields better results compared to the selection of the variables based only on the estimations of the methods themselves.

In addition, it must be noted that no set of variables or at least one variable was found to be common in all sets provided by all methodologies. The fact that the selection process was conducted differently for each method may explain the above finding. Therefore, care should be taken when examining the selection process of the most important variables. Even if a methodology can select an adequate number of important variables, one must further examine if the use of these variables yields better results. For instance, in the present study, NN, LDA, and LR have in all cases selected an adequate number of important variables; the use of these variables, however, did not provide better results in most cases. In terms of the relative importance of the variables, the ratio "interest expenses/net sales" and the variable "most recent year with default events" have appeared as important ones in several cases in most methodologies. Finally, the experiment conducted indicates that the use of balanced samples does not lead to the development of more efficient models (only the performance of RBF PSVMs models was consistently improved).

According to the results obtained, we conclude that the development process of a credit rating system requires the thorough examination of all parameters that can affect its efficiency. Since there are still interesting topics to be examined, important future research perspectives appear in the field of credit risk estimation. The Basel Committee on Banking Supervision is continuously revising its provision, aiming to provide the banking industry with sounder practices for assessing credit risk. The move from the standardized IRB approach to the advanced IRB approach requires banks to estimate the LGD and EAD parameters themselves. However, research still needs to be conducted regarding the assessment approaches of these parameters.

Moreover, the Basel Committee on Banking Supervision (BCBS, 2005) reported that credit institutions can calculate their unexpected losses using whichever credit risk model best fits for their internal risk measurement and risk management needs. However, effective loan portfolio management techniques are still an unresolved area (Altman and Saunders, 1998).

Furthermore, as Krahnen and Weber (2001) stated, a bank should develop different credit rating systems, applying each of them to evaluate homogeneous firms. In the present study homogeneity has been expressed on the basis of the firm's operational sector. Alternatively, systems that evaluate firms with similar financial records can be developed (i.e., systems that will evaluate firms with similar level of sales, liabilities, assets, etc.). Finally, the process of estimating the average probabilities of default of each grade implies a further specialized research topic. The properties of mapping techniques and scoring model-based methods employed for this reason have not been systematically examined (Carey and Hrycay, 2001).

References

1. Aas, K. The Basel II IRB approach for credit portfolios: A survey. Norsk Regnesentral, Norwegian Computing Center, 2005.
2. Altman, E. I. 1968 Financial ratios, discriminant analysis and the prediction of corporate bankruptcy. *Journal of Finance*, 23:589–609.
3. Altman, E. I., and Saunders, A. 1998 Credit risk measurement: Developments over the last 20 years. *Journal Banking and Finance*, 21:1721–1742.
4. Basel Committee on Banking Supervision. The new Basel capital accord. Bank for International Settlements, 2001a.

5. Basel Committee on Banking Supervision. The internal ratings-based approach: Supporting document to the new Basel capital accord. Bank for International Settlements, 2001b.
6. Basel Committee on Banking Supervision. The new Basel capital accord. Bank for International Settlements, 2003.
7. Basel Committee on Banking Supervision. International convergence of capital measurement and capital standards. Bank for International Settlements, 2004.
8. Basel Committee on Banking Supervision. An explanatory note on the Basel II IRB risk weight functions. Bank for International Settlements, 2005.
9. Baesens, B., Gestel, T. V., Viaene, S., Stepanova, M., Suyken, J., and Vanthienen, J. Benchmarking state-of-the-art classification algorithms for credit scoring. *Journal of the Operational Research Society*, 54:627–635, 2003a.
10. Baesens, B., Setiono, R., Mues, C., and Vanthienen J. Using neural network rule extraction and decision tables for credit-risk evaluation. *Management Science*, 49:312–329, 2003b.
11. Balcaen, S., and Ooghe, H. 35 years of studies on business failure: An overview of the classical statistical methodologies and their related problems. Department of Accountancy and Corporate Finance, Ghent University, Working paper, 2004.
12. Bishop, C. M. *Neural Networks for Pattern Recognition*. Oxford University Press, Oxford, 1995.
13. Black, F., and Scholes, M. The pricing of options and corporate liabilities. *Journal of Political Economy*, 81:659–674, 1973.
14. Breiman, L., Friedman, J. H., Olshen, R. A., and Stone, C. J. *Classification and Regression Trees*. Pacific Grove, CA, 1984.
15. Calindo, J., and Tamayo, P. Credit risk assessment using statistical and machine learning: Basic methodology and risk modeling applications. *Computational Economics*, 15:107–143, 2000.
16. Carey, M., and Hrycay, M. Parameterizing credit risk models with rating data. *Journal of Banking and Finance*, 25, 197–270, 2001.
17. Comptroller of the Currency Administrator of National Banks. Rating credit risk. *Comptrollers Handbook*, 2001.
18. DeLong, E. R., DeLong D. M., and Clarke-Pearson D. L. Comparing the areas under two or more correlated receiving operating characteristic curves: A nonparametric approach. *Biometrics*, 44:837–845, 1988.
19. Desai, V. S., Crook, J. N., and Overstreet, G. A. A comparison of neural networks and linear scoring models in the credit union environment. *European Journal of Operational Research*, 95:24–37, 1996.
20. Doumpos, M., and Zopounidis, K. *Multicriteria Decision Aid Classification Methods*. Kluwer Academic Publishers, Dordrecht, 2002.
21. Doumpos, M., Kosmidou, K., Baourakis, G., and Zopounidis, K. Credit risk assessment using a multicriteria hierarchical discrimination approach: A comparative study. *European Journal of Operational Research*, 138:392–412, 2002.
22. Esposito, F., Malerba, D., and Semerano, G. A comparative analysis of methods for pruning decision trees. *IEEE Transactions on Pattern Analysis and Machine Intelligence*, 19:476–491, 1997.
23. Fernades, J. S. Corporate credit risk modeling: Quantitative risk rating system and probability of default estimation. Banco BPI, 2005.
24. Freed, N., and Glover, F. A linear programming approach to the discriminant problem. *Decision Sciences*, 12, 68–74, 1981a.

25. Freed, N., and Glover, F. Simple but powerful goal programming models for discriminant problems. *European Journal of Operational Research*, 7:44–60, 1981b.
26. Fritz, S., and Hosemann, D. Restructuring the credit process: Behaviour scoring for German corporates. *International Journal of Intelligent Systems in Accounting, Finance & Management*, 9:9–21, 2000.
27. Frydman, H., Altman, E. I., and Kao, D. L. Introducing recursive partitioning for financial classification: The case of financial distress. *Journal of Finance*, 40:269–291, 1985.
28. Fung, G., and Mangasarian, O. L. Proximal support vector machines classifiers. In *Proceedings of the 7th ACM SIGKDD International Conference on Knowledge Discovery and Data Mining*, F. Provost and R. Srikant, Editors. ACM Press, New York, pages 77–86, 2001.
29. Gordy, M. B. A risk-factor model foundation for ratings-based bank capital rules. *Journal of Financial Intermediation*, 12:199–232, 2003.
30. Hand, D. J. *Discrimination and Classification*. Wiley, New York, 1981.
31. Huang, Z., Chen, H., Hsu, C.-J., Chen, W.-H., and Wu, S. Credit rating analysis with support vector machines and neural networks: A market comparative study. *Decision Support Systems*, 37:543–558, 2004.
32. KMV. Private firm model: Introduction to modeling methodology, 2001.
33. Krahnen, J. P., and Weber, M. Generally accepted rating principles: A primer. *Journal of Banking and Finance*, 25:3–23, 2001.
34. Lee, T.-S., Chiu, C.-C., Lu, C.-J., and Chen, I.-F. Credit scoring using the hybrid neural discriminant technique. *Expert Systems with Applications*, 23:245–254, 2002.
35. Liu, Y., and Schumann, M. The evaluation of classification models for credit scoring. Institute fur Wirtschaftsinformatik, Working paper, 2002.
36. Liu, Y., and Schumann, M. Data mining feature selection for credit scoring models. *Journal of the Operational Research Society*, 56:1099–1108, 2005.
37. Malhotra, R., and Malhotra, D. K. Differentiating between good credits and bad credits using neuro-fuzzy systems, European Journal of Operational Research, 136:190–211, 2002.
38. Mangasarian, O. L. Multisurface method for pattern separation. *IEEE Transactions of Information Theory*, 14:801–807, 1968.
39. Marais, M. L., Patell, J. M., and Wolfson, M. A. The experimental design of classification models: An application of recursive partitioning and bootstrapping to commercial bank loan classifications. *Journal of Accounting Research*, 22:87–114, 1985.
40. Masters, T. *Practical Neural Network Recipes in C++*. Academic Press, New York, 2003.
41. Merton, R. C. The pricing of corporate debt: The risk structure of interest rates. *Journal of Finance*, 29:447–470, 1974.
42. Moody's Investor Services. RiskCalc^{TM} for private companies: Moody's default model, 2000.
43. Moody's Investor Services. Systematic and idiosyncratic risk in middle-market default prediction. A study of the performance of the RiskCalc^{TM} and PFM^{TM} models, 2003.
44. Moody's KMV. The Moody's KMV EDF^{TM} RiskCalc^{TM} v3.1 model, 2004.

45. Min, J. H., and Lee, Y.-C. Bankruptcy prediction using support vector machines with optimal choice of kernel function parameters. *Expert Systems with Applications*, 28:603–614, 2005.
46. Noble, P. A., Almeida, J. S and Lovell, C. R. Application of neural computing methods for interpreting phospholipid fatty acid profiles on natural microbial communities. *Applied and Environmental Microbiology*, 66:694–699, 2000.
47. Ohlson, J. A. Financial ratios and the probabilistic prediction on bankruptcy. *Journal of Accounting Research*, 18:109–131, 1980.
48. Ong, C.-S., Huang, J. J., and Tzeng, G.-H. Building credit scoring models using genetic programming. *Expert Systems with Applications*, 29:41–47, 2005.
49. Patuwo, E., Hu, M. Y., and Hung, M. S. Two-group classification using neural networks. *Decision Sciences*, 24:825–845, 1993.
50. Piramuthu, S. Financial credit-risk evaluation with neural and neurofuzzy systems. *European Journal of Operational Research*, 112:310–321, 1999.
51. Rumelhart, D. E., Hinton, G. E., and Williams, R. J. Learning internal representation by error propagation. In D. E. Rumelhart and J. L. Williams, Editor, *Parallel Distributed Processing: Explorations in the Microstructure of Cognition*. MIT Press, Cambridge, MA, 1986.
52. Saunders, A., and Cornett, M. M. *Financial Institutions Management*. McGraw-Hill, New York, 2003.
53. Smith, K., and Gupta, J. Neural networks in business: Techniques and applications for the operations researcher. *Computers and Operation Research*, 27:1023–1044, 2000.
54. Srinivasan, V., and Kim, Y. H. Credit granting: A comparative analysis of classification procedures. *Journal of Finance*, 42:665–683, 1987.
55. Standard and Poor's. About CreditModel, 2006a.
56. Standard and Poor's. CreditModel technical white paper, 2006b.
57. Stone, M. Cross-validation choice and assessment of statistical predictions. *Journal of the Royal Statistical Society B*, 36:111–147, 1974.
58. Subramanian, V., Hung, M. S., and Hu, M. Y. An experimental evaluation of neural networks for classification. *Computers and Operations Research*, 20:769–782, 1993.
59. Treacy, W. F., and Carey, M. Credit risk rating systems at large US banks. *Journal of Banking and Finance*, 24:167–201, 2000.
60. Vapnik, V. *The Nature of Statistical Learning Theory*. Springer-Verlag, New York, 1995.
61. Vellido, A., Lisboa, P. J. G., and Vaughan, J. Neural networks in business: A survey of applications (1992–1998). *Expert Systems with Applications*, 17:51–70, 1999.
62. West, D. Neural network credit scoring models. *Computers and Operations Research*, 27:1131–1152, 2000.
63. Zopounidis, C., and Doumpos, M. Building additive utilities for multi-group hierarchical discrimination: The M.H.DIS method. *Optimization Methods and Software*, 14:219–240, 2000.

Index

acquisitions, 329, 385, 432
Akaike's information criterion, 40
American options, 246
arbitrage, 82, 187
ARCH model, 203
asset liability management, 281

bankruptcy risk, 407, 432
Basel Committee, 457
BearSpread strategy, 260
binomial tree, 331
Black–Scholes formula, 336
bond markets, 201
bond pricing, 160
Bretton Woods system, 245
Brownian motion, 158

capital adequacy ratio, 284
capital asset pricing model, 302
capital budgeting, 301
capital requirements, 462
certainty equivalent, 7
chance constrained programming, 285
closed-end funds, 82
combinatorial optimization, 26
complementarity conditions, 37
computational intelligence, 100
concentration risk, 464
conditional value at risk, 249
continuously compounded return, 107
cost–benefit analysis, 301
covariance matrix, 235, 236, 363
credit institutions, 457
credit networks, 348

credit rating, 459
credit risk, 457
critical line algorithm, 17
currency options, 246
currency risk, 245

default, 457
deterministic models, 285
diffusion model, 159
discriminant analysis, 26, 431, 468
diversification, 246
dividend yield, 82
downside risk, 49, 246
dynamic programming, 286

efficient set, 9, 51
EGARCH model, 203
ellipticity, 49
equilibrium conditions, 368
European options, 254
evolutionary computation, 104
excess returns, 205, 232, 437
exchange rates, 231, 245
exchange traded funds, 67
expectations hypothesis, 175
expected utility, 247
exposure at default, 462

financial engineering, 201, 343
financial equilibrium, 343
financial institutions, 49, 457
financial networks, 343
financial optimization, 25, 345

490   Index

financial planning, 289
financial trading, 102
fixed-income instruments, 67
forward contracts, 246
forward rates, 157
futures contracts, 69
fuzzy linear programming, 326
fuzzy logic, 312
fuzzy numbers, 312

GARCH model, 203
generalized cross-validation, 215
genetic algorithms, 100
genetic operators, 103
genetic programming, 99
goal programming, 285
graph theory, 344
growth models, 233

hedging, 246

index derivatives, 68
index tracking, 31
informational asymmetry, 385
integer programming, 25, 304
interest rate models, 163
interest rate risk, 293
intermediation, 359
internal rate of return, 301
investment projects, 303

jump process, 232

Kalman filter, 176
kernel function, 439, 474
knapsack problem, 304

LIBOR market, 190
LIBOR rate, 191, 208
likelihood function, 211, 236
linear programming, 6, 28, 473
logistic regression, 470
loss given default, 462

managerial ownership theory, 406
market makers, 76
market risk, 248
martingales, 175
maximal predictability portfolio, 34
maximum expected utility, 8

maximum likelihood, 210, 236
mean-absolute deviation model, 31
mean-variance analysis, 50, 346, 347
mean-variance portfolio, 3, 61, 346
membership function, 312
mergers, 329, 385, 432
Merton's model, 465
Monte Carlo simulation, 239, 293, 304
moving average, 110
multicriteria decision aid, 306, 348, 431, 475
multiobjective programming, 12, 285, 325
mutual funds, 30, 67

net present value, 301
network optimization, 346
neural networks, 100, 201, 435, 471
nondominated set, 3

optimal capital structure, 387
outranking methods, 307

Poisson process, 232
portfolio insurance, 246
portfolio management, 247
portfolio optimization, 25, 247, 343
portfolio selection, 3, 5, 346
power law, 233
price dynamics, 234, 240, 362
pricing kernel, 164
primary market, 74
principal component analysis, 171
protective put options, 246
put option, 246, 332
put-call parity, 337

quadratic programming, 42, 347, 438

radial basis function neural networks, 212
random walk, 238
real options, 331
replication strategies, 84
risk aversion, 8, 347
risk management, 282
risk premium, 168, 231
risk-neutral valuation, 187
riskless asset, 53
riskless interest rate, 108
ROC curve, 449, 476

safety first, 4
scenario generation, 257, 289
secondary market, 74
self-organizing map, 222
semivariance, 54
short sales, 105
short selling, 4, 76
simulation, 49, 239, 293
spot market, 254
spot rate model, 160
stationary point, 368
stochastic models, 285
stochastic processes, 157
stochastic programming, 6, 246, 288
stochastic volatility, 173
supply and demand curves, 344
support vector machines, 438, 474
systematic risk, 87

technical analysis, 99
Tikhonov regularization, 212
tracking error, 77
trading rules, 99
trading strategies, 259
transaction costs, 29, 107, 360
treasury bills, 108

utility function, 4, 50, 285, 306

value at risk, 49, 249
variational inequality theory, 349
volatility clamping, 258
volatility curve, 178
volatility spillovers, 201

yield curve, 170

zero coupon bond, 159

Printed in the United States of America